P9-DUP-967

EVE

EVE

HOW THE FEMALE BODY DROVE 200 MILLION YEARS OF HUMAN EVOLUTION

CAT BOHANNON

ALFRED A. KNOPF
NEW YORK
2023

THIS IS A BORZOI BOOK
PUBLISHED BY ALFRED A. KNOPF

Published in the United States by Alfred A. Knopf,
a division of Penguin Random House LLC, New York.

www.aaknopf.com

Knopf, Borzoi Books, and the colophon are registered
trademarks of Penguin Random House LLC.

Library of Congress Cataloging-in-Publication Data

Names: Bohannon, Cat, author.
Title: Eve : How the female body drove 200 million years of human evolution / Cat Bohannon.
Description: First edition. | New York : Alfred A. Knopf, 2023. |
Includes bibliographical references and index.
Identifiers: LCCN 2022060219 | ISBN 9780385350549 (hardcover) |
ISBN 9780385350556 (ebook) | ISBN 9781524712570 (open market)
Subjects: LCSH: Women—Physiology—Popular works. |
Women—Evolution—Popular works. | Sex differences—Popular works.
Classification: LCC QP81.5.B64 2023 | DDC 613/.0424—dc23/eng/20230616
LC record available at https://lccn.loc.gov/2022060219

Illustrations by Hazel Lee Santino
Jacket image: Chronicle/Alamy
Jacket design by Gabriele Wilson

Manufactured in the United States of America
First Edition

For my children, Leela and Pravin—
nothing has changed my understanding of time
like the small, beautiful breaths you take, every day.

CONTENTS

EVE

INTRODUCTION

We did this. Conceived
of each other, conceived each other in a darkness
which I remember as drenched in light.
I want to call this, life.

—ADRIENNE RICH,
"ORIGINS AND HISTORY OF CONSCIOUSNESS"

Elizabeth Shaw has a problem. The director Ridley Scott has impregnated her with a large, vicious alien squid. Aboard the spaceship *Prometheus*, she has to find a way to abort her uninvited guest without bleeding to death. Shambling to a futuristic surgery pod, she asks the computer for a C-section. "Error," it says, "this medpod is calibrated for male patients only."

"Shit," said a woman behind me, "who does that?"

What follows is a gruesome scene involving lasers, staples, and writhing tentacles. As I sat in a darkened theater in New York in 2012 watching this prequel to *Alien*, I couldn't help but think, *Yes, who does that? Who sends a multitrillion-dollar expedition into space and forgets to make sure the equipment works on women?*

Actually, modern medicine often does precisely that. One-size-fits-all doses of antidepressants are given to men and women, despite evidence that they may affect the sexes differently. Prescriptions for pain medications, too, are considered sex neutral, despite consistent proof that some may be less effective for women. Women are more likely to die of heart attacks, even though they're *less* likely to have them—symptoms differ between the sexes, so women and their doctors alike fail to catch them in time. Anesthetics in surgery, treatments for Alzheimer's, even public education curricula suffer from the ill-conceived notion that women's bodies are just bodies in general—soft and fleshy, and missing a couple of significant nether bits, but otherwise, just the same as men's.

And of course, nearly *all* of the studies that produced these findings include only cisgender subjects—in the world of scientific research, there's been very little attention to what happens in the bodies of people assigned one or another sex at birth who then go on to identify differently. In part, that's because there's a massive difference between biological *sex*—something wound deep into the warp and weft of our physical development, from in-cell organelles all the way up to whole-body features, and built over billions of years of evolutionary history—and humanity's *gender identity*, which is a fluid thing and brain based and at most a few hundred thousand years old.*

But it's not just that. The fact of the matter is that until very recently the study of the biologically female body has lagged far behind the study of the male body. It's not simply that physicians and scientists don't bother to seek out sex-specific data; it's that until all too recently *the data didn't exist.* From 1996 to 2006, more than 79 percent of the animal studies published in the scientific journal *Pain* included *only* male subjects. Before the 1990s, the stats were more disproportionate. And this is hardly unusual—dozens of prominent scientific journals report the same. The reason for this blind spot concerning female bodies, whether we're talking about basic biology or the nuances of medicine, isn't just

* I know some people still struggle with this idea, but most of the scientific community agrees that biological sex is fundamentally separate from human gender identity. The belief that the sex-typical features of a person's body inevitably assign them a gender identity and behavior to match is sometimes called "biologism" or, more broadly, "gender essentialism" (Witt, 1995). The thing about gender essentialism is that it is a natural extension of sexism. Societies that form deep cultural beliefs about what one or another gender "should be" also tend to believe that a person is one of two genders from birth depending on how their body looks. Those societies then strongly reinforce those beliefs through various rules for each gender, ranging from the sort of fine, irritating cognitive grit of social exclusion to incredibly violent punishment of "rule breakers" and everything in between.

sexism. It's an intellectual problem that *became* a societal problem: for a long time, we've been thinking about what sexed bodies are, and how we should go about studying them, in entirely the wrong way.

In the biological sciences, there's still such a thing as the "male norm."* The male body, from mouse to human, is what gets studied in the lab. Unless we're specifically researching ovaries, uteri, estrogens, or breasts, the girls aren't there. Think about the last time you heard about a scientific study—some article about a new window into obesity, or pain tolerance, or memory, or aging. More than likely that study didn't include any female subjects. That's as true for mice as it is for dogs, pigs, monkeys, and, all too often, humans. By the time a clinical trial for a new medication starts testing on human subjects, it might not have been tested on female animals at all. So, when we think about Elizabeth Shaw screaming her sci-fi head off at the misogynistic medpod, we shouldn't just feel terror and pity and disbelief. We should feel recognition.

Why is this still happening? Aren't the sciences supposed to be objective? Gender neutral? Bound by the empirical method?

When I first found out about the male norm, I was flabbergasted—not because I'm a woman, but because at the time I was a PhD candidate at Columbia University studying the evolution of narrative and cognition—brains and stories, to put it simply, and their 300,000-year history. I'd taught and conducted research at a number of the top-tier institutes of learning and science in the modern world. As such, I thought I had a pretty good overview of the state of women in the academy. While I'd seen some sketchy stuff, I'd personally never experienced sexism in the lab. The idea that much of biological sciences still rested on the "male norm" was the furthest thing from my mind. Though I am a feminist, mine was more a feminism in practice: simply *being* a woman doing quantitative research was, to me, the revolutionary act. And honestly, the biologists and neuroscientists and psychologists and biophysicists *I* knew, from the people I collaborated with to the people I drank with, were some of the most cosmopolitan,

* In the scientific literature, this is also called "male bias."

liberal, clearheaded, intelligent, and frankly *good* people I'd ever met. If I'd been one to gamble, I'd never have thought of them as the sort who'd perpetuate some systemic injustice, much less one that undermines their science.

But it's not entirely their fault. Many researchers default to male subjects for practical reasons: it's difficult to control for the effects of female fertility cycles, particularly in mammals. A complex soup of hormones floods their bodies at regular intervals, whereas males' sex hormones seem more stable. A good scientific experiment aims to be *simple*, designed with as few confounding factors as possible. As a postdoc in a Nobel laureate's lab once told me, using males "just makes it easier to do clean science." The variables, in other words, are easier to control, thereby making the data more interpretable with less work, and the results more meaningful. This is especially true for the complex systems involved in behavioral research, but can even be a problem with basic things like metabolism. Taking the time to control for the female reproductive cycle is considered difficult and expensive; the ovary itself is thought of as a "confounding factor." So, unless a scientist is *specifically* asking a question about females, the female sex is left out of the equation. The experiments run faster, the papers come out sooner, and the researcher is more likely to get grant funding and tenure.

But making such decisions to "simplify" is also prompted by (and perpetuates) a much older understanding of what sexed bodies are. It's not that topflight scientists still think female bodies were made when God pulled a rib from Adam's side, but the assumption that being sexed is simply a matter of sex organs—that somehow being female is just a minor tweak on a Platonic form—is a bit like that old Bible story. And that story is a lie. As we've increasingly learned, female bodies aren't just male bodies with "extra stuff" (fat, breasts, uteri). Nor are testicles and ovaries hot swappable. Being sexed permeates every major feature of our mammalian bodies and the lives we live inside them, for mouse and human alike. When scientists study only the male norm, we're getting less than half of a complicated picture; all too often, we don't know what we're missing by ignoring sex differences, because we're not asking the question.

After being struck by the stubborn reality of the male norm, I did what researchers like to do: I dug into the databases to see how big a problem it was. And, well, it's huge. It's so huge that many papers don't even mention that they used only male subjects. I often had to email the authors directly and ask.

Okay, maybe it's just mice, I thought. Maybe this is only a problem with animal studies.

Sadly, that's not the case. Thanks to regulations established in the 1970s, clinical trials in the United States, for one, are actually "strongly advised" not to use female subjects who "could be of childbearing age." The use of pregnant subjects is all but verboten. While on the face of it, that may seem perfectly sensible—no one wants to mess with our kids—it also means we've been continuing to steer the ship in a fog. The National Institutes of Health managed to update some of these regulations in 1994, but loopholes are regularly exploited: as of 2000, one in five NIH clinical drug trials still wasn't using any female subjects, and of the studies that did, nearly two-thirds didn't bother analyzing their data for sex differences. Even if everyone actually followed the new rules, given that it usually takes more than ten years for drugs to move from clinical trial to market, 2004 was the first year any new drug approved for sale would have been tested on significant numbers of women. Drugs that were released before the new regulations took effect are in no way obliged to go back and redo their clinical trials.*

And so, the majority of subjects in clinical trials continue to be men, just as the vast majority of animal studies use male subjects. Meanwhile, women are *more* likely to be prescribed pain medica-

* Similar problems appear in legal guidelines in much of the industrialized world including Canada, the U.K., and France. The good intention to *protect* pregnant women and their potential children dropped much of the female sex out of medical research for a very long time. Recent legislation in a number of countries has boosted the numbers—for instance, now NIH-funded studies in the United States have to justify *why* they're not including women in a clinical trial if they fail to—but there remain enough loopholes in the system to drive all the elephants in a three-ring circus through (Geller et al., 2018; Rechlin et al., 2021). Some journals have taken up the charge—*Endocrinology*, for instance, now demands the methods sections of papers be explicit about animals' sex (Blaustein, 2012). But most peer-reviewed scientific journals haven't made such rules.

tions and psychotropic drugs than men—drugs that haven't been tested on nearly enough female bodies. Since dosage is usually based on body weight and age, if there aren't specific recommendations for women coming from the research, doctors have to rely on anecdotal knowledge* to figure out whether a prescription needs to be "jimmied" for a female patient.

This is particularly problematic for painkillers. While recent research has demonstrated that women require higher doses of painkillers in order to feel the same level of pain relief as men, that knowledge isn't currently built into dosage guidelines. And why would it be? Official guidelines are generally based on the results of a drug's clinical trials. For many painkillers on the market today—for example, OxyContin, released in 1996—clinical trials didn't rigorously test for sex differences, because they weren't required to do so. In many cases, they were legally encouraged *not* to do so, because the trials occurred before the NIH rules changed. OxyContin has since gone on to become one of the most abused painkillers in the world, one commonly prescribed to women suffering from endometriosis and uterine-related pain. Pregnant women addicted to such drugs are warned not to go off them too quickly because the stress of withdrawal might abort the fetus. (These women are typically put on methadone.) Others begin their addiction during pregnancy, sometimes after well-meaning doctors prescribe painkillers to relieve their pain, unaware that the patients are pregnant (or about to be). One study released in 2012 shows that the number of infants born addicted to opiates had tripled in only ten years, in part due to mothers

* Sometimes regulating agencies catch up, but it takes a while. For example, in 2013 the U.S. Food and Drug Administration finally issued guidance instructing doctors to prescribe lower (essentially, half) doses of zolpidem (for example, Ambien) because women seem to clear the drug from their bloodstream more slowly than men (U.S. FDA, 2013). At that point, zolpidem had been approved for medical use for twenty-one years. The original approval letter indicated that dosage should be "individualized," but made no comment about sex differences in dosage, stating "the recommended dose for adults is 10 mg immediately before bedtime" (U.S. FDA, 1992). They did offer that "elderly, debilitated patients, and patients with hepatic insufficiency" should receive "an initial 5 mg dose" (ibid.). Perhaps women should be considered hepatically insufficient, then?

becoming addicted to drugs like OxyContin. That number is still on the rise.

According to a recent report from the American Academy of Pediatrics, many mothers didn't realize these drugs could harm their infants. They simply felt pain, asked their doctors for help, and the doctors gave them a prescription. But unlike the doctors' male patients, these women probably took more of the drug, and more often, because they weren't feeling the relief they'd expected, or the relief they felt wore off too soon: *That worked for a little while, crap, better take more, ugh, didn't work as well this time, better take more* . . . Most clinical studies show that across multiple drug types, women metabolize drugs more quickly than men.* This finding is usually shrugged off when it comes time for medical guidance, though. And unfortunately, addiction to pain medication becomes more likely the greater and more consistent one's dosage. In other words, women who take OxyContin are more likely to do precisely the sort of thing that will make their bodies addicted to it: front-loading pills to the point that their bodies "norm" a certain level of drug in their system. If drugs like OxyContin had been properly tested on women during clinical trials, doctors would have better guidelines for dealing with these patients' pain, and fewer newborns would begin their lives as drug addicts.

It's important to remember that "drugs" aren't just the pills we stash in our medicine cabinets. Ask yourself this: Is it really acceptable that we only bothered to test sex differences for *general anesthesia* in 1999? Turns out women wake up faster than men, regardless of their age, weight, or the dosage they've been given.

* While a lot of attention is given to the fact that women's bodies tend to be smaller, the reason we metabolize drugs differently may actually have as much to do with our livers. One recent study comparing biopsies of male and female liver tissue showed thirteen hundred genes whose mRNA expression was significantly influenced by sex; of these, 75 percent showed higher expression in females (Renaud et al., 2011). It's not, in other words, just a matter of how much drug distributes through how much body mass, but how the cells in a sex-typical liver go about their day. And "day" matters here, too: livers, like the rest of the body, have a circadian rhythm, and female mammals are especially sensitive to our long-evolved relationship to the sun (Lu et al., 2013). More on daylight and why it matters in the "Perception" chapter.

(I don't know about you, but I'm not fond of the idea of waking up during surgery.) And that study didn't even set out to discover sex differences. The researchers simply wanted to test a new EEG monitor during anesthesia. The study used patients who were already scheduled for surgery, and four different research hospitals were involved, so, unusually, there were loads of subjects—both women and men. The EEG monitor did prove useful, in the end, but that turned out to be far less interesting than the results in women. It seems only then did the scientists go back and analyze their data for sex differences. In other words, they didn't really ask the question. They realized, after the fact, that they should have asked the question.

Not asking the question is dangerous. I'm all for simple experiment design, but who in their right minds would call that "clean science"?

At the same time that I was learning how dire the problem of the male norm is, I started finding new research into the female body that wasn't getting nearly enough attention. Scientists don't often read outside their specialties, but my field of research required I read regularly in at least three different disciplines (cognitive psychology, evolutionary theories of cognition, and computational linguistics), and I had to stay abreast of the latest literary scholarship, too. But even for me, it was pretty unusual to start digging around in anesthesia journals, in metabolism studies, in paleoanthropology. But I was driven to keep asking the question, What about women? What changes when you ask, "What's different about the female body? What might we be missing?"

For example, why are women fatter than men (to put it bluntly)? As a twenty-first-century American woman, I'd spent altogether too much time thinking about my fat, but I hadn't the faintest clue that my adipose tissue is actually an *organ*, much less that it evolved from the same ancient organ as my liver and most of my immune system.

Let me give you an on-the-ground example. In 2011, *The New York Times* published an article about liposuction. It seems that women who have liposuction on their hips and thighs do grow back some of their fat, but they grow it back in different places. Basically, your thighs may stay thinner, but your upper arms will soon be fatter than they were before. It was a cute article. A bit of fluff, really. But unlike the majority of plastic surgeons, I'd guess, I'd just been reading the latest research on the evolution of adipose tissue—specifically female adipose tissue.

As it turns out, women's fat isn't the same as men's. Each fat deposit on our body is a little bit different,* but women's hip, buttock, and upper thigh fat, or "gluteofemoral" fat, is chock-full of unusual lipids: long-chain polyunsaturated fatty acids, or LC-PUFAs. (Think omega-3. Think fish oil.) Our livers are bad at making these kinds of fats from scratch, so we need to get most of them from our diet. And bodies that can become pregnant need them so they can make baby brains and retinas.

Most of the time, female gluteofemoral fat resists being metabolized. As many women know, these areas are the first places we gain weight and the last places we lose it.† But in the last trimester of pregnancy—when the fetus ramps up its brain development and its own fat stores—the mother's body starts retrieving and dumping these special lipids by the boatload into the baby's body. This specialized hoovering of the mother's gluteofemoral fat stores continues throughout the first year of breast-feeding—the most *important* time, as it happens, for infant brain and eye development. Some evolutionary biologists now believe that women evolved to have fatty hips precisely because they're specialized to provide the building blocks for human babies' big brains. Since we can't get enough of those LC-PUFAs from our daily diet, women

* For example, the fat deposits around your heart behave differently than the ones under your chin, and their structure is a bit different, too.

† As such, these are also popular sites for women to get liposuction. Tummy tucks rank a close second. The so-called Brazilian butt lift combines the two and makes it worse, typically by sucking fat out of women's stomach deposits and reinjecting that same fat into their buttocks. That's particularly risky because women's buttocks are especially full of blood vessels, which is precisely where you don't want to inject a bunch of lipids, risking a fatty embolism, wherein fat breaks into the bloodstream, migrates to somewhere really vital like the heart or lungs or brain, and causes a blockage.

start storing them from childhood forward. Other primates don't seem to have this pattern.

Meanwhile, we found out just a few years ago—again, someone *finally* asked the question—that a human girl's hip fat may be one of the best predictors for when she'll get her first period. Not her skeletal growth, not her height, not even her day-to-day diet, but how much gluteofemoral fat she has. That's how important this fat is for reproduction. Our ovaries won't even kick in until we've stored up enough of this fat to form a decent baseline. When we lose too much weight, our periods stop. We also learned—again, this is recent research—that while taking supplements can up a breast-feeding woman's LC-PUFAs, the vast majority of what the baby's getting is coming from her body's fat stores—particularly her big fat butt.* Most women's bodies begin preparing for pregnancy in childhood, not because it's a woman's destiny to be a mother, but because human pregnancy sucks, and our bodies have evolved ways to help us survive it.

But every year, nearly 190,000 women undergo liposuction in the United States alone. As reported in various medical journals since 2013, there seems to be something about the violent disruption of women's tissue during liposuction that prevents fat from recovering at the surgery site.† I suspect that the new fat that

* This was discovered by giving breast-feeding women a specially marked supplement that could be tracked via isotope. By sampling the mother's breast milk, researchers were able to trace which of the fatty acids in the milk came from the supplements and which must have come from elsewhere. Other studies have established that variations in pregnant women's diets can modify some, but not all, of the LC-PUFAs in the mother's bloodstream and the newborn baby's cord blood, which is often used as a measure for what the mother is transmitting to the child via the placenta in late pregnancy, and it also seems to matter what type is given (Brenna et al., 2009).

† Maybe it's the stabbing: in the most common sort of liposuction, the target area is typically flooded with a solution that helps loosen the adipose tissue, then repeatedly punctured with a hollow needle called a cannula that sucks out a mix of the fluid and local cells and supportive tissue. For the record, most people are happy with the result, and in a properly licensed clinic it can be essentially safe. The issue here isn't whether *any* liposuction should occur; it's whether we should be treating subcutaneous adipose tissue as fundamentally nonessential, and its surgical removal as having no effect, particularly in women of reproductive age. More

accumulates on women's underarms post-liposuction is not the same kind of fat that was sucked from their thighs and buttocks. So I have to ask: With a violently disrupted store of LC-PUFAs, which may or may not be able to do quite what it did before, what happens if that body becomes pregnant?

I should not be the first person to ask this question. At some point during the many decades we've been "cosmetically" sucking out women's body fat as if it were as simple as getting a haircut, someone should have asked this question. Someone should have already run the study. No one has, much as I did try to get something going after I read that *Times* article.

But back then, I was a grad student in a department that didn't have the right sorts of freezers for storing the breast milk I intended to analyze—milk I'd meant to gather from a bunch of women in Manhattan who'd had liposuction years earlier and were now breast-feeding their children.* So I sent some emails to scientists at other labs. Everyone agreed that someone should do the study. Eventually someone will. Meanwhile, women keep undergoing liposuction, and no one has the foggiest clue if it matters which long-evolved depot of fat they destroy. As with huge swaths of modern medical science, female patients and their doctors are basically crossing their fingers.

Will everything be fine? Maybe. The maternal body is surprisingly resilient: battered on all sides, evolved to be so battered, and somehow, improbably, still alive. Human breast milk, as I've since come to learn, is also remarkably adaptive. All mammalian milk is. Making babies the way we do is a messy, dangerous business. It sucks, in fact.† But hey, it's *always sucked*, so the system has some fail-safes.

---

deeply, what's at stake is whether the ways we think about what might "affect" the female body take into account the deep history of mammalian evolution—that what we are is made of how we got here.

* There are rather important rules about how one handles human tissue in the sciences. Also, my little freezer in an apartment on the Upper West Side didn't exactly have consistent temperature control. And I had roommates.

† It's possible the best illustration of the verb "to suck" would be a diagram of the female human reproductive system. More on that in the "Womb" chapter.

. . .

While the majority of scientists still effectively ignore the female body, there's a quiet revolution in the science of womanhood brewing. In the last fifteen years, researchers in all sorts of fields have been discovering fascinating things about what it means to be a woman—to have evolved in the ways we have, with the body features we have—and how that could change the way we understand ourselves and our species as a whole. But the majority of scientists don't know about this revolution. And if scientists don't know about it—because they're not reading outside their field, and their field is still permeated by the male norm—how is anyone else going to piece it together?

You know that feeling when you realize that something needs to be done, and you're not sure you're the right person to do it, but damn it, *somebody should*? That was me in a crowded movie theater watching Ridley Scott exorcise his latest "mommy issue" in the form of a sexist medpod.* The lady in the row behind me felt it. I felt it. And I bet every other woman in that room felt it, too. For my part, it was like a kind of vertigo. I'd had the same feeling when I read the *Times* article about liposuction, the one that made casual fun of women for their newly fat arms. I was pretty sure neither the writer, nor the authors of the research paper the writer was reporting on, nor the women who had undergone the procedure knew that our adipose tissue and our livers and our immune systems all came from the same primordial organ, called the "fat body." That's probably why all three share so many properties: tissue regeneration, hormonal signaling, deep responsiveness to shifts in local environments. The ancient fat body is the reason you don't need to transplant an entire liver into a patient who needs one: a little lobe and you're good to go—the whole thing will regrow in situ. Adipose tissue famously regenerates, too. But unlike the liver, the separate fat depots in our bodies seem to be geared for different jobs, each intricately linked to the digestive, endocrine, and reproductive systems. This is why people who do research on adipose tissue have started call-

* For the record, I'm a huge fan of his work.

ing it an organ system: that's not a bit of fat under your chin but a small, barely visible part of your *fat organ.* Our subcutaneous fat does different things from the deep fatty deposits around our hearts and other vital organs. The fat on a woman's butt might be more important for her possible offspring than the fat under her arms.

We don't know when that started, exactly—most mammals have special fatty deposits near their ovaries and hindquarters—but we do have a rough guess as to when our ancient ancestors split off from fruit flies, which, by the way, still have the ancient "fat body": 600 million years ago. Thinking about that timescale for too long will give you vertigo, too, but at least it's a more *useful* sort. It gives you a reason as to why it's hard to "get rid" of one's fat: if adipose tissue is a body-wide organ system that has regenerative properties that go back 600 million years, maybe lopping off a piece of it in one spot naturally triggers a self-protective response that effectively "regrows" it elsewhere. And like anything that terribly old, there are bound to be younger, newer features laid on top: specialized regions, for instance, that don't grow back. Functionality that gets lost.

Bodies are basically units of time. What we call an individual "body" is a way of bounding a series of cascading events that follow self-replicating patterns until finally entropy sets in and enough goes wrong that the forces that keep you from flying apart at the seams finally let go. Species, in a way, are also units of time. But what's unusual about the body, when you start to think about it this way, is that your basic digestive system is radically old. Your brain is not. Your bladder is a workhorse, doing essentially the same job it's been doing for hundreds of millions of years—keeping the waste products of your many millions of cells' ongoing metabolism from poisoning you to death. It's not your bladder's fault that the mammalian uterus evolved to squat on top of it like Quasimodo. That only happened about forty million years ago. Actually, if we're talking about the gravity problem, that was only four million years ago. Before then, our ancestors had the good sense not to walk on two legs, smooshing all our long-evolved organs on top of one another in our trunks (not to mention generally screwing up the spine).

In 2012, when I got home from that movie theater, I realized we needed a kind of user's manual for the female mammal. A no-nonsense, hard-hitting, seriously researched (but readable) account of *what we are*. How female bodies evolved, how they work, what it really means to biologically be a woman. Something that would get the attention of both women in general and scientists. Something that would tear down the male norm and put better science in its place. Something that would rewrite the story of womanhood. Because that's exactly what we're doing in the lab now when we study sex differences. We're building a new story. A better story. A truer story.

This book is that story.* *Eve* traces the evolution of women's bodies, from tits to toes, and how that evolution shapes our lives today. By piecing this evolution together and connecting it to recent discoveries, I hope to provide the latest answers to women's most basic questions about their bodies. As it turns out, those basic questions are producing some truly exciting science: Why do we menstruate? Why do women live longer? Why are we more likely to get Alzheimer's? Why do girls score better at every academic subject than boys until puberty, when suddenly our scores drop through the floor? Is there really such a thing as the "Female Brain"? And why, seriously why, do we have to sweat through our sheets every night when we hit menopause?

To answer those sorts of questions, we have to make one very simple assumption: we *are* these bodies. Whether we are in pain or joyful, abled or disabled, in sickness or in health until death do we disassemble, our bodies and the brains they contain are quite simply what we are. We are this flesh, these bones, this brief concordance of matter. From the way we grow our nails to the way we think, everything we call human is fundamentally shaped by how our bodies evolved. And because, as a species, we are sexed, there are critical things we should be thinking about when we talk about what it means to be *Homo sapiens*. We have to put the

* Or at least the best I was able to do, from one little desk with access to a massive library and a small army of thankfully patient scientists and scholars willing to walk me through all the things I didn't initially understand.

female body in the picture. If we don't, it's not just feminism that's compromised. Modern medicine, neurobiology, paleoanthropology, even evolutionary biology all take a hit when we ignore the fact that half of us have breasts.

So it's time we talk about breasts. Breasts, and blood, and fat, and vaginas, and wombs—all of it. How they came to be and how we live with them now, no matter how weird or hilarious the truth is. In this book, I aim to trace what we're finally coming to understand about the evolution of women's bodies and how that deep history shapes our lives. And there's no better time for it: in laboratories and clinics across the world, scientists are now coming up with better theories, better evidence, better questions about the evolution of women. The last twenty years has seen a revolution in the science of womanhood. We're finally rewriting the story of what we are and how we came to be, chapter by chapter.

HOW TO THINK ABOUT 200 MILLION YEARS

So how exactly does someone go about writing the story of nearly every woman, everywhere, ever?

As long as you're willing to get a little dizzy, it's fairly straightforward. This is how the evolutionary history of women breaks down: Roughly 3.7 billion years ago, on the thin crust of our lonely little planet wobbling around its yellow star, there were isolated microbes. Between 1 and 2 billion years ago, eukaryotes appeared—single-celled organisms with a nucleus. (Think amoebas.) Then, through a scrambling up of many branching trunks on our evolutionary tree, the subphylum Vertebrata appears. The earliest fossil records of vertebrates—that is, animals with spines—date to 500 million years ago. Vertebrates still represent only about 1 percent of all living species.* Thus, the majority of what you and I call evolution—what we're debating about endlessly in litigation and fitful bursts on op-ed pages and conflict-

* Twenty-two percent of the world's species are egg-laying beetles. Seriously. In the history of life on Earth, beetles do really, really well.

ing textbooks in far-flung communities, this thing that has caused so much trouble—represents only 13 percent of the time there's been any life on Earth at all.

Once you start thinking about deep time, you quickly realize that human bodies are new because *all* bodies are pretty new. It really wasn't that long ago that we had thumbs on our feet instead of big toes. So to realize that how women's bodies evolved must shape how we experience our lives today isn't a stretch—it's a fact. Each of our bodies' features has its own evolutionary story, and we're still in the thick of it. Evolution works by building cheap upgrades on existing systems. Once one body feature is in place, that newly changed body interacts with its environment, and those interactions influence the rise of other features. Those new features lead to more changes, which often loop back and change the first feature: milk leads to nipples, and the caretaking habits involved in being a nursing mother help enable the development of the placental uterus. The placental uterus then influences our metabolism and the needs of our offspring, so breast milk starts to change. Breast milk changes, and eventually birth canals turn into petri dishes for the bacteria that help newborns digest sugary milk. In essence, the kid is coated on the way out with handy bugs that coevolved with our breast milk.

You see, evolution is a little like P. T. Anderson's film *Magnolia*, or Paul Haggis's *Crash*, or Iñárritu's *Babel*. You can't really follow it unless you're willing to pay attention to more than one major character. It's a complicated narrative, with a lot of whimsy and accident and things that seem to be unimportant at first but turn out to be vital. It's not a bildungsroman. But unlike oversimplified stories of our origins, it's true. Unraveling how each of our features really came to be gives us a better picture of what women are: one half of a very young, complex, and fascinating species.

That's the real problem with origin stories like the one in Genesis: our bodies aren't one thing. There's no *one* mother of us all. Each system in our body is effectively a different age, not only because the cellular turnover rate *differs* between cell type and location (your skin cells are far younger than most of your brain cells, for instance), but also because the things we think of as distinct to our species evolved at different times and in different places. We don't have one mother; we have *many*. And to each Eve, her particular Eden: We have the breasts we do because mammals evolved to make milk. We have the wombs we do because we evolved to "hatch" our eggs inside our own bodies. We have the faces we do, and our human sensory perception along with it, because primates evolved to live in trees. Our bipedal legs, our tool use, our fatty brains and chatty mouths and menopausal grandmothers—all of these traits that make us "human" came about at different times in our evolutionary past. In truth, we have *billions* of Edens, but just a handful of places and times that made our bodies the way they are. These particular Edens are often where we speciated: when our bodies evolved in ways that made us too different from others to be able to breed with them anymore. And if you want to understand women's bodies, it's largely these Eves and their Edens you need to think about.

And so each chapter in this book will follow one of our defining features all the way back to its origins—its Eve, or sometimes Eves, and their Edens, from the damp swamps of the late Triassic to the grassy knolls of the Pleistocene. I will also examine current debate around how the evolution of those features shapes women's lives today, considering the current science around each thread of the story.

Though I'll have to move back and forth in time to encompass all this, each trait will appear in the book in roughly the same order it *first* appeared in our evolutionary lineage. As such, each chapter builds on the last, moving forward in time and consequence, just as our bodies built later models of themselves on previous incarnations. Without those furry milk patches of our Eve of milk, we might never have evolved the fatty breast. Without the use of tools necessary for gynecology, we might never have evolved the sorts of societies that could support the *childhoods* that built our massive human brains. Without large, complex social groups that could support the elderly, enabled in part by gynecology, we might never have evolved to have menopause. Each evolutionary accident builds on prior accidents; each new feature depends on the circumstances that make it useful enough to outweigh its cost.

Once I established this as the order of my "manual," the way I went about choosing each feature for my chapters was fairly simple: I looked to our taxonomic address, the organizing principle that biologists use to determine what an organism is. Taxonomy outlines our relationship to the rest of life on the planet according to the features we share with others. Women, like all human beings, are *Homo sapiens.* Because we are mammals, we make milk. Because we are placentals, we have a uterus that gives birth to live young. Because we are primates, we have big eyes with color vision and ears that can hear a wide range of sound. Because we are hominins, we are bipedal and now have giant brains. And so on and so forth, climbing the evolutionary tree. As I examined each feature of our history, I then asked myself whether it had a particular story for *women:* Are there ways this trait affects us especially? Is there new research that's challenging our assumptions about this trait, and thereby, about all of humanity?

The most common way evolutionary biologists think about how traits work is by thinking about the last common ancestor of a trait we share with other species. Therefore, I located—or tried to locate—an Eve for each trait. For bipedalism, *Ardipithecus*—we just found her in 2009. For milk, a weird little weasel-beast that lived under the feet of dinosaurs!* By looking for an Eve, I often discovered surprising new research in paleontology and microbiology that challenged yet more assumptions about women's bodies.

Along with all of this, I invite you to think of yourself: to think about where your body comes from, how the evolution of biological sex shapes it—whether you identify as a man, a woman, or another gender—and how those stories are embedded in humanity's everyday life. In her essay for Annie Leibovitz's book *Women,* Susan Sontag wrote that "any large-scale picturing of women belongs to the ongoing story of how women are presented, and how they are invited to think of themselves." As such, it raises

* Because the deep and dark earth likes to keep her secrets well hidden, not everything has a known or obvious Eve: either we haven't found those fossils yet, the trait doesn't lend itself well to the fossil record, or we simply haven't fully discovered how to interpret the fossils we already have. But in every case, if I don't have a name for a beastie who directly falls in line, I look for an *exemplar* species or genus: a creature whose body and time and place we know a decent amount about, and whose history can teach us something about what our *true* Eves might have been like.

"the question of women—there is no equivalent 'question of men.' Men, unlike women, are not a work in progress." In scientific terms, Sontag is wrong: there is no stopping point for evolution. All of our species continues to evolve. But in the way she meant it—in the sense that looking at women begs a "question of women" while looking at men begs no question—she's absolutely right.

Why talk about the evolution of women, if it hadn't been neglected? Why focus this camera on the female form, unless it were still, amazingly, uncommon to do so? There is no more fundamental "picturing" of women than asking a reader to think about all women, everywhere, ever. And I am. I really am asking us each to look at women's bodies and think hard about how they shape what it means to be human.

THE EVES

"Morgie"—*Morganucodon.* 205 million years ago. Eve of mammalian milk. Initially found in Wales, but since found as far as China—this was a widely spread, highly successful creature. She was a bit like a cross between a weasel and a mouse. She's not assumed to be our direct ancestor, but an "exemplar" genus; our true lactating Eve was probably a lot like her.

"Donna"—*Protungulatum donnae.* 67–63 million years ago. Eve of placental mammals (not marsupials, not monotremes—but creatures with the sort of womb that humans have). Seems to appear right around the asteroid apocalypse that wiped out all the non-avian dinosaurs, but her line may stretch back into the Cretaceous. This Eve is highly specific and named, determined by extensive comparative fossil and genetic analysis. She's basically a weasel-squirrel.

"Purgi"—*Purgatorius.* 66–63 million years ago. An ancestor of primates and, by extension, our treetop-born primate sensory array. She is the Eve of primate perception: the reason women sense the world the way we do. Her fossils were found in the Fort Union Formation of Hell Creek, deep in the badlands of northeast Montana. So close to Donna she was basically contemporary. A monkey-weasel-squirrel.

"Ardi"—*Ardipithecus ramidus.* 4.4 million years ago. The first known bipedal hominin. There is an excellent fossil, only recently recognized. She is a big jump, both in time and in evolution, from the squirrely Eves that came before her.

"Habilis"—*Homo habilis.* 2.8–1.5 million years ago. She is the Eve of simple tools and associated intelligent sociality. A prolific tool user, Habilis coexisted in Africa with *Homo erectus* for half a million years. Her fossils were found at Olduvai Gorge in Tanzania.

"Erectus"—*Homo erectus.* 1.89 million–110,000 years ago. Erectus was a better tool user, was highly migratory, and had a big braincase. She is the Eve of more complex tools and more complex intelligent sociality. We'll look to her for one of the origins of our more humanlike brain (and perhaps at least some of the childhood that builds it).

"Sapiens"—*Homo sapiens.* Roughly 300,000 years ago to the present.* Eve of human language, human menopause, and modern human love and sexism.

* The precise start of our species remains highly contentious. Very few assume that the earliest hominins had true human language, modern-type menopause, or modern-type social rules around sex and gender; however, nor do many assume these traits predate our species. As with all things in the world of paleoanthropology, having more fossils from humanity's deep past would be phenomenally helpful.

OTHER PLAYERS

"Lucy"—*Australopithecus afarensis.* 3.85–2.95 million years ago. Many australopithecines are associated with tools, and the general assumption is that most if not all were early tool users of one stripe or another. Given that today's chimps are known to use tools, it would be odd to assume that ancient ancestors like Lucy didn't at least do the same, if not even more intelligently. *Australopithecus* are both among the best-known hominins (more than three hundred individual fossils have been found so far) and the longest lived of all the hominin species—in other words, theirs was a body plan and lifestyle that worked well for a very long time. Found in Ethiopia and Tanzania. Lived in trees and on the ground, fully bipedal.

"Africanus"—*Australopithecus africanus.* 3.3–2.1 million years ago. Her fossils were found in southern Africa, and it's unknown if Africanus is a descendant of Lucy's species. She had a bigger braincase than Lucy, and smaller teeth, but otherwise she was still pretty apelike, though bipedal.

"Heidelbergensis"—*Homo heidelbergensis.* 790,000–200,000 years ago, though she may stretch back to 1.3 million years. Probable ancestor of Neanderthals, Denisovans, and *Homo sapiens* (or she at least has a common ancestor with these), according to genetic research, with divergence around 350,000 to 400,000 years ago. The European branch led to Neanderthals. The African branch (*Homo rhodesiensis*) led to *Homo sapiens.* Heidelbergensis carried on, meanwhile, and died off just before *Homo sapiens* was officially on the scene. This was the first species to build simple shelters of wood and rock. She had definite control of fire and hunted with wooden spears—the first known hunter of large game (as opposed to scavenger). She lived in colder places and showed evidence of adapting for those problems. As the name implies, her fossils were first found in Germany, and later in Israel and France as well.

"Neanderthals"—*Homo neanderthalensis.* 400,000–40,000 years ago. Neanderthals coexisted with *Homo sapiens* as they spread through Europe, and the two interbred.* Anthropologists have found tons of fossils and living environments; this was a successful species. Former assumptions of Neanderthals have now been overturned: they are known to have had a complex culture, including burials, clothing, fire, and tool and jewelry making, and might even have been capable of language. Their braincases were shaped differently but don't seem to be *smaller* than *Homo sapiens'*—in fact, sometimes they were larger (which may correspond to their larger, robust bodies). They seem to have developed more quickly than we did, however; their childhoods were shorter.

"Denisovans"—Presumed to be *Homo denisova* or *Homo sapiens denisova*, though not yet formally described. 500,000–15,000 years ago. This Eve is known only from three teeth, a pinkie bone, and a lower jaw found in a cave in Siberia and through comparative DNA sequencing. Denisovans are known to have lived at least 120,000 years ago, with the longer time stretch inferred by sediment analysis and DNA research. Thought to be a small population, Denisovans lived in Siberia and eastern Asia, including at high altitudes in what is now Tibet, potentially passing down a gene that continues to help populations in these regions succeed at that altitude. DNA research establishes that many modern humans—particularly Melanesians and Indigenous Australians—share up to 5 percent of their DNA with these ancestors, implying that like the Neanderthals, ancient humans probably interbred with them. All this interbreeding, in fact, makes the "species" boundaries between these later hominin groups rather blurry.

* I, for one, have loads of Neanderthal in my genome, as do most people descended from recently European folk.

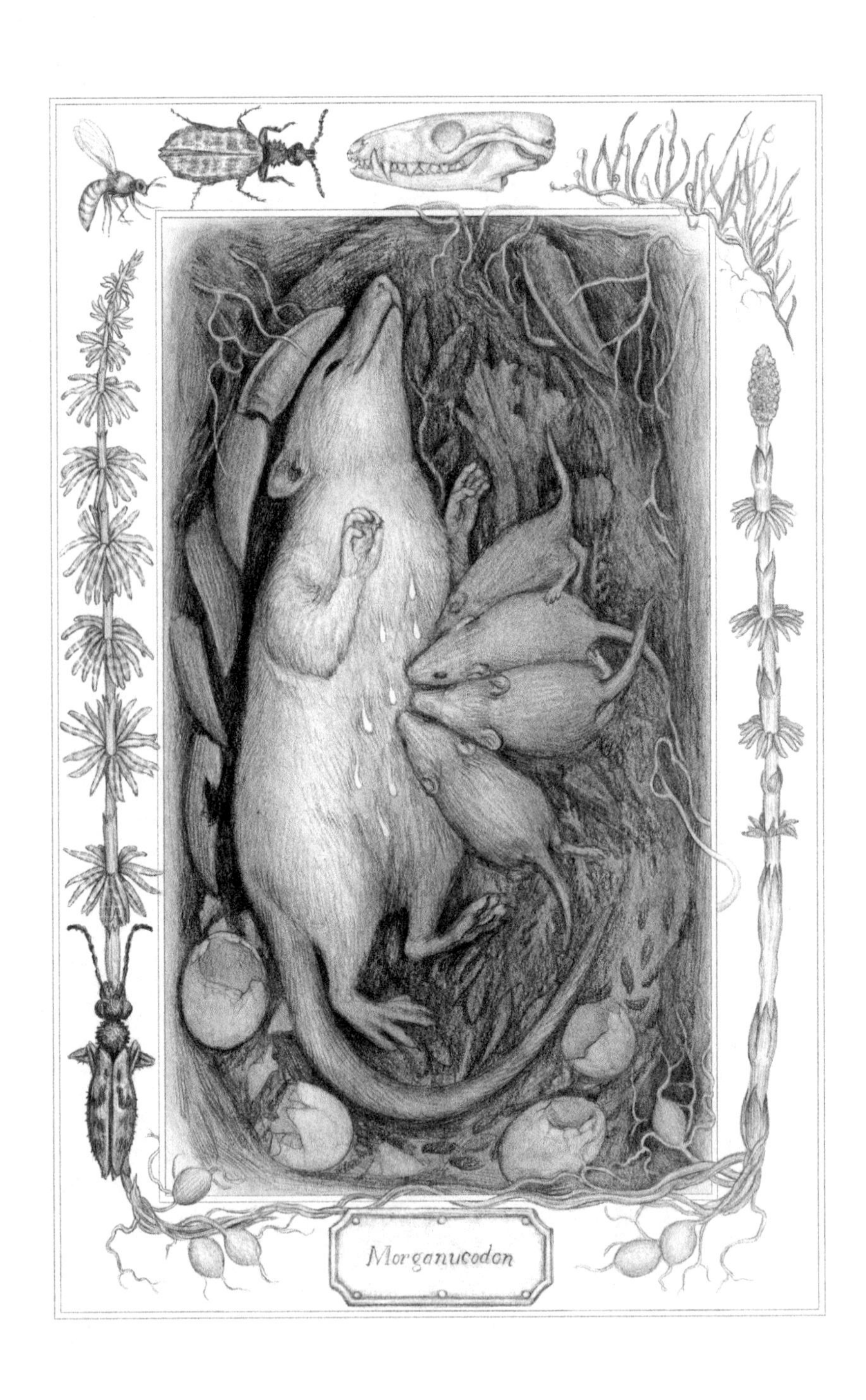
Morganucodon

CHAPTER 1

MILK

No sooner had the notion of the Flood subsided,
Than a hare paused amid the clover and trembling bellflowers and
said its prayer to the rainbow through the spider's web.
. . . .
Blood flowed in Bluebeard's house—in the slaughterhouses—in the
circuses, where God's seal made the windows blanch. Blood and milk
flowed together.

—ARTHUR RIMBAUD, "AFTER THE FLOOD"

Got Milk?

—ADVERTISING CAMPAIGN FOR
THE CALIFORNIA MILK PROCESSOR BOARD, 1993

There in the soft grass, in the wet crush of evening, she was waiting: furred body shirred with drops of rain, no bigger than a human thumb.

We call her Morgie. Little hunter. One of the first Eves.

She waited at the mouth of her burrow because the sky was still pale—streaky threads of photons refracting through clouds, the deepening blue beyond. She waited because her cells told her to, all the little clocks in her clockwork, and her eyes, and her whiskers twitching in the air, and the temperature of the dirt under her footpads. She waited because there were monsters in the world, and they waited for her, too.

When the night was good and dark, Morgie risked it, skittering along the ground, searching for her prey—insects, some nearly as big as she was. She heard them before she saw them: the high-pitched hum of their wings, the wheezy tapping of their legs. Quick as that, her skinny muzzle snapped. She loved the sweet crunch of its chitinous body, loved the little dribble of fluid down her chin. She licked it off and resumed the hunt. Never safe to stop. Jaws everywhere. Claws and teeth. The thing that looked like a tree could be a leg; the wind in the ferns could be hot breath. So, she ran, and hunted, and ran, and hid, the wet air heavy as a fist. She flitted over the feet of dinosaurs like a grasshopper hopping an elephant's toe. She felt their low bellows not as a sound so much as an earthquake.

This was life every night for *Morganucodon:* she who lived under giants.

When she was tired, she returned to her waiting place, fleeing the gray dawn. She crawled down her tunnel like a lizard, belly dragging over the familiar earth, paws pulling her forward into the close dark of home. The burrow was warm with the soft, radiating heat of her pups, all piled together. Their breath stank of old milk. Scraps of their leathery eggs mildewed gently in the dirt, along with urine and shit and dried spit, the smells mingling in the damp hole she'd dug for her family. A place safe from the monsters that thundered above. Safe enough.

Exhausted, she settled in. Her pups woke, blind and chirping, and swam across one another toward her belly, where beads of milk sweated out of her skin. Each pup jockeyed for the best spot. They slurped her wet fur, faces soon coated in milk. She stretched out on her side, whiskers finding the one closest to her head. Lazily she rolled him over on his back, nuzzling his unrolled ears, his thin eyelids, still closed. She dragged her raspy tongue down his belly to help him defecate, which he couldn't yet do on his own.

The milk and the crap and the egg scraps in that dark little burrow—these are the origins of breasts. Morgie is the real Madonna. Creatures like her nursed their young in a dangerous world, not only to feed them, but also to keep them safe.

To put it in the simplest terms, women have breasts because we make milk. Like all mammals, we nurse our young with a cloyingly sweet, watery goo that we secrete from specialized glands in our torso. Why human breasts are high on our chests, rather than near our pelvis, why we have only two of them instead of six or eight, and why they're surrounded, to varying degree, by fatty tissue that some people find sexually appealing are all questions we'll get to. But at the heart of things, human beings have breasts because we make milk.

And as far as the latest scientific research can determine, we make milk because we used to lay eggs and, weirdly, because we have a long-standing love affair with millions of bacteria. Both can be traced back to Morgie.

WHICH CAME FIRST, THE CHICKEN OR THE EGG . . .

Jurassic beasts tramped above Morgie's burrow every day. Meat eaters as big as lorries ran around like ostriches on steroids. Some, in fact, looked like ostriches on steroids. Loch Ness–style plesiosaurs lived in the seas. With all the big niches in the ecosystem taken, most of our early Eves evolved underfoot, which is hardly the place you wanted to be 200 million years ago. Even the earth was dangerous: the supercontinent, Pangaea, was starting to break up. Tectonic shifts tore Morgie's world apart. Water rushed in to fill the widening gaps, birthing new oceans with the hiss of lava hitting water.

Still, Morgie was an incredibly successful species. Her fossils have been found everywhere from South Wales to South China. Where there could be a Morgie, it seems, there was. She was adaptable. Resourceful. And she had a lot of kids. The geneticist J. B. S. Haldane* liked to say that God had an inordinate fondness for beetles, for he made so many of them; eating them was a successful strategy for insectivores like Morgie. For God so loved the beetles, and the furry, warm, heart-fluttering Eves who ate them.

* If you've ever heard of a "clone," that was Haldane. He was also the first person to compose a scientific paper in a forward trench—specifically in France during World War I. One of his co-authors, sadly, was killed in battle, so Haldane apologetically submitted for publication early, given the man's inability to collaborate further (Subramanian, 2020).

But it wasn't just the surfeit of beetles that made Morgie so successful. Unlike the Eves who came before her, Morgie nursed her young.

Once they are born, newborn animals face four essential dangers: desiccation, predation, starvation, and disease. They can die of thirst. Something can eat them. They can starve to death. And if they manage to dodge all of those, they can still die from bacteria or parasites overwhelming their immune systems. Every mother in the animal world has evolved strategies to try to protect her offspring, but Morgie managed to combat all four by dousing her kids in stuff made of her own body.

When we talk about breast milk, we usually describe it as a baby's first food. The last thing you want to do is underfeed a baby, because a newborn needs fuel to build new fat and blood and bone and tissue. As a result, we usually assume newborns cry for milk because they're hungry, but that is and isn't true. The most important thing infants need after they are born is water.

All living creatures, mammal or not, are mostly made of water. While the adult human body is 65 percent water, newborns are 75 percent. Most animals are essentially lumpy donuts filled with ocean. If you wanted to describe life on Earth in the simplest terms, you could say we're energetic bags of highly regulated water.

We use that water to transport molecules between cells, between organs, to splice molecules and build new ones, to fold proteins, to cushion our various lumps, to move nutrients and waste in the right directions. Our very DNA maintains its shape surrounded by carefully ordered molecules of water. An adult human can go without food for up to a month, but without water we die in three to four days. Any biologist will tell you that the story of life is really the story of water. Our earthly cells evolved in shallow oceans, and they never got over it.

So newborn Earth animals need water as soon as possible. Fish drink constantly from the second they hatch. On land, slaking a newborn's thirst is trickier. Some newborn reptiles are small enough that they can drink water droplets and absorb mist through their skin. Some seek out puddles and streams. Others, like newborn sea turtles, head straight for large bodies of water. But mammals seek the ocean in their mother's abdomen; human breast milk is almost 90 percent water.

Over time, ancient land mammals like Morgie evolved to slake their hatchlings' thirst with milk. There are a number of advantages to this adaptation. For example, the newborns don't have to move: the water comes to them. Pups of burrowing animals can stay in the safety of a small burrow a lot longer than creatures that need to get to water. Also, milk isn't just water but a balance of water and minerals and other useful stuff. Too much straight water all at once can be dangerous to very young mammals, and even grown human beings. There is such a thing as water poisoning, which causes all sorts of nasty side effects: Brain swelling. Delirium. Eventually, death. Our babies shouldn't even be given water until they're six months old. If they're thirsty, they should just drink more milk or formula.*

There were other advantages in replacing water with mother's milk. Water is an ideal medium for transmitting disease. That's why you're supposed to cover your mouth when you sneeze: tiny droplets of saliva and mucus hurl away from your mouth and nose at more than thirty-five miles an hour, each drop full of viruses and bacteria. That's why people started wearing masks in public in 2020: most airborne diseases actually "fly" from host to host in tiny droplets of fluid that have aerosolized. Either you breathe in a tiny droplet or a droplet lands on something you touch that makes its way to your face, where the moistness of your mouth, nose, and eye surfaces helps it replicate. Larger bodies of water are almost always host to millions upon millions of bacteria, some of which can be dangerous pathogens. Thus, controlling exposure to water and finding ways to ensure that drinking water is clean are two of the better strategies for maintaining the health of any animal.

Think of Morgie's body as the Jurassic world's best water filter. Tiny, fragile newborns are especially susceptible to patho-

* Very ill babies who can't keep milk or formula down are sometimes given a special mix of electrolytes, minerals, and water, like Pedialyte, to keep them hydrated until they're able to digest the good stuff again.

gens, in part because of their small size and in part because their newly independent immune systems are still developing. Morgie's milk might have contained whatever pathogens she happened to be carrying, but it wouldn't have introduced anything new to her pups. Her immune system could fight the good fight, until her pups were old enough to fight for themselves.

Scientists think milk evolved to solve both the desiccation and the immunological problem in one go. But how it started—how the very first droplets of milk actually formed—that's where the story takes an unexpected turn.

Like all the early mammaliaforms, Morgie laid eggs. And like many reptiles' today, hers were soft and leathery. When you crack a chicken's egg into a pan, you're actually tapping through a structure evolved by dinosaurs: a hard shell that prevents the liquid inside the egg from evaporating.* The eggs of most reptiles and insects, including the haphazard lineage that led to early mammals, were soft. There are a number of advantages to that strategy. For example, hard eggshells are primarily made of calcium. Like anything a body tries to build when making babies, all that calcium has to come from somewhere. Morgie was about the size of a modern field mouse. If she had tried to lay a chicken-style egg, it would have leached the calcium out of her little bones and teeth.† Even now, animals that make hard-shelled eggs are known to seek out calcium-rich diets before reproducing. (Chickens in industrial egg-laying farms often suffer from osteoporosis, their fragile leg bones breaking under the weight of their own bodies.)

But small leathery eggs, like Morgie's, can dry out before the

* Chickens are, after all, scientifically classified as "avian dinosaurs"—the direct descendants of Jurassic monsters. Hard-shelled eggs seem to have evolved three separate times in the dinosaur family tree (Norell et al., 2020).

† Modern human women are likewise advised to eat a calcium-rich diet when pregnant; it takes extra to build all those little bones. Pregnant women's bones and teeth are known to leach their own stores into the bloodstream; this can have serious effects for teenage mothers, whose own bones are still growing. If the diet doesn't provide enough for both mom and baby, she may be likelier to face dental work and osteoporosis down the road.

pups are ready to hatch. So Morgie didn't just need to keep her clutch warm; she needed to keep it wet.

There are a few different ways to do this. Modern sea turtles, for example, find a nice patch of damp sand, above the tide line, and bury their soft eggs in a shallow pit, coating each one in a thick, clear mucus they secrete during the birthing process. If you're a more attentive sort of mother, you might still use the mucus trick, but you'll also hang around and periodically lick the eggs or secrete some more goo onto them. That's what the duck-billed platypus does. One of the last living mammals that still lays eggs, the platypus first digs a wet den, then lines it with soggy vegetable matter. Crawling into the center of that damp pit, she lays her clutch directly on her body and folds her tail over them. She waits there, curled around her eggs, until they hatch. Platypus eggs also have an extra mucoid layer that persists until birth and is especially dense in antimicrobial material.

Morgie needed to keep her eggs moist, but she also needed to keep them from becoming festering breeding grounds for water-borne bacteria and fungus. Most scientists assume her egg mucus contained a host of antifungal and antibacterial material as the sea turtle and platypus mothers' mucus still does.

When today's leather-egged offspring are ready to hatch, they use a specially evolved tool (usually a sharp "egg tooth" that later falls out) to puncture the shell. Then they also lick up some of the egg-coating goo. Their first meal, in fact, is from the wet side of the eggshell. In all likelihood, this was the first mother's milk: an egg-moistening mucus that Morgie's grandmother secreted out of specialized glands near her pelvis. When her pups hatched, some of them licked up a bit of this extra stuff, which gave these offspring a serious evolutionary boost. By the time Morgie came along, these glands had evolved to secrete a goo containing more water, sugars, and lipids. Eventually they became "mammary patches" with specialized bits of fur over them that helped channel the gunk into the pups' eager mouths. Even today, newborn duck-billed platypuses lick milk from sweaty milk patches on their mother's stomach; she doesn't have nipples.

. . .

Early mammalian milk was probably a lot like modern women's colostrum: a thick, yellowish, sticky-sweet ichor, super dense in immunological material and protein. For the first few days after a woman gives birth, her milk is incredibly special—a hot shot of immune system for her newborn baby. New mothers can find colostrum alarming, since it looks a bit like pus, but within a few days it converts to the bluish-white stuff we're used to calling breast milk. Most mammals have this pattern: first colostrum, and later a thinner, mature milk that is richer in fat. Each of those fat globules is surrounded by a membrane that contains xanthine oxidoreductase—an enzyme that helps kill a ton of unwanted, dangerous microbes.

But colostrum is especially dense with immunoglobulins: antibodies tagged to respond to pathogens that the mother's body knows to be dangerous. In fact, before we discovered penicillin, cow colostrum was commonly used as an antibiotic.*

Despite its obvious benefits, human women throughout history have mistakenly believed colostrum to be rotten milk, or what they called beestings. Some even avoided giving it to their babies. In the fifteenth century, Bartholomäus Metlinger wrote the first European textbook for pediatrics. Despite the German's own lack of breasts, he didn't hesitate to mansplain women's milk and what to do with it:

> The first 14 days it is better that another woman suckle the child as the milk of the mother of the child is not as healthy, and during this time the mother should have her breast sucked by a young wolf.

I can't imagine where he thought each new mother would find a young wolf. But any recommendation that babies not be given colostrum as a matter of practice was, and is, dead wrong. A mammal's lactation pattern—from thick, yellow, protein-heavy colostrum to thin, white, fat-heavy milk—is specially geared toward a newborn's development. Timing is everything here. The four dangers—desiccation, predation, starvation, and disease—are dif-

* It's also used to make a particularly sweet Indian cheese.

ferently dangerous according to a timetable. In a burrow, desiccation is the first danger, both for the eggs and for the newly hatched. Starvation comes quite a bit later, since a body can always eat a bit of itself to survive.* Predation is also a later problem, especially if the baby doesn't need to leave its underground bassinet for a while. But disease is a big deal right off the bat. Colostrum doesn't just boost the kid's immune system by injecting antibodies; it's also a reliable laxative, which is also crucial to building a baby's immune system.

On top of the thick yellow stuff coming out of her nipples, a new human mother might also be startled by what's coming out of her little darling's behind. Meconium, a baby's first poop—actually first few poops—is thick, tarry, and alarmingly green-black. It doesn't smell like much, thankfully, because it's mostly broken-down blood, protein, and fluid the fetus ingested inside the womb. But it's important that the stuff comes out fairly soon, and the laxative properties of colostrum help hurry that along—so well, in fact, that the intestines of a newborn drinking colostrum are wiped relatively clean. Which is precisely what needs to happen.

Before babies start to digest the food that will give them energy, they need to line their intestines with bacteria to help them break that food down. Mammals coevolved with their gut bacteria, because it takes a village.

Friendly bacteria—present in the mother's milk, in her vagina, and on her skin—rapidly colonize a newborn's intestines. Think of a new neighborhood: whatever group moves in first has a big influence on how the place evolves. Because of the relative lack of competition, those early bacterial colonies thrive, reproducing themselves all along the intestinal walls. Initial colonies in newborns' guts also have ways of communicating with cells in the tissue of the intestine. Toll-like receptors learn, like a neighborhood watch, which types of bacteria should be catered to and which

* This is part of why human newborns usually lose weight in the first weeks after they're born: they gobble up their own fat reserves until their mother's milk converts from colostrum to mature milk and they're able to take in—and digest—a proper meal.

are dangerous. The earliest occupants have a profound influence over these receptors. That's one reason why preemie babies in the NICU are usually given donated breast milk and concentrated colostrum if the hospital can get it: their immune systems can be dangerously compromised without it.

Colostrum doesn't just clear the path for the early bacterial settlers. It also contains bacterial growth factors that help those colonies gain a foothold. A growing neighborhood might need a combination of public services and small-business loans, but for intestinal bacteria it's a hot dose of 60-sialyllactose. That's an oligosaccharide, one of the special milk sugars our breasts make for our babies. The early bacterial colonizers of newborn intestines—that is, *Bifidobacterium*, *Clostridium*, and *E. coli* (the good kind)—really like this stuff. It's their nectar of the gods. It helps them not only grow and reproduce but develop complicated biofilms: connected colonies of bacteria that, instead of just loosely floating around, adhere to the intestinal wall. Once they've set up shop, these bacteria help the newborn digest the milk their mothers are feeding them. What's more, we've recently discovered that oligosaccharides themselves can help deter dangerous pathogens from adhering to intestinal walls. Unable to find a comfortable, noncompetitive spot, the unwelcome invaders drift through and eventually get pooped out.

And that's one of the most surprising discoveries about breast milk. Just in the last decade or so, scientists have come to realize that maybe its nutritional value isn't the biggest deal off the top. Milk is really about infrastructure. It's city planning. Some combination of a police force, waste management, and civil engineers.

There's one last point to make against the idea that mammals' milk evolved mostly for nutrition. It turns out a significant portion of our milk isn't even digestible.

Modern human milk is mostly water. Among the things that are not water—proteins, enzymes, lipids, sugars, bacteria, hormones, maternal immuno-cells, and minerals—one stands out. The 6′-sialyllactose that colostrum delivers to newborn baby guts is not the only oligosaccharide in breast milk. In fact, the third-largest solid component of milk comprises oligosaccharides. These complex, milk-specific sugars aren't even *digestible* by the human body. We don't use them. They're not for us. They're for our bacteria.

Oligosaccharides are prebiotics: material that promotes the growth and generally ensures the well-being of friendly bacteria in the intestines. Prebiotics also promote certain kinds of activities among these bacteria: for example, the sort of activity that annihilates unfriendly bacteria. Commensal digestive bacteria have a complex and irreplaceable role in your digestive and immune systems, the features of which we're only starting to understand. But without prebiotics, they're up shit creek without a paddle. (Prebiotics are not the probiotics you've likely heard of—bacteria like *L. acidophilus* that the human body naturally contains. Eating fistfuls of probiotics on their own is a little like planting your garden without fertilizer, or maybe even without soil. You need prebiotics to make the whole system work.)

These special milk sugars are the target of an entirely new industry in the United States: that of lab-processed, powdered, and/or concentrated human breast milk, drawn from women who are sometimes paid handsomely for their donations. Nonprofit milk banks don't pay the mothers who donate their breast milk, because they see themselves as providing a service for patients who need such breast milk for medical reasons. These for-profit companies, on the other hand, dehydrate milk they've *purchased* from human mothers and then sell the product to hospitals, hoping to profit from providing the extra hit of oligosaccharides preemies need to start their young lives. At a cost of up to $10,000 for a few short weeks, daily doses of concentrated human breast milk product can help these little patients gain weight and develop a mature immune system more quickly.*

Other biotech companies are trying to create human-type

* The ethical questions surrounding how these women are paid are a bit less straightforward. For instance, one company—Medolac—was roundly criticized by an advocate group for African American women in Detroit, because it was believed that the company was specifically targeting poor women for donation (Swanson, 2016). If those women felt pressured to donate more milk than they really had as "extra," it could cause their own babies to suffer.

oligosaccharides on their own, eliminating the need for women's breast milk. It's unclear whether it will be more financially viable to create the sugars de novo or source them from paid donors or whether there'd even be a market for these sugars outside human infants. Scientists are feverishly trying to figure out whether they might be part of a medical treatment for patients with Crohn's disease, IBS, diabetes, or obesity, for instance. But we simply don't know if the adult microbiome would benefit from the same sorts of prebiotics that infant intestinal colonies do. Technically, the bacteria are the same bugs. But how they interact with infant intestinal walls, and how those walls help "teach" the infant's immune system in a critical window of development, is at the cutting edge of current knowledge. We know that mammals' milk coevolved with mammalian guts. We know our bacteria matter to our well-being. But precisely how and why and when? Ask again in twenty years.

Still, humans are not known for behaving rationally when it comes to our own bodies. Some bodybuilders, for example, currently buy human breast milk on the black market, erroneously believing it will help them build muscle—even though human breast milk has far less protein than cow's milk and protein is what muscle tissue is primarily made of. If getting swole were the goal, it'd be far cheaper and more effective to buy and drink a quart of cow's milk.

Two hundred million years before there was ever such a thing as pseudoscience, much less a supplements aisle, Morgie squatted in her little burrow, half-drugged with the smell of her sleeping young, a rush of pleasant feelings flooding her brain. And deep in the warm dark of her intestines, her bacterial colonies did what they always do: ferment sugars, help her body absorb minerals, and co-regulate her immune system. And maybe that's the thing. If milk's original purpose wasn't feeding our young, but solving the water and immune problems, and then evolved those nutritional properties after the fact—a wonderful door prize, if you like—then it's safe to say that the story of milk isn't just about us. It's about what "us" should mean.

After all, giving birth isn't just when *you* reproduce. It's also a key moment for the bacteria in and on your body: the construction of an entirely new environment that's especially suited for their survival. The ways your bacteria aid in the process might even fall under the umbrella of what biologists call "niche construction." In the simplest terms, niche construction is the way in which organisms change an environment to better suit their children and grandchildren. A beaver, for example, creates a dam that widens and deepens the watercourse it blocks into a pool, changing that ecosystem to better suit the beaver and its offspring. Different sorts of fish thrive in these deeper waters, and different sorts of riverine birds—even different strata of microorganisms: deeper waters, dammed by the beaver, are a very different ecosystem from a beaver-less creek. And so, some scientists argue, the beaver's children inherit both their parents' genetic material and a changed environment.* There's an intimate two-way relationship between the evolution of our genes and the inherited, changed environments that the expression of those genes produce.

So how are our digestive systems and gut bacteria like beavers and their dams? Put it this way: the main road through our organismal city runs from mouth to anus. What's inside your digestive tract is, technically, outside you, though bacteria are so interwoven with our intestines' function it's hard to say where the intestines stop and the bacteria begin. Destroying all the bacteria in a person's intestines can be life threatening. Hospital patients on industrial-grade antibiotics are famously prone to *C. difficile* infections, which are very hard to get rid of. Until recently, these patients would have no choice but to suffer repeated bouts of exhausting diarrhea, and even risk dying. The best cure for it, as we've only learned in the last ten years, involves pumping a brown slurry of a healthy person's poop into the patient's intestines. Some

* Some even say the dam should count as an "extended phenotype," given that specific behavioral outcomes produced by the beaver's genotype are the things that build the dam and, critically, the success of those genes' propagation depends on the dam (Dawkins, 1982/1999). Thus, much as one's body traits are the "phenotype" of one's genotype, the beaver's dam is an extra-organismal extension of that phenotype. It's important to know where to bound these arguments, of course; not everything an organism produces should count as an extended phenotype.

feel better in a couple of days. Many are cured entirely within a week.*

Here's the thing: a beaver's river doesn't usually up and die eighty or so years after the dam is built. Human intestines do. So, if our gut bacteria are in the business of passing on their genes, they're going to evolve in ways that help their descendants colonize the intestines of their hosts' babies. In mammals, milk is one of the key ways that happens. Our milk changes depending on our environment and the sorts of things we eat, which makes sense, given that breast milk is one of the first ways we protect our children, and it needs to be responsive to both local resources and local dangers. You can see that responsiveness in individual species, too; chimps, for example, have markedly different breast milk in the wild than they do in zoos (as do human women with differing diets). But what remains consistent in human milk, no matter where we are and what we're eating, is the extraordinary number of oligosaccharides we stuff in there. In fact, human milk has the most, and most diverse, oligosaccharides of all our primate cousins', probably because unlike other apes modern humans have had to deal with cities and high-speed travel.

Cities are bacterial cesspools. Humans are not just social primates; we're *super* social. By living in such close quarters, day and night, human bodies regularly encounter an onslaught of foreign bacteria. Pathogens can easily jump from host to host, moving through a large population like wildfire. What's more, because we've invented technologies that manage to haul our bodies (and their bacteria) across land and sea so quickly, each population at each new port of call has to confront whatever bacterial guests we happen to bring with us. Some scientists think our milk sugars are so different from other primates' because they evolved to help our gut bacteria handle our crazy human lifestyle. They may even provide clues to specific infections our ancestors had in the past: not only do our special milk sugars feed friendly bacteria, but they

* Do not do this at home. Right now, the FDA approves FMT (fecal material transplant) for *C. diff* infections only. It's in clinical trials for all sorts of other things, from obesity and IBS to lupus and rheumatoid arthritis. No one knows if any of those treatments will pan out. Meanwhile, the best advice still stands: don't put things up your butt unless you really, really know what you're doing.

can also trick unwanted pathogens to bind to them instead of to an infant's intestines, and then send them into the diaper.

Our guts are, in essence, as social as our brains—or at least as influenced by our disease-prone social nature, and that history has pressured our milk to change, too. Forget about the Paleo diet: modern *Homo sapiens* have already adapted to urbanization and the bacterial challenges that come with it.

MILK IS PERSONAL

When domesticated cats cozy up to their owners/roommates/ Known Food Providers, they'll often push their forepaws against the owner's body—left paw, right paw, left paw, right paw. When kittens nurse, they do this same motion: kneading their mother's belly on either side of the nipple, helping push out the milk into their waiting mouths. Animal behaviorists think this action is something older cats do when they're content and bonding, that the body motion is just so ingrained in them from birth that, even sans nipple, their paws go to work as a part of a familial pleasure circuit. They'll do it when they feel good. They'll do it when they want to feel good. They'll do it when they feel bonded to another being. And maybe they'll do it when they're bored.

Human babies don't nurse from a string of teats the way cats do. Maybe that's why our infants don't display this push-push pattern. What babies do have is the ability to suckle. And they are able to do that because women have nipples.

Except the platypus and the echidna, all mammals living today have teats: raised, porous, nubbly patches of skin under which highly evolved milk glands get to work when mothers need to nurse their children. At some point before marsupials and placentals arrived—somewhere between Morgie's 200 million years and marsupials' 100 million years—the Eve of nipples was born. On her holy chest were not just a few sweating patches of fur but thickened bumps of skin that helped her kid latch on.

The modern human nipple is a thicker nub of skin on a woman's chest surrounded by a flattish patch of darker skin called the areola. The average nipple has fifteen to twenty small holes that are connected via tubes to the milk glands in the breast. When a female mammal becomes pregnant, the tissue around the nipple becomes engorged with blood and new tissue as the milk glands gear up for production. The skin becomes darker and redder. Veins swell. New capillary branches feed the growing tissue. For many mammals, this is when their nipples first become apparent to an outside observer, as the teats swell past the fur of the female's underbelly, following two long lines from armpit to groin. For humans, whose nipples are generally not covered by hair, others will spot the change in shape and size.

From a waste-management point of view, it's obvious why nipples evolved. Though Morgie's sweaty milk patches probably did have "mammary hairs" that helped guide the milk into her pup's mouths, that system had a lot of slop. Inevitably milk would be wasted. Since it takes a lot of energy to make it, having a more specialized access port to the milk glands seems like an easy product of evolution. Controlling for slop wasn't the only waste-management feature of the nipple. While the mammalian body does produce a bit of milk of its own accord—pregnant women "leaking" at various inopportune moments during business meetings or on the subway or in a particularly emotional argument with one's partner—it's nothing compared with what it does in response to suckling.

For nippled mammals, the majority of milk is a "co-produced biological product." That means that while it's the mother's body that produces it, the infant's mouth is the thing that triggers the mother's body to do so. What's more, the infant has a significant role in the type of milk that the mother's body makes. There are a few different mechanisms involved, but the most important are these two: the let-down reflex and the vacuum.

Contrary to popular belief, a nursing mother's breasts are not full of milk. They're swollen, sure, sometimes to the point that they resemble fleshy water balloons, but they're full of blood, fat, and glandular tissue. There's no bladder in a breast that holds a sloshing cup of milk that empties as the baby nurses and then gradually fills up again, ready for next time. Even a dairy cow's udder isn't the bag of milk you might think it to be; like us, a cow's udder is a visible mound of mammary tissue, along with a few nipples.[*] The ductwork of a nursing human breast can hold, at most, a couple of tablespoons of milk at a time. It's the act of suckling that normally triggers a breast's "let-down reflex"—a cascade of signals that tell the milk glands to kick up production and dump fresh milk out the front door.

It's a lot like what your mouth does when it comes to saliva. Chewing your way through a typical meal produces about half a cup of spit. But you don't have half a cup of spit in your mouth at all times, ready to go. Your salivary glands get the signal to start amping up saliva production when you smell something tasty, and most especially when you start chewing.

When an infant begins to suckle, the nerves in the breasts send signals to a mammalian mother's brain. In response, the brain tells the pituitary gland to produce a lot more of two specific molecules: the protein prolactin and the peptide oxytocin. Prolactin stimulates milk production. And oxytocin helps squeeze the milk out of the glands into the waiting ducts, which are then emptied by suction from the baby's working mouth.

These two molecules have roots tied to the evolution of milk itself. Some of those roots go even further back than Morgie. Prolactin has been around since fish evolved. In fish, it seems to be mostly tied to regulating salt balance. Moving up the evolutionary chain, prolactin has a number of functions in the immune system. Nowadays, it's also tied to sexual satisfaction: no matter your gender, the more prolactin you have in your body after sex, the more satisfied and relaxed you feel. This may be because prolactin counteracts dopamine, which your body produces in buckets when you're sexually aroused. Likewise, if you have too much prolactin in your system, you're more likely to suffer from impotence.[†]

* And like us, dairy cows tend to produce the most milk overnight and first thing in the morning; most mammals' milk production is tied to a diurnal cycle of hormones. That's why a farmer's first task of the day is milking the cow: a cow with swollen udders is going to be especially cranky if you don't tend to her fast, along with more in danger of developing a mammary infection and/or losing her milk supply. (I got mastitis twice. Hideously painful. I've never had more sympathy for cows than when I had to nurse my children.)

† This is true of both male and female bodies. Many lactating folk find their sex drives, and general sexual satisfaction, plummet while breast-feeding. There are many reasons for this, but not all are "psychological." Prolactin is one obvious

Oxytocin also evolved to serve multiple purposes. This little peptide has garnered tons of attention lately because of its association with emotional bonding. Some of the science around oxytocin is good, and some of it is so tainted with stereotypes of femininity that we might as well dress it up in a frilly pink tutu: "Oxytocin makes you love your baby." "Oxytocin makes you love your man." "Monogamous men make more oxytocin than men who're going to cheat on you." While oxytocin does seem to be associated with a number of psychological states in various mammals, and higher levels of oxytocin are associated with more pro-social behaviors, there are simply too many other factors that produce these things to treat oxytocin as a solo player. Also, while human beings behave more altruistically toward members of their own group after a dose of oxytocin, they also act more defensively and aggressively against people they perceive as being *out* of their group—so it's hardly the angel of our better nature. And no one really knows what oxytocin is doing in the brain: Does it make us interpret others' social signals differently? Does it just make us pay more attention to faces? Does it simply make us feel warmer toward known things (like people we know) than toward unknown things (people we don't know)? In the end, the only thing we're *dead* sure oxytocin does is make certain kinds of tissues contract.

When you have an orgasm, oxytocin tells the muscles in your pelvis and lower abdomen to rhythmically contract. This is true for both men and women. For men, these contractions help shoot sperm out of the urethra—and also happen to pulse the muscles in the buttocks and anus, making them more likely to fart. For the woman in mid-orgasm, muscles in the uterus and vagina will pulse, and the anus and the buttocks and upper thighs will often come along for the ride. Sometimes those uterine contractions are so powerful they don't entirely stop after the event is over, and she'll experience rather painful aftershocks, like menstrual cramps (which, by the by, also involve the oxytocin pathway, helping the uterus rhythmically and sometimes painfully contract in order to

---

factor. Estrogen and progesterone play a role, too. Vaginal tissue tends to suffer a bit in the breast-feeding woman, typically becoming more dry and fragile. That can make postpartum sex painful, even after birth injuries have had time to heal.

slough off its old lining). When a woman goes into labor, oxytocin is a major player. It's so important for childbirth, in fact, that it's listed by the WHO as one of the world's "essential medicines."

Similarly, when a baby suckles and the pituitary gland up-regulates oxytocin, a nursing mother might also experience a deep sense of contentment and social bonding with her baby. Post-orgasm men and women tend to feel that, too, to varying degrees. We don't know when, exactly, oxytocin's "contraction" function became tied to the mammalian brain's "social bonding" and "feel good" signals, but now they tend to be coupled.

When a human baby suckles, it wraps its mouth around the mother's entire areola, the flesh of its lips splayed out in a kind of lamprey-like O. In response to being touched, the nipple contracts into a fleshy forward-jutting pyramid. When the kid properly latches on, the base of the pyramid rests on top of the baby's toothless lower gum, its tip extending all the way to the back of its mouth. And then the cheeks contract, sucking all the air out of the mouth, creating a vacuum around the nipple that helps pull the milk, freed by oxytocin, into the baby's throat. The tongue and muscles of the lower jaw roll front to back, massaging the nipple from base to tip, squeezing all that vacuum-hoovered milk out of it. Some of the milk can splash up into the lower sinuses and bubble out the baby's tiny nose, but most of it goes down the esophagus, swallowed in between gulps of air. The mechanics of the whole thing are quite the production.

Suckling is not something a newborn mammal always knows how to do. Though the "rooting" instinct seems universal in mammals—the way that a baby will start nudging its head around, looking for a nipple, when it comes near a large, warm, soft surface—latching on is quite a bit harder. Some babies wrap their lips just around the tip of the nipple's pyramid and can't form a good vacuum. Some get the vacuum part down, but don't move their tongue and jaw the way they're supposed to. Some appear to become so frustrated by it all they don't even bother, leading both the baby and the mother, exhausted, to cry.

And cry she might, poor Morgie's daughter, for her nipples may dry out and crack and bleed, sucked and gummed raw by a child who can't figure out how to feed. (My firstborn damaged my nipples so badly in the first twenty-four hours that they bloomed with black-purple bruises, alarming even the battle-hardened nurses assigned to my care.*) Latching can be such a problem, in fact, that a crop of "lactation consultants" have sprouted in hospitals to help new mothers teach their babies how to do this odd, recently evolved thing with their mouths. Most figure out how to do it. Eventually. But in evolutionary terms, the breast knows how to milk better than the mouth knows how to suck.

Luckily, the nipple evolved one useful compensatory measure to help withstand the learning process. Some nipple holes connect not to milk glands but to Montgomery's glands, which produce a greasy substance that coats the nipple and helps prevent the skin from being totally destroyed by insistent gumming. When a woman is pregnant, the Montgomery's glands swell and make the nipple look a bit "bumpy." For some of us, those little bumps are visible all the time. Like milk glands themselves, the Montgomery's glands probably evolved out of primitive sebaceous glands that naturally proliferate in the skin. But instead of producing the usual skin oils, the Montgomery's glands came to pump an industrial-grade lubricant that could withstand the kind of chafing a nursing baby inflicts.

It's the vacuum, though, that really changed the breast game—being able to seal a kind of docking station between the mother's body and her offspring's. Once that evolved, milk stopped being something the mother's body made on its own and started being something the mother's and baby's bodies make together. As the rhythmic rolling of the baby's tongue and jaw move the focus of the vacuum back and forth, a kind of *tide* forms between the breast and the mouth. In that rolling wave, the milk flows up over the top while, on the bottom, the baby's spit is being sucked back into the mother's nipple, in a kind of evolutionarily purposeful backwash. Lactation scientists call this the "upsuck." And that's where things get really interesting.

* He didn't have a tongue tie, for what it's worth. He just decided to chomp instead of suckle. It took weeks to recover. Meanwhile, I formed an intimate relationship with a breast pump, and he developed an intimate relationship with silicone nipples. This is fantastically common for new mothers.

The nipple itself is packed full of nerves to help detect that vacuum, which starts the chain reaction of oxytocin for the let-down reflex. That's why, for example, modern women can use a breast pump. Just about any vacuum will do to trigger milk production. But what breast pumps obviously can't do is inject salivary backwash into the nipple. Lining the mother's milk ducts, from the nipple all the way to the glands, are an army of immuno-agents. And depending on what happens to be in baby's spit that day, the mother's breasts will change the particular composition of her milk.

If a baby is fighting an infection, for example, various signals of that infection, from actual infectious agents like viruses and bacteria to more subtle indicators like the stress hormone cortisol, will be present in the baby's spit. When that spit gets sucked up into the mother's breast, the tissue reacts and her immune system will produce agents to fight the pathogen. Her milk will carry them into the baby's mouth, providing extra soldiers to combat the infection and help the baby's own immune system learn what it needs to fight. In response to raised cortisol, the milk glands and surrounding tissue will also bump up the dosage of immuno-agents in the daily brew, and it may also send down the line a number of signals to soothe the child. Some of those signals are hormonal—stuff to directly counteract the inflammatory properties of cortisol. Some of them are nutritional, with added knock-on effects to change the baby's mood. For example, milk produced by a breast that's nursing a child who's stressed tends to have differing ratios of sugars and fats, providing extra energy to help the baby's body fight off any potential invasion. It can also work as an analgesic, damping the baby's pain response and helping it rest; after all, quite a lot of healing happens when we're calm and asleep. These sorts of responsive features seem to be true across Mammalia, the particular magic potion varying from species to species—different bodies need different sorts of breast-borne chicken noodle soup—but the overall principle holds true.

The resulting effect is so powerful that when many babies grow up, their brains still associate milk-related signals with healing and comfort. Eating fat-dense and/or high-carb foods, especially if they taste sweet—the sort that many humans tend to seek when feeling stressed or lonely—produces an analgesic effect in a number of different mammals. For rat and human alike, "comfort food" can dampen the body's pain response, a kind of grown-up breast substitute.*

The evolution of mammalian nipples provided a new, vacuum-sealed transmission point between mother and child. It was a way for them both to make milk together and to communicate with each other. Communication, in fact, is such a deep feature of mammalian nursing that it's not just a matter of the nipples; the ways and occasions in which mothers nurse their babies are also shaped by the things we want to "say" to each other. Mother felids tend to rumble and pant; apes hoot and lip smack. The majority of human women favor cradling their infants and nursing them from our left breast, which also happens to line our baby up with the side of our face that is more expressive. No, really—and other primates do this, too. Among humans, the muscles on the left side of the face are slightly more adept at social signaling, and 60 to 90 percent of women preferentially cradle infants toward the left of the body's midline, with the baby's head more exposed to the left side of her face. This preference is strongest in the infant's first three months of life, precisely the period when new mothers are nursing more often throughout the day. This is true across many human cultures and historical periods.

Meanwhile, the right hemisphere of the adult brain is largely responsible for interpreting human social-emotional cues, and it receives those signals dominantly through the left eye. So the mother's left eye carefully watches the infant's face, interpreting

* Unfortunately, eating sugary foods also tends to produce a sugar crash shortly thereafter, which can feel considerably less comforting. Emotional "pain" maps in the brain in strikingly similar ways to physical pain, and aspirin, ibuprofen, and even Tylenol can work pretty well on that, too. According to a few recent studies, taking a common over-the-counter painkiller *before* you encounter negative events can greatly influence how emotionally pained you'll feel (Mischkowski, Crocker, and Way, 2016). It may not help as much after the fact, unfortunately—a good deal of the pain of remembering has to do with your emotional state at the time of encoding—so if you happen to know that you're about to break up with your boyfriend, take a couple of ibuprofen or Tylenol. It takes about thirty minutes to kick in. Unfortunately, it may also reduce your empathy for your partner's pain, so do with that what you will (ibid.).

the baby's emotional state, while the infant gazes intently up at the most expressive side of the mother's face, learning how to read her emotions and respond—something that human beings spend huge portions of their childhoods learning how to do.

MILK IS SOCIAL

When Morgie came back from hunting each dawn, she was stressed. Of course she was: she lived in a stressful world. But if her environment had been more dangerous than usual that night, or if she was hungrier than normal, her body would have produced a higher dose of cortisol. And when she rolled onto her side to nurse her pups, her milk would have contained similarly higher levels.

Milk with a lot of cortisol tends (at least in rats and mice and certain kinds of monkeys) to produce baby personalities that are less risk seeking, and those traits seem to persist through the individual's lifetime. These individuals explore their environment less. They're less social with other members of their own species. They react more skittishly to unknown stimuli. They like to play it safe. Babies with low-cortisol milk, on the other hand, explore more. They're more social. They spend more time playing with their den mates. And when they grow up, their personalities tend to have similar features. While many things go into building an individual's personality, at least among species we're able to study in the lab, what's in the milk they drink is, all on its own, a strongly predictive factor.*

But before we blame our stressed-out mothers for all our social anxieties, let's think about the evolutionary reasons for this pattern. Being social takes a lot of energy. If the milk you're

* Whether that's also true for deeply social beings like humans is unclear, and presumably genetics also come into play. But if personality is something that's built by a suite of influences over one's lifetime, and milk is already known to be an influencer in other model mammals, it would be foolish to discount it in humanity. Rather, milk—particularly its obvious signaling components, like cortisol—is one of many pathways of formative communication between the mother's and the infant's bodies.

drinking—which as a baby is *all* you're drinking—has fewer sugars in it, or if you're able to nurse less often than you'd like, you have less energy to spare. You're going to want to conserve the energy you have for growing your young body into something that can survive to adulthood. Spending that energy on a bunch of roughhousing and time- and energy-intensive socializing is unwise. If you live in a very dangerous world, a fact you're "learning" through your mother's cortisol levels and other milk content, it's probably good to be a bit fearful.

Higher-cortisol milk also tends to be protein heavy, which in principle helps an infant build a lot of muscle, good for running like hell toward safety. Sugar-heavy milk, in contrast, is great for building adipose tissue, creating a comforting energy buffer, and for fueling a growing brain. Brains are, after all, supercomputers that run on sugar. Being social takes a lot of brainpower—a lot of sugar energy. Even now, many *Homo sapiens* who have convinced themselves that a low-carb diet is a good idea feel sort of sluggish as a result, with brains in a fog.*

Still, it's not true that the best scenario is milk with *no* cortisol. A low and consistent amount of cortisol in a mother's milk helps her offspring later in life. If you lace a mother rat's drinking water with low levels of cortisol, her offspring will perform better on maze tests, have better spatial recognition, and generally be less stressed out when faced with challenges than young rats whose mothers didn't drink water dosed with cortisol.

There haven't been many studies that directly test the relationship of a human mother's cortisol levels while nursing and her baby's temperament; also, many children's temperaments change over time (they do get past the terrible twos). But one study did

* A number of papers have debated the benefits and detriments of the so-called ketogenic diet on the brain. I have no intention of giving diet advice, but at least when it comes to the typical diet of our closest cousins—chimps and bonobos—it's clear they don't live on a quivering meat pile. As opportunistic omnivores, they do quite well on a range of diets, but each of those diets has a lot of fruit and vegetable matter, with a smattering of meat and bugs and nuts and what have you as it comes along. The human gut has evolved significantly since the Eve of chimps and hominins, but it'd be wrong to assume that our ancestors were eating a diet significantly different from that of other opportunistic, omnivorous apes.

find that when a breast-feeding mother's cortisol levels were raised above a certain threshold, she would be more likely to rate her child as "fearful" or timid. But women with higher cortisol levels who were bottle-feeding their babies didn't describe them as fearful. Some degree of change in the breast milk seemed to be producing behavioral change in the infant.

So, do we want our babies to drink "stress milk" or don't we? The answer seems to be that we want milk with *just* enough cortisol and other materials, in the right balance, at the right time. Think back to the rats: A little cortisol makes rat pups learn *better* than the pups who didn't ingest any extra cortisol. Overdose them with cortisol and they freak out. That makes sense. Researchers think that to a certain degree mildly challenging environments inoculate children against the upcoming stresses of adulthood. So maybe it's better to have a mother's milk "demonstrate" a moderately dynamic and challenging environment. But if a woman is stressed out all the time, with cortisol levels through the roof, her kids might likewise be more fearful, hesitating to explore new territory and learn new things. In other words, our bodies teach our children about the world, not just by actively showing them their environment, but also by what we put in their mouths. Caretaking mothers have long evolved to take advantage of every pathway available to prepare their offspring for their looming independence. Because we're mammals, the nipple is one of our first lines of communication.

Mothers' bodies tailor milk's contents for the needs of their offspring through a complex communication system between mouth and breast. Babies' personalities are shaped by its particular makeup, are soothed by its fats and sugars and hormones, their guts purged and recolonized by friendly bacteria. Milk is something we *do* as much as something we make. It has evolved to be social.

To be fair, milk doesn't do the whole job. For example, mothers in many cultures use their spit to wipe away a bit of schmutz from the kid's cheek; it's so common, in fact, that this may be a basic human behavior. Continual exposure to the mother's more robust immune system, whether through spit or milk or breath or skin contact, should, in principle, help the child's own immune system develop and learn how to respond to its environment. This is also true of exposure to a father's spit, and a big brother's spit, and the spit of any other adult who has physical contact with the child. It's just that babies actively ingest breast milk on a regular basis, so it's safe to assume that the mother's body is in greatest molecular "communication" with her offspring.* Human infants drink about three cups of breast milk a day in their first year of life. That's clearly a greater opportunity for biochemical signaling than nearly any other pathway.†

And what of men's nipples, then? They clearly aren't doing the heavy lifting here, so why do they still have them?

We tend to think of men's nipples as "vestigial," but that's not quite right. First, "vestigial" is a term that implies an evolutionary leftover that no longer serves any purpose. But the body hates to waste. We have very few vestigial traits. Even the appendix, long thought to be vestigial, is now believed to have an important function in maintaining the health of the large intestine's microbiome. A grown man's nipple can, under the right circumstances, deliver milk. It's not nearly as good at it as the adult female nipple, but it can do it. Seriously. Men can—inefficiently, and with difficulty—nurse a baby.

There's a group of people who live in Congo who call themselves the Aka. In this tribe, gender roles are remarkably fluid. Men and women both hunt. Men and women both care for children. Given the demands of the day, a woman may cook and watch her child while the father hunts. If net hunting is done, rather than spears, they may well do it together, baby in tow. On another day, the woman will hunt and the man will watch the child. Aka men are either holding or within arm's reach of their children more

* On average worldwide, mothers also spend more time in physical contact with the child. But given that *Homo sapiens* are among the only species that regularly adopt the offspring of unrelated parents, these lines of physical signaling between children's bodies and their caretakers shouldn't be thought of as something that *only* happens in genetically linked mother-child relationships.

† Dad, you're great, but I think we're both pretty happy about the fact that I've never ingested a pint of your body fluid. It's okay. We communicate in other ways.

than 47 percent of their time. Pregnancy doesn't seem to change this ratio, either; one Aka woman was known for hunting well into the eighth month of her pregnancy. And after she gave birth, the father still traded responsibilities day to day, not only the general child care, but suckling the child at his breast.

Presumably most of the Aka men do not lactate; the anthropological study didn't mention seeing it, though it's historically known to have happened in many other cisgender men. But even if some do, it's true they don't produce as much milk as women do. The point is that suckling a baby is seen not as an emasculating thing in their culture but just as a feature of the day. As the vast majority of parents know, if your infant is fussy, one surefire trick is popping a nipple in its mouth. When they don't offer their own, American women generally offer a pacifier. Aka men use the one they have built in.

But if you want to know how hardwired milk production is among *Homo sapiens*, you need only look at trans women: people born with XY sex chromosome patterns, but who identify as women. Trans women who want to breast-feed their children generally follow the same medical treatments given to XX people* who have either adopted or used a surrogate to have children. The most common protocol involves taking high-dose hormone pills to trick their bodies into thinking they're pregnant for roughly six months. After that time, they change their pill regimen in order to mimic the sorts of changes bodies experience after giving birth.† They don't produce as much milk, and not all are able to produce any, but many can.

* I use "XX people" here not to avoid the cis terminology, but rather because there are some genderqueer people with two X chromosomes who don't identify as trans and likewise desire to breast-feed a child that their bodies didn't give birth to. They, too, regardless of their genetic background, would need to follow this hormonal protocol. When I refer to "post-birth women" elsewhere in the book, I do so because while some trans men do choose to give birth, the majority of people who give birth are cisgender women and, more important, the studies that undergird the claims I'm making about these mothers have been conducted overwhelmingly on cisgender woman subjects.

† They'll also take a drug called domperidone, which interferes with dopamine receptors and, among other effects, helps stimulate the production of prolactin (Wamboldt, 2021).

It's not clear whether this protocol really mimics the hormone changes (and their cascading effects) that women experience while giving birth. For example, during labor, pregnant women experience a huge rush of oxytocin, which not only triggers contractions of the uterus but also stimulates the milk glands. It's also true that the placenta produces and stimulates the production of a number of hormones and neurotransmitters, including human placental lactogen, which may have a critical role in the production of colostrum. Generally speaking, the milk that people produce after this treatment is remarkably similar to the milk that a post-birth woman produces after about ten days. It's mature milk, not colostrum.

Even with hormone treatments, endless nipple tweaking, and mechanical suckling, many men and trans women will not be able to lactate. Not all postpartum women with their relatively giant mammary glands and nipples automatically produce milk, either. For various reasons, some women's bodies just don't.

So, it's probably not the case that men retain nipples in order to be backup lactation specialists. Instead, men have nipples largely because women have nipples; getting rid of male nipples might mean effectively rewriting the program for basic mammalian torso development in the womb, a costly and dangerous process with great risk for mutations. Why mess with it? Mammary tissue and nipples are hardwired to respond to hormones, so it's relatively easy to change what they do during puberty. As a result, the majority of human fetuses develop nipples.*

What isn't clear is why female breasts have so much extra fat. The shape of human breasts is largely determined by the placement of

* Some of us even get extra—a third or fourth or more. These supernumerary nipples are usually no bigger than a mole and typically follow the V-shaped "nipple lines" along the torso, with most popping up somewhere between the groin and the armpit. Roughly 5 percent of human newborns have them, and extra male nipples are slightly more common than female. Why the male fetus is more likely to "glitch" in this way is unclear. Males are also more likely to have them along the left side of the torso.

large fat deposits, woven through and around the mammary tissue. But while that adipose tissue probably plays a role in both the content of milk (breast milk has a lot of fat in it) and its tailoring (adipose tissue probably helps generate at least some of the immunological content that gets dumped into milk), we also know that there's a huge range in human breast fattiness and shape. From what studies have shown, big, fatty, pendulous breasts are no more likely to make higher-quality milk than "skinny" teacup breasts, nor are they more likely to produce more milk to any significant degree. So long as the nursing mother is healthy and well fed, her milk is quite likely to be fine, regardless of how much fat she's got in her breasts.

We also know that breasts develop in response to hormones, not only in bodies in female-typical puberty, but also in bodies experiencing fluctuations in hormones in general. Many boys will develop proto-breasts as they hit puberty, only to have the fatty lumps shrink back into their widening chests as puberty progresses. Obese males, too, may develop additional breast tissue—not only fat, but also mammary tissue—likely because adipose tissue, on its own, triggers greater production of estrogen in the human body (that's true in other mammals, too). We also know that many trans women taking heavy doses of daily estrogen will develop fattier, female-typical breasts. But suckling an infant doesn't seem to require extra fat deposits around the milk glands.

So why are women's breasts so fatty? Why are they shaped the way they are?

Many people erroneously assume they evolved in this fashion because male *Homo sapiens* were more likely to mate with females who had fatty breasts. Witness, for example, the wild proliferation of breast augmentation surgery: If men didn't like looking at large breasts, why on earth would women choose to go under the knife? And given that, why not assume this is the reason breasts got so big in the first place?

The first obvious signal that breasts may *not* be sexually selected is the wide diversity of perfectly functional breast sizes and shapes, from teacups to watermelons. Breasts are typically smaller on one side than the other, and asymmetrically placed—

for most of us, only slightly, but for others, very noticeably.*
None of this affects milk and nursing capabilities. But somewhere between our split from chimpanzees (anywhere between five and seven million years ago) and now, the hominin body plan added a bunch of adipose tissue to female chest walls.

We have no idea when, within that two-million-year time span, this happened.† We don't know which genes control breast size and shape, so scientists can't do an analysis for genetic mutation rate. The breast, like all soft tissue, doesn't survive in the fossil record. The only reliable evidence we have for when human beings had fatty breasts is actually a work of art called the Venus of Willendorf. Carved from a piece of stone, it depicts an enormous human woman, with a large stomach and huge breasts. There you go: thirty thousand years. By that point, at least, we'd evolved to have human-type breasts, rather than the fluctuating mounds of our primate cousins.

Given that we haven't any true sense of when these sorts of breasts evolved, it's even harder to know whether they came about as a reproductive signal to males. We do know that among today's *Homo sapiens*, small-breasted women regularly give birth to perfectly healthy babies and can make plenty of milk, and there's no evidence that large-breasted women have more babies (or even more sex) than other women, nor do they make more milk. Among studies that try to parse modern heterosexual male desire, hip-to-waist ratio is a better predictor for whether men will find

* The left is generally a bit larger. This could be a functional feature, given that both human and some nonhuman primates tend to prefer cradling (and nursing) infants on the left—more mammary tissue could mean slightly more milk production, depending on the density of the breast, which would clearly be useful—but given that features on the left side of the face are likewise slightly wider and/or more prominent, and the scrotum in most primates tends to house a slightly larger left testicle, deeper developmental patterns in the body's chirality may simply make these things turn out the way they do, with any knock-on "perks" showing up quite after the fact. Perks, and costs—the left breast is also more likely to develop cancer.

† That hasn't stopped folk from trying. For instance, a Polish group is dead convinced the human-type breast is tied to a rise in meat eating and subcutaneous fat in general, with any add-on benefits coming after the fatty breasts were already there. That would put the fatty breast evolving around the time of *Homo ergaster* (Pawłowski and Żelaźniewicz, 2021).

a woman attractive than the size of her breasts, and this is true across multiple human cultures.*

But there's another knock against that theory: large breasts aren't a reliable sign of fertility. In fact, women's breasts are at their largest not when a woman is most likely to be ovulating but when she is menstruating, already pregnant, or breast-feeding. Not only is she less likely to be receptive to sexual advances at these times, with breasts often sore and sensitive to touch, but her male admirers would have no luck sowing their oats. Being sexually attracted to large, swollen breasts does not, by and large, have an immediate evolutionary payoff. Large breasts can, however, advertise an estrogen-heavy phenotype, particularly when combined with a relatively narrow waist, which may be good for carrying babies in general. Like any plump female feature, they're also pretty good flags for being healthy and having a ready food supply.

One of the more popular theories about the development of the modern human breast is that the shape—like a teardrop, with a slightly uptilted nipple—is easier for our flat-faced babies to suckle. After the human brain grew and the nose receded, babies would have had difficulty nursing from a flat chest. Their little noses would have been squashed, making it hard for them to breathe. Or so the theory goes. But all you actually need to fix that issue is a little uptilt, not a lot.

Others think it was a two-legged problem. As we started walking around, carrying our infants in our arms, we needed breasts that could reach their mouths in several positions. This is an attractive idea for a number of reasons, not least of which is the fact that large breasts don't look like teardrops when they're not

* I was not able to find a single study that attempted to replicate these findings among trans men, who often self-identify as heterosexual. A long-standing assumption that trans men are specifically and exclusively attracted to, and have sex with, cisgender women has now been undermined by a proliferation of studies in the field (Sevelius, 2009; Bockting, Benner, and Coleman, 2009; Iantaffi and Bockting, 2011; Katz-Wise et al., 2016), but of attraction patterns among better-studied queer populations, it's well known that queer folk attracted to femme women tend to find similar features attractive, including a low hip-to-waist ratio (Cohan and Tannenbaum, 2001).

stuffed into a bra. Large breasts that have never seen a bra and have nursed one or more children tend to look like long, deflated balloons. Think of the tugging of gravity and endless suckling. That is what mature female breasts *evolved* to look like.

I'm saying not that modern human breasts aren't sexual show traits today but rather that the original driver of their evolution might not have been sexual selection. Even among traits that are sexually selected for, the result isn't always beneficial. For example, there's no clear evolutionary reason why *Homo sapiens*' male genitalia are the way they are.

Put it this way: The average vagina is only three to four inches deep. When a woman is sexually stimulated, hormonal changes tense the ligaments holding the uterus and cervix in place. This makes them rise relative to the vaginal opening as the vagina expands its depth considerably. But a six-inch aroused vagina does not accommodate a seven-inch erect penis. In other words, there's nothing usefully *adaptive* in a long human penis when four to six erect inches will do the job. In evolutionary terms, that is probably why the *average* erect human penis is still only a bit over five inches long.* And yet, in a number of studies, heterosexual women rate pictures of men with longer penises as more attractive. There is, in other words, a disconnect between the human penis as a sexual show trait and its functionality.

Meanwhile, there is the issue of a man's badly protected, sparsely furred scrotum. It's probably *not* the case that mammalian testes evolved to dangle on the outside in order to keep sperm cool. The original reason they dropped out of the abdomen might have had more to do with running. It was a locomotion problem. Morgie had a sprawling pelvis, with legs that jutted out to the side, the way an alligator's do. But her descendants had a more upright pelvis, like a dog's. And once her grandsons were trying to gallop around with femurs rammed vertically into the hip sockets, they were putting a lot of pressure on the lower abdomen. The

* Coming in under the average vagina's depth is useful: you don't bump into the cervix, and there's a bit of "wiggle room" for depositing sperm without the risk of dragging the majority of it back out as you withdraw. Other mammals with penises frequently follow this model. We'll talk more about vaginas later.

"galloping theory" of scrotal evolution holds that the fragile male testes were pushed out of the male torso because running and jumping and bounding basically hurt.* In much the same way, the evolution of the human breast probably had to do with its general function, and was only secondarily a show trait.

But that hasn't stopped theorists from writing their exuberant stories. Some of those stories go way back. For example, thanks to Hippocrates, European anatomists were convinced, well into the seventeenth century, that all women had a vein connecting the uterus to the breasts that existed for the sole purpose of transforming "hot" menstrual blood into "cool pure" mother's milk. Even Leonardo da Vinci, a careful anatomist, drew veins in his diagrams that connected the uterus to the breasts. Despite conducting multiple autopsies and finding no such vein, each anatomist believed it was there. It was called the *vasa menstrualis*—which should probably translate as "the emperor's new clothes."

Still, the idea of the *vasa menstrualis* was probably born of careful observation. After all, women don't menstruate much when they are pregnant, and breast-feeding women tend not to menstruate for a while after giving birth. So, they stop losing one kind of liquid from one part of their anatomy and begin to pour a different liquid out of another. Any reasonable person can see why they drew the conclusion.

But the idea of Leonardo drawing in a *vasa menstrualis* that he couldn't even see, simply because he steadfastly believed it should be there, as did everyone else at the time, is the sort of thing that keeps me up at night. You see, the ideas that human beings have about reality—what it's made of, how it works, how we all fit into grander schemata—can change fundamentally. Sometimes, those changes are so dramatic and so far-reaching that it becomes nearly impossible to understand the world the way we did before. In the history of science, the germ theory of disease was one of those paradigm shifts: knowing that infections aren't the result of a miasma or an imbalance of body fluids or godly punishment but

* From what I've heard from men, running on two legs with dangling testicles isn't all that great, either; it's just perhaps more advantageous than the alternative, which is having one's testicles crushed by pressure in the lower abdomen.

are instead caused by bacteria and viruses. Still, even after scientists had discovered the germ theory, our understanding of what the human body was made of was so deeply entrenched that it took a long time to accept it.

I know there are ideas about human biology that we hold right now that will ultimately prove to be deeply incorrect. Of course, we don't know what they are; they are the "unknown unknowns." If I had to place a bet, I'd say the human microbiome and emergent properties of complex systems are going to form the foundation of a paradigm shift in biology: in multiple fields of study, we're in the process of unraveling the boundaries of what individual organisms are. But again, by definition, the people who live and think and work before, and even during, a paradigm shift are largely in the dark.

The only reason this doesn't drive me completely batty is that there are little tricks one can use to try to identify at least some of our blind spots.* Here's a good place to start: anywhere you see scientific assumptions that seem suspiciously *cultural*—tied, in other words, to recent human ideas about the way things are, rather than to numbers—you can dig a little deeper.

For example, there's a long-standing assumption that cities came to be because of the discovery of agriculture. More food, we assume, allowed populations to grow, and those greater populations would remain in place to tend to the processing and storage and distribution of that food. Urban specialization easily followed: a certain class of people would tend to the growing of the food, and another to its storage, and more to the building of shelters and the healing of the sick and—maybe the most popular human occupation—doing none of these things, but instead attending to invisible gods and/or learning. It's not true that specialization required cities—modern hunter-gatherers have specialized roles in their societies—so let's say ancient cities took those skills and ran with them.

* If this sounds hyperbolic, think of it this way: it's true that as a researcher I have a relentless need to know, but more important, as a person who prefers to think the reality I perceive is actually a suitable representation of the world and how it works, it's more than a little disturbing to think everyone, everywhere is presently getting some unknown feature of reality profoundly wrong.

All of that makes perfect sense. But I also know that we often forget how buggy human reproduction actually is. And we tend to forget that because we have cultural assumptions about femininity; most people think it's easy for human women to make babies. It isn't. We're not like rabbits. Our reproductive systems aren't even as reliable as most other primates'. Morgie had a much easier time laying her eggs and sweating milk into her fur. That means lots of behavioral factors come into play to let human populations rapidly expand. So, let's concede that agriculture was crucial to the rise of cities. But then let's ask the other question: not just who's feeding the adults, but who's feeding the *babies* in this growing population, and how that affects how they are being made in the first place. After all, female bodies are the literal engineers of urban populations.

Agriculture might have helped an abundance of bodies to come together, exploiting all those urban niches, but we should also assume that new problems arose from such close contact: widespread infections, for one, the legacy of which we can see in the oligosaccharides of human milk. We also know that from the dawn of recorded civilization human beings have employed wet nurses. These women, paid or enslaved to breast-feed others' babies, enabled population booms. In fact, human cities may be Morgie's greatest legacy. Without wet nurses, city life might never have taken off the way it did.

I'm not the first to make this argument, though it's largely been hidden in academic journals, read by only a handful of academics and scientists. It runs like this: While agriculture might have allowed more humans to live in one place, the problems associated with population density should have provided their own checks against exponential population growth. This is part of why the first human cities, which arose somewhere between four thousand and seven thousand years ago, are assumed to have been not much bigger than towns—sometimes as few as a couple of hundred people, or as many as three thousand.* Agricul-

* Or at least that's the range for ancient Jericho eleven thousand years ago, depending on whom you ask. To put that in perspective, the U.S. Census Bureau currently defines "small towns" as anywhere with a population fewer than five thousand.

ture demanded a lot of acreage, which presumably kept most of the "suburbs" of these city-towns fairly spread out (if they even existed). Those who lived in the more crowded, urban centers suffered increased mortality and reduced fertility through disease and anemia, and in these circumstances more young people in their reproductive prime died in violent conflicts caused by social friction. The bigger a city gets, the more the pressures of urban living can blunt the growth of its population.

And yet, somehow, big cities did come about. Cases of explosive urban growth are documented in the earliest of written human records. And in some of those ballooning cities, urban women regularly employed wet nurses to feed their children.

Let's do a little math. Among today's African Ju/'hoansi hunter-gatherer tribes, women regularly nurse their children for up to three years and have a mean birth interval of 4.1 years. These women have an average of four to five children over their lifetime. In the middle of the twentieth century, North American Hutterites—a rural religious group who do not take birth control and who wean their children before they reach one year of age—had a mean birth interval of 2 years and gave birth to more than ten children.* Women who do not nurse their children at

* As of 2010, Hutterite women have dramatically fewer children—now only about five. Though this might be due to changes in breast-feeding or birth control practices (White, 2002), it can also be tied to social intervention: Hutterite women used to marry early, around age twenty or twenty-one. Now it's common for them to wait until their late twenties (Ingoldsby, 2001). That decreases the birthing window and clearly results in fewer babies. This is a big part of why women in many industrialized countries are having fewer babies, too: not simply the advent of birth control ("waiting for motherhood"), but that population-wide most babies are still born to married couples ("waiting to be a wife"); reducing the time spent in wedlock naturally reduces the number of babies that marriage will produce. These standards have shifted over time, and there are known outliers; for instance, as of 1990, 64 percent of babies born to Black mothers in the United States were out of wedlock, where it was only 24 percent in 1965 (Akerlof, Yellen, and Katz, 1996). We should assume complex social issues drive that difference. The mass incarceration of American Black men is a huge one (Western and Wildeman, 2009). Less access to birth control and sex education is another, compounded by fatalism and a distrust in government-driven medical advice (Rocca and Harper, 2012). We can assume lots of things drive changes in marriage and birthrates. But nevertheless, if you look across *enough* social groups and, particularly, if you look at global statistics, the trend holds: where you find later marriage, you'll find fewer babies.

all—for example, British women in the 1970s who chose not to breast-feed—have a mean birth interval of 1.3 years.

Breast-feeding, in other words, is a very predictable sort of birth control. It is an imperfect birth control, with a far lower success rate than our more modern interventions (condoms, hormones, bits of copper inserted in the womb), but nonetheless breast-feeding is Nature's Pill. Morgie didn't have the energy to nurse more than one set of pups at a time; it would have been suicide not to space out her pregnancies. For this reason, the genetic mutations that allowed birth spacing were favored. Once primates evolved to have fewer offspring at a time, that evolutionary legacy had a strong hold. Generally speaking, our ovaries stay quiet while our breasts are at work.

So, imagine what happens to a city's population when you have a large percentage of its mothers employing wet nurses. In principle, that would reduce a woman's mean birth interval significantly, from 4.1 years to as little as 1.3 years. Gestation takes about 9 months. You'd be pregnant pretty much all the time.

Meanwhile, wet nurses wouldn't be pregnant as often as you, but because breast-feeding is an *imperfect* ovulation suppressor, they wouldn't be entirely un-pregnant, either. Many would have their own children, some of them born immediately before yours and some born while yours are still nursing. Many women are perfectly capable of nursing more than two children. These so-called super producers—we can assume such women would be more likely to find steady employment as a wet nurse—could be nursing three or four kids at a time without much increase in infant mortality. It's not hard to imagine how the population of an ancient city could explode under such circumstances.*

If marriage is delayed long enough, you'll also find an increase in the percentage of babies born to unwed mothers, but the associated decline in the total number of births still holds. Whether women are choosing to wait to become a wife *specifically* to wait to become a mother is another matter; we should assume that varies both from culture to culture and from woman to woman. Women's decisions are complex. The range of a woman's fertile years, however, is more consistent.

* For the stats inclined, here's the quick and dirty: If only one-tenth of the female population were outsourcing their milk production, it would take twenty years for the city to double. If the tradition continues, growth is naturally exponential. That's assuming one wet nurse for every woman who employs her, and assuming

Remember, too, that the people who are having so many kids are (whatever would count as) upper- and upper-middle-class women. If their kids grow up with enough resources to employ their own wet nurses (of course, the wet nurses' kids wouldn't), that would presumably increase the proportion of the city's population that is using wet nurses, accelerating growth. Eventually, either the city would have to absorb more wet nurses from the surrounding rural areas or some sort of rebellion against wet nurses would topple the ridiculously fertile ruling class.

Right. In this imaginary world, where only wet-nursing influences a city's population, it would seem Hammurabi had a hell of a lot of babies on his hands. No wonder regulations for wet-nursing made it into his written law. Of course, in the real world, many other things kept urban populations in check: famine, disease, floods, violence. For example, in eighteenth-century France, where large swaths of the middle classes employed rural wet nurses, not just the rich, many infants farmed out to the countryside died, presumably of disease or neglect. It became such a problem that a nationwide regulatory agency called the Bureau des Nourrices was conceived to help protect the infants and look after the interests of both mothers and wet nurses. It stayed in business until 1876, and the French continued to employ wet nurses through World War I. In the United States, African American women regularly nursed the white babies of the American South throughout slavery, through Reconstruction, and in some cases all the way up to the mid-twentieth century (there was no

---

the wet nurse doesn't become pregnant herself while employed. If you allow even two women for every three wet nurses—which could easily have been the case—the city's population would double after only ten years. That's because the average delay in childbirth for a woman who's breast-feeding is more than double that of women who aren't breast-feeding, so for every employer and wet nurse, you get an extra two babies. For two women nursing their own kids, you'll get two kids every 4.7 years. For every urban woman and her corresponding wet nurse, you'll get three kids—three and a third, to be precise, born an average of 1.3 years apart. You might even get four kids, if the wet nurse gives birth right before she's employed—twice as many children, therefore, as a group that doesn't use wet nurses. If you allow that even one-tenth of the wet nurses become pregnant while employed, suddenly the city is doubling after eight years.

bureau to regulate that, of course; it was one of the many degradations of slavery and continued racist exploitation).

Remember Babylon? That massive, terrifying city, so loathed by ancient Hebrews? Around 1000 BC, its population was roughly 60,000. Meanwhile, the denizens of the Golden City of Jerusalem under King David (same era) numbered a measly 2,500. While some women famously nursed *other people's* children, Hebrew mothers were in the habit of nursing their own, as the sacred texts exhorted them to do.* Babylon had wet nurses. Their gods were more urban. Time and again, ancient wet-nursing cities saw their populations swell and heave against their walls: Mohenjo-daro, 50,000; Thebes, 60,000; Nineveh, 200,000. Ancient Romans formed organizations to regulate the practice; Roman families would solicit the services of wet nurses in the city square at the Columna Lactaria.

And so, Morgie's legacy was both boon and bane for the rise of *Homo sapiens.* Ancient cities had major overpopulation problems, and these problems bled into their origin stories. It seems, for instance, that the tale of Noah and the ark wasn't originally about sinful humans; it was about urban overpopulation and birth control.

* There's some disagreement here, even within sacred texts. In the Talmud, breast-feeding is seen as a service to one's husband, much as spinning wool or making his bed. But if a woman brings *two* maidservants with her to a marriage—if she was wealthy enough to have two slaves, in other words, who came with her to the man's house when she married, much as money or cattle or any other property—then she could choose to give her baby to a wet nurse. On the other hand, any woman who gives birth is considered *meineket* for two years—literally, a "nursing woman"—and falls under a special protected class of women who cannot remarry, but also doesn't have to do things like ritual fasting if she feels too weak to do so. Moses's mother famously served as wet nurse to her own river-abandoned baby in the Pharaoh's court, hired as his wet nurse when he (symbolically) refused to feed from an Egyptian breast. In both the Talmud and the Torah, breast-feeding is repeatedly praised as a good thing and recommended for as long as two to four years. That continues to be the common practice in many different global Jewish communities, in no small part due to religious tradition and cultural support. For his part, the Prophet Muhammad had three wet nurses as a baby, and he held special consideration for "milk brothers" fed from the same breasts—a common practice in those communities at the time; one could have all sorts of new "siblings" by having shared the same wet nurse.

Among scholars who spend their lives studying such things, it's generally agreed that the Hebraic flood myth didn't originate among the ancient Hebrews. The earliest account we have of such a myth is from Sumer. Situated between two rivers in an otherwise arid land, Sumerian cities depended on irrigation canals and a regular cycle of flooding and ebbing to fertilize their crops. But when the flooding got out of hand, cities could be destroyed. There are other cultures with flood myths from around the world, but the Sumerian one has enough in common with the story of Noah and the ark to be the obvious precursor. And it is surprisingly bound to women's reproduction.

As the story goes, the Sumerian gods were lazy. They didn't like to do all the annoying work of growing food and making clothes for themselves. So, they gave the work to Man. But human cities grew so quickly they irritated the gods. One god, Enlil, famous for copulating with actual hills and begetting the seasons, woke from his sleep because a nearby city was overpopulated and so noisy the banging and babble punctured his dreams.[*] Royally pissed off, he decided to wipe humans from the face of the earth with a flood. If it weren't for the intervention of another god, it would have worked. But that god tipped off a man named Utnapishtim—the Sumerian Noah—and told him to build a boat on which he should put his wife and plants and mating pairs of all the animals. When Enlil sent his terrible flood, Utnapishtim and his family survived. Later, when a raven they'd sent from the boat didn't return, they knew the waters had receded.[†] They quickly reseeded the city with their children.

But soon the place became overcrowded again. That's when

* In other versions of the story, the noisy urban folk were lesser gods, and only after they were silenced were human beings created. But the idea of overpopulation, noise, and general irritation leading to a mass punitive genocide still held.

† Not a dove—which makes perfect sense, because the clever corvids quickly adapted to coexisting with urban human populations and are, even now, a commensal species, much like the rat and New York's pigeons. For what it's worth, in both the Torah and the Christian Old Testament, Noah originally sends out a raven. The dove came later. The Quran is entirely uninterested in the birds. No mention.

Enlil and the rest of the gods stepped in. Aside from inventing mortality, to set an upper limit on the human problem, they also set down a bunch of edicts about birth control and sexuality so there would be fewer births. Women were categorized into sacred temple prostitutes with special knowledge of herbs and birth control; wives, who would be okay for sex and reproduction; and "forbidden women," who were off limits when it came to sex. Other Sumerian cuneiform tablets lay out advice for the best herbs and methods for both aiding and hindering fertility.

So, a story born in ancient cities beleaguered by too many people, and arguably about the dangers of urban overpopulation and the benefits of birth control, is adopted by the mostly nomadic ancient Semitic tribes who didn't use wet nurses as often. And those tribespeople repurposed it as a story about urban wickedness (Noah and the ark), thereby generally screwing women for the next three thousand years.

But that's all pretty recent history. *Homo sapiens* have been around for 200,000 years. Mammals, 200 million. Scuttling as she did in and out of her little burrow in the early Jurassic, we can't really ask Morgie for an apology, nor does she owe us one. In general, I'd say mothers everywhere owe us far less than we like to think they do. And we owe them more.

Full of the immune system's soldiers, a mother's milk extends the protective borders of her body to envelop her children. But like many of the things we do to protect our babies, it's costly to make and costly to give. Breast cancer is common and deadly precisely because mammary tissue evolved to strongly respond to hormonal changes; wherever you have a bunch of cells proliferating and changing and reverting, you're likely to find cells that go rogue. It's not just us, either: dogs, cats, beluga whales, sea lions, *all* animals with mammary tissue are known to have mammary cancer.*

* In fact, North American jaguars in zoos have a cancer profile strikingly similar to that of women who carry the *BRCA1* gene mutation, with increased risk for

While 1 percent of breast cancers occur in men, breast cancer is a full 30 percent of *all* cancers in women.

Unfortunately, it's also the second leading cause of cancer death for women. Because mammary glands evolved from skin along the torso, human breasts are now stacked right on top of our heart and lungs, run through with blood vessels and lymph tissue; there's a terribly good chance that a breast cancer will metastasize before we even notice it's there.* Breast cancer deaths have been going down lately, largely because we're getting better at finding them and treating them before they find their way out of the breast. But the incidence of breast cancers hasn't been going down at all. There's still a one-in-eight chance that I, as an American woman, will develop breast cancer at some point in my lifetime, and those stats are similar worldwide.† Having breasts and making milk isn't just *socially* expensive, in other words; it is, all on its own, a dangerous affair.

But that's motherhood for you—even for women who aren't mothers and never will be. The legacy of mammals' evolution on the female body prepares us for these feats, with varying degrees

both mammary and ovarian cancers (Munson and Moresco, 2007). They, too, have noted mutations in the *BRCA* genomic sequence—researchers compared it with the same sequence from domestic cats to look for changes—though, as is also true in human women, no one is entirely sure what those mutations actually do in the body or why they lead to a greater risk for female reproductive cancers. It largely seems to have to do with cellular repair. Men who carry these mutations are also eight times more likely to have cancer than the normal population (Mano et al., 2017). They still don't get breast cancer as often as women, but these men are also much more likely to get prostate, skin, colon, and/or pancreatic cancer.

* As a woman with "dense breasts"—a higher ratio of mammary tissue to fat that both makes them harder to properly image with standard ultrasound and increases my risk of developing breast cancer—I can also report that they generally feel a bit lumpy when I do a self-exam. If you, dear reader, are in the same situation, this was the analogy I was given: you're looking for raisins in the oatmeal. And if you find one, generally speaking, it doesn't move as easily when you poke around at it, unlike the other lumpy bits.

† Obesity strongly raises one's risk, as do some known genetic mutations, but whether obesity or specifically *abdominal* adiposity is the central driver remains unclear (James et al., 2015). Still, the best risk reducers for breast cancer remain the same: get screened regularly and learn how to screen yourself at home. Above all, take your own body seriously; if you're worried something's wrong, talk to a doctor.

of cost. From the immune system to the intestinal flora to the fat and mammary tissue and reproductive organs, the female mammal is born ready to brace for impact. Preparing to make milk is part of it. Preparing to make womb-grown *babies* is quite another.

Morgie didn't have to do that; live birth came later. The reason giving birth and recovering afterward are so stupidly hard these days is that we give birth to live young. Human beings are placentals. And for that, you can actually blame an "act of God." Milk started under the feet of dinosaurs, but live birth took hold in an apocalypse.

Protungulatum donnae

CHAPTER 2

WOMB

Then the second angel blew his trumpet: and something like a great mountain, burning with fire, was thrown into the sea, and a third of the sea became blood.

—REVELATION 8:8

It was cold. The ash fell like snow for years. When it stopped, everything was dead. The great beasts as tall as trees, and their hook-toothed predators, the lake things and the river things. But she survived, like other creatures who were small enough or burrowed deep enough to hide. The tiny. The minute. Those easily forgotten few.

The ones who could live off the dead did well, too, fed by massive corpses drifting down to the ocean floor—the death of ten million leviathans. Their bodies fell down, down, down to the tables of the spindly and sucker-mouthed, who feasted like kings until they, too, grew quiet. But on land, soon there were tender shoots, and insects, like manna from heaven. And our Eves rejoiced then—or however much a half-starved, apocalyptic weasel-rat can rejoice.

We don't know whether it was a comet or an asteroid. Most think it was an asteroid. We're pretty sure where it hit: there's a crater, 110 miles across and 12 miles deep, half buried in the water off the coast of a place we now call the Yucatán. The asteroid was

6 miles across, and it hit Earth with a force that exceeded a hundred teratons of TNT: more than one billion Hiroshimas.

And so, there is such a thing as the K-Pg boundary, a strange shift in the fossil record between the Cretaceous and the Paleogene periods. If you dig anywhere in the world, you will find a thin layer of clay from this moment, soused with iridium—star stuff, very rare in Earth's crust, but common to asteroids and comets. When that enormous rock hit, the force of the impact threw iridium-rich fragments and dust into the air and carried it in clouds around the globe. Those clouds blocked the sun, but it didn't get cold right away. First, the world caught fire. The sheer energy of the impact launched molten debris, hot ash, and other ejecta into the sky, and when it fell back down, the planet ignited like so much tinder; wildfires burned across the continents, pulsing heat over the course of many days. The ash from the fires joined the dust clouds, which spun up into the sky in massive fire tornadoes and sweeping plumes. The sky grew darker. The ash fell. And when the fires finally went out, it got cold. Cold and quiet.

The clay layer dates to about sixty-six million years ago. So do the rocks in the Yucatán crater. Before that time, the world was full of all kinds of dinosaurs. After, mostly birds. And us, or rather what became us. Along with some lizards. Amphibians. Frogs and beetles and dragonflies and mosquitoes. We are the descendants of the survivors, of whatever managed to adapt.

There is no holocaust, no natural disaster, no great terror in the history of humankind that can compare with the apocalypse we call Chicxulub. It's fair to say that it's unimaginable. We know that there was ash. We know, for many years, it was very cold.

And we know that there, somewhere in the ashfall, is the reason women have periods. In the middle of one of life's worst disasters, the placenta took hold. Ancient mammals gave birth to live young.*

* For the record, Chicxulub isn't the worst thing that's ever happened to life on our planet. In terms of death, that probably falls to the Permian extinction event, popularly known as the Great Dying. About 250 million years ago, 96 percent of all species died. No one knows why. Best theory going is oxygen depletion—climate change, triggered by Siberian volcanoes pumping out too much CO_2. But

THE TRUTH IS WE SHOULD HAVE MORE VAGINAS

Since the age of Morgie, mammals have nursed their young. For some species-specific period of time after newborn mammals exit their mother's uterus, they suckle and slurp their way through early development.

But somewhere in deep time, after the dawn of milk but before the Chicxulub apocalypse, mammalian bodies started veering off the main road. Instead of laying their eggs, some number of ancient creatures started incubating them inside their bodies. Some of them became the marsupials, while others became eutherians like us—the placentals.* We didn't just keep our eggs warm in there; the entire female body became a gestation engine.

I'm not sure it's possible to sufficiently explain how insane this is. The majority of multicellular animals lay a clutch of eggs. Some of us let them loose in the ocean in a free-floating stream. Some of us tuck them safely away in a sticky glob. Some stay with the eggs, guarding them until they hatch. Others skip town. In other words, what animals do with our eggs varies widely. But laying eggs is normal.

What's not normal is letting eggs incubate and hatch *inside your body*, where they can do all kinds of catastrophic damage. What's not normal is building a placenta and anchoring a developing fetus to the wall of the uterus, thereby transforming the mother's body into a kind of H. R. Giger fever dream meat factory. What's not normal, in other words, is giving birth to live young.

But that's precisely what most mammals do, along with a very small number of unrelated fish and lizards. Thanks to the world-clearing burn and freeze of Chicxulub, gestating our young inside our bodies might have been a significant part of how our Eves managed to succeed. For whatever reason, mammalian bodies

---

in terms of a disastrous event that directly shaped the evolution of all *mammalian* life, Chicxulub is the winner. Dinosaurs are still pretty pissed about it—to whatever degree the common sparrow gets pissed about things, that is.

* The easiest way to remember the difference between marsupials and placentals is that one has a pouch and one doesn't. Kangaroo, pouch. Cows and cats and dogs and mice and just about every other mammal you can think of, no pouch.

were able to fill some of the niches the non-avian dinosaurs left behind.* We spread out. Diversified. Crammed into ecosystems and competed. And all along the way, we carried our young inside our bodies instead of laying eggs like sensible creatures. This is why women's bodies are built the way they are today. It's a huge part of why our *lives* are the way they are: most women have periods, get pregnant, and give birth.

And the whole situation's pretty lousy. Both gestation and birth are far more taxing and dangerous than anything egg layers have to face.† Being able to pull it off requires jury-rigging not only the female reproductive system—those organs that used to pump out eggs and the tubes that carried them—but huge portions of the immune and metabolic systems, too. It's not a simple fix. Giving birth to live babies is a big deal.

Like any deal, there are pluses and minuses. The known advantages? It's great not to worry about having to tend your nest of eggs. That means you can spend more time looking for food in a wider area.‡ You also don't have to worry as much about keeping your eggs at a certain temperature, since your warm-blooded

* No one's really sure why, but many have their theories. It's possible our Eves had faster growth rates, or maybe a slightly better ability to burrow, or were more diverse in what they were able to eat. Or maybe something about gestating young inside their bodies made them better at keeping their young alive than the external egg layers. A good many paleontologists lean on a combination of the burrowing and diverse diet—hiding from the fires and the cold, being small enough not to need much food, and being able to eat anything that could, after an apocalypse, count as food.

† There are always exceptions to the rule. Many salmon species, after swimming upstream to their spawning grounds, lay their eggs and promptly die, their bodies fertilizing the waters for their future hatchlings. Not all salmon mothers die upstream, but many do, and many more die trying to get there. This seems less a product of the egg laying, however, than of the mass migration. There are other species of egg layers—especially among insects—who have evolved to live only briefly during their reproductive periods. For example, some fireflies hatch from their cocoons to find that they *have no mouths*, so whether or not they manage to reproduce, they'll soon starve. In nature, the horrors of motherhood know no bounds.

‡ Though *you* might get swallowed by a postapocalyptic serpent, you at least have the chance to run away. Eggs can't run. That means a mother can do only so much to protect them when she's not at home.

body is already built to keep your organs at a fairly steady heat.* You can also regulate your eggs' bacterial environment a bit better, along with the level of moisture, and all the useful things your body already does for your vital inner parts.

But creating a body that can do all those things for gestating little ones also means making some big sacrifices. For example, nowadays, we have only one vagina. More would have been handy. Most marsupials have at least two; some have three or four.

In case you don't have a vagina yourself, or are otherwise unfamiliar, here's the lowdown: Like all placentals, the vast majority of human women have one vagina.† It's a muscular, mucous tube that's normally only about three inches long and sort of collapsed in on itself. It's what biologists call "potential space"—something that can expand to accommodate intrusion but doesn't normally hang open. Most women's vaginas end at the cervix: the brief neck of the uterus, an organ that's normally only about the size of a woman's fist that can expand to the size of a watermelon when the woman is pregnant. The human uterus is shaped a bit like an upside-down pear, flanked by a couple of tiny fallopian tubes that end in fringed bits right next to the ovaries, which themselves are typically the size of a large grape. These are the leftovers of the old egg-laying system: a place to make the eggs, some tubing for those eggs to roll down, a pouchy gland to secrete materials to create an external eggshell (that gland is what turned into the uterus), and a way for the final product to roll out.

A diagram of women's reproductive organs is usually drawn like a capital T, with the fallopian tubes extending out on either side. But in the cramped, well-occupied space of a woman's lower

* Echidnas and platypuses have more variable body temperatures than marsupials and placentals, but no one knows if that represents the basal mammalian state for our ancient Eves—birds are warm-blooded, after all, and presumably the ancient dinosaurs they evolved from were warm-blooded, too. Some even think species may evolve to have live births specifically to benefit from using their bodies to *carry* their internally gestating babies to warmer places—not just early mammals, in other words, but cold-blooded creatures like certain sharks, who seem to migrate to warmer waters during pregnancy (Farmer, 2020).

† Some few of us are born with a divided vagina or a small, closed-off portion of a second vagina that never fully developed. Trans women are usually born without a vagina.

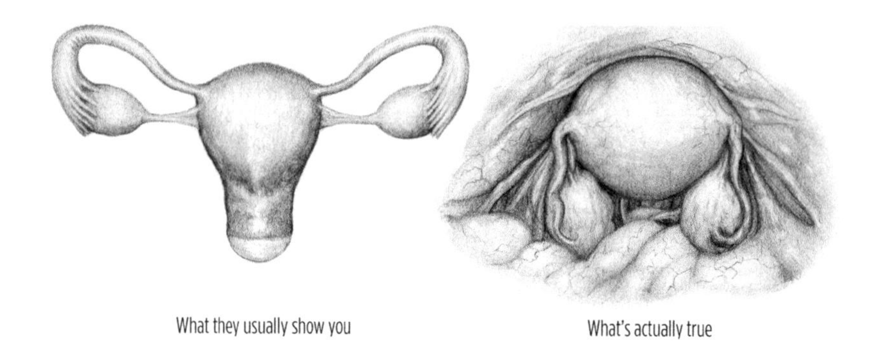

Female pelvic anatomy: it's a tight fit.

abdomen, the ovaries are actually tucked in tight to the uterus, and also smooshed close to the bladder and large intestine, and the fallopian tubes don't extend so far. That's why, if you've ever had an ultrasound of the area, your technician might not have been able to find one of your ovaries, which are often obscured by the uterus or bladder or part of the bowels.* It's pretty crowded in there.

Morgie's egg-laying pelvis would have been tightly packed, too. In terms of development, the biggest difference between us and her is how the female fetus evolved to grow a separate vagina, urethra, and rectum. Doing *that* was likely one of the most important steps in becoming placentals like us. Marsupials had to do it, too, but the way they built their new nether regions might have been just different enough from our Eves' solution to limit their options.

* This is also why ovarian cancer is so dangerous: not only do the ovaries regularly undergo strong hormonal changes and cell turnover, which makes them more prone to cancer in general, but they're small and tucked against other organs. By the time ovarian cancer is diagnosed, it has often spread around the lower abdomen, with tumors popping up on the intestines, uterus, bladder, kidneys, and/or liver. It's also true that ovaries regularly pain women—benign cysts are common—and many of us have learned to ignore little aches and pains down there. Though ovarian cancer usually arrives after menopause, one in seventy-eight people with ovaries will be diagnosed with it at some point in their lives (SEER, 2021). Don't let this fact make you too anxious, but do tuck it away somewhere in your brain. As always, if something's bugging you, please talk to your doctor.

One way to trace the path from Morgie to live eutherian birth is to simply move forward through time. Roughly 200 million years ago (that is, shortly before Morgie came to be), the mammalian line split into three: the monotremes, the marsupials, and the placentals. Monotremes are named for their *one* outgoing passage: a single (mono) hole at their bodies' back end, called the cloaca. (More on that in a minute.) The biggest difference between the monotremes and everybody else is the obvious fact that they still lay eggs, pushing those eggs out of that single hole.

The marsupials have two holes, generally—the "urogenital sinus" and a rectum. They give birth to barely developed little jelly-bean babies, which promptly crawl into an external pouch, where they suckle from a teat, serving their time until they are ready to come out.

The eutherians, with our three-holed female pelvic plan, give birth to live, relatively vulnerable babies that nurse for a variable amount of time, usually in one sort of safe place or another, like a den or nursery or burrow or wherever we can manage.*

What divides these three mammal groups is how our bodies changed over time to accommodate baby making and baby raising. All mammalian offspring will nurse, as Morgie's did, but how developed our infants happen to be when they start nursing varies from species to species. How long a baby nurses, when that child finally becomes "independent," and how much the mother has to do to get her offspring to maturity—all this varies, too.

Compared with humans, most marsupials are born (that is, exit the uterus and head to the pouch) at a point of development that would be roughly seven weeks into a human pregnancy—*incredibly* underdeveloped, in other words. Their forearms are strong, which helps them crawl into the pouch—using ultrasound, researchers have been able to watch wallabies practice climbing in the womb a few days before being born—but their rear limbs are often little better than buds. Once in the pouch, most marsupials

* Some rare placentals retain a two-holed plan, and a very rare few retain a cloaca, though it's unclear whether those animals represent a line that "failed to advance" to the more common three passages or reverted to an older system at some point, keeping the trait because it didn't harm them enough to matter. Earth-born life is a messy business, and categorizing living things accordingly allows for mess.

essentially fuse their tiny mouths with a nipple, maintaining that close connection with the mother's body as they grow. Think of the marsupial nipple as a lesser umbilical cord: there's still that two-way communication (remember human babies' "upsuck"?), but the mother's body doesn't have to do quite as much for a kid in a pouch as it would for one in the womb. For example, if a joey happens to die, it's quite a bit simpler to "cut the cord."

Mouse pups are a bit further along developmentally when born, but they, too, are pink and hairless, their eyes fused shut, their ears rolled back against their tiny skulls. More advanced mammals have varyingly independent newborns. Cats and dogs are born squiggly and incompetent, but they grow rapidly as they nurse and sleep, nurse and sleep. Others, like giraffes, are born essentially able to live in the world right from the start, which is good, given that they drop a full six feet from the laboring mother's vagina to the ground. The force of impact is what breaks the newborn giraffe's umbilical cord and natal sac and jolts its lungs into action, causing it to gasp at the air shortly after it lands. About an hour from that rude awakening, the newborn is usually able to stand.

Getting a body ready for *that* kind of arrival in the world is the essence of the eutherian story: giraffe mothers are pregnant for fifteen months, and those pregnancies are very taxing. Marsupials, meanwhile, barely notice they're pregnant because so much development takes place in the pouch instead. For mice, a bit more happens in the womb, but still quite a lot in the nest. So moving from egg layers to baby havers (diverging, in other words, from monotreme-like Eves into the creatures that eventually became today's marsupials and placentals) was, for the mother, a question of how much you give, and when. You can't actually separate fetal and juvenile development (what the kid does) from female reproductive plans (what the mom does) because the two are intrinsically linked. They evolve together. Biologists call this "maternal investment"—an umbrella term for all the things a female has to do (physiologically and behaviorally) to make reproductively successful offspring and what it's gonna cost her. Costs are distributed over time in differing severities, depending on the species and its milieu. Will she "spend" more (energy, resources, time) making eggs?* How much will that expense deplete her? Will she spend it shaping the environment her eggs hatch in, or the environment the hatchlings mature in? Will that put her body at greater risk? All the answers depend on chance, but that chance is deeply shaped by the mother's environment and her species' body plan. None of the answers come without cost. Every strategy has risk. But those risks, when they survive evolutionary churn, tend to have payoffs in babies that make it to the point of making more babies. So giraffes are pregnant for fifteen months. By the end, the mother is pretty exhausted. But her babies arrive able to walk.

Changing one's body plan comes with a similar risk calculus. A big part of the evolution of milk had to do with protecting newborns from bacteria. With the development of the three-holed body plan, our ancient Eves had to evolve ways of protecting the birth canal from contamination with bacteria from feces.

Like Morgie, today's monotremes don't have a separate vaginal opening. Like birds and reptiles, they lay their eggs via the cloaca, a single exit from the pelvis to the outside world. Just behind the cloaca's purse-string opening, birds and reptiles have some version of a cloacal sinus: a pouch of varying size that the ureters, uteri, and large intestine dump their various products into. Thus, when the platypus lays her leathery eggs, she pushes them out through the same exit she uses to get rid of urine and feces. It's an efficient system.

But they still have to safeguard their offspring from the bacteria in their body waste. And, in fact, most egg layers have a folded

* Like human beings, other species have "opportunity costs" around maternal investment: a bit like a person who might sacrifice the opportunity to take a higher-paying job because she's already busy working a lower-paying job and doesn't have the time to look for a new one, an animal that spends a ton of time building a nest isn't spending that time finding food for itself, or finding new mates, or even finding a better location for a nest. The fact of needing to build any nest, in other words, takes a portion of that animal's life. This is, of course, a large part of why so many egg-laying species have *males* that do the nest building, or at least contribute—a naturally attractive trait in a partner for an egg layer. How well a male contributes to this task may even shape how many eggs she lays (García-López de Hierro et al., 2013). Time is not, for human or sparrow, free.

bit of tissue inside the cloaca called the uroproctodeal fold, which handily shifts one way or the other to help protect the urinary system and reproductive tract from exposure to bacteria from the colon. But it's a flap, so it's not perfectly sealed. That means eggs often get a little bit of poo on them on the way out—which is to say, they're exposed to intestinal bacteria. For a baby in an egg, tucked neatly into that protective shell, no big deal. But when you get rid of the shell, that sort of system could lead to all sorts of bacterial overgrowth on your newborns, whose immune systems may not be ready for it. So if the evolution of live birth came *first*, building a more reliable separation between feces and the birth canal should have quickly followed—it doesn't take that many generations of babies who survive a bacterial hellscape, unlike their poo-dosed cousins, for a trait like that to become the norm.* And indeed, both marsupials and eutherian placentals like us have a separate rectum and urogenital sinus: mammals who give birth to live offspring generally keep their babies away from their butts.

Though not all fetal development maps onto evolutionary time, you can actually get a window into how this might have happened by looking at what mammals do in the womb:

At first, eutherian embryos grow a cloaca, and the ureters dump straight in there, as does the neck of the bladder and the lower intestine and the oviducts—two of them, of course, one for each ovary. Then a fleshy wedge starts forming between the intestines and the bladder, extending down until it finally forms the back wall of the vagina.† Meanwhile, the ureters shift up and

* It could have happened the other way as well. For example, having your ureters regularly exposed to bacteria from the lower intestine could make you prone to bladder infections. It doesn't seem like such a big deal for reptiles and amphibians and birds, who all still have a cloaca, but if some early mammalian Eve had any sort of advantage during a local outbreak of intestinal flu—for example, a more permanent, fibrous septum between the colon's dumping place and the ureter—it's not hard to imagine how that'd be selected for. And once the poo door is separated from the pee door and the egg drop, the reproductive system could have been freer to do something silly like keeping one's offspring inside until birth.

† Recent research indicates this "descent" is more a product of differing growth rates between different regions of the cloaca rather than an active downward thrust (Kruepunga et al., 2018). But the resulting division remains wedge-like, does take shape over time in this way, and still forms a fascinating window into the developmental differences between monotreme, marsupial, and eutherian back ends.

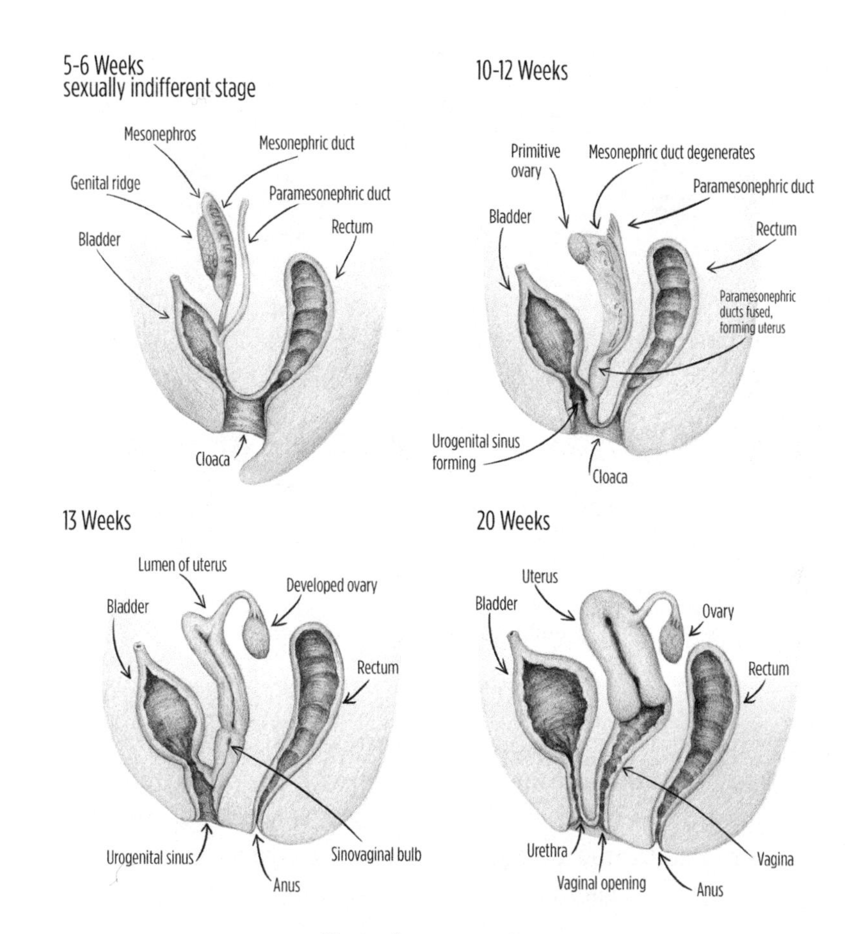

From cloaca to vagina

connect with the bladder instead of the ancient cloaca, and the urethra extends down from the bladder to the front of the old sinus (in females) or all the way out to the tip of the penis with a connection to the vas deferens (in males) to serve as a way to get pee and semen out of the body. In the human embryo, the cloaca is present by the fifth week in utero, and then subdivides into two separate passages (urogenital and rectal) during the sixth and seventh weeks, with the full three-way division largely completed by week twenty.

Note that the human embryo is, in many ways, "sexually indifferent" up to week seven or so, at least in terms of these urogenital

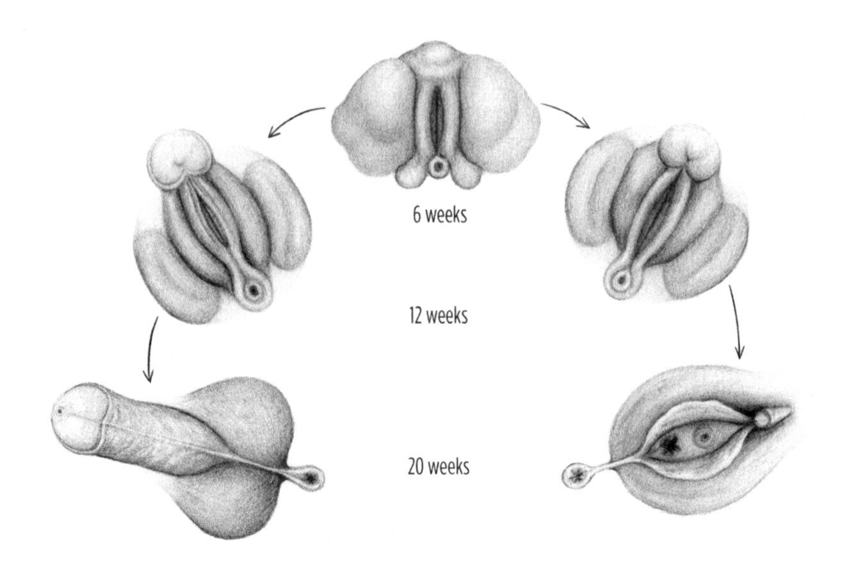

How embryos turn a hole into genitals

parts: the formation of a penis and testes and that long male urethra doesn't really start until after then, with the penis bud largely indistinguishable until week twelve. Even at week twelve, there's still a hole along the bottom of the penile structure that looks a heck of a lot like a vaginal cleft—the developmental remains of that urogenital sinus—which won't seal fully, with a penis and glans in place, until week twenty. That's also when the side swellings that will become either the labia, in females, or the scrotum, in males, have moved into place, and the little genital bud is pushed forward into the glans in males and, in females, forms the basis of the external part of the clitoris. In many adult men, you can still see the remains of how that cleft sealed: a little line, sometimes even a tiny ridge of flesh, that runs along the underside of the penis, straight down the middle of the scrotum, and all the way to the perineum and anus. This is a visual reminder of how that man managed to turn his ancient cloaca into a penis, scrotum, and anus.*

* It's also a useful reminder that what gender essentialists seem to find so essential—the presence of one or another sort of sex-typical genitalia—is a rather

Though the Eves that came between the dawn of placentals and us have long since died, the monotreme arrangement is a decent model for what their arrangements must have been like: the echidnas and platypuses retain their cloaca, and the males still have their testes inside their body instead of hanging out in a scrotum. They also have ureters that dump right into their cloacal sinus.

Modern placentals go through a complicated dance of development—flaws in urogenital development are among the most common birth defects human beings suffer. Some babies are born with a cloaca, though that's very rare. The closer you get to the current model in our evolutionary past, the more likely you are to find mild malformations—maybe a hymen that covers too much of the vaginal opening, maybe a urethra that's kinked or cramped somehow.* Less common are subdivided vaginas, betraying that deeper past: two uteri, two cervixes, and two birth canals. The penis sometimes has issues with the urethra being blocked, divided, or partially open. The testes sometimes fail to descend properly into their scrotum after birth, or sometimes the hole through which they drop fails to close properly, allowing a loop of bowel to slip outside the abdomen. The labia and the clitoris, too, have diverse development,† with some clitorides responding to greater androgen signaling in the womb and developing into proto-penises. Some of these conditions require surgery at a young age—certainly nobody wants a baby boy to die because part of his bowel necrotized from being pinched off in a hernia.

---

small difference in fetal development that takes only a handful of weeks and frequently goes astray. Our evolutionary path is littered with glittering difference, and so it is with bodies in any population. In life on Earth, diversity is a feature, not a bug.

* Though we attach cultural value to the human hymen, it's probably just an awkward leftover of urogenital development. Lots of mammals have them: elephants, whales, dogs. While it's vaguely possible *some* selection maintains the human hymen in terms of "virginal" confirmation for particularly choosy males, it's prone to break from all sorts of normal living well before any penetrative sex and many girls are born without one. Having too much of a hymen is actually a threat to a girl's health. The likelier reason for why human females have hymens is simply that our reproductive system is buggy.

† Diverse evolution, too—female genitalia are far more diverse than the male's in mammals, despite how much attention the penis havers get (Pavlicev et al., 2022).

Others require no surgery at all—there's nothing life-threatening in a proto-penis.

Thus, making the modern placental vagina meant rearranging the lower pelvis, including how it grew a fetus from conception to birth. We still needed to pee, poop, and push our offspring (or semen) out of our bodies. We still needed to make sure urine and feces didn't go careening into our abdomens, so growing the tubing and fleshy divides *correctly* mattered. And once we got rid of the eggshell, live birth required a birth canal that would not only physically accommodate whatever size baby you're trying to push out but also shield the path from things your baby's body might not be ready for.

Before the asteroid hit, mammals were already working on these problems. But from what we can tell from the fossil record, the long winter after the impact killed off more of the marsupials' ancestors than the placentals'. If that hadn't been the case, the majority of us might now have two or three vaginas.

As I mentioned before, marsupials—every single living species—have at least one vagina for each uterus. Those vaginas connect into a short, central out passage—that "urogenital sinus." Those vaginas are how the sperm gets where it needs to go. Many marsupials also have an extra birth canal or two for the fetus to crawl out of the mother's body, up along her belly fur, and into her pouch. The marsupial penis coevolved with these respective scabbards, as penises always do (more on that in the "Love" chapter), which is why possums and kangaroos have a forked penis to match their mates' two vaginas.*

So think of vaginas as a specialized gene delivery system: up go the sperm, down come the offspring. Simple. But, again, the abdomen is crowded. For many marsupials, like the kangaroo, the ureters go *between* the female's three vaginas toward the bladder.

* The monotreme echidna, despite having a cloaca and thus having to *evert* it to push the penile tissue out, still somehow managed to evolve a four-headed penis. Two of the four heads hang back during an erection so that the next time the wayward bachelor happens on a willing mate, the penis can switch hitters and erect these other two, like a game of sexual whack-a-mole.

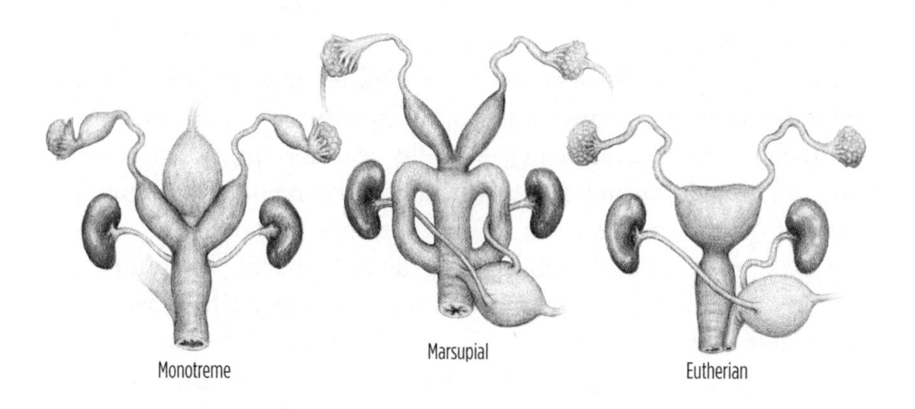

Mammalian vag plans

That means she can't give birth to anything bigger than a jelly bean or she'd tear her ureters. If internal bleeding didn't kill her, she'd quickly die of poisoning from the nitrous waste normally tucked safely away in her kidneys and bladder.

In other words, somewhere along the evolutionary path from the cloaca to the vagina, the marsupial body plan became self-limiting. For the Eves of placentals like us, the ureters either weren't a problem or, by happy chance, stopped being a problem as they evolved. That meant giving birth to larger babies wouldn't kill us by ripping our tubing to shreds. It might kill us in *other* ways, but not that one.

Still, there are always trade-offs when it comes to new body plans. In principle, the more recent a feature, the more likely it is to fail—as true of smartphones as body parts. The walls of our modern placental vagina are, being rather new, a kind of "poorly tested product." If marsupials are any clue, it's likely that the supporting structures for the relocated urethra, which is now just behind the vagina's front wall, evolved more recently than the structure that divides the back wall and the rectum. In human women, these structures aren't as robust as one might hope: as many as one in ten women suffer from urinary incontinence after a vaginal birth. Modern human babies are so large at birth, and have such large *heads*, that the process of "crowning" (moving the head down the birth canal) can be traumatic for the vaginal walls and their surrounding tissue structures. After a difficult birth weakens the deep tissue between the bladder and the front vaginal wall, many women suffer a prolapse, in which their bladder partially falls into the vaginal cavity.* Physical therapy to strengthen the pelvic floor muscles helps most women recover, possibly by retraining the local nerves to appropriately respond to an urge to pee, and possibly by making the layer of pelvic floor muscles *thicker* and thereby propping up the wayward flesh above that muscular shelf.

The other evolutionary problem with human birth and bladders, of course, is that we stand and sit upright. That means there's a lot of downward pressure on the vagina from the organs in the pelvis. Our Eves' bladders presumably hung forward in their fleshy nest, toward the front of the belly, so there would have been no additional strain from gravity as the vagina recovered from the trauma of birth. But now, given our upright posture, the bladder naturally puts pressure on the front vaginal wall. If that wall is weakened by birth, the bladder is liable to fall down and through. It's just physics.

Having a vaginal birth is the biggest risk factor for bladder prolapse in women. The second-biggest factor is menopause; as the hormone balance in a woman's body shifts, the lowering estrogen levels naturally loosen the vaginal tissue and the surrounding pelvic floor muscles. Many women will have surgery to tighten the tissue and repair the prolapse. If the prolapse is significant

* Weakening of the wall between the vagina and the rectum is also common and leads to a particularly feminine sort of constipation. Most of these injuries aren't a *full* tear, but a tear through one or more layers of the vaginal tissue or supporting structures—so not a gaping hole. Though that can happen and is called an obstetric fistula (more on that in the "Tools" chapter). These injuries can also mess with the delicate arrangement of nerves that control the local sphincters, which can present their own problems. Even absent such issues, tears in the vagina and surrounding tissue are par for the course when giving birth, particularly for the first time. Having done it, I can report that most OBs stitch things up pretty quickly once the baby is out. For me, it honestly felt like a bit of *tugging*. Everything was already so destroyed down there that the nerves weren't sending many discernible signals to my exhausted brain. The road to recovery after the birth, however, was surprisingly long and painful.

enough—for instance, part of the bladder actually falls out of the vaginal opening, or the uterus does, cervix slumping down past the vaginal walls—some women may even have the vaginal opening surgically closed to support their organs. Naturally, this involves being willing to never have vaginal intercourse again, but many older women are willing to accept the trade-off.

You might be about to accuse me of reinforcing a stereotype about older women's lack of interest in sex, but only 25 percent of women of any age reliably experience an orgasm during vaginal intercourse. When you control for whether those women also experience any clitoral stimulation during sex, the numbers drop further. Despite their obvious evolutionary function as birthing tunnels and receptacles for sperm, the vagina isn't the center of most women's sexual satisfaction, old or young—that remains, hands down, the clitoris. If you use a No. 4 camel-hair paintbrush to stimulate a female rat's clitoris, she'll happily return to the place she associates with it, over and over and over. She'll emit a series of subsonic squeaks while she's there—a quiet lover she's not—and both her brain and her behavior will show evidence of reward seeking and pleasure, and if you do it near an almond-scented pad, she'll solicit sex from an almond-scented male later. Female rats who experience clitoral stimulation also show lowered stress and better general health than rats who don't have that sort of stimulation. In other words, clitoral stimulation is good for a lab rat's health, much as it seems to be for human women.

Birds, poor things, do not have a clitoris, and most bird and lizard cloacae don't have the same sort of nerve sensitivity.* Honestly, you probably wouldn't *want* a heck of a lot of sensitive nerve endings inside the place you use to push out eggs—and indeed modern vaginal walls are similarly nerve dull. Most male birds

* Some female birds do show evidence of pleasure during sex. Far more data exist on masturbating males than females, unfortunately. The one bird we know of that does have something like a clitoris is the *male* weaver bird, who has an additional "fake" penis that doesn't deposit sperm. He will, if properly stimulated, shudder and curl up his scaly toes. We can't say he's feeling the same things as human women do when we orgasm, but from an outsider's perspective it looks like it (Winterbottom et al., 2001). He certainly wouldn't be motivated to fake it.

don't even have a penis.* Ninety-seven percent of birds have sex through a "cloacal kiss," wherein the female lines up her poochy, inside-out cloaca with the male's open cloacal slit, whereupon he forcefully ejaculates, shooting the stuff directly onto/into her. She then pulls her cloaca back inside her body with a feathery shuffle—a modern dinosaur adjusting her skirt. Sex for most birds is a brief affair; it's the mating rituals that are elaborate.†

Squamates (scaled reptiles, such as snakes and lizards) do tend to have penises—usually a Y-shaped thing called a hemipenis—that they keep deflated and tucked inside their cloacal opening and evert for sex as needed. In fact, it seems all amniotes have descended from an ancient Adam that had an erectable penis. But the dinosaurs that became birds inactivated the gene that allows for an embryo's penile development. You can see it in the egg, actually: a little tip of flesh that rapidly shrinks back into the bird's fetal body as it grows.‡ The running theory is that mate choice made it useful to get rid of the penis: when a female chicken doesn't choose to usefully present her cloaca to a randy rooster, there's simply nowhere for him to put his sperm. Most of what's left of the dinosaurs, in other words, evolved away the penis altogether, because giving females what they wanted (having bodies that necessitated a female's *willingness* to have sex) proved better for their survival. Or so the theory goes.§

After all, whenever you find a penis-having species, you should know that the penis *coevolved* with the species' vagina. It's not simply that they are useful to deliver sperm directly into the female's

* Cocks, in fact, do not have one. Human language often misrepresents reality.

† And the pair-bonding rituals, if they have them, like the extensive grooming and nuzzling you'll find in certain parrots.

‡ Tuatara, another type of reptile, also got rid of the penis in a similar manner. While the penises of the world are wildly diverse, they're all modifications of one basic, ancient evolutionary innovation. The amniotes that don't have them today evolved from ancestors that got *rid* of them for one reason or another.

§ It's also true that growing a penis is an error-prone thing: 1 in every 125 boys are born with one or another sort of penile defect today, the most common of which is a misplacement of the urethra, which may betray the human penis's own evolutionary past (Paulozzi et al., 1997; Bouty et al., 2016; Gredler et al., 2014). So it could also have been better to have *no* penis, somewhere in the dinosaur past, than a bunch of defective ones.

reproductive tract, increasing the male's chance of passing on his genes, but the female is also interested in the *right* penis—the one she particularly wants—to do the job. Or lack of one, given the bird penis that vanished likely as a result of the positive influence of female choice. Since many species have forced copulation, vaginas have also evolved a number of ways of setting up little foldy bits of tissue that can close off or open up depending on the female's willingness to be impregnated by that male. The "rapier" the species, the more likely the female will have such a thing: duck vaginas are notoriously foldy. (More on that in the "Love" chapter.)

If placentals had retained multiple vaginas, we would have also had to deal with an irritatingly complicated phallus, which might have worked against us in the long course of evolution that recently produced humans. Human males have some of the only penises in the world that lack a baculum (a small support bone), relying entirely on turgid tissue to support their effortful thrusting. This has led to a number of broken penises,* not to mention the extremely common (but evolutionarily severe) problem of erectile dysfunction.

But humans' relatively simple heterosexual mechanics nonetheless helped us avoid other problems. For example, a female rhino has such a convoluted vagina that the rhino male evolved a two-and-a-half-foot-long penis shaped like a lightning bolt to match it. A long time ago, people in China glimpsed the lightning bolt penis (or perhaps witnessed the typical two and a half hours of mating the rhinos have to go through just to make the darn things work) and erroneously believed that rhinos' physical prowess could be transferred to humans. Rhino horn—illegally

* Though human penises have no baculum bone to break, the outer sheath surrounding the erectile tissue of the penis can rupture, typically by being struck or being bent too far while the penis is erect. This is both hideously painful and a medical emergency that typically requires surgery. The most common way it happens is when, during a particularly vigorous bout of sexual intercourse, the penis slips and strikes the perineum, causing the penis to buckle. Thus, the evolution of live birth not only led to the creation of a vagina separate from the anus but also produced a common *hazard* that, in the human reproductive system, can potentially render a male unable to bear children. Especially vigorous sex, in other words, is a sign not of a man's virility but of his recklessness.

poached, dried, and ground into a powder—continues to fetch a high price for poachers on the black market. That's why most rhinos are now endangered species; thanks to that complicated vagina, zoos have a difficult time impregnating them to increase their dwindling numbers.*

So, rhinos got complicated vaginas and are going extinct at our hands, marsupials kept their multi-vag but are mostly isolated to Australia, and placentals like us, with our simple, single vaginas, have spread all over the world. From there, we built the modern placental uterus—or rather, uteri.† Like today's marsupials and a majority of rodents, our Eves originally had two.

HOW TO TURN YOUR BODY INTO AN EGGSHELL

In 2017, a group of American researchers did what no one thought was possible: they rigged up a mechanical uterus that could bring baby lambs to term. They called it the "biobag." Videos of a *Matrix*-like contraption soon popped up on the world's news websites: a pale, fetal lamb barely contained in a translucent sac of artificial amniotic fluid, tubing pumping blood and waste in and out of its body, little hooves delicately kicking in their alien pool. I imagine the video might have struck viewers as a kind of clarion call: *The end of pregnancy, rejoice!* In reality, the biobag works only for part of the third trimester. In other words, if it works for human babies, not just lambs, it's meant to be an improvement on what a NICU can offer, supporting preemie babies in a way that better mimics the mother's pregnant body.‡ No one has

* Never mind that the horn isn't even horn, but tightly compressed hair with a calcium-rich core, or that the horn has absolutely nothing to do with rhinos' sex organs.

† For the grammar nerds: it does have a Latin root, so it is formally "uteri," while "octopus" is from the Greek and should be "octopodes." American English allows for both "uteri" and "uteruses." Personally, I like "uteri." It feels wonderfully sci-fi.

‡ The big innovation here wasn't the fluid but being able to "plug in" to the preemie's bloodstream via the umbilical cord, thereby letting the lungs develop a bit longer without having to breathe air. In the womb, the fetus inhales amniotic fluid throughout the end of pregnancy, a critical part of fetal lung development for land

ever invented a true external uterus. To do that, they would have to invent an entire mechanical mother, because placental animals like us use our *entire bodies* as an eggshell.

The eutherian uterus evolved from the "shell gland"—a muscular, oozy organ that secreted all the stuff necessary to produce an eggshell. Each shell type evolved to serve each species' needs until the babies were ready to hatch. It's a fairly straightforward process: the egg matures in the ovaries, rolls down a little tube, gets fertilized, and develops a shell in a muscular sac that spooges various materials over it to get it ready for the outside world. Meanwhile, the mother's brain—likewise evolved in its particular environment—prompts her to perform various behaviors that help her eggs make it all the way to hatching. Once you stop laying eggs and instead give birth to live young, you don't get rid of those other needs; it means you need to find a way to turn the mother's body into a combination of eggshell and nest.

That's a tricky prospect. Not only do you need to find a way to let the kiddo respirate, but you need to find a balance between providing enough resources for the full length of gestation—however long your species' progeny takes to "hatch" and become an independent offspring—and, you know, not completely destroying your body in the process.

In a sense, milk-producing species such as Morgie's already had a leg up. Because lactating species were already accustomed to more intensive caretaking after their babies hatched—providing nutrients and water and immuno-goods from their own bodies through milk—they didn't have to entirely change their behavior and physiology around reproduction. In the beginning, all they really had to do was move the egg nest inside and devise a nondestructive path out when it was time for their offspring to be evicted. After that, motherhood was largely the same as it ever was.

Of course, in the wild, both mother and offspring are terribly vulnerable, often near starvation, until the pups are developed

---

animals. Very premature babies' lungs have oxygen forced into them in the NICU. They would die without it, but it does damage their lung tissue.

enough to stop nursing and start foraging for themselves.* And unless she's stored up a tremendous amount of fat or shelf-stable food in a burrow, the mother is still going to have to leave the nest to go get more food; a pregnant mom is a hungry mom, and a nursing mom is even hungrier.

It's not hard to see why the egg-laying strategy has worked so well for so long. Even among caretaking egg layers—some dinosaurs among them—letting the egg do its own thing for a while would have been a tremendous relief for the mother's body.

So how did we get here? Who was the Eve of eutherian placentals—the mother of our collective womb?

Like finding the Eve of milk, tracking down the placental Eve is tricky given that soft tissue, like breasts and uteri, doesn't survive in fossils. With milk, we had genetic clues: specific genes that code for necessary proteins in egg laying, and others that code for making proteins in milk. With them, we could track down a general range of time when the egg laying probably stopped, and likewise find the origins of milk. But there are so many genes involved in wombs and placentas (most of which we haven't even isolated yet) that it's still pretty hard to narrow down where we break from the marsupial plan and start being placentals. Most paleontologists rely on studying the general pattern of bones among today's living marsupials and placentals to help them theorize.

There are some things we're pretty sure about. For instance,

* This is also true among the poor and oppressed today: never forget that more than 50 percent of Indian women—a nation of more than 1.3 billion people—are currently suffering from anemia and malnutrition, due in no small part to local traditions where young women eat last, after the father, children, any men in the extended family, and finally older women (Hathi et al., 2021; Coffee and Hathi, 2016). When the young woman is pregnant, malnutrition becomes especially severe and harms the fetus, further stunting the upcoming generation of one of the world's most important economies and, by far, the world's largest democracy. This won't be the last time I say this: humans are mammals. If we want to invest in humanity's future, we have to feed human mothers, and we have to feed them well. It would also be nice if we'd stop abusing and killing women in general, but let's start with the food.

because of genetic dating methods, most estimate that the ancient placenta probably evolved anywhere between 150 and 200 million years ago—a long time before the asteroid, in other words. The placenta is the organ that lets embryos attach themselves to the mother's uterus without being wholly destroyed by her immune system—a pretty important feature for live birth. It's derived from the same membranes that surround embryos in eggs, but evolved into a big, fleshy, alien docking station between the mother's body and the growing embryo.* Not everyone who gives birth to live offspring has a placenta like ours. Roughly 70 percent of all living shark species give birth to live pups (and were among the first on the planet to do so). But only one group of them—the ground sharks, or Carcharhiniformes—evolved to use placentas. Those placentas are relatively shallow affairs, compared with the highly invasive things so many eutherian mammals produce. The sharks that give birth to live pups but don't make placentas use a range of strategies to keep them fed in utero: secreting a thick mucus from the uterine walls that the pups can munch, firing unfertilized eggs down the fallopian tubes to waiting hungry mouths, or even having the earliest-hatching (and thereby largest) pup *eat* siblings in the womb, leading to a rather violent bit of in-family cannibalism.† Tawny shark embryos are actually known to swim between the mother's two uteri in search of a meal, and even, occasionally, poke their little heads out of the cervix and through the cloacal opening to take a look around. If you aren't attached to the uterine wall and your next meal lies in the other uterus, you might as well swim in search of your lunch.

But we don't descend from sharks. As for dating *our* ancient Eves, researchers found a fossil in 2011—*Juramaia sinensis,* or

* If you've never seen a human placenta, the internet is waiting for you. But a warning: it's bloody and huge and *extremely* horror show. Guillermo del Toro would be proud.

† While that cannibalism strikes us as horrific, consider that the mother shark gets to do a lot less in terms of tricking her immune system and robbing her own resources to grow her offspring than we do with our deeper, more invasive placentas. And because "survival of the fittest" continues to be true in much of the natural world, the competitive shark embryo may have genes that likewise aid its fitness for life outside the womb.

"ancient mother," a squirrel-like thing that ate a bunch of tree bugs roughly 160 million years ago in what became northeastern China. Because she had teeth that were more like our teeth than those of marsupials, most think *Juramaia* is the oldest known Eve of the eutherian line.

Still, a lot happened between 160 million years ago and the asteroid apocalypse, and even more happened (rather quickly) after the world burned. It's very hard to know what the mammalian placenta was doing all that time, or why ancient marsupials and *our* Eves remained head-to-head in terms of dominance throughout the Jurassic era, despite their bodies' differences. Of *Juramaia*'s many descendants, we simply don't know how many evolutionary paths were dead ends: Were her children among the ones that went on to survive the apocalypse, or weren't they?

Shortly after the turn of the millennium, an international group of paleontologists and comparative biologists assembled a massive database of morphological features from all known living and extinct mammalian species. Then they used complex computation to trace everything they could think of backward through evolutionary time: from whence comes this particular jawbone, from whence those curious toes, from whence (importantly, for our purposes) these sorts of pelvic bones. About forty-five hundred characteristics, all told. They found that the last, true Eve of today's eutherian mammals was almost certainly an arboreal insect eater, about the size of a modern squirrel, who spent most of her life climbing trees and snatching bugs from their high perches. She lived roughly sixty-six million years ago. Like many of our true Eves, we have no fossil that's *for sure* the one. But we do have a creature with all the right traits dated within a useful margin of error that the researchers call *Protungulatum donnae*.

Let's call her Donna.

Jubilant, the researchers even commissioned a rather adorable portrait of her for the paper:

Her eyes, beady but blithe, shine black in the high forest light, where she stretches forward to snap up an insect. Her nose is large, her whiskers short, and her tail long and bushy tipped. There she is: our womb's many-times-great-grand-rat.

So, Donna, the Eve of the modern eutherian uterus, had her

Donna, our womb's great-grand-rat

toe pads in the right spots.* She had a fondness for the sweet crunch of live insects, which she'd catch with the cone-shaped, jagged teeth that lined her delicate, narrow maw. Her ears, set close to the hinge of her jaw, were furred, as was the rest of her. Unlike Morgie, her legs didn't splay to the side like a lizard's, but instead ran more vertically from pelvis to ground.

For eutherians, the alteration in the pelvis is pretty crucial. In order to fit a swollen uterus, you need a pelvis that is more of a

* Some still think *Juramaia*, or an Eve similar to her, is the likelier candidate, in large part because there's a whole camp of scientists who like to rely on molecular dating and another camp that prefers to rely on known morphology—in other words DNA versus the Stuff DNA Makes. Both camps have problems. The DNA camp makes a lot of assumptions about how long it takes for DNA to mutate and for mutations to spread throughout a population. The SDM camp makes a lot of assumptions about how long (or short) the oldest "branches" of a taxonomic tree might have lasted before changing. Donna is the favored candidate of the SDM camp, essentially, and *Juramaia* (or some contemporary of hers) is a candidate for the DNA camp. These ancient weasel-squirrels are very similar creatures, so the debate is really about *when* rather than *what*. The *when* matters here because of the asteroid—most paleobiologists assume that the apocalypse was the big driver for why ancient mammals proliferated and diversified and essentially took over huge portions of the planet. In that case, Donna is our Apocalypse Eve.

bowl shape. Rather than scrambling along with our bellies dragging on the ground like alligators, we evolved in such a way that our torsos naturally lifted higher so that the upgraded pelvis could support a pregnant placental uterus.

Uterus, not uteri: the authors assume she had a single, horned uterus, and handily provided an illustration of that, too, alongside a sketch of her various cuspids, and skeleton, and the flattish tadpole-like features of her partner's sperm (let's call him Dan). When Donna and Dan mated, she gestated her big fetuses in her fused uterus just long enough to produce something like newborn squirrels, hairless and blind, which arrived in the world through her (presumably single) vagina.

Since Donna is the lovely lady squirrel from whence all extant, placenta-having, non-marsupial mammals evolved, she's the one we can blame for the *modern* placenta, single womb, and vagina. But she's not the only model of mammalian uteri. Mice and rats, for instance, still have two separate uteri with a cervix for each.* Elephants and pigs (eighty million years ago) have a partially divided or "bicornuate" uterus, with its upper "horns" more or less separate and some portion of the bottom fused, but they have a single cervix. Basal primates like lemurs also have that sort of uterus, but more derived primates have the fused, pear-shaped arrangement that we do. Since our Eves split off from lemurs around thirty-five million years ago, that means the semi-divided uterus hung around in our Eves' bellies for a good, long time.

Though not all developmental snafus are true atavism,† you

* If that's confusing, remember that the paper has Donna appearing roughly sixty-five million years ago. Our Eves split off from the rodent lineage around eighty-seven million years ago. Different traits appear in the body at different points in history: the placenta appears in ancient uteri well before Donna gets here with her semi-fused womb. After all, even marsupials have a placenta; it's just smaller and shallower than ours, which makes sense, since they only need to grow to the size of a jelly bean before they transfer out to the pouch.

† That is, a reversal to a basal or ancestral state. Many oddities in the development of the Müllerian ducts—the two fetal tubes that turn into female reproductive organs—are pretty darn atavistic, but something like hypertrichosis (that is, "werewolf syndrome," where a patient has long hair growing all over the face and body), however primitive it may appear, is less so. Because the Müllerian ducts develop alongside the Wolffian ducts in the embryo and seem to interact during the process, uterine malformations are often associated with renal problems—

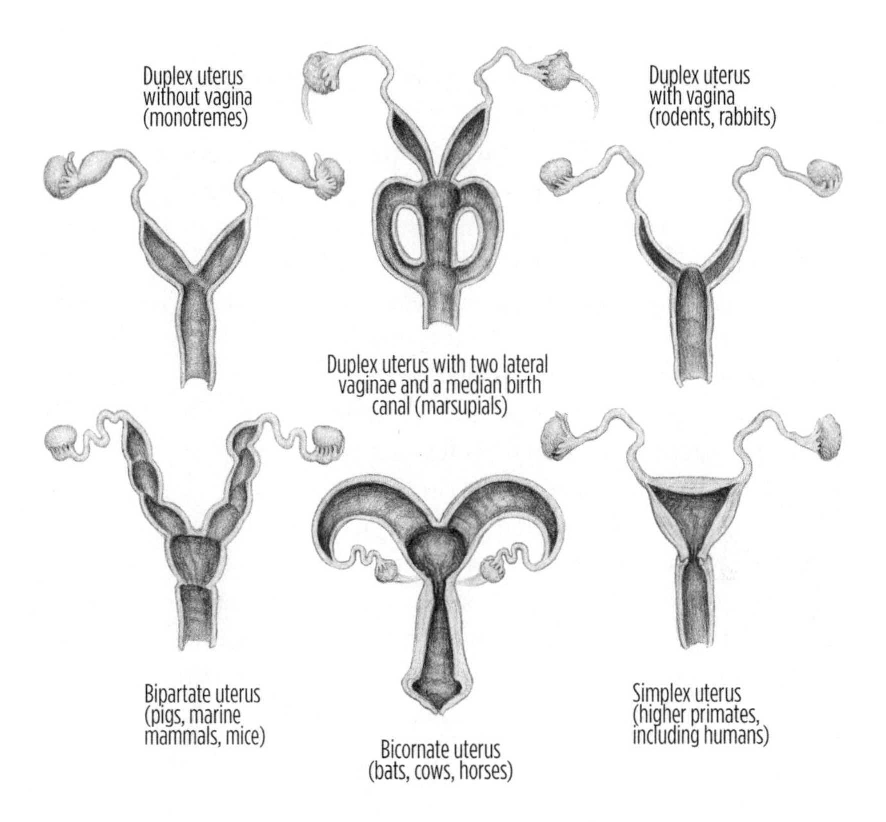

Timeline for mammalian uterus evolution

can still trace how this evolutionary history might have played out by looking at the wombs of women today. Roughly 1 in every 350 human girls are born with two uteri and cervixes at the end of their normal, single vagina—a glitch in the developmental programming that clearly harks back to our evolutionary past. Even more commonly, 1 in 200 women are born with a "heart-shaped" uterus, wherein the upper half of the uterus is split in two. Roughly 1 in 45 girls are born with a "septate" uterus, wherein a fibrous wall separates the upper part of the uterine cavity from the lower,

---

including the rare failure of an entire kidney to form. It's becoming more common to screen for kidney issues when uterine malformations are found, and some in the medical community are calling for the reverse to be true as well (van Dam et al., 2021).

and 1 out of every 10 girls are born with a uterus that has a slight "dent" in the top—a wobble, if you like, in the modern outline of the human womb.

Each of these abnormalities is tied to malfunctions in girls' fetal development, and the most common glitches are probably tied to more recent developments in our evolution. It's been a very long time, for example, since our ancestors had two uteri, but not as long since our uterus was partially fused. Much more recently, there was probably that minor, fibrous wall, and that leftover "dent" at the top was probably the last to go, given that one in ten of us still have it. The little dent doesn't seem to negatively affect pregnancy outcomes, so it's safe to assume there isn't a lot of evolutionary pressure to get rid of it.

I haven't yet mentioned the 1 in 4,500 girls born every year *without* a uterus. Since the male-to-female birth ratio is about 1.7 to 1 and roughly 133 million babies are born annually, that means more than 14,000 baby girls are born without a womb every year. The vast majority of those girls aren't trans, and for the ones who aren't, being born totally without a womb has to involve some

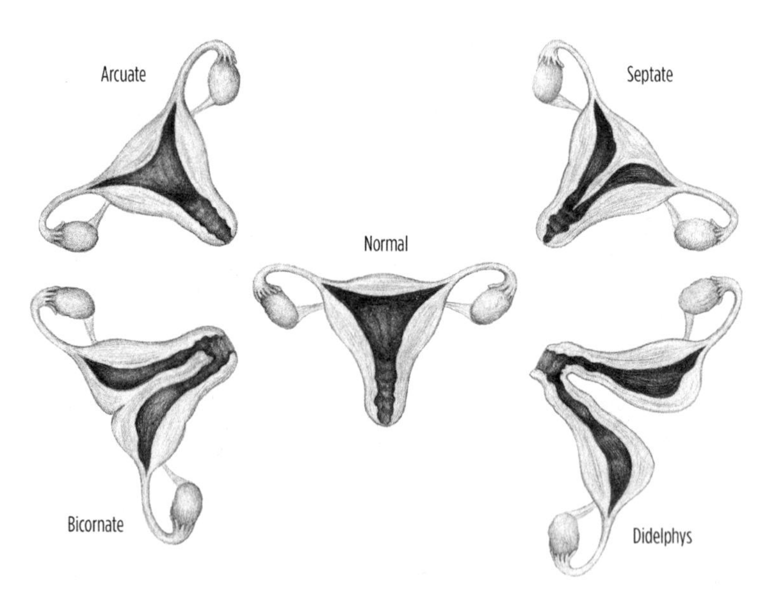

Human uteri today

dramatic detours from our genetic and/or developmental past.[*] For example, sometimes a genetically male fetus (XY or XXY) doesn't respond to androgens in the womb in the usual fashion and so is born with the external appearance of a girl. Such people usually grow up identifying as a girl, but later find out they have two testes where their ovaries would normally be. That's a pretty cool mutation, but also an evolutionary cul-de-sac, since they can't pass on the trait to their offspring.[†]

The mutation is too rare for us to think of it as a major part of the human reproductive plan—this wouldn't be a part of humanity's potential eusociality. A quick primer, if you're not familiar: with eusocial species, not every individual has a chance to reproduce, but a caste of asexual members is useful, even essential, to the group's success, most typically by helping care for the young. Ants are some of the most famous eusocial sorts of creatures, with their female, childless workers and giant egg-laying queens.[‡] Bees are eusocial, too. And while it's a more popular arrangement among social insects, even mammals have eusociality—most famously the naked mole rat. Cooperative breeding and child rearing that seem a heck of a lot like eusociality also show up in many mammalian species, like the meerkats. Human child rearing is already highly cooperative, and perhaps homosexuality—wherein, barring social pressure, an individual does not naturally produce his or her own children—is a strong case for human eusociality. The latest numbers estimate that as many as 20 percent of humans are homosexual, and given that the majority of scientists think that

* To put it simply, we *know* why most trans girls aren't born with a uterus: most of them have a functional *SRY* gene on their Y chromosome and, like other such babies, went on to develop male sex organs in a typical pattern. We haven't the foggiest idea why so many cisgender girls are born with wonky sex organs—I mean, we don't know the exact *mechanisms*—but given our evolutionary history, it's obvious why there might be so many fail points along the developmental path.

† Because of the high risk of testicular cancer in these patients, they usually have a gonadectomy after diagnosis. Given that amenorrhea (no periods) is what leads them to the clinic in the first place, they're usually teenagers at that point, though allowing them to finish puberty first might be useful for a number of reasons (Barros et al., 2021). It's possible advances in IVF will someday allow these women to have genetic offspring of their own, but to my knowledge, that hasn't happened yet.

‡ The males largely exist to deliver sperm in most ant species.

homosexuality is a trait present from birth, those sorts of numbers indicate that whatever part of homosexuality is classically heritable can't have been too strongly selected against. In highly social species like ours, the benefits of having extra hands for child rearing—hands that aren't busy taking care of genetic children of their own—might have outweighed the evolutionary pressure against homosexuality. Homosexuality has been observed in countless species, mammal and bird alike.

One complicating factor in this is that, frequently throughout recorded human history, oppressive social pressures have forced people who might otherwise prefer not to have sex with the opposite gender to do so anyway—usually some measure of "god" is invoked—thereby passing on their genes.* So while homosexuality might not be something that's selected *for* in classic evolutionary terms, it nonetheless exists commonly in the population for a number of reasons. And in the end, it doesn't seem to have much effect on the reproductive success of the species as a whole: witness our multibillion-strong global population. What's more, given that our species' child rearing is deeply collaborative across local communities, the commonality of people who, for one reason or another, had no easy ability to produce children of their own might even have made our ancient offspring that much more likely to survive to adulthood.

If anything, the sex drive is what's really fundamental to evolution. Most mammals are sexually oriented in one fashion or another. Like chimps and bonobos, *Homo sapiens* are an especially promiscuous species. Changing the gender of the target? It happens. Truly asexual people are probably the ones who are *really* rare.†

* It's impossible to know if this occurred before humanity started keeping (readable) records, but let's just say cuneiform isn't when we started treating each other badly.

† As all sorts of atypical sexualities become more socially tolerated, the number of publicly self-identifying asexuals will rise, but it's probably safe to assume those numbers would be much smaller than other orientations, and the latest research on the subject carefully qualifies that it may have many different underlying mechanisms (Bogaert, 2015). It's also true that one's *desire* for sex, in general, can vary drastically over the course of one's lifetime. But sexual orientation is different from fluctuations in desire, and our understanding of the biological underpinnings of

Donna, living in her squirrely body in ancient gingko-like trees, probably wasn't asexual. Her fused, horned uterus was regularly obliged to produce her tree-born babies—otherwise we'd never have descended from her. But given her relatively small size, there also wouldn't have been a lot of pressure to make her uterus merge into the single, pear-shaped organ women carry today. That probably didn't happen until her descendants got a bit bigger.

We might even be able to see this sort of evolution in action by looking at species living today. Generally speaking, the biggest mammals in the world usually have a single, fused uterus, with a single cervix leading to a single vagina, and they likewise have one or two offspring per pregnancy. The smallest? Two uteri, two cervixes, and a litter. Domesticated cats have a strongly bicornuate (two-horned) uterus, shaped like a Y. Bigger cats like the Bengal tiger have a bicornuate uterus as well, but the two uterine horns curiously *curl* down toward the body of the uterus, like a court jester's hat, and the bottom is a bit more fused. They usually give birth to one to three large kittens, while domesticated cats give birth to four to six. Smaller rodents usually give birth to six or more pups, while the largest, the South American capybara, usually gives birth to four.*

If that's true—if evolving a larger body size means making bigger babies, which subsequently means having *fewer* babies—it makes sense. As a reproductive strategy, it's less risky when you're

---

sexual orientation will likewise deepen with time and provide more nuance to what it means to be sexually "oriented" in one way or another.

* It's true that some domesticated dogs give birth to up to ten pups at a time, but wolves in the wild usually have only four to six. There are multiple evolutionary pressures on dogs: first, domesticated dogs are often deliberately bred, with the genes for desired traits clearly benefiting from whatever the breeder prefers, and the more pups a prized bitch can produce, the better for the breeder's wallet. Second, only 30 percent of wolves in the wild survive their first year. If the litter's bigger, they're even likelier to die; larger litters of eight to twelve tend to whittle down by the summer, with roughly the same number (one to four) surviving to the following winter. That's probably because after they're done drinking milk, pups eat regurgitated meat from the older wolves, and any given environment supplies only so much prey for the parents. Thus, the fact that domesticated dogs can have so many puppies *and* have them survive to adulthood has a lot more to do with what humans do than what canids' bodies evolved to do before we intervened.

big to have fewer babies because, as a big creature, you're less likely to die from predation and/or being stepped on. But it also makes sense for placentals to adopt this strategy to save their own health, because making big babies also means being pregnant *longer*, with all the accompanying risks. The bigger we get, the bigger our babies, and the more likely we are to have a smaller number of them. Imagine being pregnant with the same fetus for two years. That's what elephants do. And you really wouldn't want to have more than one elephant placenta in there. (Elephant twins are extremely rare.)

So, the *real* Eve of placentals, the one that predates Donna, might or might not have had a fused uterus with two horns (though she probably had some manner of modern vagina and cervix, given the timing of it all). But as her descendants experimented with larger body sizes, they would have had bigger offspring, which made a more fused uterus necessary. As fetuses grew in size, their placentas became more penetrative—it takes a lot of energy to build bigger bodies, so the more a fetus can draw from its mother, the better off it is . . . so long as it doesn't kill her. Each evolutionary step toward a fused womb and greedier placentas makes sense: it's a dance between what the mother's body needs and what her hungry offspring need, with each accommodation skirting just on the edge of killing one or both of them.

That means, at least in part, our Eves gave birth the way they did because they got *bigger* than ground squirrels. And their uteri and placentas adapted accordingly, not only to accommodate the heft of their children, but possibly also as a way to help the mother endure more taxing pregnancies.

By the way, there's been one more attempt to make a mechanical uterus: in 2021, a lab managed to keep mouse embryos alive and developing normally in a spinning vial soused with a complex, amber-colored fluid that carefully controlled the exchange of oxygen and carbon dioxide. The spinning part was important: they had to prevent the developing embryos from attaching themselves to the walls of the vials, as they would in their mothers' uteri. So they grew tiny mouse placentas in the oxygen-rich liquid, like little disks, floating and twirling, attached to the amniotic sacs, which also floated, and their tiny hearts grew until they beat well on their own—until, in fact, they could no longer survive without a blood supply (the amber liquid could do only so much). And while the placentas *looked* perfectly normal, they couldn't have been exactly right. Because a normally grown, living placenta is made of *both* the mother's and the child's bodies. It has two plates that fuse together to make that singular organ: one side that's always hungry and one side that's trying to protect itself from that hunger.

WHAT TO DO WHEN YOUR KIDS ARE TRYING TO KILL YOU

At some fairly flexible point in her adolescence—anywhere between eight and eighteen—the female *Homo sapiens* arrives at menarche: the rite of passage that involves uterine blood and tissue leaking out of her vagina for an average of three to seven days.* If she happens to be an American *Homo sapiens*, the girl will probably have awkward conversations about purchasing tampons or diaper-thick pads from the local drugstore. The girl may or may not also suffer from menstrual cramps—deep, grinding pain from uterine contractions as the organ sloughs off its unused lining—or headaches, or mood swings, or food cravings, or breast pain, or acne, or any of a host of fun new additions to her young life. Over time, she'll also encounter the joys of being told that her unhappiness about one thing or another is down to "PMS" and that, as a woman, she's far too "emotional" to handle the sorts of life challenges usually assigned to men.

What her parent probably won't tell her—because relatively few *Homo sapiens* know this—is that she should be impressed by the fact that she menstruates like this at all. There are only a handful of species in the world that do. Among the descendants of

* I know—nine is *young*. The start of menarche has been slipping downward in industrialized countries (Bellis, 2006; Winter, 2022). Many fear it's due to hormones in our food and other endocrine-disrupting chemicals in our environment, while others think it has to do with the rise in childhood obesity: girls who are overweight or obese by age seven are significantly more likely to start their periods sooner (Ghassabian et al., 2022; Diamanti-Kandarakis et al., 2009; Jacobson-Dickman and Lee, 2009; Freedman et al., 2002).

Donna that automatically build up and shed their uterine lining as we do, the vast majority simply reabsorb it.

The girl would probably not be impressed by this.

But it's true. Shedding menstrual material out of a vagina is super rare. And we've only just come up with a good theory for why we do it.

Every month, the interior lining of the human uterus thickens. This is the endometrium, a layer of hillocked tissue thick with blood vessels, ready to nourish a freshly fertilized egg when it rolls down the fallopian tube and gently tumbles onto its soft bed. From there, the thickened endometrium—repurposed from a shell builder in deep evolutionary time—will create a network of blood vessels to nourish the growing placenta, and the pregnant woman will glow with satisfaction and eat chocolate-pickle ice cream and all will be right with the world.

Or at least that's what I learned in my eighth-grade health class—a thick white curtain drawn down the middle of the room, with the girls on one side, learning about their vaginas, and the boys on the other, learning about their penises.*

In the class—taught, as I remember, by a person who had no particular training in anatomy or medicine or human sexuality, for that matter—I learned that menstruating was just my body's way of getting ready for a baby, that the endometrium was a lush cushion of baby love, and the fact that I suffered through menstrual cramps was punishment for not getting pregnant often enough.

But we shouldn't fault my teacher, given that this theme still runs through the scientific literature on the human uterus. In doing research for this book, I learned that I'm having too many periods because I'm not pregnant or breast-feeding as often as my ancestors would have been (and that's bad), that not being pregnant often enough or early enough could put me at greater risk for certain cancers, that delaying pregnancy into my thirties could make my babies deformed (or at least cognitively challenged), and

* No, really. This happened. Americans are very, very, very uncomfortable with children knowing anything about sex. We also have some of the worst teen pregnancy and STD rates in the industrialized world. Yes, these two things are related to each other.

that—as if there weren't enough salt in the wound—European women who fall pregnant in their twenties are less *happy* than women who become pregnant later in life, but experience far fewer physical consequences from becoming mothers, which can make a person miserable all on their own. If all of that is really true, maybe it's not so wrong that people call menstruation a curse.

In the 1990s—an era when many Americans spent quite a lot of time thinking about AIDS—some researchers thought that maybe human menstruation was a kind of anti-pathogen mechanism that dumped tissue infected with sex-delivered invaders once a month. That idea has since been abandoned since the vagina doesn't seem particularly *less* loaded with foreign bugs after menstruation.

And then there is the behavioral camp. A number of scientists think maybe women's periods evolved as a social signal: that because one or another ancient hominin male could ostensibly *see* when a female wasn't fertile, there'd be a brief respite in the sexing that could, say once a month, let the females do other things.

Never mind that many human men, much like their fellow apes, have no problem having sex with women who are obviously not at peak fertility: already-pregnant women, breast-feeding women, women who are clearly menstruating, postmenopausal women, and even women who are visibly *ill* will all experience sexual advances from men at one or another clearly infertile point in their lives. Also, some women experience an enhanced sexual drive while menstruating. Like the two species of ape we're most closely related to—the aggressively horny chimpanzee and the sociably horny bonobo—when it comes to sex, human apes are generally good to go, regardless of the fertility status of the female.

Certain members of anthropology and biology departments in the 1980s and 1990s wondered, *What's with all those women synchronizing their periods when they live together? That must have an evolutionary advantage, right?* One ambitious fellow (published by Yale University Press, no less, in 1991) decided this meant that ancient women somehow evolved to go on collective *sex strikes* by synchronizing their periods, thereby enabling/encouraging men (less distracted by the pressing desire to screw) to go out and hunt and forage. This, the author theorized, was the root of all human culture. In effect, he argues that humans build cool stuff like the Pyramids and rocket ships because women get periods and therefore don't have sex for a set number of days per month.*

Menstrual blood has taken on all sorts of cultural significance throughout human history, most of it bad. But to think that evolutionary processes would produce such a deeply significant mutation as external menstruation just so guys would be less horny for a while misconstrues what the uterus actually has to go through to make babies.

Refocusing on that simple fact—what the uterus does, rather than what men may or may not think about it—has led to a much more promising theory.

The endometrium has two parts: the basal layer and the functional layer. The basal layer, clinging to the muscular interior of the uterine wall, isn't shed every month. We shed only the functional layer, which is produced by the basal layer. When the right amount of estrogen rises in a woman's bloodstream, the basal layer of the endometrium starts building up the functional layer that tops it, forming a spongy mass of mucous tissue and coiled blood vessels, riven by deep, narrow canals and tipped with a waving fringe of cilia.

If a fertilized egg manages to hook onto the endometrium's functional layer, it will start building a placenta. The functional layer of the uterus will then rapidly transform into what's called the decidua, a thick buffer between the mother's body and the growing embryo. Meanwhile, digging down into the decidua, the embryo will start building its part of the placenta. That's right: the placenta is actually made of *both* embryonic tissue and the mother's tissue—one of the only organs in the animal world made

* I'm simplifying here. But accurately. Notably, one study of Dogon women in Mali found *no* evidence of menstrual synchrony among women, despite the total lack of (modern) birth control, artificial lighting, or cultural squeamishness about sex, often pointed to as causes for our excess periods. These women had eight to nine children over the course of their lives and roughly 100 to 130 periods all together (Strassmann, 1997). By comparison, the average American woman today is likely to have around 400 (ibid.). So we may not actually sync our periods, but we Western folk *are* having roughly four times as many menstrual cycles as our ancestors would have.

out of two separate organisms. One half is built from the blueprints in the embryo's genetic matter. The other half, the placenta's "basal plate," grows out of the mother's decidua. Two fleshy landscapes, one organ.

If no fertilized egg is in the picture, the mother's ovaries trigger a rise in progesterone after she ovulates, and the "functional layer" of the uterus breaks down and gets sloughed off. The uterus even helps out with minor contractions. If they're severe enough, women experience these contractions as "cramps." I have a distinct memory of lying on my bed as a fifteen-year-old, green with pain, punching my stomach to make it go away. It worked, as I remember. Or at least it was a different kind of pain.*

The fact that menstrual material comes out of the vagina isn't the most interesting part. The question is why the uterine lining starts building up before it knows a fertilized egg is barreling down the fallopian tubes toward it. Among Donna's descendants, this trait is exceedingly rare. Yet it has evolved independently three different times: once for higher primates, once for certain bats, and once for the elephant shrew.†

Why would the trait arise in such radically unrelated species? Does it serve some purpose? Is there anything, in other words, for human women to be grateful for in our otherwise-unwelcome monthly uterine awareness program?

Not really. It turns out the mammalian uterus isn't a lush pillow—it's a war zone. And ours may be one of the deadliest. Human women menstruate because it's part of how we manage to survive our bloodsucking demon fetuses.

The fetus has long evolved to hoover massive quantities of blood and other resources through the placenta. The mother's body, meanwhile, has longer evolved to . . . survive. We mammals

* I strongly advise against this technique.

† The elephant shrews stand alone as the menstruating few among Afroinsectiphilia, that weird clade of not-rodents that includes things like the aardvark, shrews, golden moles, and cloacal tenrecs. A few biologists I know are oddly reassured that, at the very least, the tenrec doesn't externally menstruate. No one knows what to do with them. They seem to be a lot like very ancient mammals: nocturnal, low body temperature, insectivores, cloaca, but they don't lay eggs. Their teeth are unusual, too.

aren't like salmon. We don't tend to die right after we lay eggs. We actually need to live at least long enough to breast-feed our offspring. And for social mammals—especially creatures like us, who often have lifelong supportive relationships with our children—the benefit of a parent's survival to our offspring greatly outlasts their gestation period.

The uterus and its temporary passenger are, in fact, in conflict: the uterus evolving to protect the mother's body from its semi-native invader, and the fetus and placenta evolving to try to work around the uterine safety measures. If a certain set of genetic mutations makes the offspring generally *stronger*, slightly better developed, and better nourished when it exits the mother's body, those genes will be selected for. If it kills the mother, of course, it loses the war. Likewise, if the mother's self-defense mechanisms are too strong, they kill the baby and she won't pass on her genes. When the stakes are this high, each "healthy pregnancy" is a temporary détente: a bloody stalemate that lasts, in our case, roughly nine months.

Like many other mammals with highly invasive placentas, our apelike Eves evolved a strategy for survival. Instead of waiting for a bomb to land, we dig our defenses early. We build up our linings on a regular basis, long before they are needed to protect the mother against the never-ending hunger of a human embryo.

If it sounds as though I'm describing human motherhood as a kind of horror movie, you're not entirely wrong. I love my children and wouldn't trade them for the world. But I did risk my life to have them, as do *all* women who have children, some more obviously than others. We seem to be driven to assume that being pregnant is innately good for us—that fetuses give us a "glow," that they calm us down, that pregnancy is a *healthy* state for a woman's body. One can, indeed, have a perfectly healthy pregnancy, and most women do, but being pregnant can also make a woman deeply unwell.

For example, in 2014 an American woman was at the salon when she felt a deep, painful pressure on her back. Because she was in her third trimester, she assumed this was just another fun feature of being pregnant, like the farting or food cravings. But when the pain spread to her chest, she contacted the hospital—

good thing, because she has no memory of what happened after: not the creak of the ambulance door, nor the ride to the hospital, nor the concerned faces of surgeons. She doesn't remember the emergency C-section, immediately followed by open-heart surgery. It turned out that her pregnancy had caused her blood pressure to skyrocket, and while she sat there at the hairdresser, the body-wide burden of her lovely fetus had managed to tear a twelve-inch fissure down her aorta, from which she was rapidly bleeding to death. The doctors were astonished she'd made it to the hospital alive.

Largely because modern medicine is amazing, she and her new baby survived. Afterward, she told reporters (who'd somehow gotten wind of the "miracle delivery" on the operating table), "I was just happy that I was alive and our daughter was alive. . . . I think that the baby saved my life." Of course, it was the baby who nearly killed her. But that's no way to start a relationship with your child.

Preeclampsia—a disorder that plagues more than one in twenty pregnancies in the United States, which is what this woman had—is characterized by spikes in blood pressure that cause knock-on effects in the mother's other organ systems (for example, her kidneys, which start to have difficulty filtering out excess protein in the blood). Thanks to new research and rising awareness, most pregnant women with preeclampsia will go on to give birth to healthy babies. When their aortas aren't dissecting.* The problem with preeclampsia is it can progress from mild to severe very quickly, and scientists aren't entirely sure why.

A number of different risk factors seem to be involved. For example, being obese greatly increases a woman's risk of preeclampsia, as does having a history of hypertension and/or dia-

* The risk of aortic dissection, while rare, is strongly associated with late-term pregnancy in women of reproductive age. In fact, I'll let this report from the U.K. speak for itself: "Perimortem caesarean section is an important part of the resuscitation of a pregnant woman. Ambulance crews should not delay this by prolonged attempts at resuscitation in the community before transferring the woman to hospital" (MBRRACE-UK, 2016). In other words, if a pregnant woman has collapsed and needs CPR, the goal should be getting that baby out of her as soon as possible, because the pregnancy may well be why she's about to die.

betes, all of which increase risk for heart problems in general. But there are risk factors more specifically tied to pregnancy: for example, being a mother over the age of thirty (but especially over the age of forty) or being a mother pregnant with multiple fetuses. Although deaths from the disorder in the developed world continue to be rare, preeclampsia diagnoses are on the rise in the United States, due in part to the rise of in vitro fertilization among older mothers. It's not uncommon to have more than one embryo implanted in a mother receiving IVF treatments; some fertility clinics are in the habit of upping the woman's chances of successful implantation by trying with multiple fertilized eggs at a go, and then either culling the excess or, in the famous case of Octomom, simply letting the whole bunch ride. The mothers-to-be, meanwhile, are so excited to be successfully pregnant at all they may not adequately consider the consequences of carrying twins or triplets: for example, the significantly higher risk of pregnancy complications that naturally arise when a human body tries to gestate more than one fetus.

Preeclampsia is the most common of these complications. While only 5–8 percent of standard, singleton pregnancies will suffer from preeclampsia, one in three women pregnant with more than one fetus at a time will develop the condition. This seems to be regardless of whether they are carrying identical twins—which usually share a single, somewhat larger placenta—or fraternal twins, as is most often the case with IVF, each with their own placenta.

What is clear is that the placenta lies at the center of the problem. Researchers have managed to isolate two proteins that placentas produce that seem tied to women with preeclampsia. Normally, these proteins help increase the mother's blood pressure just enough to help deliver a bit more blood, more often, to the placenta to supply the fetus with what it needs. But in certain concentrations, these proteins narrow the blood vessels too much, which starts the hypertension cascade of preeclampsia. Whether through genetic predisposition, some response to the uterine environment, or a combination of the two, producing too much of these proteins puts a mother more at risk.

But a third protein also plays a role, which may be the best illustration of maternal-fetal conflict of all so far—PP13 (that's placental protein 13). Until recently, we weren't really sure what the protein does, just that mothers who go on to get preeclampsia usually have rather low amounts of it.

After implantation, the placenta sends cells called trophoblasts into the uterine lining. These trophoblasts attack the mother's uterine arteries in order to try to gain more nutrients for the growing fetus. Naturally, the mother's immune system tries to kill these trophoblasts and often does.* But the human placenta has evolved some sneaky ways to get around her defenses.

In 2011, a group of researchers in Haifa, Israel, examined placentas from normal pregnancies aborted before fourteen weeks. These were *young*, frontline placentas. Initially, the scientists just wanted to determine if there were varying concentrations of PP13 in the placentas. But they noticed something odd. All around the maternal veins in the uterine lining—the *veins*, mind you, not the arteries—they found necrotic tissue: dead and dying cells. And not just a little bit. A lot.

Veins carry waste away. The placenta wants more nutrients to come *toward* it, which is what arteries do. So why on earth would a war be going on around the veins and not the arteries?

One word: distraction.

In large animals like *Homo sapiens*, the immune system usually works at two levels: global and local, with an emphasis on the local. At the body-wide level, you might get a fever when your system is waging war; most bacteria have evolved to function within a certain range of temperatures, and turning up the ther-

* One recent paper indicates the evolution of the placenta may be tied to the maternal body's preexisting strategies to prevent cancer metastases—the patterns of gene expression in the uterine lining that allow a "window" of implantation, and otherwise resist invasion, are similar to what the body does to prevent wayward cancer from setting up new tumor sites (Mika et al., 2022). That may also be why ectopic pregnancies are so common for human women. When super-invasive placentas like ours manage to land in a spot that isn't properly resisting and controlling growth (like the fallopian tube), the embryo will just go about its business setting up shop (ibid.).

mostat is still a pretty effective way to kill them off.* But, except for things like fevers, healthy immune systems work by "focusing" on the areas in which they are needed.† If there's a lot of inflammation in one area—and inflammation is generally what happens when tissue is being attacked—the immune system will fortify its efforts there. Such a focus often means paying less attention to other areas. That's the feature of the mother's immune system that the fetus hijacks by way of PP13. As the lead researcher put it, "Let's say we're planning to rob a bank, but before we rob the bank, we blow up a grocery store a few blocks away, so the police are distracted." They surmise that the placenta produces PP13 to inflame tissue around uterine veins so that the arteries are left relatively unprotected. That way, the trophoblasts can do their thing and the placenta can set up its arterial supply of nutrients while the mother's immune system is busy fighting all those distracting skirmishes around the veins.

This is what goes on as PP13 wages its war during a normal, healthy pregnancy. Maybe preeclampsia is what happens when the placenta starts *losing* the war and brings out the nukes.

One of the most common effects of preeclampsia—which may speak to its underlying cause—is that the placenta doesn't get enough blood. Less severe cases are often associated with low birth weight: no surprise, if the fetus hasn't been getting everything it needs. Infants whose mothers have preeclampsia often struggle to thrive in the womb. In other words, preeclampsia may be the result of the tide turning in the normal battle between the fetus and the mother's body. As a result, the placenta gets desperate, which in turn prompts a larger response from the mother's body, and so on and so forth until the whole situation gets out

* It's also a pretty effective way to kill our own cells off, too, which is why if you ever have a fever of 104 degrees or higher, you should seek medical attention. Boiling too long in your own juices can cause brain damage.

† We know what the alternative looks like: anaphylactic shock and cytokine storms can kill you. That's why kids who are deeply allergic to peanuts carry EpiPens. It's also why so many people died during the 1918 flu—the body's immune reaction turned deadly—and why, many suspect, so many people died in the initial stages of the COVID pandemic. Many of the drugs used in 2020 to treat COVID-19 had to do with dampening the body's immune system.

of hand. The struggling placenta sends out more of those blood-pressure-altering proteins. Maybe it sets off too many of those smoke bombs of PP13 near the uterine veins, sending the mother's immune system into overdrive, increasing inflammation, which drives up her blood pressure, too. There are many scenarios in which an imbalance in the maternal-fetal conflict—a conflict that every eutherian pregnancy naturally involves—could produce problems like these. In severe cases, women with untreated pre-eclampsia can progress into full-bore eclampsia, which can cause seizures and kidney failure.

In a healthy pregnancy, you don't want the fetus to win *or* lose the war, because either way can kill you. What you actually want is that uneasy nine-month stalemate. Women's bodies are particularly adapted to the rigors of pregnancy not simply so we *can* get pregnant but so we can *survive* it.

Some think that these adaptations put women who never become pregnant more at risk for illness than those who do. But recent research undercuts that theory: women who never give birth are less likely to develop autoimmune diseases than women who have given birth at least once. Meanwhile, a number of studies have come out in recent years indicating that if you've managed to become pregnant and give birth in your twenties, your risk for certain kinds of cancers is lower than a woman who's never been pregnant.* One possible reason is that the down-regulation of the mother's immune system during pregnancy may somehow keep women's innately more aggressive immune system in check. Chronic inflammation is a known risk factor for many types of cancers, so the theory holds, and maybe being pregnant—particularly being pregnant more than once—is a good way of "turning down the heat."†

* But if you've ever given birth, your chance of being diagnosed with breast cancer is actually slightly *greater* than if you hadn't done so (Nichols et al., 2019). The latest research shows your risk peaking at about five years after the birth, lasting for more than twenty years, and isn't improved by whether you choose to breast-feed.

† There are also problems with interpreting these studies as solely causal—for example, women with autoimmune disorders also tend to have fertility problems, so it may be that the autoimmune issues are causing the lack of pregnancies, not the other way around. Ditto for the cancer: bodies that are already genetically

We shouldn't, however, think that this means it's healthier for all women to become pregnant. Pregnancy is inherently dangerous and can have crippling long-term side effects on women. The *safest* thing for a woman's body is to never be pregnant at all. But when we do choose to have children, at the very least, evolution has managed to provide us with a suite of tools to be able to endure it.

And most of us will. Most women do have at least one child, and that pregnancy is usually straightforward. Nearly all women suffer from muscular tears and immunological snafus and a host of other problems during and after their pregnancies, many of which can and do lead to disability and death. Again, medicine helps us with those. Not everything is curable, but most is manageable. Having a wonky hip or lower back pain is certainly better than tearing a hole in the walls of your vagina, but even those terribly common tears can be repaired. What's more, women who benefit from modern gynecology usually don't *die* becoming mothers anymore.

That includes most women in the industrialized world. If you're a pregnant woman living in a malaria-prone country, you have a very different relationship to risk. Pregnant women with malaria are three to four times more likely to suffer from the most severe forms of the disease, and of those who do, 50 percent will die. Ever wonder why the Centers for Disease Control is located in Atlanta? Malaria. The entire reason the United States built the CDC is that malaria was rampant throughout the American South. Malaria was finally eradicated in the United States in 1951. That wasn't very long ago.

Some argue that getting rid of malaria did more good for American women than universal suffrage. Some say it had a bigger effect than *Roe v. Wade*. Nowadays, in the United States, only 0.65 out of every 100,000 legal abortions will result in the woman's death, while 26.4 American women still die for every 100,000 live births. Before *Roe v. Wade*, 17–18 percent of all maternal deaths in

---

prone to certain cancers may also have issues with the early stages of pregnancy. Since this is a hot area of research right now, you can expect the field to have better answers in the next ten to twenty years.

the United States were due to illegal abortions—that stat was as true in 1930 as it was in 1967. Meanwhile, as many as one in four maternal deaths in today's malarial countries are directly tied to the disease. During our worst outbreaks, the same was true in the United States.

Isn't it wonderful for women to live in a place where *both* ways to die have been basically eradicated? What a thing, to choose to be pregnant, in a place where it's significantly less likely to kill you.*

Donna certainly never got to choose. Our Eves had a long way to go before anything like conscious choice would come into play. First, they needed bigger brains. To do that, they had to become primates.

* In the United States, unfortunately, the risk of maternal death has been going *up* lately, quite unlike every other industrialized nation that isn't presently at war, and that was true even *before* the COVID-19 pandemic, which made even more pregnant women and new mothers die (Hoyert, 2020). More on this in the "Love" chapter.

Purgatorius

CHAPTER 3

PERCEPTION

New organs of perception come into being as a result of necessity. Therefore, O man, increase your necessity, so that you may increase your perception.

—JALĀL AD-DĪN AR-RŪMĪ, 13TH CENTURY

In college, I worked as a model at the local art school. For a few hours a week, I was the girl who stood nude on a raised dais while teenagers attempted to draw what they saw with awkward lines on canvas. It's true: I was a professional naked person. Easy way to make money.*

I posed in a big drafty prewar building with huge windows in what used to be the fancy part of town. But the rich people were all gone now—fled to the outskirts of the city, as they usually do. Instead of carriages and servants outside the windows, now there were weeds and rats. And artists. Artists love these broken places.

* Easy for someone like me, that is: mostly cisgendered, able-bodied white women do figure modeling. There are many ways to sell one's body; those of us who do usually do it in ways that we've seen other people do it. I knew of only one male model at the art school, though my brother had been a model when he was in college before me, which is where I got the idea. He'd also done some medical studies for pay, and I followed suit, though I was always more broke than he was, which meant I always needed to work more than he did. This wasn't solely because of our differing genders, but it's also not an irrelevant factor. More on that in the "Love" chapter.

They're cheaper, for one. They also make time feel slippery, as if the past were always there, ready to be repurposed: fresh coat of paint, don't mind the ghosts.

Classes ran for two to three hours, which meant I was grateful for the little space heater near my feet. About halfway through each session, all the students would go outside for a smoke and I got to put on my robe. I'd walk among the easels, watching my body take form—here a leg, there a torso. One pattern always held true: at the start of the semester, the male students—only the guys—drew my breasts too big. And I don't mean a little out of proportion, but *huge*. Then, a few weeks in—and this happened time after time—they would start to shrink, as the guys learned to draw what their eyes saw and not the cartoons their brains had made.

At this point, you probably have questions.* I did, too. For instance, did those boys really see me differently at first than the girl students? Were their eyes drawn to breasts through some ingrained hetero-ness or gender mess, or was it simply that the girls, who had breasts of their own, were used to seeing them? Padding around that room in a robe and bare feet, I remember wondering, *Do men really see the world differently than I do? Do I live in a different sensory reality from the men around me?*

These are hard questions to answer. Perception is made of two things: the brain and the sensory array—this thing we call a *face*, which is really a tight mound of bone and flesh we mammals hang our primary sensors on: the eyes, ears, nose, and mouth. Sight, sound, scent and taste. To understand human perception, we'll have to think back to where our face was made. The Bethlehem, if you like, of the male gaze: that crèche of ancient forests. Because those students trying to trace my body on canvas weren't just mammals but primates.

* Questions relevant to this book, I mean. If you're wondering if I was uncomfortable being naked, maybe a little, but I can also say it's terribly empowering to be a young woman facing down a group of young men actively judging your naked body and realize that you *don't care*. Take that, nightmares.

TWO ROADS DIVERGED IN A YELLOW WOOD

After the asteroid—when the land was scorched, and the ash fell, and everything froze, and Eve's children hid in their burrows trembling in the long night—the landscape began to change. The first plants to return were the ferns. We know this from their fossilized remains, delicately fringing over razored shale, just above the irradiated ash line of the K-Pg impact winter. We also know this because we've seen something a bit like it happen more recently.

The day after Mount St. Helens exploded in 1980, much of the surrounding land was destroyed. Rockslides and lava took out some of it; boiling rivers took out more; and any life-form unable to flee—for instance, trees—either burned or choked in the ashfall. Then, under that thick layer of fertile ash, things began to grow again. Among the first to return were the ferns, their hairy little heads poking out of the dead land, shaggy tufts of primordial life clumped in with dust and mud, cooling lahars and decomposing bodies.

Like moss and fungus, ferns are happy to sprout from a fallen tree, sluiced ash, wet dirt under the carcass of a dinosaur. They're the nomads of the plant world. After the ferns came the ant colonies, building their vast underground cities, fueled by the economies of the dead. The ants broke up the hardened soil, aerating the compressed earth, allowing bacteria and fungi to thrive. After the fungus and ferns and ants came the creatures that eat ants, and eventually the predators that prey on the eaters of ants. The trees came back later, carefully at first, many of their tender shoots mowed down by returning animals. But it didn't take long for Mount St. Helens to look much as it had before the eruption, save for a thickened undergrowth and a lake covered with shattered trees. And, of course, the mountain was a thousand feet shorter.

In the ancient world of early mammals, things didn't go back to the way they were before. They couldn't. The Mount St. Helens eruption quieted in less than a day; Chicxulub was an apocalypse. But something else was different, too, something more fundamental. A new sort of plant life had quietly evolved in that Jurassic Eden. These were the angiosperms: flowering plants. And they were getting ready to take over.

Before the asteroid, our planet's forests were massive conifers and ferns.* But out of the ash, in place of those ancient forests, fruiting trees and their canopies formed brand-new ecosystems. Flowering trees produced, at regular intervals, vast bounties of fruits on their terminal branches—fat bulbs of sweet and sugary flesh. Fruits. Bugs. Moss. New things that ate the fruits and bugs. New things that ate the new things.

It was those fruits, ripening high above the forest floor, that gave rise to the Eve of human perception: *Purgatorius*, the world's earliest known primate.

Purgi appears in the fossil record roughly sixty-six million years ago, precisely when angiosperms started filling in the smoking holes left in the old conifer forests. Scientists found her little bones on Montana's Purgatory Hill in the 1960s, and more of her many sisters throughout the Fort Union Formation: broken jaws, fractured ankles, scatterings of teeth. From what we can tell from the fossils, Purgi looked like a freaky monkey-squirrel, roughly the size of a modern rat. She had a long, bushy tail, a medium-length nose, two beady eyes—the usual features of our early Eves. But unlike Donna, the Eve of the modern uterus, Purgi had hinged, rotating ankles, which were especially good for climbing trees and skittering along branches. And quite unlike Donna, she'd eat just about anything she could get her paws on: berries, fruits, tender leaves, bugs, seeds. If she were alive today, she'd probably eat our garbage. We'd complain about Purgi stealing all our birdseed, rooting through our trash cans, making nests in our attics.

The mammals Purgi evolved from were mostly insectivores like Morgie and Donna. But Purgi also ate fruits. We know this because her teeth were specialized for both crunchy, chitinous things (bugs) and squishy plant stuff. From looking at her ankles, we also know she spent a lot of time in the trees—hunting new, specialized insects in those new forest canopies of ancient fruit bearers. As those bugs went about their business, carrying tree

* In the heavenly garden before the apocalypse, there wasn't anything remotely like an apple. The very first fruiting tree was probably some sort of gingko (Zhou and Zheng, 2003). The angiosperms did pave the way for later primate intelligence, however, so you could still call them the trees of knowledge.

sperm to waiting she-flowers, predators like Purgi went about the business of eating them and, while they were at it, some of the sweet fruits, too. Like most of her primate descendants, Purgi was an opportunist: she probably preferred certain foodstuffs, but she was open to new things. And her teeth evolved accordingly.

We haven't yet found her full skeleton in the dust fields of paleontology, so we don't know for sure if she did what so many modern primates do: cling to branches with her hind paws and use her forepaws to manipulate food. But many tree mammals do that today. Some scientists even think primate hands and posture evolved from sitting upright in trees, using their forepaws to delicately handle food. You can see this behavior in other arboreal mammals, too, from opossums to raccoons to squirrels. Living in trees does certain things to a mammal's body: You have to be able to hang on. You have to have good balance and depth perception. And if you're munching on more complicated stuff than insects, you might need to use your forepaws to eat.

Purgi was a near contemporary of Donna's. We don't know their precise relationship. We do know that Donna and her placental uterus were up in the trees not long before Purgi and her relatives gave rise to later primates. The arrival of angiosperm forests profoundly shaped the evolution of tree dwellers, just as tree dwellers shaped the evolution of those trees. They pollinated flowers. They ate fruits. They pooped out seeds. And far below, in the dim light of the virgin forest floor, those seeded droppings grew yet more fruiting trees.

And so, at the dawn of the Paleogene, there were fruit trees, and there in the leaves was Purgi, each aiding the other's success. She had many children. Some of her relatives continued on as the plesiadapiforms: ancient primates that did very well in their time but whose genetic branch withered and shrank and finally fell off into extinction. Other members of Purgi's family became today's typical primates: big-brained and flat-faced, most of whom are *still* in the trees.

It's those faces that concern us now. We're primates, too, which means we evolved from creatures that adapted to live in trees—most especially the terminal branches, where Purgi and her kind needed a gift for acrobatics to be able to eat, along with a sensory array that could handle this new environment. We needed eyes that could see when fruits were ripening and distinguish when leaves were young and nutritious and tender. We needed ears that could hear our children in a loud, leafy landscape high above the ground. And while we wouldn't use them nearly as much to find food as our foremothers did—a sweet fruit's scent doesn't always travel far—we needed noses that could handle a sex life in the canopy. Adapting to those needs changed our sensory array. But was it different for males and females? And if so, is it still different for humans today?

EARS

When you first visit a tropical rain forest, the dominant emotion is usually surprise—not over the beauty of the place, nor over how hot it is. The biggest shock is that it's bloody *loud.* On any given day, it's louder than a Rio street carnival. The insects thrum and buzz at screaming decibels, their wings and legs rubbing a frenetic jazz. The frogs bellow. The birds caw. And the monkeys, the howler monkeys, like the horns of hell, day and night they call.

Life bursts at the seams here, overstuffed and overcrowded—the place on Earth with the greatest diversity of land animals. Given that the rain forest is a profoundly vertical space, life on the ground is only the first in a series of riots up into the canopy. There's an abundance of food and an abundance of predators and parasites ready to kill you (for you are also food). Before Bangkok, Hong Kong, and New York, *this* was the city that never sleeps. And it's the closest we have today to the place where primates evolved.

In the rain forests of Brazil, you can hear the short, eerie siren of the white bellbird—and of course you hear it, because the damn thing hits 125 decibels. To put that in perspective, screeching brakes in New York subways peak below that level.* Howler

* Union Square station has been measured at 106, but that includes the roar of multiple trains, stupidly sound-reflective tile, and half a dozen buskers with portable amps (Gershon et al., 2006). Smaller stations are generally quieter.

monkeys can reach 140 decibels—per monkey. They usually roar in a chorus, and they're not the only beasts roaring.

To handle communication in that kind of din, the aural part of your sensory array is going to have to be adept at separating important noises from non-important noises. When our Eves went into the fruiting trees, their ears had to change.

THE ORIGINS OF THE BASS CLEF

Primates are able to hear much lower frequencies than many other mammals. And the best theory going for why we can is our move into the forest canopy. It's actually a physics problem: when you're at ground level, you can bounce your sound waves off the earth, doubling your signal strength. When you're in the trees, the ground is too far away to amplify your vocalizations. But that's not the only problem that resulted from our Eves' relocation to the trees.

If I were to yell at you from across an empty room, you wouldn't have any problem hearing me. But if the room were full of junk, you'd have a harder time. That's because not only has the path between my mouth and your ear been obscured, but the stuff between us also absorbs some of the energy of my sound waves. Now add dozens of others yelling just as loud as me. That, dear friends, is the forest canopy: leaves, fruits, branches, moss, trunks, and many other screaming bodies between you and the ear you're trying to reach.

Animals generally adapt to a soundscape in one of two ways: they tweak their pitch range, or they boost their volume. Primates did both: they evolved to both hear and produce lower pitches, and they found ways to get louder. By lowering the pitch, they automatically gave themselves more distance, since the lower the pitch of a sound, the longer the sound wave, and the longer the wave, the farther it travels. You've probably experienced something like this yourself. For example, in my old apartment in Brooklyn, I regularly heard the booming woofers of some distant car's sound system. In the summers, when the air was humid and my windows were open, I could even feel the bass vibrating my rib cage. It was hard to tell which song was playing—the music's higher frequencies were being absorbed by buildings and bodies, degrading over the urban distance between me and the car. But the bass? The bass passed right on through.

So too goes primate evolution: as their lifestyles changed, our tree-borne Eves needed those lower frequencies to cut through the sonic clutter.

In a sense, when it comes to our sensory array, we're talking about the evolution of the primate social network. At first, all we had was the primate equivalent of "yo." We could boom our brief, specific yo's through the canopy, ears specially tuned for the voices of our friends. In that way, we could establish our territory, find mates, and even make new friends. Eventually, the system could carry more complex messages. We could say things like "Yo, I'm here!" "Yo, where are you?" "Yo, awesome buffet happening up this fig tree!" And even more important, we could shout, "Yo, you're sexy!" "Yo, I'm sexy!" and "Yo, shit, tiger!"*

Larger primates have lost some of the higher end of our range, but we haven't lost all of it. Humanity's high end, around 20 kHz, vibrates at twenty thousand times per second, which is comparable to many other mammals our size. But most people find that pitch disturbing, and we're not great at discerning what's being communicated in that range. Dogs, on the other hand—who evolved mostly from ground-based mammals—can hear pitches quite a bit higher than we can. That's why the "silent" dog whistle works—it produces a sound close to 50 Hz that humans can't hear. If we built a "primate whistle" that dogs couldn't hear, it would sound like whale farts.

Up in the canopy, the ears of Purgi and her fellow Eves became specially tuned to the pitches that traveled best over the crowded, leafy distances that mattered. Modern human ears inherited those changes—many primates living today have them, in fact. We're able to produce and hear sounds at greater decibels and lower pitches than is typical for animals of the same size. Even male

* Cross-canopy calls may lie at the root of human language, setting a primate-friendly stage for later brain development. The calls of one species of monkey even seem to have a primitive grammar. More on that in the "Voice" chapter.

gorillas, which spend most of their time on the forest floor, have a fantastic low *rumble* when they want to get their point across. That rumble can really carry. But up in the canopy of the South American rain forest, howler monkeys can be heard three miles away.*

Among primates, females and males have slightly different hearing. That might be because the males don't need to hear everything the females need to. It's not that they have different *ears*—like a hi-def stereo, the equipment is largely the same. Rather, the tuning is a bit different, and that's still true for men and women today.

BABIES BOOM BOOM BOOM

For the record: babies do *not* pitter-patter. Babies boom boom boom. For a brief time, I worked on this book in the basement of a friend who'd brought forth into the world a small boy named Rex.

Rex was two. Like most children his age, Rex thundered across the floor with the force of stampeding bison. That is, bison also capable of high-pitched, siren-like wails that erupted without warning and poured through the floor like hot panic. His cries filled me with dread. I froze. My heart raced. I *could not* stop listening. Sometimes I even broke into a sweat.

It's unclear whether I was more or less aware of Rex than the average man might have been. I didn't grow up with small children around and hadn't hung out with many toddlers at that point. I remember thinking, Maybe you get used to how loud they are if you actually live with them.

And yet years later, when my own son arrived, my body responded the same way. If anything, even more intensely, given that my breasts ached each time he cried and leaked milk into my

* It's true that some savanna creatures have evolved ways of optimizing for distance. For instance, a bull elephant in rut can rumble his way to a listening female six miles away. But she's not only using her ears; she's also listening with her feet. The low, 20-Hz call he's making creates a corresponding seismic wave in the ground (O'Connell-Rodwell, 2007). Lion roars have a similar effect, presumably for the same reason (Pfefferle et al., 2007).

dress. This is a very common reaction for breast-feeding women: babies cry; boobs leak.

I don't think it was the baby noise in general. I think it was the crying. From what physiology labs have been able to determine, men's and women's ears respond differently to different pitches. Female-typical ears seem to be specially tuned to the range of frequencies that correspond to baby cries. Both men and women can hear and differentiate between noises in a certain range of pitches. Most can hear both bass notes and the high end of a violin. But generally speaking, men's ears seem to be better tuned to lower pitches, while women's ears are more sensitive to higher pitches—usually those above 2 kHz. That just so happens to correspond to the standard pitch of a baby's cries.

Now, if you're a female primate, there are obvious evolutionary advantages to being able to hear your baby well. So, while the entire primate line might have shifted the bottom end of their hearing downward—presumably to correspond to long-distance, low-band communication through the forest canopy—being the primary caretakers, females would particularly need to retain their ability to hear their higher-pitched offspring. Via pathways that are still mysterious, female-typical hearing became tuned to these higher pitches. Most women can hear them better than men even in noisy places. And while typical masculine ears tend to lose their higher range as they age, women's ears are better at hanging on to those pitches. Importantly, our better ability to hear the very upper end of the human register is also tied to hardwired emotional response: baby cries alarm women more than men. It's not that men *can't* hear the kid crying, but that for many adult men, their ears snip off the upper end.

Making a sound with vocal cords doesn't just produce a single note. Like playing a stringed instrument, when you sing something, your vocal cords produce harmonics.* Though it's harder to discern, this is also true when you speak. These upper registers are called overtones. If you sing the note A at 4.4 kHz, your throat

* Your vocal apparatus *is* like a stringed instrument. It's also a little bit of a phonograph, a clarinet, and a set of bagpipes with weird bellows. But the larynx itself is a flexible box of damp strings.

produces overtones of 8.8 kHz, 13.2 kHz, 17.6 kHz, and so forth.* But the higher you go in your register, the more "piercing" or disturbing a sound is. So, while both men and women might hear a baby screaming at 5 kHz, a woman is much more likely to hear the highest overtones at 10 kHz and 20 kHz, presumably making the cry more alarming to her.

That panic does produce some useful outcomes. For example, one recent study had subjects listen to a recording of a baby crying or a more neutral noise. Then the subjects had to play a game of whack-a-mole. The ones who'd listened to babies crying were faster and more accurate in their mole-whacking efforts—they were, in other words, more alert and focused after being exposed to the sound. Women showed this result more robustly than men.† The evolutionary advantages are pretty clear. If you are tuned to the sound of a baby crying, you'll probably be better at taking action to make it stop: flee with babe in arms, fight off predators, shove useful bits of fruit or a nipple into its mouth.

This difference in perceiving registers has very real consequences. It's not just about the babies. Men are also far more likely to suffer common types of hearing loss than women, with those higher pitches the first to go. That's probably because these shortwave sounds are greatly diminished by the time they make it all the way down the ear canal to the cochlea, which means the human ear has to "work harder" to be able to focus on them. Also broken hair cells in the cochlea—damage that usually accumulates

* If you've ever played a stringed instrument, you might have done this: on a guitar, for instance, pressing a string down between the frets produces a standard note, while lightly tapping the string at the right intervals produces a harmonic. Tuvan throat singers famously manage to sing two notes at once, though that isn't happening in the larynx—they're adding strength to some of their vocal overtones by manipulating the back of the throat and mouth above the voice box, making the overtone more audible alongside the primary note. It's a difficult thing to do. Professional throat singers often sweat when they perform because it takes tremendous concentration and muscle control.

† Tellingly, these studies failed to assess whether the female subjects were less *happy* after having heard these sounds than the men. Personally, while I'm happy enough for the evolutionary door prize of being better at solving problems when my babies cry, I'd also readily trade that skill for feeling less stressed. But then, my children's survival doesn't currently depend on my feeling stressed. Presumably, our ancestors' survival did.

with exposure to loud noises and repeated use over time—make the entire apparatus less able to flexibly sense and respond.

Though this sort of hearing loss can be sudden, the more usual trajectory is a gradual slope, with the ability to hear at the highest frequencies gradually declining from age twenty-five on. It's so predictable, in fact, that most men over the age of twenty-five are unable to hear noises at 17.4 kHz or higher, which led to the invention of an alarm in the U.K. specially targeted at young people. It's called the Mosquito. It blares out a horrifying whine at 17.4 kHz precisely, and can be cranked higher than 100 decibels, so that shopkeepers can use it to disperse loitering groups. The assumption is that people over twenty-five won't be chased away by it, because they can't hear it, whereas young troublemakers will. The device is controversial, but largely unregulated. Interestingly, it also targets women.

I'm in my thirties and I have absolutely no problem hearing 17.4 kHz. (I've tested sounds at that pitch. It's horrible.) Adult men are nearly twice as likely as me to be "protected" from this high-frequency alarm, thanks to their sex-typical hearing loss. Middle-aged and older men also have more trouble following a conversation in a crowded soundscape, especially if it involves a lot of higher-pitched sibilants. That also means they have difficulty hearing women's voices, with their characteristic higher pitches, but retain the ability to hear men's voices and other low, rumbly things. Because social power is typically assigned to men as they age, women's voices are literally not being heard by men in power.

Of course, there are other day-to-day insults to women that stem from sex differences in hearing. Have you ever been enraged by the whine of a computer monitor and tried to explain what's bothering you to your husband or father or other significant male in your life, and the guy cannot hear what on earth you're talking about?

Modern computer screens tend to emit high-pitched sounds, starting around 30 kHz, well out of the range of human hearing. Computer *fans*, however—the ones cooling down the blazing-hot processors—generate a high-pitched whine of their own that bothers women's ears more than men's. Blame it on the sex of the designers and testers: back in the day, televisions and computer monitors used cathode-ray tubes, which regularly buzzed at an obnoxious 15.73 kHz. But with these departments largely staffed by men, no one noticed it before they hit the sales floor. What made the sound was the transformer in the back of the machine, whimpering like a mad mosquito while it strained against magnetic forces.*

These sorts of things continue to plague women—the periodic hum of electricity in fridges, the overtones of ice machines, the tinny buzz of a vacuum when its filter is too full. But it's not just technology. We're also more likely to hear the high squeaks of mice making a home in our walls. We're not crazy. We really can hear these things.

What's unclear is *why* women as they age retain their hearing better than men do. The assumption among scientists has been that women have fewer high-volume, ear-damaging jobs—like driving a jackhammer into concrete. It is a significant factor, but not enough to explain it all. Even among men and women who work in high-volume environments, men are the likelier patients in hearing clinics later, and—in a pattern that's usually the opposite for men—they go to the clinic sooner than their female coworkers.

So do men's ears age a titch more rapidly than women's? In other words, is this an ear problem or a global repair problem? Both men and women are born with roughly twenty thousand hair cells in the cochlea of each ear. Aside from the membrane detaching, the most common cause of hearing loss has to do with hair cells breaking and dying. After eighty, both men and women suffer from hearing loss equally. But before seventy, men are more than twice as likely to have hearing loss as women are. So why? Do women's hair cells somehow repair themselves better? Do we have other compensatory mechanisms? It's fine that our ears might be more tuned to crying babies, and there are plenty of evolutionary arguments to be made about that. But for women

* By the way, if you do get a high-pitched whine from your computer, simply adjust the refresh rate of your monitor downward. You can do this in the control panel. You'll still get good image quality, but you should have less of that whining sound to contend with. I like to think of it as making a display setting "female friendly."

to *retain* the ability to hear those higher pitches over time is a bit curious. There's some support for the repair model; as I'll discuss in the "Menopause" chapter, women's bodies do seem to be a titch better at fixing themselves than most men's bodies. But we don't have solid answers yet. Given current research, you can expect a little more light on these questions over the next ten to twenty years.

In the end, maybe a good chunk of it will come back to behavior. For one thing, when women and men are exposed to equally loud environments, on average women will feel more distressed by them. That distress may lead women to try to escape the noise more quickly than men will. After all, it's not just what your sensory array can *do*. It also matters what you do in response to what it reveals to you.

AMPLIFIERS

In 2015, my boyfriend was addicted to *Fallout 4*, a video game set in postapocalyptic Boston—a rather boring place to spend the end days, in my opinion. For about two months, my home was filled with the noise of radioactive zombies, robots, and explosions. I have pretty good speakers. The treble carries. My woofers can really roar. It became a war in our apartment, with the boyfriend cranking up the volume to total immersion levels and me asking him to please turn it down. We finally agreed that he could leave the soundtrack on (a solid mix of mid-century American pop) if he muted the weapons noises.*

True in the home, true in the lab: men can function more happily in noisier environments than women can. Maybe some of that has to do with the range of overtones female-typical ears can hear. But if we think back to that metaphor of a hi-def stereo system, it probably also has something to do with the amplifier.

Ears aren't passive receivers; they also make their own sound.

* You may be wondering why he didn't wear headphones. I wondered that, too. He did wear them a couple of times, but always with hilarious amounts of resentment. I was, to be fair, the first girlfriend he'd ever lived with. Adulting is hard.

Deep in the cochlea of the inner ear, the hair cells *snap* in a series of tiny clicks called otoacoustic emissions (OAEs). Every time a sound cascades down from the eardrum and middle ear, the hair cells in the cochlea wave and snap, boosting the signal.* The pace and volume of these movements increase and recede like a tide.

Women's OAEs tend to be both stronger and more frequent than men's—so predictably that acoustic researchers describe inner ears as "masculinized" or "feminized." Some think these patterns might be tied to why females of many primate species seem to be more sensitive to noise: if the cochlea boosts the sound signals in female ears more than in male ears, that could, in principle, make the experience of hearing loud things feel *louder* for females. And it's not just true for human beings—even marmosets have feminized OAEs, ever so slightly more dominant in the right ear, just as most human girls do.

This right-sided quirk in females isn't isolated to the ears. For example, the length ratio of the pointer finger to the ring finger for most human girls is lower than for most boys, and that difference is more pronounced on the right than on the left. You can see similar hand/paw differences in other female primates. But what makes this "girl" trait is complicated: human women with complete androgen insensitivity syndrome (CAIS)—that is, girls born with an XY chromosome but whose bodies don't respond to androgens, so they develop a female-typical body—still tend to have male-typical OAEs and the male-typical digit ratio on their right hands. That means something more complex must be driving these differences than just exposure to more male sex hormones in the womb.

Curiously, if you happen to be born with two X chromosomes and also happen to identify as gay or bisexual, you're also more likely to have "masculinized" OAEs. But if you've got two X chromosomes and identify as heterosexual, your OAEs will likely be similar to most females. But it's not as simple as some

* The physics of this mechanism are both complicated and highly contested throughout the world of hearing research—no one knows *entirely* why ears do this, or even precisely how. It's known to be driven by the action of outer hair cells in the cochlea and that listening for the *absence* of OAEs is a good predictor of certain kinds of hearing loss.

sexual orientation binary: XY men's OAEs nearly always fall in a male-typical curve, regardless of their sexual orientation.

This isn't enough to be able to assume that lesbianism or XX bisexuality is something that happens when a typically female fetus is exposed to higher-than-usual levels of androgen in the womb. In male-female fraternal twins, however, the sister is more likely to have masculinized OAEs than if she hadn't shared a womb with her brother. The same thing can be said of sheep and other mammals—this sex difference, in other words, seems to be something fundamental in how the mammalian body builds its ears in the womb. If you castrate a sheep later in life, for instance, the OAEs don't change their sex pattern. For the twin human girl, that change to female-typical ear function implies the androgens coming off her brother in the womb are what shifted *her* ear development. But since CAIS women have male-typical OAEs, there may be other pathways to this sort of ear development, too.

None of this tells us why a woman might be queer, of course, nor does it shed much light on the deep complexity of human sexuality.[*] All that can really be said is that queer women's ears often behave slightly differently than straight women's do, and that the ears of gay and bisexual men don't reflect a similar difference.[†] As with all things sex related, it's very tempting to treat these data like a smoking gun. Many scientists have come to believe that among the undoubtedly numerous biological roots of male homosexual-

* For example, the original OAE study—to my knowledge, and I presume to the lab's knowledge as well—didn't examine trans subjects, nor did they rate their subjects' degree of attraction to same- or opposite-sex partners on a complex sliding scale (McFadden and Pasanen, 1998). Subjects simply self-identified using categories common in the 1990s in Texas, where the study took place: heterosexual, homosexual, bisexual. These data were supported by a couple of extra questions from Kinsey's famous sexual fantasy survey, and in the rare cases where they weren't sure, the lab basically asked the subjects about their relationship history. Since many of the queer subjects were recruited by contacting local gay organizations and publishing an ad in gay magazines, it's safe to say that many subjects had self-identified as queer well before they put their ears to the test at the university. Whether any of the straight-identifying subjects later graduated from college and came out of the closet, I can't say. There is always a little noise in the data.

† These data come from queer women who've already passed puberty, but because mammals in general maintain these patterns throughout their lives, there's no reason to assume OAE patterns would have "flipped" from infancy in the lesbian ear.

ity there may be a kind of "hyper-masculinity" driving the system. This theory holds that, contrary to the "sissy" stereotype that gay men still endure, the underlying reason they're into guys is that they're somehow physiologically *more* masculine than the typical hetero man.

Straight women's ears may be better tuned for a world of needy primate babies because they calibrate their instruments more regularly than men, and women in general—whether straight or not—simply have more sensitive instruments for hearing than males, with those abilities better preserved over time. But if we think of Purgi's tree-borne face as a sensory array—one ready-made for both sensing and *communicating* with her babies—then maybe we should rewind the tape. Hearing isn't the only thing we do with our children. Though the first sound of a baby's cry helpfully confirms that the child is alive, mammals do something far more ancient with children when they're born: We smell them. We put our faces close to our pups, whatever species they may be, and we breathe them in.

NOSE

Long before we could see, before we could hear, before we could *feel* anything at all, we could smell and taste. This is olfaction: our ability to sense chemical gradients. From the very dawn of life, single-celled animals needed to be able to distinguish chemicals in the water around them and sense their concentration. *Are we getting nearer to food? Is that toxin getting farther away?* The more mobile we became, the more important it was to be able to track the various chemicals in our environment.

But our single-celled ancestors didn't have sex to reproduce. We do. Once *sex* happened, male and female olfaction started to diverge, with each species' "nose" (or olfactory organ of whatever sort) tailored to the sex-specific needs of its carrier.

Hundreds of millions of years later, Purgi's mammalian nose lifted in the cool, dry air of dusk. She smelled the moss on the bark, the ripening fruit, the musk of a male on a nearby tree. Her body was more complex than earlier mammals', and so was her social life. But like our most ancient ancestors, she was essentially smelling and tasting food, sex, and danger.

That's still true of humans today, and we're doing it basically the same way—except now our chemical sensors line the wet tubes of our nasal passageways and the tiny, spongy nubbins on the surface of our tongues. But the nose is the major player here. Taste is massively compromised when we're unable to smell.

Our hearing and sight sensors don't require as much room in our heads as our olfactory system, which takes up a good third of the volume of our faces. Because olfaction involves molecules rather than waves of light or sound, and there are millions of different molecules in the air we breathe, being able to smell something requires a big, wet, warm surface area lined with sensors.

That our noses can make sense of the chemical world around us is impressive. Think of the difference between English and Chinese. In the English alphabet, there are only 26 characters that we combine to produce a narrow range of sounds. Chinese script, however, is not phonetic. There's a different

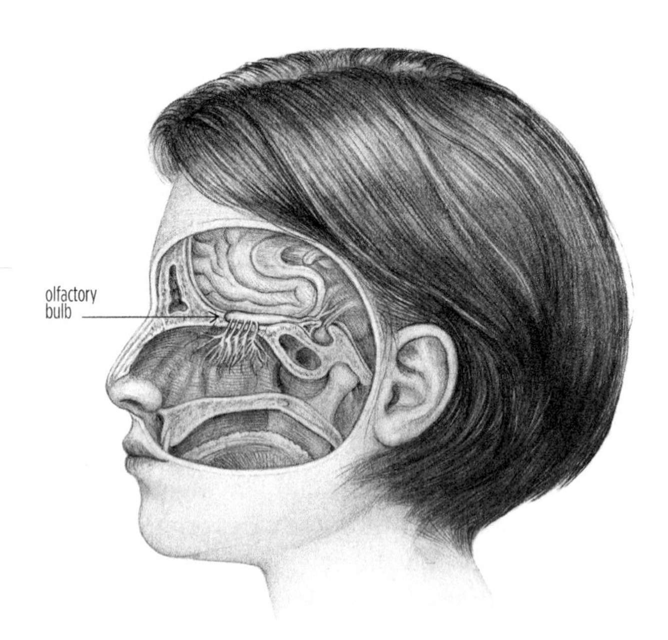

Human olfactory system

symbol for *every word.* You're talking about 106,230 Chinese characters.*

In the alphabet of our olfactory sense, there are roughly four hundred known receptors in the human nasal tract, and roughly a thousand known genes for odor receptors in mammals, though the majority aren't functional in the human body. Even setting aside the nonfunctional ones, these genes constitute as much as 2 percent of the mammalian genome—a truly massive number. So what do they build? Essentially, a bunch of receptors shaped a bit like catcher's mitts; it's surprisingly accurate to say you "catch a whiff" of something. But each gene for an odor receptor builds *one* type of catcher's mitt, and each mitt binds to only one molecule of the right size and shape. Since the air is full of an absurd number of molecules, any of a host of which might be important for us to smell, it's easy to see how a genome might get clogged with such information.

But thankfully, odors tend to activate multiple receptors in the nose. That's because most odors are a combination of different chemicals. So even with so many of our olfactory genes turned off, human beings may not catch the full complexity of a scent, the way a dog does, but we can still get the gist. In the intensely complex world of invisible stuff floating around in the air, our noses are still able to tell the difference between the scent of an orange and that of a grapefruit.

Or rather, the female nose can do this—men's noses aren't as good at that kind of granularity. Both women and men have those four hundred receptors, but women live in a more particular olfactory world.

THE SCENT OF A MAN

It's impossible to overstate the importance of the nose to the life of a mammal. It tells you where's safe and where's not, what's good

* Still, if you know only 900 characters, you'll be able to read about 90 percent of a Chinese newspaper.

to eat and what's poison, who's nice to have sex with and who might kill you instead. It can even tell you if a tiger has eaten a member of your species recently—useful to know if you're on the menu. This information and these olfactory skills naturally influence your behavior. For example, you can deliberately mask your own scent in order to avoid predators; predators can mask their scents in order to better hunt. Among the best-studied mammals, mice and rats, olfaction is so important to the animals' lives that researchers can radically change their behavior by changing what their environment smells like.

That's especially true for sex-specific smells. For rodent males, the scent of other males' urine can cause stress or interest, depending on the situation, while the actually-banana-scented pee of a pregnant female *really* gets them going. For rodent ladies, the scent of male urine piques their curiosity. Female mice and rats *love* sniffing a male's pee-soaked bedding. They'll seek it out. You can train a female rodent to prefer a certain spot in a cage or maze just by making it smell like male pee. Even after that spot stops reeking of male, the female will tend to keep hanging out in the place she's learned is Boy Town.

This is attributed to male pheromones: volatile compounds, awash in a male's saliva and also produced by tiny glands on his rump where they get mixed with his urine. Most mammals seem to have this scent-based social signaling system. Pigs also excrete pheromones in their saliva. In dogs, it's their saliva, pee, and rump sweat. When it's the season for sex, male mammals tend to rub and pee on everything around them to mark their territory, broadcasting social signals far and wide. Male goats, in a display I can only hope was never part of hominin history, actually urinate on themselves, spraying a thick, musky pee up their belly all the way to the chin. As any goat breeder will tell you, it's about the most disgusting and instantly recognizable stink a person could ever encounter. It contains putrescine and cadaverine—two organic compounds that corpses produce when they decompose. One can only assume the goat ladies like the sickly sweet "smell of death."

Until recently, the scientific community assumed that human beings don't have pheromones anymore. That's because we don't have much of an accessory olfactory system, a particular cluster of sensors and nerves in most other mammals that runs through the upper palate of the mouth, around and through the nose, hitting a peculiar little bunch of flesh called the vomeronasal organ, and up along a specialized pathway toward the parts of the brain that handle sex and socializing. Rodents have such a system. Monkeys have it, too. Even Purgi probably had it. But human beings and other apes do not.

The theory for why we lost it is that Purgi and other early primates slowly evolved to become more visual and less scent-driven. Maybe that's because, for primates at least, life in the canopy made it harder for them to distribute social stink than creatures on the ground. Whatever the reason, the further along primate evolution you go, the flatter the face. The eyes move forward. The nose shrinks. Maybe you even turn off a bunch of your olfactory genes, as is the case in the human genome. Eventually, you start knowing the world by seeing it, rather than smelling it.

Purgi's sensory reality in those ancient angiosperm forests wasn't the same as the world her descendants experienced. That isn't only because the forests themselves changed. In order to adapt to life in those fruiting forests, the sensory array of the primate Eves and its corresponding brain architecture shifted to such a degree that for those Eves, the Self would have become fundamentally different in its relation to the World.* Once ancient primates evolved into apes, the olfactory system had massively degraded. What's left of humanity's vomeronasal organ is just a

* The Self is something a brain makes. Change the brain in deep ways—and its ways of sensing and relating to the environment around it—and you must inevitably change the Self. Whether or not Purgi had a Self in the philosophical sense, maybe ask Peter Singer, but I'll take the easy way out here and simply say that as a functional feature of a well-developed mammalian brain, the Self—surely built from that brain's cumulative sense of its own body and its relation to the world, memories of experience, and the general buzz of activity that constantly reconciles these categories of information with one another—seems both useful and a bit of a shrug. Of course the Self changes as the body changes. In fact, you and I simply cannot comprehend what life was like for creatures like Purgi, because we don't have Purgi's body and brain. One reason that's true is that our sensory array is radically different. Another reason is that our brains are obviously different. But each time we try and fail in that imaginative exercise, we get a little closer to parting the veil around our own body-bound set of experiences to see the world as it is, and our own bodies as they are in that world.

tiny bit of flesh that usually ends in a blind tube toward the floor of our nasal sinus. Though it may still be connected to our endocrine system in some ancient fashion, it has none of the obvious nerves or general connectiveness present in other mammals.

Still, there may be another way being smelly helps you get laid. Some of humanity's most odorous parts are the crotch and the armpits. Perhaps because it's harder for researchers to ask subjects to give them their dirty underwear rather than a dirty T-shirt, most studies on the social influence of smell are armpit based. It may also be because the sorts of smells the armpit produces seem stronger than those from a healthy crotch. I have distinct memories of riding in taxis and buses in Marseille, in Istanbul, in Cairo and Dalian and Nairobi—any of the cities I've been where deodorant isn't a given—in a thick miasma of manpit. It almost feels wrong to call it an odor. It cloaked. It suffocated. It actively wrestled with the lower parts of my brain. Sweet, sharp, tangy, pungent, as heady as old cheese, and as musty as some long-forgotten cave, it was unmistakably male. I know the smell of a woman's pits. I know the metal tang of old menstrual blood, unwashed hair, the masking odors of too much perfume. But absolutely nothing a healthy female body gives off rivals the impression of a ripe male underarm.

Maybe, just maybe, that impact was so great not just because men's armpits are strong smelling but because I'm a female who's sexually attracted to males.

There's one human hormone that scientists have been studying as a potential male pheromone. It's called androstadienone (AND), a volatile steroid that's present in nearly all men's sweat.* It is structurally similar to the pheromone in male pigs' saliva, the scent of which literally makes females spread their legs and prepare to be mounted. In humans, not so much. But there are some effects: when you put some AND on the upper lips of heterosexual women (this was a real thing—scientists diligently

* It's present in women's sweat, too, but at far lower concentrations. That makes sense given that it's a steroid compound that's derived from testosterone. Adult men tend to have fifteen times more circulating testosterone than women of reproductive age.

swabbed highly concentrated man-pit on female undergrads' upper lips using a Q-tip or, when specifically studying AND, a concentrate made from boar testicles[*]), they're more likely to find certain guys sexually attractive, more likely to enjoy talking to men at speed-dating events, more likely to show particularly high activation in their hypothalamus, and more likely to have higher levels of cortisol in their saliva.[†] Results like these tend to be more robust if the woman is nearing ovulation, which suggests her sensitivity is about being able to sniff out a good mate—though without a transvaginal ultrasound and a battery of blood tests, ovulation is a tricky thing for most studies to really nail down.[‡]

The odor of man-pit also seems to have some play in sexual orientation. If you test gay men for AND reactions, you'll find similar activity in their hypothalamus as in heterosexual women; lesbians show no such reaction. In a less direct test, researchers wafted smelly T-shirts under the noses of gay men, straight men, and heterosexual women (going for the whole pit-smell, rather than just AND). Gay men particularly liked the smell of other gay men, as it turned out, but less so the straight guys, whereas women preferred the stinky pits of gay men to those of straight

* The big reason pig farms castrate male pigs is that otherwise, once they reach puberty, their testicles pump out androstenone, which becomes concentrated in their adipose tissue and can make the meat taste like sweat and piss. It's called boar taint. Not all humans have the right genetic makeup to be able to taste or smell it thoroughly (Keller et al., 2007), but when we do, it's universally unpleasant. Human males also have testes making this stuff from puberty on, and our adipose tissue also has a complex relationship with androgens (Mammi et al., 2012), but thankfully we're not in the habit of eating a well-marbled man steak.

† I should mention one quick note: whether these results are tied to AND particularly or instead to another chemical, or even to a complex interaction between various components of pit juice, remains unclear; scientific publication tends to be biased toward positive results, and anything tied to cultural notions of human behavior, like sexuality triggers, is particularly vulnerable to editorial bias toward flashy findings (Wyatt, 2015). As with all research, some of these experiments are better conducted than others.

‡ Most labs settle for asking whether a subject is on birth control, when her last period started, and how long her menstrual cycles usually are. It's not so accurate, as any woman who's gotten pregnant using the "rhythm method" will attest, but in a pinch it'll do. Most women of reproductive age ovulate about fourteen days before they start their next period.

men. And male-to-female transsexuals showed similar hypothalamus activity to heterosexual women.

I found considerably more studies about women's scent preferences than men's. I don't know if that's because male scientists are particularly curious about What Women Want. Among studies on men, there's the now-famous bit about men tipping strippers more if they're ovulating—they do, the effects are reproducible, and they go away if the woman is on birth control—but that may or may not be scent related. (It's hard to say what you're smelling, exactly, in a strip club.) Men also prefer the smelly T-shirts of ovulating women, don't like the pit smells of menstruating women and women who are less immuno-compatible as much, and almost universally dislike the smell of a woman's tears, regardless of her reproductive status.

It used to be that these sorts of studies were amusing to other scientists, but largely dismissed because in some cases the sample sizes were too small and in others, the effect too tiny. Those problems continue to plague some of the research on human pheromones. But as the literature grows, and more and more people are subjected to scientifically specific armpit scenarios, the picture's starting to look more persuasive. Though we're not as driven by pheromones as other mammals, the human nose may play some role in our sex lives.

Now, whether using deodorant—a recent human practice—eliminates that influence, simply reduces it, or otherwise changes the signal is unclear. Some scientists, giddy with fresh data, have gone so far as to claim that deodorant and birth control pills are screwing up our built-in compatibility sniffers, making our offspring more prone to genetic disorders. I'm unconvinced. So many other factors go into human mating—one's physical appearance, job, cultural background, regionality—that the sniff test would seem less influential. What's more, our hominin ancestors presumably had fewer mates to choose from: ten or twelve local suitors, rather than, oh, *most of the user base* for any given dating app.* As a healthy, modern American woman living in a large

* Hetero-compatible female users of most online dating platforms famously receive more contact from men (both requested and unsolicited) than men receive

urban area, I have an actual *million* potential guys to choose from to father my potential offspring, all of whom generally benefit from modern medicine, which allows them to far outlive most of the crummy genes they might carry. I can only assume having such a diverse array of sperm on the shelf is more influential on my offspring's chances of avoiding genetic catastrophe than whether I like the smell of a guy's armpits while I'm ovulating.

And yet! My long-evolved female scent superiority still holds, and labs are finally getting a sense of the mechanisms that might make it so.

THE (LADY) NOSE KNOWS

It's one of those things everyone who works in human olfaction simply accepts: a woman's sense of smell is more sensitive than a man's. Women are better at detecting faint scents, telling the difference between different sorts of scents, and, once they catch a whiff, correctly identifying what it is. Though you can find some of these differences in newborn baby girls, it's especially true of grown women around ovulation and pregnancy, and lessens in women after menopause. That's why most olfaction researchers think female sex hormones may play a role. Because this female advantage is present in a number of other mammalian species, too, it probably was true for Purgi. We don't know exactly why. But just as olfaction originally evolved to sniff out sex, food, and danger, most evolutionary theories for the feminine nose still fall under those three categories.

Being able to smell a man covers two of them: he's fairly useful for sex, but he can also be dangerous. While males in other species do a lot of smelly social signaling, the females of many species are often a titch better than the males at smelling such signals.* When you think about it, that's pretty odd, not only because

from women. That is, so long as the women are white. Being nonwhite significantly reduces one's candidate pool, and being a Black-identified woman on a dating app is, quantifiably, the worst (Rudder, 2014).

* In many human neighborhoods, scattered deposits of dog urine form an invisible social media platform, letting each passing pup know who's around, who's

male social mammals waft stink at each other all the time, but also because most female mammals aren't ready to reproduce at all times. Except rabbits, who ovulate in response to sex itself, most lady mammals only blast their pheromones to let guys know they're ready when they go into estrus. Yowling tomcats in alleyways. Stallions pawing the ground. A majority of male mammals can smell when a female is reproductively viable.

Since men and women have roughly four hundred different types of odor sensors each, men should, in principle, be better at it, given that their nasal passages are slightly bigger than your average woman's. Human puberty builds a bigger nose in boys in order to provide the oxygen they need to run their larger muscle mass. A typical teenage male will grow a nose about 10 percent bigger than a typical girl his size. The resulting adult male nostril sucks more air, and more odor molecules, into his nasal traps. And yet women are still better at detecting diluted scents—fewer molecules of the scent, in other words, in any given local quantity of air.

Something is making a woman's odor receptors function better. To tease out why that is, we need to look to sex differences in the underlying nasal tissue. And we also need to look upstream to the brain. That's because discerning scent is about both detection and deduction: catching enough of the scent to generate sufficient signal and then comparing it with prior knowledge.

In 2017, a lovely little mouse study provided one useful window into how this might work. Mice still have a vomeronasal organ, but they also have olfactory sensory neurons (OSNs) just as we do—neurons that physically contact odorants through the chemical "traps" in the nose and then transmit information about them to the olfactory bulbs of the brain. When female mice smelled something, their OSNs responded more broadly and transmitted information more quickly to the brain than the male OSNs did. But when the mice were *neutered*, a funny thing

dominant, even what dinner's been like lately. This is a large part of why dogs on a walk insist on smelling every last thing. When you tug on the leash to rush them along, you're interrupting the conversation. If the walking dog is *female*, she's probably smelling what male dogs had to "say" to her *and* other male dogs, and when she pees in on the conversation, she's probably advertising her own reproductive status as well (Cafazzo et al., 2012).

happened: the females became slower and less nuanced, while the males became *more* nuanced and faster. That means both sets of sex hormones seem to be at play in a mouse's nose: the estrogens enhancing OSN performance and the androgens somehow *suppressing* or interfering with their smelling abilities. And because human OSNs seem to be structured similarly to other mammals', those same hormonal influences are probably at work in our noses, too.

It's hard to suss out whether this particular strength is selected for by evolution or just a handy by-product of other traits. For instance, it's hard to imagine why being worse at smelling things would ever be useful. But in humans, it's widely known that a woman's sense of smell heightens around ovulation, and it's not hard to imagine why *that* might be adaptive. After all, ovulation is an important time for a female mammal to be discerning. Since it's more costly for us to get pregnant and give birth than for other female animals, we need to be fairly careful about which male gets to do the job.

But it's not enough to be better at sending data from the nose to the brain. What the brain is able to do with it is what really makes the difference here. As any pregnant woman will tell you, it's not that she couldn't smell the cleaning fluid in a restaurant bathroom from where she's sitting at her table in the dining room before carrying a child. It's just that now, smelling the stuff produces a wave of nausea and negative *emotion*—a strong signal and a strong response—which means being seated at a table too close to the bathroom isn't going to work, thank you very much.

Being pregnant might have changed her ability to sense the smell, perhaps due to changes in blood flow in the nose. But the real reason she needed to change where she was sitting, while her male companion wasn't bothered, was that her baseline ability to smell the toilet started at a different point. Her olfactory bulbs are simply built differently than his.

In most of the brain, neurons are wired dendritically—that's the classic picture of a neuron, with those spiderweb sorts of long arms that reach out and form synapses with other neurons to create action chains. In the olfactory bulb, however, signals are more diffuse. An activated cell tends to radiate the information out in all directions to nearby cells. In that sense, the wiring of the olfactory bulb is less about sparking a *chain* of events and more about creating a *ripple* over a pond.

In 2014, one lab thought it might be a good idea to see just how many cells were in women's olfactory bulbs versus men's.* Though the sample size was relatively small—only so many cadaver brains to go around—the results were clear: women's olfactory bulbs have massively more neurons and glial cells than men's do, even controlling for size. More than 50 percent more. Women's are simply more dense. And given the way olfactory bulbs process signals, density might have a large effect on overall function. The density, and thereby strength, of any given signal is enhanced. The ripples spread faster over the pond. And given that women have the same number of odor receptors as men do, the primary site for how women's olfactory system differs from men's might be here in the bulbs.

Given how primitive olfactory bulbs are, this difference may be present from birth. At the moment, there's no way to know for sure, but I wouldn't be too surprised to hear a lab decided to dump some newborn mouse brains into a scientific Vitamix in the near future, just to have a look-see at the numbers. Given that differences in olfaction form such an intimate part of our lived experience, the idea that sex differences might differently wire an ancient part of our mammalian brain is always going to be an attractive target.

A MEAL YOU'LL NEVER FORGET

Pregnant women, menstruating women, and ovulating women are famously prone to food cravings and food aversions. The usual stereotype is that we're hunting for something fatty, salty, and/or sweet. In America, chocolate is popular. So is mac and cheese.

Evolutionary scientists tend to think our food cravings are

* They basically took olfactory bulbs from cadavers and shoved them in a blender. Then they used a machine to differentiate all the different types of cells and count them. It's wonderfully Frankensteinian.

instead tied to nutritional deficiencies—that our bodies, under a unique set of stressors while ovulating, menstruating, or during pregnancy, simply "know" that we need to eat one sort of thing or another and prompt us to seek out those foods.

There's some support for this. For example, pregnant women sometimes suffer from pica: the uncontrollable urge to eat things like dirt or hair or pencil shavings. The placenta sucks a lot of iron out of a pregnant woman's body, and women who have pica also tend to have iron deficiencies. We don't know yet if this is a causal relationship, but topsoil can be high in iron. Of course, if you get an intestinal blockage from your new dirt-eating habit, that's not something we'd call evolutionary *fitness.* Also, many cravings we experience don't seem tied to immediate nutritional needs. Steak, while fatty, isn't a stereotypical PMS craving, despite its high iron content, something you'd think you might crave when you're shedding a lot of blood.

Likewise, the cravings for ice cream and pickles or other odd combinations of food aren't necessarily a good thing for a pregnant woman (though they may not do much harm, either). Even as her desire for specific foods might intensify, her negative responses to scents and tastes also increase. If anything, among pregnant women, food aversions are more common than cravings, as well they should be: you need to eat, but you also need to stay alive. Though most of us would prefer to never feel it, nausea is one of the most important sensations a body can produce. It's right up there with pain. Your body evolved to motivate you to learn valuable lessons: if you manage to survive being poisoned by something you eat or drink, it's only sensible that your body would do whatever it could to make sure you don't eat or drink the damn thing again.

Part of what's so interesting, then, about a pregnant woman's nausea is how *powerfully* her taste and scent preferences can change. Some of the nausea is simply a result of basic indigestion: a pregnant woman's hormones also tend to slow down her intestines, making her feel bloated and generally nauseated. So some of those waves of nausea might just be a lousy side effect of feeling backed up. But that might not be enough to explain the powerful changes in her sense of smell. For example, previously loved foods can smell absolutely disgusting. The scent of a cigarette, previously innocuous, can be like someone farting directly in your face.

A pregnant woman's nausea, in other words, is more than mere tummy trouble, and may be strongly tied to olfaction. Her emotional responses to the world are often on high alert, too. Sartre's *Nausea*? Paltry. A bored Frenchman. Try a pregnant woman who's puked twice in one morning, nibbling saltines from a plastic sandwich bag on the subway from Brooklyn to midtown. She can smell every single dead thing that's ever been in that car.

But you have to eat, especially with that fetus hoovering nutrients from you like a crazed Dyson. So what on earth could be the advantage of these new, random associations of stimuli with nausea? Why doesn't all this nauseating instability in her olfactory system simply kill her off?

Avoiding death is the goal, actually. Most would argue that avoiding toxins is well worth a bit of nausea and starvation, and toxins are particularly deadly when you're pregnant. For example, most humans aren't fond of bitter tastes. It just so happens that most of the most highly toxic foods in the world tend to taste bitter. Cyanide famously tastes and smells of bitter almonds—in fact, almonds would still be dangerous today if ancient farmers hadn't managed to breed the cyanide out of them.* The plant world hosts a nearly unending list of likely-to-kill-you dishes, each more bitter, metallic, or sour than the last. And plant-eating mammals' sense of taste has evolved accordingly, with females typically more sensitive to bitterness than males. After all, when it comes to passing down your genes, placental females are *always* eating for two. Because their bodies do so much of the heavy lifting when it comes to reproduction, the death of a female is always going to be far costlier for the species' local fitness than the death of a male. So if having a nose that's better at detecting threats and sex gives females an edge at the survival game, it benefits the species as a whole. If it happens to help a woman find tasty food, too, well, so much the better. You need a lot of calories to make babies the way we do.

* We'd have done the same for oak trees, but their toxins are more complex than almonds', so acorns are best left to squirrels.

Purgi, as one of the early mammals living in the trees, would have used those dense female olfactory bulbs to add fruits to her diet of bugs and leaves. Fruits taste better and are better for you when they are ripe. If she was close enough to them, her sensitive nose would have helped her discern the choicest morsels. But she needed her eyes to spot ripe fruits from across the forest canopy and to plan a safe route to get to the table.

EYES

Standing on my modeling dais, I could smell the old electric heater burning dust at my feet, the thin, distant wisps of turpentine, the cigarette smoke on the students' clothes. I could hear the scratching of palette knives mixing paint and the swish of brushes against canvas. But the boys weren't hearing or smelling me all that much. And it wasn't just because they were boys. It's because they were primates. And modern primates are really all about eyes.

As the art students looked at me, trillions and trillions of photons bounced off my flesh and streamed toward their primate faces. The minuscule muscles of their irises contracted, widening the pupils to let in more light. As photons pummeled the backs of their eyes, their retinas sent information about my contours to the optic nerve, which carried a surge of data toward the visual centers of their brains.

Think of the difference between an old dial-up modem and today's broadband. Sensing the world through sound waves is all well and good, but without echolocation you're not actually learning that much.* The nose, too, is good at telling you about nearby chemicals, but probably isn't going to help you climb through a tree canopy. But eyes! Eyes can give you the equivalent of a million trillion gigabytes of information a second. They'll tell you what things are, and where they are, fantastically quickly. So long

* The reason bats and other echolocators are so good at using sound waves to build a model of the world is that they are able to *send out* sound and then hear what bounces back—hence the "echo." Ears, by themselves, are more passive—they rely on sound already being transmitted from elsewhere.

as you have the processing power to make sense of a data stream like that, you're in business.*

PARALLAX

When you think about primates, you probably think about monkeys and apes. And when you think about monkeys, you probably can't help but picture their faces: short, squashed noses and big, binocular, stereoscopic eyes, usually ringed by orbital bone, sitting right on the front of the face.† Even my two-year-old can recognize a crude drawing of a monkey: ears to the side, short nose bridge, and big, forward-facing eyes. By and large, that's how the primate sensory array evolved: over the eons up in the trees, our noses shrank, our eyes moved forward, and the visual centers of our brains exploded. If you line up fossilized skulls in chronological order, you can see the eye sockets move toward the front of the head. And as this happened, the size of the visual processing portions of the brain increased dramatically.

If you want to optimize how you interact with your local environment, where you place a pair of sensors matters. Because lungs constantly suck in new air, the best way to orient an olfactory sensor is to place it in the path of that river of odor-laden air—it makes sense for our nostrils and their corresponding olfactory bulbs to be smack in the center of our faces. The ears, meanwhile, are best placed on either side of the head so they can hear sounds radiating from both sides of the body—better for triangulating how far away a sound might be and what direction it's coming from. Eyes use similar strategies, but generally speaking, which

* Any decent gamer will tell you that a top-flight monitor is garbage without a computer that can quickly process visual information, and a superpowered graphics card is garbage without a monitor that can actually display that much rapid detail. The only fun thing about mammalian brains and eyes, in this regard, is that the brains of born-blind folk are fantastically good at repurposing what would have been the "visual cortex" for a range of other purposes, which your Nvidia obviously can't do. More on brain plasticity in the "Brain" chapter.

† You might also be aware that lemurs and bush babies and other weirdos are *also* primates, but unless you're a primatologist, their faces aren't the first that come to mind.

ones they use depends on what sort of creature you are: predator or prey.

In mammals, there are essentially two strategies for eye placement. Prey animals usually have their eyes on either side of the head. Think of deer, rabbits, small birds: by having eyes on the sides of their heads, they're able to keep watch for predators over an incredibly wide field. What's directly in front of them matters a lot less than spotting the lion in the grass. Meanwhile, predators—dogs, eagles, snakes, cats—generally have their eyes on the front of their heads. While this produces blind spots at the far left and right of their visual field, it greatly increases the amount each eye's visual field overlaps. That overlap—the parallax—makes it a lot easier to see how far away something is from you in space. It's also easier to make out fine-grained features of items in that overlap zone. Having a large parallax means we can see farther away, in

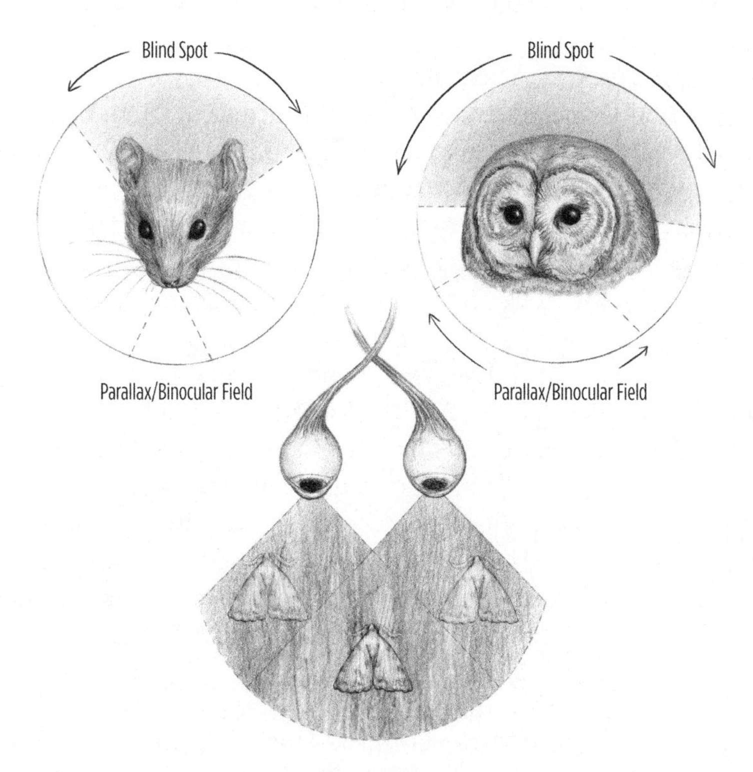

Parallax and stereopsis

greater detail, and are better able to judge the distance between ourselves and faraway objects.

When it comes to primates, where our eyes are located may be more complicated than the needs of predator versus prey. That's because as the primate line evolved, we started to change both *what* we ate and *when* we ate.

Let's start with the food on the menu. If we assume our most ancient primate Eves, like Donna and Morgie, ate mostly bugs, then all they needed to be good at was catching bugs. But what if those bugs became really good at hiding?

For example, if tree-based insects evolved to freeze and camouflage—staying very still, their bodies matching the mottled green of leaves or the dark striations of bark—then it's harder for distant predators to find them. But not so if that predator has two eyes on the front of its head: stereopsis gives you really good 3-D vision. With forward-facing eyes, you might be able to see that camo'd bug even if your nose is confused by other scents and the bug is being super still. Also, if you live in a massively 3-D space, like a tree canopy—where up and down matter as much as back and forth and side to side—and you're trying to catch bugs that keep flying away from you, your ability to judge depth and direction suddenly matters a lot. Your brain might have to get bigger, too, since processing a lot of 3-D visual data takes a lot of computational firepower. Indeed, when paleontologists measure primate fossils' skulls, the more stereoscopic the eye placement, the bigger the brainpan.

Like those of the Eves before her, Purgi's eyes were much more like a rodent's or weasel's, placed on either side of her head. Our earlier insectivore Eves mostly used their impressive hearing and smell to find their prey. Purgi likely also found insects to eat by listening for the high-pitched, delicate taps and thrums of their wings and smelling the distinctive odors of their bodies. But as she and her primate relatives established themselves in the canopies, many of them evolved to be more binocular. And that might be because a good portion of the ancient primate line was trying to eat in a 3-D space at *night*.

Binocular, stereoscopic vision is a convergent trait that has evolved a number of different times. Owls and bats, both predators, move through the air at night, and both have eyes on the fronts of their faces. Not *all* predatory birds have binocular vision, nor do all insectivorous mammals. The defining circumstance is hunting at night, when it's that much harder to see things, so being able to utilize a parallax is important. In this line of thinking, maybe primates' eyes slowly moved forward because it's hard to catch insects in the treetops at night.

So they twitched and skittered in the nighttime canopy for hundreds of thousands of years. The bugs got better at hiding. Our ancestors got better at finding them. Predator and prey body plans competed with each other in their slow, evolutionary dance. The more time passed, the more our ancestors started eating other things: leaves and fruits, particularly. So even if our binocular vision did initially evolve in service of following insects in a 3-D space, that predatory advantage rapidly fell to the wayside. The bigger brains, however, stayed.

And those brains and their corresponding forward-facing eyes became rather useful for our new diets. With our wider parallax, we were able to use our forepaws to manipulate leaves and fruits and seeds close to our faces, with much greater clarity and precision. When it comes to eating insects, you're not gently turning a bug over in your palm, checking to see if it's ripe, all the while carefully trying not to detach it from its stem (in case it's not ripe, and it'd be better to wait a bit).

Consider the raccoon. Not a primate, but another relatively clever tree-based creature who is an opportunistic eater, much like humans. The raccoon uses her forepaws to carefully manipulate foodstuff. She's not a predator. She doesn't hunt. But her eyes are located decidedly to the front. Like our ancient placental Eves, she's also usually nocturnal. The raccoon will, however, convert to a diurnal lifestyle if food in her territory is more plentiful in the day. It's not her *usual* way, but like most opportunists she's flexible. But as it does for human beings who work the night shift, changing her natural rhythm can cost her: the circadian cycle is embedded in nearly every body system in mammals. Important hormones peak and ebb at different times of day. The way we digest food, the way we repair injuries, even what sorts of *cognition* we're better at can change according to what time it is. Some of these signals are internally bound and are usefully flexible: if you fly across multiple time zones, for instance, you'll suffer less and recover faster from jet lag if you adjust your mealtimes to the new schedule before you go.*

But other things seem to respond directly to the sort of light that hits your retinas: your eyes, in other words, help your entire body "understand" what time it is, and the clockwork of your internal machinery responds accordingly. This became only more true as primates became more visual creatures. What our bodies do with eye signals influences pretty baseline stuff. For example, women who work night shifts famously have trouble with fertility: It's not just the general stress, nor them not being home in the evenings, that can make their sex lives a bit more complicated to coordinate. It's also that the intricate timing of their ovaries' cycles is tied to a circadian rhythm. When a woman's egg is developing during those first two-ish weeks of her cycle, progesterone peaks in the morning, estradiol peaks at night, and luteinizing hormone seems to have a slow rise that peaks somewhere in the afternoon. All of these need to maintain their proper rhythms and relative balance for normal egg development and ovulation to happen. The complicated conversation that the brain, ovaries, and uterus constantly maintain can be disrupted by divorcing oneself from the normal rhythms of a sunlit day.†

Men who work the night shift have similar metabolic and immunological problems as women do, but it doesn't affect their

* Effectively, you're "tricking" your body into thinking it's another time by shifting when your blood sugar and cortisol rise and fall, both of which normally peak a bit after eating a proper meal. There may also be useful psychological effects, since when a person eats, for many of us, is a deeply normalizing thing: it *feels* like the evening when a person who's used to eating dinner in the evening eats dinner, and it can feel like bedtime if the usual amount of time has passed between dinner and getting into bed.

† This may also be part of the reason that breast-feeding affects ovulation: not only is the hormone pattern of a breast-feeding woman different from a woman who's returned to a menstrual cycle, but at least in those first few months of a newborn's life, breast-feeding women are awake a *lot* at night. This produces no small amount of stress, of course, which is already an ovulation disrupter, but it may also screw around with the body's circadian rhythm. For my part, I can easily say that by month two of my son's life, I had very little sense of the difference between day and night.

fertility as greatly. Testosterone usually peaks in the morning, but in male bodies it's tied less to the eyes' light exposure than to sleep: it rises during sleep cycles, and falls off after waking. So, if men make themselves sleep during the day, then their testosterone will simply shift accordingly, and their testes' production of sperm will similarly adjust to the new normal. Because it's so much cheaper and easier for the mammalian body to make sperm, there's just less to screw up by turning men into night owls.

In evolutionary terms, changing from a day dweller to a night creature is dangerous for a placental species' fitness. The reverse should also be true. You're not just changing habits—you're messing around with base code. And yet we opportunists are known to do it: not often, not always, but if it *benefits* us, then yes. Once upon a time, some of our Eves were opportunistic enough to make the switch. It changed a lot of things in our bodies. But first, and most obviously, it changed our eyes.

TECHNICOLOR

Most paleontologists assume that our early mammalian Eves were largely nocturnal insectivores, skittering about in the safety of moon shadow. As the tree canopy evolved with all its fruits, and insects evolved to take advantage of it, the insectivores naturally followed their prey up into the trees. At first there was no reason to change over from their nocturnal lifestyle. Bugs were out at night, after all. Why subject themselves to the dangers of daylight predators? Why risk being seen? You'd need a *really* good reason to stop falling asleep at dawn. But at least one of Purgi's granddaughters started going to bed earlier, and earlier, and earlier, until finally our primate ancestors were fully diurnal: daytime dwellers who slept at night. The reason for that was, in all likelihood, fruits—that fantastic food supply in the canopies of the angiosperm forests that usefully advertises its readiness by color.

Most mammals are color-blind—unable to differentiate between red and green. Their world is more blue-gray, or even sepia. This is how color vision works: special receptors on our retinas, called opsins, respond to different wavelengths of light; longer waves skew red, while shorter waves are bluish. The retina takes these different color wavelengths and "mixes" them in the underlying nervous system. One receptor activates for blue, and another for red, and the brain sees purple—so long as you have those two different receptors. If you don't, you'll just see variations of blue. Most placental mammals are dichromatic, meaning they have two primary types of color receptors: blue and green. If you don't have a red opsin, you simply can't differentiate between red and green very easily. Which doesn't matter when you're nocturnal—there's not a lot of red and green going on.

Birds can all see red. Most fish can see it, too. But not cats, not dogs, not cows or horses, not rodents, not hares, not elephants or bears. Their worlds are red-less. Even the bulls of Pamplona can't actually *see* the matador's red cape, nor the traditional red stripes and jackets of the bull runners, streaming through the city streets like some freak bovine death gang. The bulls aren't aggressive because they see red, which probably looks sort of dark brown to them, or maybe even black. The bulls are aggressive because they are treated like crap. The red, that's just for us.

Since kangaroos and other marsupials are trichromatic, we think the change to dichromatism happened around our placental Eve, Donna. She or one of her daughters lost her red color receptor in the long, dark night of the forest. Being fully nocturnal, Purgi probably couldn't see red, either.

The genes responsible for our red-green color vision arose by gene duplication roughly forty million years ago, right around the time a bunch of proto-monkeys floated on a land raft across the Atlantic Ocean and created a new monkey kingdom on the North American continent. Land rafts are precisely what you might imagine: a floating mass of earth and vegetation. Because the tectonic plates that hold Africa and South America were closer at that time, and because so much of the world's oceans was bound up in Antarctic glaciers, the sea was narrower and shallower than it is now. Scientists assume that primates, living as they are wont to do in trees near good sources of water, were caught up in storms along the African coast and got tossed—possibly along with their trees and the earth bound to their roots—into the ocean, where currents swept them across the sea. Astonishingly, many survived. From these storm-tossed creatures descended the howler monkeys, the spider monkeys, the capuchins. The New World monkeys. They're the only ones left with prehensile tails. Most of them are also still color-blind.

But back in Africa, primates became increasingly frugivorous and foliage-friendly—away went the insect diet and in came the tender young leaves and ripe fruits. These primates became the Catarrhini: Old World primates, a select group of monkeys and the apes, some of whom would eventually evolve into humanity.* To eat all those tender green and ripe red things in the daytime required a retina with a red opsin. The genes for creating that opsin, as luck would have it, are located on the X chromosome.

If you have two X chromosomes, as most women do, it's incredibly unlikely that you'll end up being red-green color-blind, whereas roughly 10 percent of men are. If red-green color vision was obviously selected for in diurnal primates, why was it located on the X chromosome?

It's possible this type of color vision was more advantageous for the primate Eve than for her consorts and sons. Perhaps being more efficient at spotting more nutritive foodstuffs (extra-sweet berries, extra-tender young leaves) made a real difference in pregnancy and breast-feeding. If Purgi utilized the same sex-specific parenting strategies as many living primates do, foraging for herself and her infant offspring, then the survival of the young depended far more on the female than the male. In other words, there was *more* pressure to see red and green on the newly diurnal Purgi than there was on her male counterparts.

The second possibility is that Purgi foraged for food with a group, as some of today's New World monkeys do. In that sce-

* When I say "Old World" or "New World" here, I'm utilizing the common terms for Catarrhini and Platyrrhini: the taxonomic groups of primate species that are largely found in Africa and Asia versus the Americas. These terms are obviously colonial and outdated, but remain useful in their way (for example, they remind us of the ways that "New World" primates descended from their Eves, who crossed the Atlantic from Africa roughly forty million years ago).

nario, it'd be advantageous to have both trichromatics and dichromatics working together, grazing not only in daylight but in the dim light at dawn and dusk, when the dichromats would be better at finding the good stuff.

Or both of these things were true: our Eve, as the female, had the most pressure on her to be able to see red and green, but in a highly social species that did some amount of food sharing, it would have been advantageous to have some dichromats, too.

The color-blind aren't at a great disadvantage today, given that their survival doesn't depend on picking red fruits out of green foliage all day. And of course, just as our fellow primates do today, the nose can always give a clue when eyes fail you—spider monkeys smell fruit to check when it's ripe when their eyes can't figure it out, a bit like smelling a melon in a grocery store. But group living also favors group strategies: today's human sensory array is also utilized in groups, which might be a little bit closer to our evolutionary past. Mixed-sex groups of foraging New World monkeys—some of whom have recently evolved to tell red from green, and some who haven't—give us a window into what it means for social species to evolve. The groups with a mixture of color vision among their members appear to be slightly better at foraging as a group. Humans, like most of our social primate Eves before us, have bodies that work the way they do in large part because they live alongside other humans. Just as we carry the deep past in our differently ancient physical traits—some things old, some things new—our social groups carry the past, too: some things old, some things new.

PHOTO-REALISM

And so it is with perception: you can move where your sensors hang on your head and then repurpose them for new contexts, each shift evolving in lockstep in that long evolutionary dance. You can change the inner *mechanisms* of the sensors, too, to make them more or less responsive to different environmental signals, depending on your lifestyle in that environment. But changing how you sense and interact with your environment inevitably changes the brain processing all that information, which in turn drives some of the evolution of your sensory array.

When we talk about perception, it's important to suss out what is and isn't brain based. But that's a very tangled web. Attention directs perception just as perception influences attention: the sensory array and its corresponding brain centers are in near-constant communication with one another and signals go both ways. Eyes move from one focal point to another. The ears do this as well, even when you're not consciously trying to scan your surroundings. For example, when you listen to a human voice in noisy environments, the cochlea tamps down its amplifier, reducing competing signals—in effect, restaurant conversations involve more lip-reading than talking in a quiet place. The eyes, too, are built to reduce signal when needed: not only are color receptors clustered toward the center of the eye, making your peripheral vision markedly different from what your brain directs you to focus on,* but eyes regularly respond to *internal* thought, too. If you're a person who can see, when you're asked to imagine or remember a vivid visual scene, your pupils will dilate, even though you're not paying attention to the external world at that moment. When your brain is internally modeling visual information, the nerve pathways that control the muscles that contract and dilate the pupils come along for the ride. That's true of the tiny muscles that direct your eyes overall as well—which, by the way, are almost constantly in motion.

The complex interactions between the brain's attentive perception of visual information, our eyes' mechanics, and memory making in the human brain are the really fiddly bits at work here; cognitive scientists are honestly just starting to figure out how all these things fit together, much less how sex differences might come into play. But for able-bodied, highly visual primates like

* Because retinal cones are more diffuse toward the edges of your retinas, your peripheral vision is largely red-green color-blind for smaller objects (Hansen et al., 2009). We are far more able to detect *movement* at the edges of our visual field than differences in color (ibid.).

Homo sapiens, these pathways are also deeply embedded in how we understand ourselves as creatures in the world with rich, remembered experience. Think back to those teenagers looking at my naked body: the most likely reason the boys regularly drew my breasts larger than they actually are isn't simply that they were socially conditioned to do so.* It's that, for one reason or another, their eyes were literally *fixated* on my breasts more than the girls' eyes were.

Generally speaking, human eyes do two things: saccades and fixations. Saccades are the twitchy ways eyes move from one spot to another in a visual field, and when they linger on a spot, it's called a fixation. There are known sex differences in these patterns when people look at human faces—adult women tend to have more saccades that move between different parts of a person's face and eyes, whereas men tend to fixate a bit more around the nose. No one knows why. But this might be why women are famously better than men at learning new faces, and it might also be why women seem to be a bit better at accurately judging what emotion that face is conveying. We also tend to focus on the left eye region a titch more, which is likewise the side of the human face that tends to be more emotionally expressive.†

All of that has something to do with the eyes themselves and what the brain, upstream, is doing with that information as it arrives in real time, further directing the eye to move or linger. But when the eye does linger, it makes a greater impression in the brain's memory after the fact, just as it seems to make a greater impression on one's perception in real time. We're talking about the nuts and bolts of reality building. So if the boys' eyes *fixated* on my breasts more frequently than the girls', they might have

* To be fair, it's also because they simply weren't as *experienced* as artists at the start of the semester . . . but neither were the girls, presumably, so I think it's safe to rule that out as the major factor.

† This comes up in a few places in this book: women tend to favor cradling infants on the left side of the body, regardless of whether the woman is left- or right-handed, and this bias seems to be useful for social interaction, because it lets both the mother and the infant better see the more expressive side of their faces. Population-wide, nearly all humans have this habit of cradling on the left, but women do it slightly more, and mothers are still far more likely to be the ones cradling their infants throughout the first three months of those children's lives.

been more likely to perceive them as larger in relation to the rest of my body—not because they wanted them to be, necessarily, in that culturally driven cartoon sort of way that the "male gaze" renders a woman's body in social spaces, but literally in the cognitive mechanics of the thing. Consider, for example, what happens when untrained artists try to draw human faces: they forget to draw foreheads.

Because human beings tend to fixate on the eyes, nose, and mouth—which is to say, where our identifying features are located (who is this person) and also where we do most of our social signaling (what is this person feeling, what might his or her intentions be)—that also means our brains *perceive* those features of the face as more prominent than they actually are on a real human face. So the untrained artist tends to draw human faces like a Neanderthal: with low, short foreheads, big eyes, big nose, big mouth. And as the artist learns that the forehead usually takes up a full *third* of a human face below the hairline and begins to internalize ways of "correcting" his or her brain's normal interpretations of the visual field, the face on the canvas starts to look more human.

Over time, the boys in the class were not only better able to draw my breasts; they started to give me a forehead, too, which is reassuring, given that the majority of my frontal cortex is safely lodged behind it and is a large part of what makes me human. I honestly can't say if the experience of drawing me made them look at women's bodies *outside* class any differently; each of us has socially specific ways of being and interacting, and skill sets don't always transfer neatly from one scenario to another. I don't know if a naked body in an art room "normalizes" that body for the viewing mind or makes it more exceptional. But I can't help but think of the girls' eyes, too, in that room, not because they were accidentally a bit better at drawing my boobs, but because some of their eyes—specifically, their *retinas*—might have been very different from the boys'.

What the brain does to perception can be seen in the ways in which culture naturally limits girls—ways that are hard to even notice. Because women are generally born with two X chromosomes, some are actually *tetrachromats*—they see the world not in three color dimensions but in four.* Like birds, these women can tell far subtler differences between red, green, and yellow wavelengths, potentially making them able to see as many as *100 million* distinct colors: a full 99 million more than the average human being.† A tetrachromat woman's visual world should be full of fine, shining, delirious detail: the kaleidoscope of color glinting off each wave on a pond as it catches the light, the shimmering flutter of the underfeathers of a robin's articulate wing.

Or *could* be, anyway.

Except our human world isn't designed for anything greater than trichromacy, and, sadly, women who have the genetic predisposition to see all those extra colors usually don't. That's because the color receptors aren't what fundamentally decide what colors we perceive. There's a directional stream of information between the eyes, the optic nerve, and the vision regions of the brain. Some of it loops—for instance, while the eye moves through its automatic saccades, the brain directs the eye to focus on some things over others, look one way or another. The brain determines the need, and the eye adapts accordingly. If you *need* to see a certain color, and especially if you've had the habit of seeing it your entire life, you'll probably see it—so long as you have the color receptors in your retina. But if you don't? If there's no need? Then you probably won't. We honestly don't know why.

As many as 12 percent of all human girls may be born tetrachromats. They have the potential to see a world that no man will ever be able to see. To see a world most *women* don't even see. But

* Not UV light, though: from tests of human tetrachromats, it seems the fourth type of human retinal cone is sensitive to wavelengths in the middle space between red and green. The fourth bird cone is specially dedicated to UV wavelengths.

† The number is unclear, given that we're talking about very fine differences between similar wavelengths of light. But whether it's 10 million or 100, bird vision is even more sensitive than that, because each of their retinal cones also contains a droplet of colored oil that seems to help boost birds' ability to sense fine differences between different types of colors. Lizards have these oil droplets, too, and owls have notably fewer of these colored droplets than their day-dwelling peers. When light is scarce, perhaps it's better for color rods to absorb more of the light, with less fine distinction between colors—and that might have been true for ancient nocturnal placentals, too. Marsupial retinas still retain some of these oil droplets in their retinal cones, as does the platypus.

because they grow up in environments that will never ask them to use it, they'll never know that they have this ability. It simply won't develop. The strange extra cones in their retinas will lie dormant, or maybe their optic nerve just ignores them. We don't know exactly what happens to them. These girls are like secret superheroes. They have eyes like *birds*.

While men and women live, in many ways, in different sensory worlds, what we share is social context: because we're so fundamentally and deeply *social* primates, the social context of our perceived worlds influences how we interpret and act on the signals brought to us through our sensory array. Change the context, and you're very likely to change the perception. So the bird-eyed girls experience the world much the same as we do, despite their superpowers, and color-blind men live with their small handicap. Women smell things more finely and accurately than men do, but mostly we only notice that when we're ovulating or pregnant (or when a man tells us that whatever we're smelling isn't really there). The shared social context of today's dating scene largely overrides any benefit women might have gleaned from their gut reactions to man smells in our deep ancestral past. And so long as we remember that women can often hear things that men can't, we'll be better at designing auditory environments that are more inclusive for any given ear.

Ardipithecus ramidus

CHAPTER 4

LEGS

We should go forth on the shortest walk, perchance, in the spirit of undying adventure, never to return,—prepared to send back our embalmed hearts only as relics to our desolate kingdoms. If you are ready to leave father and mother, and brother and sister, and wife and child and friends, and never see them again,—if you have paid your debts, and made your will, and settled all your affairs, and are a free man, then you are ready for a walk.

—HENRY DAVID THOREAU

(some opinions on what it means for a woman to walk out of a house)

—ZILPAH WHITE, PROBABLY

DAHLONEGA, GEORGIA, 2015

The soldiers leaned into the mountain. Their lungs burned. Their muscles burned. Their eyes. Everything but their fingers and toes, which went blue in the cold as they climbed to higher altitudes, blood shrinking away from their extremities and pooling in their torsos in some long-evolved, last-ditch effort to keep vital organs alive. The team had been moving, day and night, up the mountainside, barely stopping to sleep, to eat, to speak. It was too much to ask of a body. But that was the point—war doesn't stop to ask how you're feeling.

Captain Griest paused and tore open a little tan packet containing a meager MRE—what passed for a meal, the first in thirty hours. The captain's boots were soaked in mountain. Brain soaked somehow, too—that place you get to after your body has done more than it's supposed to be able to and you know there's still more to do. Soldiers call it "the suck."

Survival becomes a matter of minutiae: the stupid little things you do to keep muscles moving. Tearing open a packet. Keeping your socks dry. That's something soldiers learned in the trenches of World War I. Wounds on legs and arms could heal, but if their feet went, they were done for.

This mountain was part of the U.S. Army Ranger School: a carefully planned barrage of trials meant to select and train elite soldiers for leadership in combat. Very few even qualify for entry; even fewer complete the course. In sixty-two arguably horrific days, soldiers suffer near-hypothermic conditions, heatstroke, near starvation, and delirium from sleep deprivation. Sixty percent of the people who quit drop out in the first week. They never even make it to the mountain. Those who do, and manage to come back down, then have to survive a simulated air raid in a festering hot swamp complete with poisonous snakes. Some soldiers are forced to stop by medical observers—the ordeals are legitimately dangerous. With too little sleep, toxic cellular waste builds up in the brain. By the end of the course, hallucinations are not unheard of.

Like training for the Navy SEALs or the Marines' Force Recon, the Army Ranger School is considered the ultimate macho test. You have to be strong enough to carry a wounded, two-hundred-pound man on your shoulders up a muddy hillside, but you can't just be strong. You have to be able to run a seven-minute mile, fully loaded, but you can't just be fast. You have to be able to do all the things that a soldier has to do in combat, under the worst conditions, over and over and over again, and never lose your shit.

Men are supposed to be best equipped to endure the mountain and ignore pain. They're supposed to be able to support each other, to show leadership, brotherhood, grit. Between the sexes, the male body is supposed to be stronger, faster, and more resilient.

Just look at the Olympics. The fastest runner in the games has never been a woman. The strongest lifter, the swiftest swimmer, the highest jumper—these bodies are always male. There are separate men's and women's divisions of most professional sports because it would be unfair, we assume, to allow the superior body of a male athlete to pummel a woman's in a competition. Except for the very rare woman with an androgen disorder, most females seem unable to compete with male physical performance.

But Captain Griest is a woman. So why on earth did she take on the mountain? Aren't women the weaker sex?

As with most things in the body, the answer is deeply rooted in how we evolved. In this case, we're really asking about the modern human musculoskeletal system. Five million years ago, our ancestral Olympics would have consisted of chin-ups, swinging from hand to hand through the trees, long periods of starvation, and fleeing our predators. We were terrible runners because, living in the canopy, we didn't need to be good at that. We didn't need to jump straight up into the air, because we had strong shoulders and limbs to pull ourselves up. We had powerful upper bodies and relatively weak lower bodies—basically the opposite of today's human anatomy. But the world changed. In order to survive, a small band of primates started walking on two legs.

ETHIOPIA, 4.4 MILLION YEARS AGO

Our primate Eves lived in the high gardens of tree canopies for tens of millions of years, noshing on fruits and bugs and tender leaves, having sex, having babies, getting into fights, having more sex and more babies. A luxury of petty bickering. Plenty of food. Some of their descendants stayed tiny; others got big; still others got big and then, weirdly, got small again. Dinosaurs spread their wings and mammals conquered. Life was good up in the trees.

But Earth never stays the same for long.

Although the continent of Africa had been covered with forests for eons, in the Miocene the planet's climate started to cool. While our primate Eves were skittering and swinging in the fruiting trees—their eyes moving forward, their hearing deepening—global weather was fairly warm and steady. But starting around 20 million years ago, region by region, things got chillier. By the time the Pliocene came around—about 5.5 million years ago—global weather had changed.

But that wasn't the only thing that changed. East Africa, humanity's sacred garden, was pushing upward as the Great Rift Valley formed. The reason the Ethiopian highlands reach more than nine thousand feet above sea level is that Africa is splitting in two. The shifting mass of the mantle underneath the Ethiopian plateau—a place now so high and so large it's called the Roof of Africa—shoved the earth up, atop a flood of lava. It's still happening. The great African separation will take millions of years, but the end is clear: East Africa is striking out on its own, lurching toward the Arabian Sea. The existing crack has started to fill with water: Lake Turkana. Lake Naivasha. Nakuru. Eventually, a narrow, shallow sea will stretch down the middle of Ethiopia, Kenya, and Tanzania.

Tearing a continent apart does a number on the weather. Up in the canopy, our primate Eves were evolving into apelike things. As the planet cooled, wind patterns shifted over East Africa's rising plateau, separating the continent's central rain forests from our ancestors' home ecosystem. By 8 million years ago, it wasn't raining as often as it used to.* Forests shrank and wide, grassy plains opened up, as fertile and treacherous as the sea. Our Eves peered out from their safety in the canopies. Most of them stayed there, their numbers shrinking, sustaining themselves on what the smaller riverine forests could offer.

* We used to think it was only 2.5 million years ago, but more recent isotope studies of soil in East Africa push the date to 6 million years at least (WoldeGabriel et al., 2001). Studies that model the impact of East Africa's uplift also date the shift from rain forests to grasses between 5 and 8 million years ago (Sepulchre et al., 2006; Pik, 2011; Wichura et al., 2015). And then there's the Messinian salinity crisis, wherein the entire Mediterranean Sea repeatedly flooded and dried out between 5 and 6 million years ago as a narrow channel in the Strait of Gibraltar periodically blocked and opened the way to the Atlantic (Krijgsman et al., 1999). Just as ancient salt farms used evaporation pools to harvest sea salt, but on a massive scale, this process at Gibraltar managed to remove 6 percent of all the dissolved salts in the world's oceans, profoundly decreasing their alkalinity, with knock-on effects for ocean species and, for our primate ancestors, screwing with East Africa's precipitation (Bradshaw, 2021). The salinity of the ocean, after all, shapes the global water cycle. So for our Eves in East Africa, this moment in Earth's history was a perfect storm.

But some of them ventured into the ocean of grass with the giant cats, the raptor birds, the hidden serpents. They went because they had to in order to find more food. And then they ran the hell home.*

"Running" is a key word here: we're the only living apes that do it. Human beings share nearly 99 percent of our DNA with today's chimps and bonobos. Most scientists estimate that our species diverged between 5 and 13 million years ago, toward the end of the Miocene and the start of the Pliocene.† Somewhere in that time, our closest cousins were learning how to walk on their knuckles, scrambling over the ground between increasingly distant tree trunks. But *our* own ancestors learned how to walk on their hind legs, and eventually they learned to run.‡ Many scientists think we actually started that process in the trees: walking on our hind limbs along larger branches as we used our hands to pluck fruits and bugs on tiny, higher boughs, especially when the trees were *shorter* and hanging was better supported than sit-

* To be fair, our forest home wasn't much safer. Take *Machairodus kabir*, a 770-pound jaguar, who liked to leap on us from above, puncture our necks with short, thick canines, and lick his paws while he watched us bleed to death (Sardella and Werdelin, 2007). Ancient Africa was no picnic (Peigné et al., 2005). But our primate Eves were already adapted to a life in the trees, fleeing all sorts of forest monsters. While it might have been hard in a shrinking forest to cope with an increasingly hungry set of meat eaters, adapting to the unknown savanna was probably harder.

† Venn et al., 2014; Steiper and Young, 2006; Diogo et al., 2017; Harrison, 2010. Again, as with most things in human evolution, this is a contentious number. For decades, we thought it was anywhere between 3 and 12 million. Then, in 2005, the number narrowed to 5 to 7 million (Kumar et al., 2005). In 2014, a genetic analysis based purely on the average rate of mutation in living chimps and humans showed that, at least today, human and chimp DNA may mutate more slowly than we'd thought, pushing the number back to 13 million (Venn et al., 2014)! But when a *species* splits is different from when DNA diverges. With a big enough population, or if our ancestors were split into two subgroups that didn't interbreed that much, a date of 7 million would be fine. Six to 7 million years ago, at least according to isotope studies, seems to correspond to when the climate of our forest Eden started to give way to a mixture of forest and grassy savanna, or at least was still forested enough to imply savanna living was not yet the norm for our Eves in that period (WoldeGabriel et al., 2001).

‡ That's the current model. It's not that knuckle walkers learned how to walk on two legs, but rather our Eves became bipedal, while the gorillas' and chimps' and bonobos' Eves went on to make use of their knuckles.

ting on branches. It was easy to take that behavior and apply it to walking upright on the ground. By 4.4 million years ago, we were doing it regularly. That's when *Ardipithecus ramidus* walked the earth—the Eve of human bipedalism—about 3–4 million years after the last common ancestor of chimps and humans.

Scientists found Ardi's skeleton near Aramis, Ethiopia, in the mid-1990s, but it took the better part of a decade to analyze the fossils and realize what they'd found: the earliest bipedal ape, the Eve of women's legs, hips, spine, and shoulders. Ardi is the best evidence we have for the root of the sex differences in men's and women's musculoskeletal system. She is the reason there are men's and women's divisions of competitive sports. She is the reason women have crappy lower backs and knees. And she is also the reason women are more likely to survive a zombie apocalypse (should you be concerned about such things).

BONES

Standing about three feet eleven, Ardi was somewhere between a chimpanzee and a really furry human, which is to say, she walked upright but still spent a lot of time in the trees. Her hands were more primitive than those of chimps, but her pelvis, legs, and feet were much more like a human's. She wasn't a knuckle walker. Her hands and shoulders weren't good for that. She moved around on her two feet, not as much as we do today, but more than someone who spent all her time in the tree canopy. When you look at a modern woman's skeleton, you'll still see a lot of Ardi.

For example, modern women's feet and knees kind of suck. Because our leg and foot joints naturally absorb a lot of the pressure of our body weight when we move, you'd think their failures would simply depend on how heavy that body is. But though women tend to weigh less than men do, we're still more prone to trouble in our feet and knees than men are. Some of that has to do with modern footwear, but not all. Even when we wear the most supportive orthopedist-recommended shoes, women's feet and knees still falter. Becoming upright was in some ways *harder* on Ardi and her granddaughters than it was on the males.

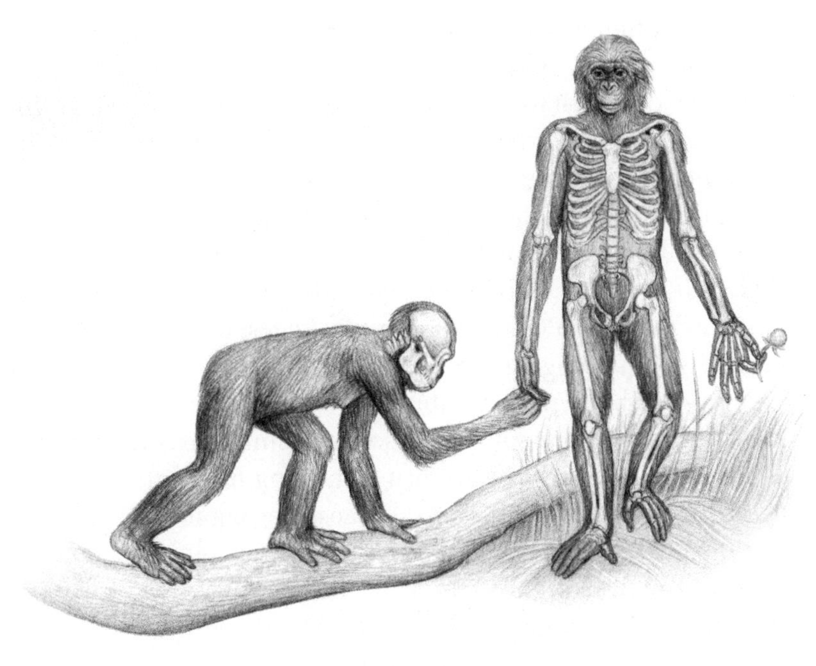

Ardi and her bones

Ardi's foot wasn't fully modern. Her big toe was set off from the rest of her toes, which allowed her to better grasp branches when she hung out in the trees. But the bones in her feet were oriented in a way that helped stabilize her when she walked upright. They were stiffer than the feet of tree-dwelling apes, which is a big part of why human beings are so prone to bunions, that painful bony lump that forms over time at the joint where the big toe begins. When we take a step, the stiff bones of our upper foot stabilize the force between our toes and our ankles. Starting at the heel, we essentially roll our weight forward, over the upper and mid-foot, onto our toes, stepping from one forefoot onto the opposite heel. We've taken something that originally evolved for *grasping* and made it a hinged series of levers for bearing weight while *walking*. Your big toe is basically a short thumb. Ardi's toe thumb was more like our hand thumbs, opposable, so she could still use it to wrap around branches. Walking for Ardi was probably a bit like walking in a snowshoe—she hadn't yet evolved the ability to roll smoothly from heel to toe.

Modern humans inherited the problems that come with any sort of bad design. Our feet are, in many ways, the biological equivalent of duct-taping your car's bumper back on when you don't have the money to send it to the body shop. But it's worse for women. Stiffening the upper- and mid-foot bones so we can walk means a lot of force is transferred from our ankles to our forefoot. All that force on the forefoot, especially the big toe joint, weakens it over time. Combine that with a female body that tends to "sway" in motion (wider hips, funky knees, more butt fat), and eventually something's gotta give. It's probably going to be the big toe joint—both the most flexible part of the foot and the one that receives the most pressure. That's what bunions are: the physical reminder of how hard it is to turn a grasping hand into a foot.

Ardi didn't develop bunions, because her big toe was set apart from the others. She also didn't wear heels and didn't spend nearly as much time walking upright as we do. Her gait was probably a bit stilted and waddling, unlike Lucy and the australopithecines after her. But as we evolved to get better at walking, we also got more bunions—especially the females among us.

It's just physics: force has to go somewhere. Our foot distributes pressure down toward the forefoot as we walk. The rest radiates back up through our leg bones, knees, hips, and spine. Unlike men's, women's femurs come into the knee joint at an angle. This was true of Ardi, too, but it's much more pronounced in modern women. Because our hips are wider than men's, our knees are somewhat closer together to help balance that differing center of gravity. That sexual dimorphism lines the pockets of orthopedic surgeons, who regularly perform significantly more knee replacements on women than men.

Consider that every pound of body weight normally puts an extra pound and a half of pressure on the knee joint when we walk around barefoot. It goes up to four times the pressure when we jump. Our bodies have evolved to mostly handle that. But modern, gendered footwear can pull the rug out from under us: in high heels, our center of gravity is tipped forward, meaning that instead of the buttocks and hamstrings, the quadriceps at the front of the thighs have to do the lion's share of the work, yanking the top of the knee upward, further compromising the joint. Over time, that can damage the ligaments in the knee, wear away at cartilage, and generally wreak havoc. It's bad for our toe joints, too: walking in heels eliminates the "roll" of normal walking and instead can mean, depending on the heel's height, a repeated slamming of all your body weight and momentum onto the forefoot. The heel of a high heel is mostly there for balance, which is precisely why stilettos work at all—we're just tiptoeing our way down urban streets like bewildered ballerinas.

But we can't blame high heels entirely for the damage done to modern women's feet and knees. There's something more subtle at play—something chemical—and Ardi probably had to deal with it, too.

In the fourteen days leading up to her period, a modern woman has a small, cyst-like structure on one of her ovaries.* This is the corpus luteum, what's left of the follicle that hatched her egg when she ovulated. In most women, the hole where the egg emerged seals over and the corpus luteum swells a bit, sending out signals to the body to increase the production of certain hormones and decrease others. This is a large part of what changes the uterine lining and sparks a host of other fun PMS symptoms, like bloating and acne and general irritation.

The corpus luteum also tells the body to produce more relaxin, a hormone that makes ligaments more flexible, loosening the muscles from their skeletal anchors. Most scientists assume this allows the uterus a little more room to grow. Normally, the uterus is anchored fairly tightly by a network of ligaments and fascia. Loosening those anchors allows it to puff up with blood and fluid in the first trimester. Relaxin also loosens the connections between the bones surrounding the pelvic area, from the hip bone to the sacral spine to the femur heads, so the lower pelvis can loosen and widen in order to carry the growing uterus in later trimesters and then widen even further for birth. Relaxin levels are highest in ovulation, the first trimester, and the last trimester—

* Sometimes it can persist after the period instead of being reabsorbed. If it grows too large, or leaks blood into the abdomen, it can be very painful. Before ultrasound technology improved, many surgeons wouldn't know at the start of an abdominal operation if they were looking for a cyst or a ruptured appendix. They just had to figure it out on the table.

when the uterus needs to start getting bigger and before it needs to squeeze a large baby through a small birth canal.

Relaxin is found in all placental mammals, both females and males, though in much more significant levels in females. But destabilizing a four-legged musculoskeletal system is a bit less damaging than destabilizing the system of a creature that's only recently evolved to walk upright. Ardi, in other words, was probably the first of our Eves to have chronic lower back pain, knee pain, and pregnancy-related musculoskeletal dysfunction. She was probably the first female to tear her ACL and the first to have a slipped disk in her lumbar spine.

If anything, when it comes to withstanding high heels, male drag queens are arguably better equipped to wear those Louboutins than women. Despite their heavier weight, their masculine body traits—straighter knee joints, more leg muscle, and lower levels of relaxin, all of which make men's knees and backs less prone to injury than women's—make drag queens less likely to suffer long-term damage from their high-heel habits.* They'll also never ovulate or become pregnant, lacking both ovaries and a uterus,† so those lower levels of relaxin *stay* lower, their spine stays nicely fixed, and their hip joints never have to widen to accommodate a newborn's head and shoulders.‡ Relaxin will also

* Which isn't to say heels *won't* ruin men's feet, hips, and back—see Steven Tyler and Prince—but it doesn't come up as much. I haven't had the chance to ask Eddie Izzard how she's fared, but one imagines all those marathons might have done a bit more to the knees than her shoes.

† There are some trans men who choose to perform as drag queens, but they're rare and don't universally consider themselves in the same category as traditional drag. Gender play is complicated. I don't know of any research into whether these men suffer from shoe-related joint problems. While gender-affirming treatments are finally available to pubescent trans people, it's true that most trans men still go through a female-typical puberty before transitioning, so I'd guess that for many trans men who perform in drag, their knees and backs would still be an issue, despite the quietus in the ovaries from hormone therapy and/or surgery.

‡ In men, relaxin is produced by the prostate, but it mostly goes into semen rather than circulating in the bloodstream, and seems to help with sperm motility (Ivall et al., 2017). This isn't exactly great for men, however, because relaxin helps blood vessels "relax" body-wide, lowering blood pressure. It also seems to help promote wound healing, probably in part due to better blood circulation at the wound site (Unemori et al., 2000).

never mess with the ligaments tying together a drag queen's foot bones—something every pregnant woman has to deal with.

However, relaxin would have made Ardi's upright frame a bit more *yoga*, if you like. Combined with lower muscle mass and more flexible joints than her male counterparts, she would have been more able to contort herself to navigate awkward spaces. Much like women today, she might have been better equipped to be nimble.

Ardi's lower spine had evolved with a slight S curve, as humans have. The spine is a bit like a spring: each time we walk, that S-shape absorbs some of the shock of impact. When the heel strikes the ground, it sends force up through the ankle to the knee and hips and spine. The knees take a lot of that. The hips, some more. The curved spine manages to absorb most of what's left. That's why we don't feel a horrible, jarring impact in our lower skulls every time we take a step. But our lumbar spine—the tiny tailbone, the fused sacrum, and the rest of the vertebrae that rise to our waist—absorbs more of that distributed force than our mid- and upper spine does. Over time, all that absorbed force compresses the cartilage between each vertebra, causes tiny micro-fractures in the bones, pinches nerves, and weakens muscles. Lower back problems are some of the most common human ailments; by the end of our thirties, quite a lot of us will have sought medical treatment for lower back pain.

And women bear the worst of it. When a woman becomes pregnant, her center of gravity should quickly change. But women's spines have evolved differently from a chimp's to compensate, keeping the center of gravity more stable by flexing the spine. That makes the human spine uniquely vulnerable, and that evolution is more dramatic in women than in men: as the uterus grows, that extra weight pulls the lumbar spine forward, tightly compressing the outer cartilage. That's why women in their third trimester seem to have a kind of swayback; their spines and pelvises have changed shape to accommodate the heavily laden uterus. Chimps and other four-legged mothers don't have to deal with that. As their uterus grows, their abdomens simply expand toward the ground. So their lumbar spine doesn't have to curve like ours, squeezing the cartilage and nerves in between the bones.

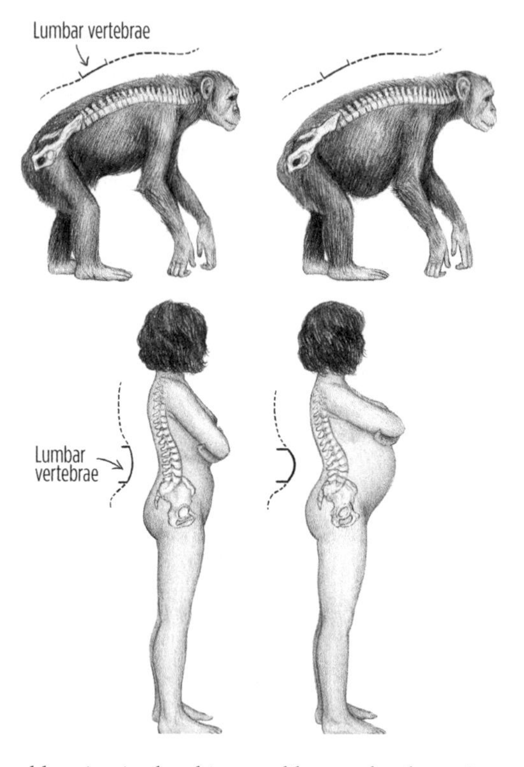

Load bearing in the chimp and human lumbar spine

In Ranger School, Captain Griest was neither pregnant nor wearing high heels. But even as she climbed in her sensible army-issue boots, taking extra pains to keep them dry, she did have to contend with her female-typical skeletal woes. If we're so prone to injury, why shouldn't men be thought of as innately stronger?

MUSCLE-BOUND

We know Ardi stood only about three feet, eleven inches. But she was probably more muscular than the average woman today, since scientists estimate she weighed roughly 110 pounds. To put that in perspective, the average adult woman in the United States today is about five feet five and weighs around 168 pounds, with a good 30 percent of that weight in body fat.* Human bodybuilders come in a bit heavier. Heather Foster, for example, a five-foot-five bodybuilding champion, reportedly weighs about 195 pounds off-season, while her weight at competitions is around 150 pounds. To imagine how muscular Ardi would have been, picture a tiny bodybuilder at her cut weight, just shy of four feet tall. Then stretch her arms, make her hands and feet a little funky, and cover her in fur.

There are three different kinds of muscles in our bodies: cardiac, smooth, and striated. Smooth muscle mostly belongs to the abdomen: intestines, stomachs, lungs. Cardiac muscle, as you might imagine, is only in the heart. Most of what we think of as "muscle" is striated skeletal muscle, which we use to stabilize and move our bones around. Unlike the other two types, these muscles are voluntary. They're also what we usually think of when we say someone is "strong."

But the "musculoskeletal" system gets its name from the fact that skeletal muscle isn't really separable from our bones. In fact, when we go through growth spurts, it isn't exactly right to describe it as just our *bones* growing. Rather, our skeletal muscles and ligaments bulk, stretch, and tug at their anchors on the bone. That tugging is intimately tied to calcification and how bone tissue grows. That's true whether it's happening in childhood, in adolescence, or during the odd bit of extra growth some people experience in their twenties.† That's why older women are encouraged

* In the U.K., that shifts to five feet three and 152 pounds, five feet three-ish and 155 pounds in Canada, and just under five feet five and 138 pounds in France (St.-Onge, 2010). Women's average weight drops to 112 pounds in Cambodia, but women there are usually only five feet tall. Still, barring rarer things like dwarfism, the modern body hews fairly closely to its norms. Our species simply doesn't have a lot of Pomeranians and Great Danes. If you feed mothers well enough, and their children after them, most of us end up roughly the same size.

† I grew a full inch between ages twenty and twenty-seven. I don't know exactly when this happened, because I only found out at the end, though it's safe to consider this a last hurrah of my already odd puberty. The same thing happened to my mother, though she's since diminished in height, as most people do in old age. Though this is rare, it's not so rare that it's outside normal human experience. Human growth is bursty, and no one actually knows the full mechanisms that undergird the onset of those bursts. The twenties is an odd time for the human

to add weights to their exercise regimen: *tugging* at the muscular anchors of our bones encourages those anchors to add more calcium, strengthening the bone. It's a simple way to counteract the dangers of osteoporosis—a disease in which bones lose too much calcium and become brittle—to which postmenopausal women are especially prone.

Modern women's skeletal muscles have evolved another 4.4 million years past Ardi's body plan. We've added a lot more fat around those muscles, for example. Our arms and hands have gotten smaller and our shoulders have narrowed. The more ground based we got, the less our upper bodies mattered. But there seem to be certain fundamentals about how muscles work—what strength really *means*—that holds true across all mammals, and especially for primates like us.

In high school physics, you probably learned that the length of a lever had a lot to do with how much potential force that lever could wield. Shorter arm, less force. Longer arm, more force. That's why the arm of a car jack needs to be long to let you apply enough force to lift the car so you can change your tire. Right. Now think of the bones of your legs. Your femur is one arm of the lever that folds at your knee. How strong your leg can be, therefore, has a lot to do with how long your bones are. The same is true for any other joint in your body: your muscles are there to support, stabilize, and pull on your skeleton. There are ligaments and fascia to connect muscles to bones, muscles to muscles, and cartilage plays a role, too. But fundamentally, a musculoskeletal system is a set of levers. Lots and lots of levers—things that pinch closed and widen, depending on the task at hand.

In a few key spots, there are also ball-and-socket joints that allow a wider range of motion, swiveling and rotating, for example, where your femur connects with your pelvis, or where your upper arm connects to your shoulder. Once upon a time, those joints had incredible range of motion for swinging our torsos

---

body in general: for example, people who've never had allergies before may develop "adult-onset allergies," typically in the twenties, to some previously benign pollen or the like. Though the human developmental plan is usually divided neatly into distinct periods—infancy, childhood, puberty, adulthood, and senescence—there's a lot of blur at the boundaries.

through trees. Humans, orangutans, gibbons—and Ardi—all have *brachiating* shoulders: a joint with a wide range of motion that lets us move, arm over arm, through the trees. Most four-legged beasts would never be able to use the monkey bars on the playground because their shoulders don't have the range of motion.

Brachiating shoulder joints are what let Captain Griest move arm over arm along a narrow cable during her physical tests. They're a big part of how she scaled sheer cliffs. And she used them again later, in the swamp trial. But compared with her fellow soldiers, this task was significantly harder for her because most modern human women don't tend to have as much upper body muscle mass as men. Somewhere in puberty, men's and women's average body plans diverge, with men's shoulders and chests broadening and bulking up, while women's hips widen and their breasts develop.

That's one of the most popular arguments for why women are weaker than men: not only are we a bit shorter and narrower, which reduces each body lever's potential force, but the muscles of our upper body don't develop the way men's do. When trans men are given androgen and testosterone hormone treatment, they do develop more upper body strength and muscle mass because skeletal muscle—especially upper body skeletal muscle—seems to be modulated by male sex hormones. That may speak to a kind of male *continuity* over millennia: the muscles of an adult human man are more like what our ancestors' were.

Today's chimps are explosive athletic performers. They're incredibly strong and agile. Even knuckle running over flat ground, they can run as fast as twenty-five miles an hour. To put that in perspective, that's just a few miles an hour shy of Usain Bolt, the fastest human alive. But chimps don't run for very long. In fact, they can't do much of anything we think of as athletic for very long before they tire. Chimps' metabolisms and muscle tissues are designed for explosive effort: to occasionally fight, to briefly chase things, or to flee into the safety of the trees when predators come calling. Their bones are heavy and they have tremendous amounts of muscle mass in their upper bodies, which do most of the heavy lifting when it comes to moving themselves around.

This top-heavy power distribution isn't just a matter of adaptation for knuckle walking. Every day is "arm day" for these brachiating primates. Orangutans, who unlike other apes still spend most of their lives in the trees, have this type of muscle distribution, though the orangs' is more pronounced than chimps', because their upper arms are much longer and therefore require even more muscle mass to control them and provide force. Human shoulder and hand anatomy still have some features left over from a brachiating ancestor: the rotating, flexible shoulder joint, the grasping fingers and thumb.

In that sense, it may be better to consider men's muscular upper bodies—along with their ability to do all those chin-ups, push-ups, and burpees—as something closer to our tree-dwelling ancestors'. Though boys and girls are relatively similar as children, adult men distribute muscle mass over their upper bodies much more than adult women. We women, meanwhile, tend to have very strong legs—as strong, for our height and weight, as men's, and in some cases stronger. In evolutionary terms, modern human women's muscle pattern changed *more* than men's.

There's a popular stereotype about how male athletes tend to be good sprinters and women athletes tend to be good endurance runners. While many women are great explosive athletes, they rarely approach the same speed as men over short distances. In feats of strength, we likewise don't generate as much force on average. Being bigger animals, men also have bigger lungs and hearts, which helps to get that extra oxygen to working muscles.

But despite all these advantages, when it comes to endurance sports, women frequently perform as well as men. Once you get to ultra-endurance distances, we even beat them. Part of it may be because women are just a bit lighter and smaller, which means it costs us fewer calories to move our bodies over the same distance. But there may be something else at work. Instead of just using carbohydrates for energy as they work, mammalian muscle cells *shift* and start metabolizing fats and amino acids, too. That switch is a lot like the "second wind" endurance athletes talk about: when you start to get tired, but then somehow you feel energized again. It's actually about firing up the mitochondria—the powerhouses in each cell. Women of reproductive age may be better at utilizing that metabolic switch. They're not only better at getting to their second wind, but once they're there, they last longer than men do. And that may be because there's something in the mitochondrial metabolism in skeletal muscles that's controlled by female sex hormones.

In the mid-1990s, a group of orthopedists ran a study comparing small samples of their patients' skeletal muscles. They found that a certain metabolic pathway (the "mitochondrial electron transport complex III") was significantly more active in their younger women subjects than in the young men.* This particular pathway has to do with using fat to give muscle cells energy. Young women's bodies are really, really good at lipid beta-oxidation: using our mitochondria to take little molecules of fat and break them down. And while all mitochondria are able to do this, having muscles that are especially good at it may be built into our female body plan. A more recent study showed that genes related to this particular sort of fat metabolism are more expressed in young women's muscle cells than in men's.

Being able to kick into that second wind, possibly with lipids as your second energy source, is incredibly important if you want to be an endurance athlete. You can sprint on your first wind. You can do a ton of explosive power performance on your first wind. But if you want to do anything for a long time, you need that metabolic second wind. Before Ardi, we weren't running or walking anywhere for any length of time. There weren't a lot of things our Eves needed to do that involved much endurance.

A number of popular science writers like to say that human beings evolved to be runners, but it's probably more true to say we evolved to be endurance *joggers* and walkers. One of the big changes that happened in hominin evolution, starting—we presume—around Ardi's time and continuing through to modern humans, is that we became "gracile": the Eves that led to humanity evolved lighter bones and different sorts of muscles. It's generally

* Boffoli et al., 1996. This activity only started to decline when women passed menopause—in other words, when their circulating levels of female sex hormones declined (ibid.). For a more recent overview of all the ways female mitochondria seem to win the race, see Cardinale et al., 2018.

assumed we did this because walking upright is calorically expensive. We curved our spines and slapped a bunch of butt muscle on our hind parts to hold it all together. But we also shaved some weight off our overall bodies and shifted our general athleticism toward stamina rather than explosive (short-lived) strength.

There's a good chance it was the females who led the charge, not just because we had a metabolic advantage, but because Ardi and her daughters might have had more need to leave the forest and venture out into that ocean of grass than the males did.

LEAVING SHORE

When you're long adapted for one environment, you need a reason to risk life and limb to venture into an environment you're not as well suited for. Ever since Darwin wrote his *Descent of Man*, scientists have been debating what on earth got us to come down from the trees. For a long time, most assumed the trees simply disappeared, forcing us onto the plains.

But the discovery of Ardi is adding nuance to that story. Between the obvious specialization of her skeleton showing that she was a part-time tree dweller and a part-time walker, analysis of the flora and fauna found around her fossils reveals she lived in a wooded environment. Further analysis of soil isotopes and pollens makes it likely that she lived in a riverine forest within a larger savanna—rich clusters of trees along the water's edge, no doubt ripe with fruits and tender stems. So why did she need to leave the trees? What happened?

A few types of arguments dominate the field. For a long time, the idea that walking was something we did to hunt was incredibly popular. We freed up our arms to carry weapons, right? We could use our brachiating shoulders to throw spears at all those grass eaters on the savanna. But chimps use spears to hunt bush babies. Right now. In the trees. Without walking on two legs.

So okay, what if we evolved not to walk on two legs but to run, because we were trying to hunt things that were really *fast* and running on two legs was the only way to catch them?

Unfortunately, it's simply not the case that a two-legged creature is faster than a four-legged creature. Cheetahs can run sixty-four miles an hour. Horses can gallop up to fifty-five miles an hour. Being two-legged actually seems to decelerate us. The fastest human can run only thirty miles an hour, and only for a few seconds.

But maybe raw speed and acceleration aren't the issue. Stamina's what's interesting in bipedal human beings—how long we can keep it up. Horses rapidly tire at their top speeds, lathering and needing to stop after only a couple of miles. Given enough time, a human could actually run a horse down. Most healthy human adults can trot at, say, five miles an hour, for hours at a time. Ultramarathoners can pretty much go for days if they get sleep breaks. Horses? They'd die.

As a result, many paleoanthropologists now argue that we evolved to outrun nimble ungulates—deer, horses, bison. We'd just keep jogging after them until they got really tired. And then maybe we'd use our brachiating shoulders to throw spears at them and then carry home all that meat in our arms.

Somewhere along our hominin path that might have been true. But since we found Ardi, it doesn't look as if that style of hunting were what *drove* the evolution of humanity's bipedalism. Ardi didn't eat much meat. Analyzing her tooth structure and enamel, scientists determined that she was primarily a plant eater.

Because Ardi is so much closer to the last common ancestor of chimps and humans than we are, maybe we'd be better off looking at modern chimpanzee behavior to figure this puzzle out. Among chimps, primatologists have observed two-legged behavior in a few scenarios. Either the chimps are trying to be impressive, or they're using one or both of their forearms to carry something (usually food), or they're wading through waist-deep water with both arms in the air.

The water theory is tempting if Ardi did, indeed, live in a riverine environment. Maybe she did a lot of wading in search of crayfish and clams. It's certainly possible, but again, given what scientists have been able to tell from her enamel, she wasn't eating a ton of shellfish. Reaching up for fruits on higher branches while standing on her rear limbs, the way modern orangutans still do, is probably a better model for how Ardi evolved with an upright pelvis. But that still doesn't explain why she came *down* from the trees to walk upright more regularly.

Here theorists usually posit a war of the sexes. As I've mentioned, today's male chimp will (briefly) rise on his rear legs when he wants to look impressive. Sometimes he hopes to impress by wagging his tiny erection in front of some females, making a general racket by using his forelimbs to swing branches about. Other times he bares his huge canines and puffs up his chest to attempt to intimidate. Sometimes these shows of strength involve leaning forward on his powerful forelimbs, flexing his biceps and shoulder muscles while he knuckles hard into the ground. Sometimes, like a gorilla, he'll even beat his chest (though this is more rare). And sometimes he'll alternate between standing up on his hind limbs and knuckling forward, baring his teeth and screaming loudly. What a guy.

So the idea has been batted around that increasingly complex social groups of early hominins, like Ardi, evolved to walk upright because guys wanted to look good for the girls and scare the other guys away. But chimps and bonobos and gorillas seem to get by with only infrequent upright displays, which is why the walking-upright-because-it's-sexy hypothesis hasn't gotten a lot of traction.

But one modification of this argument has gotten more attention, in no small part because the main scientist who's making it is the same guy who published the papers on Ardi: Dr. Owen Lovejoy, a towering figure in his field. His theory is that the changing climate of the Miocene finally meant our riverine ape ancestors had less to eat than they were used to. Thus, males started walking out into the nearby grass in order to find more food, which they'd then trade with females for their exclusive sexual attention. The females were presumably caring for increasingly needy babies, so they couldn't do their own walking and would be game for such a trade.

Sex for meat is a good argument for bipedalism. (Or sex for really good tubers, I suppose, because Ardi wasn't a heavy meat eater.) But there are many problems with it. For example, we have no idea when, exactly, our Eves' babies became so needy they couldn't simply ride along with their mothers in the search for food. We were probably still covered in fur that tiny fists could hang on to. And even if we were upright enough to carry a babe propped on a hip, we'd have the other arm free for food gathering and food carrying. In every other living ape species, food gathering for babies is the primary responsibility of the mother.

From what little we can see in the fossil record, it does seem true that hominin babies were needier than the children of earlier Eves. We don't know exactly when the neediness would have started changing hominin society, nor when (or if) it would have radically changed basic maternal behaviors.* Still, having highly dependent offspring does put more pressure on females across the board: they would inevitably be more hungry, more tired, and generally more pressed for time when it came to keeping themselves and their needy babies alive. Even in today's chimp societies, many females will barter sex for meat and other prized foods. But because chimps aren't monogamous, trading treats for sex isn't a safe survival strategy. So maybe, the argument goes, Ardi and her kin *invented* hominin monogamy to make the trade more appealing: that way, the guys would be motivated to bring home the bacon in exchange for a little lady waiting faithfully under the canopy.

It's an idea in keeping with our current sexual mores. But there are lots of ways of dealing with hungry kids, and a rapid flip from promiscuity to monogamous prostitution seems a little far-fetched.† Maybe Ardi did exchange some risky food sourcing for sexual rewards, the way many primates do today. But just like today, it was probably opportunistic: *Here's some rare and delicious food that I happen to have in my hands and am willing to share. Now can we please have sex?* But having to constantly *commute* to find enticing food makes the male even more vulnerable to sneaky sex back home. Without some kind of strict social policing of female behavior, how on earth would such monogamy even work?

If Ardi was anything like most primates today, her babies were with her most of the time. Responsible for their nutrition from breast-feeding to early childhood, she had greater food needs and

* More on this in the "Tools" chapter.

† For more on ancient hominin sexual relations, see the "Love" chapter.

a greater need to be innovative than her male counterparts. It's terribly unlikely that she was home in the trees, hanging out, waiting for a male to bring home the tubers. More than likely, she was out in the grass herself, looking for food. And when she found it, she might have needed to take it somewhere safe to eat it—not just to avoid predation, but maybe to ensure that other members of her group didn't *steal* it. That's what chimps do now with prized food—especially females responsible for young.

When we think about a trait evolving, it's always useful to ask who has the most need for such an adaptation. There's no question that both male and female primates need food and anything that influences whether you actually get the food you need will put a lot of pressure on evolutionary selection. But if females bear a greater need for food—in terms of both having pregnancies and needing to feed their offspring—then it seems safe to assume that added food pressure would be a driver in selecting for evolutionary change. In many mammals, female bodies have evolved to adapt to that food challenge by being smaller than males so that, when they're not pregnant, they need fewer calories to survive. In a changing world with increasing food variability, Ardi would have had to venture farther to get enough food, and once she got it, it would have been highly beneficial for her to be able to walk away carrying a bunch of it in her arms—doubly true if she had to carry a baby, too. In Ardi's time, single mothers were doing a lot of the same things they do now: commuting to work, dealing with the kids, scraping by on whatever they can. I think that's a more likely argument for the evolution of bipedalism than a sudden invention of the monogamous nuclear family with a sexed division of labor.

Instead of waxing poetic about ancient male hunters, we need to ask what the *female* upright ape might look like—focusing not simply on the detriments of having a female body, in other words, but on the benefits. So here's a useful thought experiment: If the female hominin were a primary driver of bipedalism, what would becoming upright mean for the evolution of our bodies?

The evidence is all around us: The food-hoarding behavior of extant chimps. The metabolic advantages of female skeletal muscle, which decline once women pass menopause. Flexibility. In most of the explosive, male-typical sports of the Olympics, women *do* fall behind. We are a bit slower over the ground. We are a bit weaker at lifting things. Our upper bodies aren't as muscular. But the most compelling arguments for why hominins evolved to walk upright aren't about short-term performance. They're about endurance. It's a question of range.

If you're an ancient hominin like Ardi, how do you greatly expand your range? How do you swim out into that sea of grass? You need to walk in order to carry stuff, yes. But you also need to *endure*. You need to be able to tap a second wind. You need to push past the wall. You need to survive in the suck.

The things that let us survive in the suck are the things that make us human. Yes, our capacity to innovate, but also our ability to endure in the worst conditions. To keep at it when we're already tired. Our ability, in other words, to not give up.

SLOW TWITCH, FAST TWITCH

Captain Griest knew people were watching. She was a member of the Ranger School's first coed class, held in 2015. The whole exercise was meant to be a one-off. The American military hadn't yet changed any policies about whether women would be allowed in forward combat positions. But the top brass had decided to allow women to try out for the Army Ranger School because they wanted to test the capacities of women's bodies. If any candidates managed to qualify for the school, it'd be interesting to see if they passed. No promises were made to Captain Griest. If she got through, there was no guaranteed position in a combat unit, just permission to wear the Ranger tab on her uniform.

The man in charge of training Griest's class—Sergeant Major Colin Boley, who'd been through fifteen deployments over fifteen years and doesn't suffer fools lightly—admitted that he didn't like the idea of a woman going to Ranger School. But he did want to see one woman make it through the physical assessment phase. Not for the advancement of women, mind you, but because then he wouldn't have to justify the school's tough standards: if a *woman* could pass, then surely the requirements should be acceptable for men.*

Out of four hundred members of the coed class, only nineteen women qualified. Most of those dropped like flies at the start of the course. By the time Captain Griest was clinging to the side of the mountain, she was one of only three women still going.

Knowing that she had a lot to prove was a big part of what got her that far. The tests she'd had to pass were designed for the male body. She'd had to use a humongous amount of upper body strength. She'd had to do a lot of things very quickly. She'd had to regularly demonstrate explosive muscular strength. But once she was on the mountain, what Griest really had to do was survive.

When it comes to large tests of endurance, the playing field between the sexes seems to even out. In fact, female bodies regularly win: Female runners typically log faster speeds in the longest ultramarathons.† Martin Strel aside, many of the world's champion long-distance swimmers are also women.‡ Some of that has to do with the fact that women have more subcutaneous fat, which is more buoyant than muscle tissue and helps with insulation. It's also a very useful store of energy when muscles use up their sugar reserves; as we've discussed, women's bodies are better at dipping into those fat backups than men's. But fat isn't the whole story. Female bodies may be innately better than men's when it comes to long-term, grueling tests of endurance. Despite less muscle overall, the muscles we do have give us an advantage.

* After the Vietnam War, the American military shifted to an all-volunteer force, and lately it's been having problems keeping up its numbers. In 2018, the army fell more than seven thousand short of its recruitment goals, and of those it did recruit, many needed waivers of existing standards to enlist (Phillips, 2018). The debate about whether to "soften" standards isn't something just the Special Forces are wrestling with; it's something the entire American military is asking right now.

† It's been quantified, in fact: while men are nearly 18 percent faster than women in 5K races, they are only 11 percent faster at marathons, 3.7 percent faster at 50 miles, roughly even as they approach 100 miles, and then women routinely outpace the men at races 195 miles and up (Ronto, 2021).

‡ I'd argue being born in Slovenia in 1954, on its own, is enough to boost a person's psychological endurance—Strel's first river swim, the Krka, was in 1992. The water ran southeast, right to the border of the new Croatia. His adipose tissue was also probably useful in this matter.

Skeletal muscles are made up of large bundles of fibrous tissue. Think of them like twitchy rope, all packed together and anchored to the bone by ligaments. These fibers are divided into two primary types: fast-twitch and slow-twitch. Fast-twitch fibers contract very quickly and generate a lot of power, but they tire easily. Slow-twitch fibers contract more slowly but are much slower to tire, with an increased aerobic capacity. Sprinters have a lot of fast-twitch muscle fibers. Marathoners have a lot of slow-twitch.

The muscles that anchor your lower spine to your lower back, your hips, the top of your buttocks—those are slow-twitch fibers. They work all day long to hold you up, fighting gravity in order to keep you from collapsing in a heap on the ground, whereas your jaw muscle is both the strongest muscle in your body and, no surprise, predominantly fast-twitch. You didn't evolve to *constantly* chew.

We've managed to learn a heck of a lot about slow-twitch and fast-twitch muscle fibers by hurling human bodies into space. As soon as astronauts leave Earth's gravity, their muscles begin to atrophy. That's why if they're staying on the International Space Station, they have to do grueling daily workouts on space treadmills daily. From studies published in the late 1990s and early years of the twenty-first century, both patients on hospital bed rest and astronauts returning from the ISS had significant muscle atrophy. But unlike hospital patients, astronauts also have *conversion*, wherein muscle tissue shifts from slow-twitch to fast-twitch. If you're not constantly asking muscle fibers to work, the way slow-twitch muscles do when we walk around in Earth's gravity, the muscle will optimize for fast-twitch fibers. This is true of both men and women astronauts, but women start from a different baseline, given that adult women's muscles tip toward slow-twitch. We don't know if that's because today's women don't usually *try* to do things that require explosive strength, or if, by nature, women's bodies are better built for the long haul. The data so far imply it may be innate: In one recent study, 75 percent of "untrained" women—that is, women who'd never undergone any sort of weight-training regimen—had significantly more slow-twitch muscle than fast-twitch. For untrained men, that balance is more even.

To know why that baseline matters when asking whether Ardi or her male companions were better at walking upright, just look down at your own legs. The muscles that run over the tops of your thighs are mostly the quadriceps: two long, bulky ropes responsible for hiking your knee toward your hips. Unless you live in a very hilly place, these don't get worked nearly as often as the muscles on the backs of your thighs—the hamstrings, which straighten the leg. Think about the mechanics of it: You don't have to lift your foot very far from the ground to walk forward. But the hamstrings and glutei maximi (your butt muscles) hoist your entire body forward over your foot each time you take a step. If you're running, that action is even more pronounced.

That difference in use also shapes what these muscles are made of. The quads tend to have more fast-twitch muscle, good for explosive movement. The hamstrings, meanwhile, tend to have more slow-twitch fibers, powering more fluid movement over a much longer period.

Soccer players, who have to constantly jog over the field but *also* kick and sprint, do a lot of explosive movement. Unsurprisingly, they usually have legs like tree trunks: they've developed the fronts and backs of their legs fairly evenly. Competitive marathoners, meanwhile, have rather spindly legs with highly pronounced buttocks and hamstrings. Sprinters have much thicker hamstrings *and* quads than most other runners, as do hurdlers who leap over obstacles—again using those fast-twitch muscles (mostly on the front of the leg) for explosive movement.

It's harder to ask your quads to do endurance exercise. You can certainly force them into it—climbing uphill a lot, for example.* But given the choice, the back side of the lower human body is better able to deal with feats of endurance than the front.

Of course, we don't have any of Ardi's leg muscles to do direct comparisons, but from what we know about our own, they probably had a similar sort of balance between fast- and slow-twitch

* On average, Swiss and San Francisco–based people usually have bigger quads than their flatland counterparts. The frequency of regular, slow hill climbing garners more endurance in those quads than, say, weighted squats—much to the chagrin of gym-prepped tech bros, who emigrate from places like Boston and New York and think their workouts will prepare them.

fibers. Or at least, more similar to ours than to a chimp's. Her shoulders were much stronger than a modern woman's. And her lower back probably didn't have as much slow-twitch fiber, because she spent much more time in the trees than we do. But to keep those legs moving on the ground in an upright position, her leg muscles were likely already moving toward endurance, and her female-typical metabolism might have given her long-distance walking skills an edge over the male *Ardipithecus.*

So at this point, it seems women are at least as good as men, and may even be innately *better,* at hard-core muscular and metabolic endurance. And there's one more thing to consider: We may be better at dealing with muscular tissue damage than men. Women recover from exercise more quickly than men do.

Whenever you use a muscle in difficult exercise, you damage it a little. That's a big part of how muscle tissue "bulks up." By putting strain on your skeletal system, you increase calcification at the anchor site and you also create micro-tears in the muscle itself. The tissue quickly inflames, flooding with blood and fluid and all the little microscopic "helpers" that repair damaged tissue. Nearby muscle cells get the signal: *Better proliferate so we can handle this the next time.* When the muscle heals, it comes back stronger, more capable, less quick to fatigue. In other words, lifting weights is a careful way of beating up your body. You can certainly push it too far—there is such a thing as serious muscle tears, and bone breaks, and you definitely don't want that to happen. But in modern, industrialized societies, underutilizing our musculoskeletal system is a much bigger problem.

Damage and healing are part of how muscle and bone do their job. That's universally true of both men and women. But how women's muscles go about the whole business is a bit different from men's.

Immediately after exercising, women lose more strength in the relevant muscles than men do. If you've ever tried Pilates, you might recognize the "jelly legs" feeling after you've done a session. But we recover much more quickly than men do. In studies from 1999 and 2001, some men took more than two months to fully recover the strength they'd lost from an elbow flexion exercise. They weren't aware of it, though—subjects *reported* feeling normal. The only way to gauge the truth was to ask them to perform the same exercise at the same weight and tension. And they couldn't. Women—radically more likely to lose strength right after the exercise—recovered much more quickly. That's a matter not of being stronger or weaker but of metabolism and tissue repair.

In the short term, men can do more "strong stuff" with their muscles than women, but it will hurt them more in the long run. Women can also do strong stuff. We may need to stop sooner than men, but once we do take a breather, we can go at it again before they can in a similar situation.

Coaches can run men into the ground, in other words, but then they have to bench them. Women, meanwhile, have to hit the bench for a rest sooner, but then we can go back onto the field.

And that's precisely what Captain Griest did, over and over and over.

WOMEN AND WAR

Captain Griest was tired. She was also up to her neck in a Florida swamp evading enemy fire, poisonous snakes a few feet away. She was in the final stretch. But she had more reason to be tired than some of her peers: she'd been "recycled." That's what the Army Ranger School calls sending candidates back to redo the part of the test they fail—a privilege awarded largely by having positive peer evaluations. So Captain Griest was tired in no small part because her peers respected her enough to let her repeat the process of systematically destroying her body in combat scenarios. To be in the swamp at the end, in other words, was, in sum, a hell of a thing.

This practice of "recycling" candidates was in place before the first coed class of Ranger School in 2015. Many of the men had been recycled, too. But her experience was particularly intense. Captain Griest started in April 2015 and finished in August, so she made it through four months' worth of grueling tests, sleep deprivation, and near starvation, not the usual two.

In a typical class, 34 percent of Ranger candidates will have to recycle at least one phase of the course. But Captain Griest faced the worst case: she had already been there for six weeks when the commander offered her the option of starting again from day one. There would be no time for rest or recovery. She had to either redo the entire physical assessment again from the very beginning or quit.

Knowing the army might not give her or any other woman another chance, Captain Griest accepted the challenge and went on to complete the entire course without any more do-overs. She even finished second out of the entire class for the twelve-mile "ruck" (a difficult hike with a heavy pack).

By the time she was wading through that swamp at the end, Captain Griest knew there was a lot more at stake than whether women became Rangers. Women being Rangers needed to be a good thing on its own merits. In life-or-death situations, the integrity of a combat unit *matters.* If someone goes down, someone else needs to step up, which means that in principle every member has to be able to do everything the group needs to do. This is why the question of integrating women into combat troops isn't just a matter of combating sexism. Lives are on the line. Captain Griest needed to keep an eye out for snakes not only to avoid danger for her own body but to keep that body available to help others.

There are things about mixed-sex groups that are very hard to quantify. For example, group "bonding" has far more to do with culture than anything physiological. There's been a lot of lip service paid to the idea of a combat group's "brotherhood"—that necessary, ephemeral social bond that lets members of a group rely on one another in life-or-death situations. A lot of people were concerned about what adding women to the front lines would do to that bond.

Captain Griest's peer evaluations show that her team had nothing but the utmost respect for her. Many said they'd happily trust her with their lives. Despite this woman having carried those men on her shoulders through crappy terrain, she made a point of dressing in the barracks separately from the men—catching a glimpse of her naked female body was still taboo. This is common in mixed-sex scenarios: when I spoke with my cousin, a former tank platoon leader and twenty-six-year veteran army officer, he said he'd seen women soldiers using ponchos to change clothes and urinate when privacy wasn't available. He also said undesired public nakedness in general tends to lower morale, but he worried that could be especially true in mixed-sex groups. The troubling idea of a naked woman's body wasn't only on my cousin's mind: When a fellow soldier who'd gone through the course with Captain Griest wrote about the experience for his peer review, he made an enthusiastic report of her battle readiness. He also took pains to mention where and how Captain Griest had changed her clothes.

But the men eventually got over the fact of her female body. They got over the two remaining women candidates passing them in the urinals, too. They'd be peeing, and the women candidates would simply walk straight by to the stalls. Brotherhood, it seems, is also made of shared stress.

So maybe it really comes down to what today's combat environment requires. What are soldiers on the front line generally asked to do? From what I've been able to learn, they need to handle odd sleep schedules. Though rations are generally on hand, they need to be able to handle varying availability of food and water. They need to be able to move equipment from one place to another in challenging landscapes. They also need to be vigilant for longer periods than normal life requires. And they need to make fast, rational decisions under extreme duress.

Some of what that list requires has to do with metabolism, body size, and musculoskeletal strength. The rest of it really has to do with psychological readiness. As Army Ranger graduates are all too willing to confess, beyond a basic physical readiness, Ranger School is meant to test the *mind:* Your grit. Your resilience. Being able to think with any amount of clarity when you're really, really tired. All the candidates who enter the course are physically fit. But not all of them have the same sort of mental stamina.

For example, Captain Griest and her fellow recruits had to haul a large machine gun up a sloppy hillside. The guy who'd been carrying it was starting to drop. His muscles were giving out. She offered to carry it for him.

Part of that had to do with her psychological resilience. And maybe another part had to do with how much was on the line as a female recruit. But she also might have been able to carry the gun the rest of the way because she was a woman. Supposedly, she even did it with a smile. When the man she relieved wrote his evaluation of her (all Ranger peers have to write such evaluations), he said he was particularly struck by how *enthusiastic* she was in that moment. There he was, completely broken, and she was practically chipper.

That's something very few of the military debates about women in combat consider: that female bodies may bring key *advantages* to combat groups. If you control for height, weight, and body fat percentage—and the simple fact that joining a volunteer army is naturally self-selecting—comparing male and female soldiers' general strength may come out a wash. But if a mixed-sex combat group has some bodies that are particularly good at explosive strength and others that are particularly good at endurance, would that group be more battle ready than a group composed of only men?

The answer would likely depend on what sort of combat scenario the group was facing, and military strategists would be better able to answer than me. But I can say that when my brother, a journalist, was embedded with troops in the Middle East, he told me how weirdly *bored* they were most of the time. Quite a lot of modern warfare has to do with simply holding an uncomfortable position. In most of today's conflicts, soldiers aren't really asked to march long distances carrying heavy loads. These days, American soldiers on the front lines mostly need to get somewhere, secure the area, stay there, and stay *awake*. They have to deal with the stress of sleep loss, monotony, muscle endurance, and the sort of neurological fallout that comes from having to be vigilant in a dangerous environment for long periods of time.

Female bodies are pretty good at that. It's not that females should *replace* males in combat roles. Rather, it may be silly not to take advantage of what female bodies could add to a group in combat situations. The point for any military strategy is to *win* with as few casualties as possible. Some advantages gained by including female soldiers in combat missions could be physiological. Others could be psychological.

When the Kurdish Peshmerga retook Sinjar from ISIS, cutting off a critical supply line on Route 47 between Syria and Mosul, women soldiers were part of the winning army. *Peshmerga*, in Kurdish, means "one who stands in front of death." Though their numbers are small compared with the men, Kurdish women are allowed to join the Peshmerga. And they have. They fight, and they win. They believe ISIS fighters fear death at their hands, worried that if they're killed by women, they won't be allowed to enter heaven. "It's a weapon for us," one female Peshmerga fighter told a Western journalist. "They don't like to be killed by us."

That isn't true—ISIS believes that all of their "martyrs" go to heaven, whether killed by men, women, or their own explosives in a suicide mission. But the idea took hold among the Peshmerga, and it emboldened them, men and women alike. They tell stories about a "tigress" sniper they call Rehana who's out "hunting" ISIS men, robbing them of paradise. *She's killed a hundred of them. Oh? I heard two hundred.* Eyes widen. ISIS, for their part, were threatened enough by the idea of Rehana to pretend they'd caught and beheaded her, posting photos on Twitter in 2014 of some stupidly grinning, dust-stained man holding a woman's severed head.

But none of these things are true. There are, indeed, excellent women snipers among the Peshmerga, and there are, indeed, women beheaded (and raped, and tortured, and enslaved, every single day) by misogynist terrorist groups like ISIS. But Rehana is a myth. It started with a photograph of an attractive Kurdish woman in military gear. It rapidly spread across Twitter. But she wasn't a sniper at all. In fact, her name probably wasn't even Rehana; it's not a common Kurdish name. A Swedish journalist did meet the woman the day her photograph was taken—August 22, 2014—and talked with her, briefly, but never got her name. This is what he remembers: The color of her eyes. Her hair. That she said she'd come to help keep the peace in Kobani, a town on the border of Syria and Turkey. ISIS besieged the city for the better part of a year, but the Kurds controlled most of it throughout the siege, which was lifted in January 2015. The journalist also learned that she'd been a law student in Aleppo, but when ISIS killed her dad, she decided to volunteer. The journalist never got a second interview and has no idea what happened to her since then. She may be a refugee in Turkey now. She may still be fighting. She may be dead, like so many others. If she's not, she has obvious motivation to remain quiet about her own fate: once ISIS *pretends* to behead you, it's safe to assume there are a number of people who would welcome the opportunity to finish the job.

Still, Rehana the sniper tigress is an effective story: one of many countermyths about women's power—tiny, brutal fairy tales—that stand in opposition to myths about women's god-sanctioned subjugation. If the women weren't there fighting, this story wouldn't have been told, inspiring the troops to fight harder, weakening the enemy's psychological reserves. It's a weapon made from the very *idea* of a woman.

And that, in the end, may be part of what ISIS (and certain American military figures) are afraid of. Maybe the debate about women in combat is not about what men's and women's bodies can or can't do—about the strengths and weaknesses of our sexually dimorphic musculoskeletal system, our metabolism, or even our psychological grit. Maybe it comes down to the idea of women's bodies in the world—what they're supposed to do, what they aren't, and how they serve as a counterpoint to the idea of Manhood.

After 162 days of mostly hell, Captain Griest completed the course. She'd carried the men. She'd carried the machine guns. She'd gone up the mountain and come down. Twice. She was, of course, ecstatic. She was also very, very tired. And more than just about anything in the world, she was probably looking forward to a hot shower to wash off the sweat and the mud. And sleep. She was certainly looking forward to sleep.

Having a woman pass the test was a huge moment for the U.S. military. By the end of 2015, Secretary of Defense Ashton B. Carter recommended that all women have equal access to combat roles throughout the military. For the most part, this move was welcomed, in no small part due to Griest's performance in the Ranger tests.* Even the Navy SEALs are welcoming women who are able to pass their qualifiers—a set of tests considered by many even more difficult than the Rangers', perhaps because unlike the Rangers SEALs have to be able to *hold their breath* underwater while performing difficult physical feats of strength and flexibility. But women will apply, and eventually some of them will pass, and then that threshold will likewise be met. For her part, in 2016, Captain Griest went on to become the first female infantry officer in the U.S. Army.

Predictably, there's still the usual sort of worry about "lost morale" in the military should many women find their way into attack forces. But recent studies have shown—including within the U.S. Marines, a group that especially protested the change—that mixed-sex combat groups exhibit high levels of group cohesion and loyalty. In fact, mixed-sex military groups' feelings of "belonging" are as high as, and in some cases higher than, single-sex groups'. What's more, the rate of sexual assault is no higher in mixed-sex groups than in male-only ones.†

* One other woman, Captain Shaye Haver, completed the course alongside Captain Griest, receiving similar praise from her peers. Lieutenant Colonel Lisa Jaster finished a few months after them. Jaster was thirty-seven years old at the time and the mother of two young children. All three women served in Afghanistan. Captain Haver served as a helicopter pilot and led the military honor guard that carried Ruth Bader Ginsburg's casket when the Supreme Court justice was lain in state at the U.S. Capitol in 2020. As of April 2020, fifty women had graduated from Ranger School.

† Same-sex rape does occur in the military and, like all sexual assaults, is underreported. The main concern here was that the presence of opposite-sex members in a combat group might make rape *more* common, given that most soldiers are heterosexual. This was not the case. That may be because, as clinical psychologists have been saying for years, human rape is often less about sex than power. The fact that rapes don't increase in mixed-sex combat groups may also simply be that being in possession of a vagina does not automatically make people with penises attracted to you. The reverse is also true: I can say, without hesitation, that I am not attracted to the majority of people with penises *or* vaginas. It's not that I'm picky, exactly—there are 8 billion people in the world. If I wanted to have sex with *most* of them, I'd have a mental disorder, not to mention the obvious problem of simply not being alive long enough to accomplish such a thing. But even of the astonishingly smaller number of people I'll actually meet or even *see* in my lifetime, they, too, are mostly people I'm *not* sexually attracted to. For most healthy

It's hard to say whether that last one is truly a win. The entire American military has a problem with sexual abuse and assault, so knowing it's evenly distributed despite some groups' regular exposure to mixed-sex teams is disheartening. But at least they can't blame it on the mere presence of a woman.

And should the machine gun start slipping, deep in the suck, in a few years a woman might be there to take it up.

---

minds, sexual desire is, by its nature, both rare in occasion and notably limited in its targets.

Homo habilis

CHAPTER 5

TOOLS

I would rather stand three times in battle than give birth once.

—EURIPIDES, *MEDEA*

THUS SPOKE ZARATHUSTRA

The Dawn of Man. A sallow light rises over the land. Stanley Kubrick's shot pulls in on a band of male hominins gathering around a watering hole.* Their bodies are lean. Their fur is long and black. No women, no children—or at least none easily discerned. The earth is likewise barren: patches of tan rock and scree giving way to dusty savanna.

The males drink the brown water, nervously scratching their fur. A neighboring band of hominins comes over a ridge. They screech and call and chase the others off.

The scene shifts to a young male hunkered alone near a skeleton. He reaches out an arm and pulls a large bone from the pile. He stares at the bone for a moment and then starts beating the

* Actually, British mimes dressed in ape costumes. The film is *2001: A Space Odyssey*, one of the most critically acclaimed films of the twentieth century. And the music is Strauss, loosely interpreting Nietzsche, in much the way we all seem to keep wandering the woods for the last two hundred years with a handful of German men.

ground, slowly at first, then furiously. Ancient Man has invented the first weapon.

The first group returns to the watering hole, chasing off their competition, except for a single opposing male, who dares to cross the water. One of the bone-wielding hominins strikes the challenger over the head. Others join in, taking turns beating the fallen body. The unarmed members of his troop look on, shocked, then run, leaving him to his fate. The primordial inventor throws his bone into the air. Kubrick traces its rise, and when at last it reaches its apex—high against the clear sky—he cuts to the future: a spaceship suspended in orbit. And *The Blue Danube* begins to play.

This is the story of Tool Triumphalism: man invented weapons, claimed dominion over his peers and the rest of the animal kingdom, and all our achievements flow from there. From bone cudgel to spaceship, from the Stone Age to now, Kubrick wasn't the only one to tell this story: the clever ape—always male—picks up something from his environment and uses it to hunt, to murder, to dominate Earth.

We still tell ourselves that this ability is what makes us *human*, what separates man from beast. We even tell ourselves this special cleverness is why we've succeeded as a species—that our golden ticket was crafted with hands that could *craft* and a brain that could design things.

And maybe that's true—but not in the way you might think, nor for the reasons most assume are the ones that matter.

LESS TRIUMPHANT, MORE TERRIFIED MACGYVER

If you had to guess which tool-inventing ancestor Kubrick was going for in *2001*, the safest bet would be *Homo habilis*, an Eve from roughly two million years ago. The face looks right. The behavior fits early hominins, too. But tools aren't unique to human ancestors. Our first tool users probably weren't male. And our most important early invention probably wasn't a weapon.

Far from some great symbol of human uniqueness, tool use is a convergent trait. Lots of intelligent problem solvers do it. They don't even have to be mammals. The octopus uses tools with its tentacles, and it's more closely related to a *clam*. Crows are avid tool users. They don't even have hands.

The early hominins Kubrick portrays mostly ate grasses and bugs and fruits and tubers. Like other primates' today, our ancestors' first "tools" were probably rocks to break open nuts and sharp sticks to dig up some kind of ancient turnip. But Tool Triumphalists, like Kubrick, want the "Dawn of Man" to be the moment we started using tools as weapons to hunt animals and beat the crap out of each other. Fine, except for one more catch: the first such weapons might well have been invented by a female.

Right now, somewhere in Senegal, a chimp is hunting. She's carrying a spear in one hand, made from a branch she snapped off a young tree, then took some time to prepare, pulling away all the leaves and offshoots, then chewing the end to a point with her powerful teeth. Her offspring clings to her back as she moves through the grass, hanging on to her long black fur. The kid's been suckling for months now. The mother is lean and hungry. She's looking for meat.

She's learned that during the day bush babies—tiny, small-brained, big-eyed primates—tend to sleep in the hollows of trees. When she finds one, she stabs it with her stick. It wakes up, snarling and scratching. It's too small and weak to be a mortal danger, but it could definitely wound her, and it might kill her offspring. Better to use a spear, which keeps it at a safe distance. She stabs the bush baby again and pulls it out of the tree only when she is sure that it's dead.

When male chimps go hunting, they sometimes use spears, but their own bodies, bigger and stronger than the females', are often weapon enough. Even if they're injured as a result, no offspring will starve. From an evolutionary point of view, their injuries aren't as costly, because males aren't caretakers in chimp society. Generally speaking, innovation is something that weaker individuals do in order to overcome their relative disadvantage. As a primatologist in Kenya told me years ago, "Women do clever things because we *have* to." She was talking about the female primates she'd observed being clever, but of course she meant human women, too. From a scientific perspective, we female primates have more to gain—and more to lose. Most of us are smaller and weaker than the males.* Given that our bodies are the ones that have to build, birth, and nurse babies, females also have more urgent food and safety needs than males. Simple tools were the easiest way to meet those needs. If the females in question were also good problem solvers—as all higher primates are—then it makes sense for females to be inventors, though that's not the picture we usually paint of our ancestors.

Habilis—"handy man," or in this case "handy woman"—lived in the grassy highlands of Tanzania between 2.8 and 1.5 million years ago. This Eve of tool making was a pinch over four feet tall, with long arms and strong legs and a brain around half the size of ours. We have no idea how furry she was, nor how fatty her breasts. But she was brainier than australopithecines like Lucy, and overall more like modern humans. She was an opportunistic eater, as we are, happily snacking on all sorts of food. Her jaws were strong, and her tooth enamel was thick, but she wasn't in the habit of cracking hard nuts or tubers with them. Why would she when she had handy stone tools to break open (and break down) tougher fare?

In the places where we've found her fossils, we've also found hundreds of stone tools. In the Olduvai Gorge in Tanzania, archaeologists unearthed so many fossils and tools that the Oldowan tool technology was named after it. The Oldowan tools are one good reason we should think of Habilis as an Eve of tools. Though chimps use tools today, and Lucy also used primitive stone tools, the Oldowan style—adopted by later australopithecines and finally by Habilis and *Homo erectus* after her—was our first advanced tool technology. Our Eves deliberately shaped these large pebbles, carefully chipping off bits of a stone at just the right angle to make axes or scrapers or awls. In the beginning, she used stones that were already pretty close to the shape she wanted, mostly river cobbles, already smoothed by water. Eventually she used rocks from miles away that, if hit just the right way, would flake into the specific shapes she was after. She could use one sort

* Though the differences are far more pronounced in nonhuman primates. More on that in the "Love" chapter.

of tool to dig up tubers, another to pound their fibers into something edible, and yet another to chop up grasses and nuts.

Habilis used the flaked-off bits, too. Longer, thinner, sometimes delicate looking, but tough as nails, such flakes let her do more delicate tasks: carving meat away from sinew, peeling fat from skin, delicately removing bitter parts of a plant to get at the good stuff. She used certain kinds of stones to cut off the juiciest steaks and others to break open bones to get at the marrow, which she sucked still warm from the animal.

If she could get to an animal that was still warm, that is. While Habilis loved a hot cut of meat, she probably didn't do much big-game hunting. Most of the animal bones scientists have found near her fossils and tools are from the beasts' extremities. She was likely a scavenger: a thief like a baboon or hyena, but much less dangerous. If some big predator had made a kill, she'd probably stay hidden until it had finished feeding, then run in to steal part of the carcass. Maybe she'd use her stone ax to hack off the lower part of a leg and then pick it up and run like hell. Habilis was by no means the top of the food chain. Like many hominins, she was often prey.

So her stone tools weren't exactly triumphant. No alien light shone in her eyes. Like the mother chimp hunting with a spear in Senegal, Habilis was simply a very smart primate using everything she could to survive. She walked through the tall grass in fear, clutching a rock ax and whatever bit of stolen meat she could find, baby in tow or even in arms.

Tool use is the first trait in this book that's purely a set of behaviors—not an organ, not neurological hard wiring, but something our Eves used their cognitive and physical abilities to *do* in order to change their relationship with the world around them.* Put it this way: paleo-archaeologists don't really care about rocks; they care about what rocks can tell us about the lives of the creatures who used and shaped them. Without a hungry person nearby, a fork is just a stick with some pointy bits—tool use,

* Most do assume tool use is a fundamental trait in the primate line, which does imply some sort of "hard wiring," but its arrival wouldn't be as obvious as the expansion of the visual centers of the brain.

in other words, is about the relationship between the object, its intelligent user, and the world in which both are situated. The study of ancient tools is always the study of ancient behavior. And for an evolutionary biologist, thinking about hominin tool use is a way of tracing changes in the habits and capabilities of all those pro-social, problem-solving hominin brains along humanity's ancestral line. Brains don't become fossils. But the artifacts of tool-using behavior can and do—particularly when they're made of rock and usefully situated near the fossilized bones of their makers, and even more so if they're near some obviously butchered bones. The reason any of us should care about Oldowan tools, in other words, is that they might be able to tell us something about the minds and social lives of our ancestors: how they made stuff, how they collaborated, how they overcame adversity.

That last one is particularly important. For every species that does it, tool use is fundamentally about solving problems. At the dawn of humanity, deep in the dry savanna, Habilis had a ton of problems. She had hunger. She had predators. Every morning she wrestled with the angels of death and disease and despair. She used her stone tools to help solve many of these problems.

But her biggest problem wasn't something she could throw a rock at. It was part and parcel of her own body. Evolution had dealt her a lousy hand.

THE HARD PROBLEM

A number of prominent evolutionary thinkers hem and haw over how it was that we hominins managed to succeed. It is an unlikely story. Aside from the usual suspects—stone tools, hunting, growing really large brains—one of the big topics is how vulnerable our babies are. They're needy not just as newborns but for an extraordinarily long time.

Therefore, in order for hominins to flourish, some kind of cultural revolution around child care must have occurred. How else, after all, would species with such needy babies survive? Chimp society is in no way prepared to deal with the sort of day-in, day-out labor involved in keeping human newborns and toddlers and kindergartners alive. The mothers would starve. The baby would starve faster. So some scientists argue for the invention of monogamy, as improbable as that is. Others say we came up with kinfolk eusociality—a kind of furry "spinster aunt." Maybe we even started alloparenting, as we still do now, with unrelated folk helping care for others' babies. Whatever the change was, many argue for it as the root of human culture: we're obviously more collaborative than chimps and bonobos when it comes to child rearing. We're also more social, if such a thing were possible, with highly specialized roles within our various human cultures to help those communities survive and thrive.

Regardless of how, exactly, our early child rearing changed, it clearly did. The thing that's usually left out of these arguments is what happens *before* our famously needy babies are born.

Species don't really get a harder problem than the one we have to deal with: We're really, really bad at reproducing ourselves—demonstrably worse at it than many other mammals. We're worse than most other primates. We're even worse than our fellow apes, whose bodies are so like our own we're called "the third chimpanzee." Human pregnancy, birth, and post-birth recovery are harder and longer for human females, leaving them significantly more prone to crippling complications. These complications can, and still regularly do, lead to the death of the mother, the death of the offspring, or both. And when these complicated reproductive processes don't kill a mother, they can render her infertile or deform the child. Most of the features that make our reproduction such a crapshoot were probably already in place by the time Habilis arrived. And they only got worse for her descendants.

In evolutionary science, a factor that directly affects whether an individual's genes are passed on is what you call a "hard selection." You can limp around on one foot. You can see with one eye. But if you can't have babies, your lineage is headed for extinction.

And yet, somehow, there are 8 billion *Homo sapiens* on the planet right now. That's not just impressive—it should have been impossible.

There are many other species that are terrible at reproducing themselves. Of the other ones that are still around, they're either sequestered in a weird little ecological pocket or sliding toward extinction: The white rhino. Giant panda. Northern hairy-nosed wombat.* That should have been the fate of the hominins: relegated to being a curiosity in some other creature's zoos.

If you want to talk about how humanity managed to survive and thrive, you need to talk about what it takes to make those babies in the first place. If Habilis sucked at reproduction even a fraction of the way we do, that was clearly the most important problem she had to solve. I propose that she did it with our ancestors' most important invention: It wasn't stone tools. It wasn't fire.† It wasn't agriculture, or the wheel, or penicillin. The most important human invention—the very reason we've managed to succeed as a species—was gynecology.

And we're still using it. We use it in every single contemporary human culture. From the records we have—and there's a surprising number, ranging from written accounts to ancient specula made of iron—we did it in every known historical culture, too. We've done it in various sorts of ways, scaffolded by various belief systems, but all human gynecological practices have some very basic things in common: They try to preserve the life of the mother and, if possible, the child. They try to prevent and treat excessive uterine bleeding. They try to prevent and treat bacterial infection.‡ They tend to guide the intensity of the mother's labor

* Each of these species is endangered because of habitat loss and poaching. But while other such species do well in captive breeding programs, these guys are going the way of the dodo. Why? They suck—I mean really, really suck—at having sex and making babies. Famously so. Rhinos of various species have various reproductive problems, all of them bad (Pennington and Durrant, 2019). The wombats in question normally reproduce only every two years, only make one baby at a time, and get stressed out by having others around (Horsup, 2005). Giant pandas seem to have largely forgotten how to have sex at all. Zoos are making them watch panda porn (Wildt et al., 2006). It only sort of works.

† Fire came into widespread use some half a million years after Habilis chipped away at her rocks (Berna et al., 2012).

‡ Though the users may not be aware that's what they're doing. That isn't meant to be patronizing in the slightest—whatever worldview one happens to have, biological outcomes are what matter here. For instance, you can preferentially use copper tools without a germ theory of disease. You don't have to *know* that copper

efforts to coincide with the dilation of her cervix. And finally, in most cultures, both contemporary and historical, they come with a wide array of techniques, pharmacology, and devices to intervene in a women's fertility: enhancing, or preventing, female reproduction when desired. Because there's no more reliable prevention of pregnancy complications than the prevention of pregnancy itself.

This continually evolving body of medical knowledge and practices is what I'm calling, for want of a better word, "gynecology."* It is absolutely essential for our species' evolutionary fitness. Without it, it's doubtful we would have made it this far.

This may be hard to accept. After all, women become pregnant and give birth every day. Some women die. Some babies die. Some women become infertile. Most of us don't. So it can't be that big a deal, right?

Wrong. The effect of gynecology is huge, especially if you're talking about taking a reproductive system like ours in its ancient state and creating enough of a population to successfully migrate across most of the planet, withstanding repeated periods of starvation as they adapted to different environments. As our populations were repeatedly hit with one ridiculous challenge after another, our Eves would have needed to regrow a viable population. That's the thing about migration and adaptation: you need *enough* of a subsequent generation to carry on your innovations, whether they're physiological or behavioral. You need enough kids, in other words, to buffer the random sprees of death that were part of the ancient hominins' changing world.

But how are you supposed to do that when your reproductive system is inherently dangerous and frequently fails?

Other primates—creatures whose bodies are to this day a lot like Habilis's—have a much easier time giving birth than Habilis

---

isn't a bacteria-friendly surface in order to notice that using copper in a birthing room seems to help new mothers survive better. The same can be said for having local traditions for serving pregnant women well-cooked food and boiled water and keeping them away from diseased members of the local community.

* Personally, I'd rather call it something like "the study and practice of how to survive the entirely stupid human reproductive system and still make it as a species," but that's too long.

would have. In the wild, a chimp female is very unlikely to die because of pregnancy-related complications. Among wild chimps, maternal death of that sort is so infrequent that primatologists haven't even agreed on a representative number. It's probably quite low.* Human women, meanwhile, hover between 1 and 2 percent. If that still seems low, remember that's the maternal *death* rate: the percentage of us who actually die because of pregnancy and birth within a narrow window. The pregnancy and birth complication rate—which, again, can readily stop a genetic line in its tracks—shoots up to a full *third* of human women. Fifty-eight percent of American women have continuing health problems associated with the pregnancy more than six months after giving birth; the rates worldwide are higher. In Nairobi, maternal complications are so common that some clinics hang large signs advertising treatments for "Fistula" in big, bold type, visible from down the road. An obstetric fistula happens in most cases because of a prolonged and difficult birth in which the baby's body puts so much pressure on the pelvic tissue that it tears a hole between the vagina and the bladder or the rectum, rendering the woman incontinent.

There are two likely reasons why human reproduction is so dangerous. First, the risk of internal bleeding. Our deeply invasive placentas can rupture veins and arteries (rare), can separate from the uterine wall before it's time (less rare), or can hemorrhage during or just after birth (still rare, but one of the leading causes of maternal death).

The second reason our reproductive system causes so much trouble is what's called the obstetric dilemma. Compared with other apes, human women have a really small pelvic opening and human babies have a really big head. When humans evolved to walk upright, the structure of our pelvis had to change, which led to a smaller pelvic opening and birth canal. For Ardi, it probably wasn't such a big deal, but for Lucy more so, and by the time

* The primary reasons chimps are an endangered species are that they compete with humans for territory and that poachers profit from chimp bodies for trophies and bush meat. While it was still legal to do so, primate research centers in the United States were very successful at breeding chimps. The problem is not with the chimp body plan but with the world chimps normally live in.

Habilis and her peers came around, it had become a real issue. It's hard to fit a watermelon through a lemon-size hole.

Births would have taken progressively longer. Today's American woman averages six and a half hours for labor. Chimps labor about forty minutes. Eves like Habilis presumably would have been somewhere between the two. While a chimp's cervix needs to dilate to only 3.3 centimeters, ours needs to get to ten centimeters. And, *man*, does it hurt. It's also ridiculously risky: six and a half hours of labored heart rate, coursing adrenaline, and downward pressure.* Plenty of time for the placenta to start detaching before it should, for blood vessels in the pelvis to strain and tear, or for a hungry pack of predators to attack you.

Once the cervix is dilated, things get even crazier. The modern human birth canal sort of *twists*, wider in some spots and narrower in others, which means a newborn actually rotates ninety degrees in the middle of the vagina while being born. That's another gift from hominin evolution: big heads need big shoulders to brace developing neck muscles. The newborn head is smooshable, thanks to all those flexible skull plates. But the wide clavicles are rigid, so the shoulders have to come through the pelvic opening *sideways* after the head's made its way out. It's push, turn, and push again.†

In other primates it's a straight shot to the finish line.‡ So, no

* Twelve to eighteen hours, if you're a first-timer.

† The peculiar shape of our birth canal is likewise a gift from evolution: not only is it narrower because of walking upright, but the pelvis itself is oddly arranged, such that the upper part of the birth canal is round, while the lower part is distinctly oval. This is likely because having a differently shaped pelvis would destabilize the pelvic floor, requiring an even greater curve to our spines to maintain stability (Stansfield et al., 2021).

‡ There are a couple of exceptions here, but they're very rare. Squirrel monkeys, for example, share the obstetric dilemma, and they are rapidly going the way of the panda (Trevathan, 2015). Nearly half of all their pregnancies end in the death of the offspring, and they give birth to only one pup at a time (ibid.). Certain macaques, however, show similar covariation between fetal skull plates and pelvic arrangement to our own (that is, big-headed women usually give birth to big-headed babies and are *also* more likely to have pelvises that better accommodate such babies), which may imply that primate birth in general involves a deep history of mother-offspring birth compromises, or at least in the strepsirrhines (Kawada

surprise, a chimp's delivery, as opposed to the labor that dilates the cervix, takes only a few minutes. Ours regularly takes as much as an hour. And if the baby gets stuck . . .

From evaluations of average skull size, neonatal shoulders, and pelvic openings in hominin fossils, it looks as if our fetuses started coming out wonky as early as Lucy. By Habilis, fetal skulls and shoulders would have been a major problem. Labor and birth would have taken longer. Gestation was probably getting longer, too: modern human pregnancies take about thirty-seven days more than you'd expect for an ape of our size. In other words, as our Eves evolved, the whole process of having babies, from top to bottom, became more dangerous and difficult.

Let's go back to that number: 8 billion human beings. If you looked only at the raw mechanics of reproduction, you'd never think the hominin line could arrive at that number. There are fewer than 300,000 chimps in the entire world and fewer than a million olive baboons, even though their bodies are better suited to rapid population expansion. But here we are. Billions of us.

It's generally true that necessity is the mother of invention. We know Habilis mothers faced obstetric challenges, so we also know they had the *need* for a solution—likely something only a very social, very smart, problem-solving tool user could come up with. The biggest clue to Habilis's potential for gynecology is actually those famous Oldowan tools. Mapping those caches—how far they spread, how consistent the tech, how often they're found with fossils—is the best way we have of tracking how early hominins were sharing complex social knowledge.

These Oldowan tool users were individuals who spent a lot of time together. Flint knapping isn't fast or easy. It's something you need to learn how to do. So Habilis probably lived in collaborative groups, desperately trying to outlearn and outrun a world full of muscled, toothy things that were all too happy to eat them. When they weren't running, they were, now and again—painfully, and with difficulty—giving birth. And they were surviving, in no small

et al., 2020). But again, macaques, whatever type they may be, don't have the same issue with maternal mortality and injury that we do.

part because of the same sort of behavior that produced their stone tools: they were working *together*.

HELPING A SISTER OUT

The arrival of midwives is one of those moments in hominin history when we can truly say, "Here is when we started to become human."

But it's hard to know precisely when that happened, since the practice of midwifery doesn't leave a neat record the way stone tools do. It's also true that in order to do something like help someone else give birth, these Eves had to become a heck of a lot less chimpy than earlier ones.

No other mammals on the planet have been observed regularly helping one another through birth. Or at least, none we know of. Two monkey species have been observed assisting in a birth, but each case seems incredibly rare. One was a black-and-white snub-nosed monkey in 2013, but it was hard to draw conclusions since it was a daytime birth and usually they occur at night. The second, involving a langur monkey, was recorded in 2014—and if it hadn't been recorded, no one would have believed it.

Chinese primatologists had observed this group of langurs for years and saw that the females generally gave birth alone. But not this time. On a rocky outcropping, an older female monkey hung around a younger mother who was clearly struggling in active labor. The newborn came out halfway. The older monkey quickly pulled the baby out of the mother's vagina, held the kid for a minute, licked it, and then handed it to its mother. This may be the first clear evidence of active birth assistance in any mammal besides humans.

As a rule, evolution doesn't produce new traits from thin air. This is as true of behavioral evolution as physiological, particularly when you're talking about our pre-lingual ancestors. If midwifery was something Habilis used to her advantage, there would have been some precursors that created a foundation to build on.

But consider how *trusting* you need to be to let someone help you give birth.[*] Our Eves would have needed a social structure that rewarded helpful behaviors. Mothers could help daughters, sure, but for midwifery to become widespread, collaboration between members of a wider social group would also have been key.[†] Collaboration *over* competition.

Once ancient hominins were regularly gathering in this way—not just to sleep at night, but during the day—they started eating together. Sharing food is a big deal for primates like us. Food sharing is a big part of chimp social bonding, too—you don't let just anyone eat your banana. Habilis already had a significantly larger brain than our earlier Eves. Many think she used all that extra brainpower for keeping track of an increasingly complex social life.[‡]

But to invent gynecology, our Eves needed a cooperative *female* society. Females needed to be able to trust one another enough to be around each other at those critical moments of vulnerability: labor, birth, and early nursing. That might have been harder than you'd think. Our hominin Eves were similar to today's great apes. Because modern humans are most closely related to the chimpanzee and the bonobo, let's compare their birthing behaviors.

In contemporary chimpanzee societies, introducing a newborn to the group is a rather tense affair. After a female gives birth, she'll wait a bit, nursing her baby in those crucial early hours, staying quiet and away from the troop. Then she'll usually try to introduce the newborn to her closest allies first. If the alpha female isn't her dearest friend, she'll put off that introduction as long as possible. There are a number of accounts show-

* Or desperate?

† Many think that's the likeliest scenario for the very first hominin midwife: the mother would still be helping her own genes pass down by helping her daughter give birth to a grandchild, so if they lived in an ape society where mothers and daughters stayed together, that would have an obvious genetic reward. Chimp females, however, tend to leave when they come of age, seeking out new troops. When they give birth, they're nowhere near their mothers. One exception to this rule is if their mothers are high ranking in their home troops—it seems the social perks of having a powerful mom outweigh the risks of inbreeding.

‡ More on that in the next chapter.

ing chimp mothers with newborns desperately trying to protect the baby as they are being chased by groups of competitive females.

And well they should. Dominant female chimps are known to kill the offspring of females with lower status. Maybe they do it out of spite or maliciousness, but from a biologist's perspective it's probably because it helps them maintain their social position. They don't just kill the baby. They may even *eat* it in front of the crying mother.

It's incredibly hard to imagine human obstetrics developing from a social environment like that. But I suspect there's an easier path. And for that, we can look to the hippie side of our primate family: the bonobos.

Just over the river from chimpanzee territory, in an area rich with easy food, the bonobo goes about her days. Unlike chimps, where dominant males are a regular menace, bonobos are both matriarchal and strongly averse to violent conflict. They do fight. In fact, they fight all the time. They just tend to resolve such conflicts with quick bouts of sex. And somewhere in the middle of all that sex, there's one strict rule in bonobo society: nobody messes with the kids. If a troop member harasses or harms a juvenile, they're quickly reprimanded by nearby adults. So, unsurprisingly, the introduction of newborn bonobos to the social group isn't as big a deal as it is with chimps. But it gets better: In 2014, researchers in Congo were finally able to witness a bonobo give birth. She went into labor in the late morning, in a nest in a small tree, with *two other females* in the tree with her.

Because the nest was high in the tree, researchers weren't able to see what went down at the moment of birth. But one female seemed to stand guard while the mother labored, looking on with interest. And at some point, the second female joined the laboring bonobo in the nest. Did she help with the delivery? We don't know. We do know that all three females shared the placenta afterward, gulping down pieces of it. Afterward, the mother didn't seem stressed out about introducing the newborn to the rest of the troop. And why should she? Despite decades of careful field research, no dominant bonobo has ever been observed murdering the offspring of lesser females or committing that sort of cannibalism.*

That's not to say they're not capable of it. It just seems their particular social organization doesn't easily lend itself to it.†

Then, in 2018, researchers gathered three more observed cases of what might as well be called bonobo midwifery—this time in captivity, where observations were naturally easier (the bonobos were used to human beings being around, and the location of the births were more predictable and visible). In each case, other females gathered around the laboring bonobo, grooming her and standing guard. In a couple of cases, females even cupped their paws under the newborn as it came out of the mother, and again they all shared a bit of placenta as a bloody reward. This is, as the researchers note, entirely unlike the behavior of the chimpanzee, whether in the wild or in captivity, most likely—they plainly note—because chimpanzee society is male dominated, whereas bonobo society has strong female coalitions and is female dominated.

So maybe, in the evolution of human gynecology, early hominins were more like bonobos than chimps. Maybe Habilis had that sort of female social structure. We can't prove it. But from what primatologists have seen among extant ape communities, a more collaborative female environment would provide the sort of fertile social ground that could allow a creature like Habilis to invent a widespread culture of midwifery.

But the dawn of midwives wasn't the only thing in play for our Eves. There was another, wider foundation they were able to build on. Human "gynecology," at each stage of its evolution, also

* In 2010, a couple of German scientists did see a bonobo female eat a baby, but it was already dead (Fowler and Hohmann, 2010). A dominant female took the dead infant from the lower-ranking mother, began to eat it, then *shared* the body with other females. When they were done, they gave what was left—just a hand and a foot, connected by a ragged bit of skin—back to the mother. She draped the strange memento mori over her shoulder and walked away.

† There was, as it happened, quite a bit of female-female genital rubbing around the bonobo births, and yet more when the placenta was shared. Bonobos do that a lot, but especially when there's food involved. It's possible that bargaining of that kind—a bit of a nutrient-dense placenta in exchange for protective and assistive behavior—could be part of how hominin midwife culture got started.

includes many types of birth control, abortion, and other fertility interventions. Female reproductive choice is ancient.

A VERY SEXY ARMS RACE

While genes go about the business of trying to perpetuate themselves, female animals are also generally trying to stay alive. When it comes to reproduction, they want the very best sperm, from partners they prefer, at a time and in the circumstances they prefer. Males, meanwhile—who as a rule expend very few resources on the business of reproduction—are also trying to stay alive, but because reproduction doesn't cost them much, they're mostly trying to get their sperm into any female they can. And that means, for all intents and purposes, male and female bodies have been at war for hundreds of millions of years.

Consider the duck: Mallard ducks are constantly raping each other. Whole groups of males will trap and gang-rape a single female. As a result, over hundreds of thousands of years of evolution, female mallards started building "trapdoor" vaginas—oddly shaped, and full of twists and folds and pockets. When she has sex with a *desired* partner, her vagina unfolds, opening the path to the waiting ovaries. When she's raped, portions of her long, winding vagina will close off, trapping unwanted sperm in a side tunnel. After her rapists run off, her body will get rid of that sperm as best as it can. Sometimes she'll even tap her beak against her lower abdomen, helping expel it from her cloaca. The males didn't take this lying down. The mallard's penis coevolved with the female's changing vagina and now has a kind of corkscrew structure—presumably to try to sidestep the trapdoors.

You can see this sort of coevolution in all animals that reproduce with a penis inserted into a vagina. These evolve in lockstep. And because female bodies generally evolve in ways that benefit their owners, male bodies tend to evolve in ways that counter those measures. Thus, raping species' genitals are in a sexual arms race: the more common it is for a male to force copulation, the more likely the female will evolve various anti-rape mechanisms to try to prevent being fertilized by her attackers' seed.

Dogs have a knot at the end of their penis that swells and "locks" a female in place for a good half hour, making it hard for her to run away before the male has ejaculated. A male cat has prominent spines along the penis that rake the vaginal wall whenever he pulls back. This raking seems to help trigger ovulation, but it also appears—for the female, at least—to be highly painful (and this is during consensual sex). Meanwhile, the dolphin's penis can actually *swivel*, feeling around its environment—a bit like a blind tentacle—before hooking itself into a vagina. The whole business can get rather violent. In the wild, gangs of dolphin males can prevent a targeted female from surfacing to breathe, exhausting and suffocating her into submission, raking her with their teeth, taking turns pushing and grasping with their J-shaped penises from whatever angle they can.

And penguins, well, penguins are famously terrible. I'll leave you to the wilds of the internet to explore that further.

ON THE EVOLUTION OF CHOICE

So there is a war. A sex war. Some of it plays out in the external sex organs. Some of it plays out in deliberate behavior. Yet more goes on in the dark—in the quiet, violent bowl of a female's ovaries and uterus.

When a pregnant woman miscarries, what's happened is what doctors call a spontaneous abortion. Humans aren't the only species that do it. Abortion is common across mammals. Some of it is really "spontaneous," and some of it is more deliberate.

If you put a pregnant mouse in an enclosure with a male who isn't the father, she'll abort (this is called the Bruce effect*). The consensus is that this capacity evolved as a response to threat since male mice will usually kill and eat pups they don't recognize as their own. From the female body's perspective, why invest energy giving birth to pups the new guy will eat? Cut your losses and abort.

* So named for the scientist who discovered it, and not the influence of people named Bruce.

Once the scientific community recognized the Bruce effect in the 1950s, researchers started finding it all over the mammalian world. Rodents do it. Horses do it. Lions seem to do it. Even *primates* do it.

But we humans don't. And that's rather telling.

We're not really sure how, exactly, female mammals who have Bruce-style abortions actually achieve their goal. But we have some clues. Among mice, it seems fairly automatic: if the pregnant female smells the urine of a strange male, she'll abort. She doesn't even have to see the guy.* But the mouse gestation period isn't terribly long—roughly twenty days—and if the pregnancy has advanced past ten days, the Bruce effect doesn't seem to kick in. Essentially, there's a kind of reproductive tipping point: if her body has already invested a certain amount of energy in the pregnancy, then she'll carry the pups to term.

It's easy to argue that, at least in rodents, the Bruce effect isn't behavioral, which makes it harder to compare it with what we usually call abortion—an act where human women deliberately and consciously choose to end their pregnancies.

But consider the gelada. On a high, grassy patch of Ethiopia, primatologists have observed a troop of geladas for nearly a decade. They're a lot like baboons: big, shaggy, smart, and highly social. Within their large societies, reproductive groups are harem based: one dominant male with a bunch of females, surrounded by roving packs of outsider males who regularly try to challenge the alpha male. If a new male manages to take the crown, a curious thing happens: a full 80 percent of the currently pregnant females

* Specifically, her olfactory system sends a signal to her brain, which changes the activity of her pituitary gland, which then has a knock-on influence on her corpus luteum. The corpus luteum shrinks, progesterone levels drop, and her uterus contracts and sheds its lining, thereby flushing out the embryos. But there's a loophole, wherein exposure to the father (familiar male) is okay for the pregnancy, while exposure to an unfamiliar male (abort!) provides the trigger, with more exposure over time attenuating these effects in varying cases (Yoles-Frenkel et al., 2022). It seems to directly involve learning via the olfactory system, and thus it's a more direct cause and effect than "she's stressed out"—the Bruce effect in mice is a more reliable abortive scenario than just raising a pregnant mouse's cortisol levels (de Catanzaro et al., 1991). But nobody's saying she made a conscious *choice* in the matter. She's still clearly a mouse.

will abort within weeks of the new male taking over. (Why not 100 percent? First, always be suspicious of perfect numbers. Biological processes are messy affairs. But also, much like mice, it seems to depend on how far along the pregnancy was when the new gelada male assumed the dominant position.)

Male geladas, like male mice, can be dangerous beasts. After taking over a troop, the new male may kill any offspring who are still nursing and may even kill the freshly weaned. That's probably because their mothers will become fertile again sooner than they would if they were tending to these infants. The sooner they ovulate, the sooner the new guy gets a chance to pass on his genes. And for the females, like mice, continuing a pregnancy that's going to end in the death of the offspring is kind of a lousy investment. In fact, among the geladas, the females who *do* abort reap a clear reproductive benefit: they're usually pregnant again in a matter of months.

But even more tantalizing, for our purposes, is the fact that no gelada male will successfully rout a dominant male without the support of that male's current sexual partners. In other words, it's not as simple as saying that the females abort out of fear of the new male; some scientists propose the females may even abort to make them better able to bond with the new guy.

Remember, these are higher primates—in evolutionary terms, just shy of being great apes. They're not aborting because of a simple biological trigger, like the scent of a male's urine. This is something that happens as a result of directly observed social change.

And then there are horses. That's where things get really behavioral. Domesticated horses are significantly more likely to miscarry than wild mares—as many as one in three. Researchers tried for years to figure out why. Was it the type of feed? Stress? The stallion's mounting style? The answer was strikingly simple. To avoid these spontaneous abortions, you have to let the mare have sex with a familiar male.

Like the gelada, a wild stallion who takes over a herd may kill any foals he has reason to suspect aren't his. Still, monogamy isn't the rule. After running blood tests on wild herds, scientists determined that roughly a *third* of foals aren't sired by the dominant stallion. That stallion does get first dibs on reproduction, but mares also have "sneaky sex" with outsider males. Then they immediately seek out the stallion to try to have "cover-up" sex with him. If they don't get the chance to have cover-up sex? That's when they'll usually abort.*

Domesticated mares are regularly stabled separately from stallions to prevent any unplanned pregnancies. But when the breeder takes a mare away from the "home herd" to have sex elsewhere, the mare will seek out the local stallion for sex as soon as she can manage it. If they're separated by a fence, she'll actually present her backside to him across the fence, tail to the side. If she manages to have the cover-up sex, she'll settle down. If she doesn't? Yep, most of the time she'll abort.

Thus, whether we're talking about the lowly rodent, lusty mares, or clever primates, we can see that social abortion—"miscarriages" that occur as a response to the local social environment, rather than any problem with the embryo itself—is a well-documented part of mammalian reproductive biology. Abortion is just one of the things that female mammals do. We don't know the ins and outs of its mechanisms yet, and they probably differ between species. But if rodents, equines, and primates have all developed some version of the Bruce effect, then we should stop thinking that human abortion is something unique. The way we do it—using human gynecology—is different, but ending a problematic pregnancy in response to social stress is something a lot of mammals do.

If anything, the fact that human women *don't* have long-evolved internal mechanisms to support female reproductive choice is what's unusual. Research has shown that a woman who's pregnant as a result of a rape won't miscarry at a higher rate than

* By the way, geladas also have sneaky sex. In fact, they're demonstrably sneaky: if a nondominant male has sex with a female, he'll do it out of sight of the dominant male, and the amorous pair will suppress their normal sex vocalizations. If the male notices he's being cheated on, he'll berate both of them in ways that are clearly punitive (le Roux et al., 2013). To my knowledge, no data exist as to whether the female is more likely to abort the way mares do if she doesn't "get away with it."

a woman who's pregnant by a partner. Apparently, 5 percent of American rapes result in pregnancies. Rates in other human communities are similar. That might not sound like a lot, but the chance of pregnancy resulting from a single bout of intercourse on your most fertile days is only 9 percent, with that chance dropping to near zero on non-fertile days.

For a little while, it did look as if human women might have a mini version of the Bruce effect, though: a woman who's having regular sex with a man is more likely to become pregnant and carry that baby to term than a woman who has sex only once or twice around the time of her ovulation. At first, researchers thought this was maybe a way to ensure the success of a local male's sperm—after all, he's more likely to help with his own offspring, right?—and reduce the chances of carrying a wayward male's baby to term. But with further research, it doesn't seem to be a built-in monogamy booster after all—so long as they don't have a sexually transmitted infection (STI), women who have sex with multiple men frequently are *also* more likely to carry their babies to term.* So it's probably immunological: being exposed to sperm regularly, whether it's with a monogamous partner or many partners, could help a woman's body "recognize" the intruding sperm and attack them less, a bit like how slightly allergic people can get used to pollen or pet dander.

Why human women have so many miscarriages after the egg implants in the womb may also have little to do with the partner. Most miscarriages occur in the first thirteen weeks of pregnancy, and even more commonly in the first eight. And most of them seem to be due to chromosomal abnormalities. That means one of two things: either the egg or the sperm already had some genetic issues, or at some point in early cellular division something went

* It also wasn't great for the practical experience of living in a monogamous couple while trying to have a baby: there's nothing that kills the mood quite like being told you *have* to have a bunch of sex. Couples actively struggling to have a baby almost universally report that the process dramatically lowers their sex drive. Ovulation predictor kits net a tidy sum in the United States. It's unclear which is worse: peeing on a little stick every morning in order to have largely undesired sex a few times a month or having largely undesired sex every two days without having to pee on anything. The latter very slightly improves one's chances.

wrong. That's not a Bruce effect. It's simply a body ending a pregnancy that would not have produced a healthy baby.*

Stress seems to have an effect on early pregnancy, too—human women who are highly stressed are more likely to abort—but it's not as predictable as the Bruce effect. After all, thousands of babies are conceived and born in refugee camps every year. I can't imagine what the word "stressed" means to a pregnant woman in the DRC right now—the chance that her fetus was fathered by a man who raped her is higher than almost anywhere else in the world. But even so, once she passes her first trimester, she's likely to carry the baby to term.

So here we are, then. Modern human beings don't have anything like the Bruce effect, which means our ancestors probably didn't, either. We do have sort of foldy vaginas, but they're not "trapdoor" vaginas, so it's also likely that we didn't evolve with a lot of gang rape going on. The human reproductive system doesn't betray a past in which competitive men regularly committed sexual violence or infanticide. Ancient hominins just weren't all that rapey. If they had been, women would probably have fancy vaginas, men would have hi-tech penises, and women would have a more reliable miscarriage response to rape and male threat.†

But that doesn't mean our Eves weren't doing everything in their power to pursue female reproductive choice. Like other mammals, they were choosy about their partners. And at some point along the evolutionary path, they also started utilizing what-

* I've had four miscarriages, to my knowledge, though at least two of them weren't due to genetic problems. In each case, my body didn't exactly help me out: one was ectopic, and I had to be hospitalized for internal bleeding; another was an "empty sac" where everything developed except for an actual embryo; a third made it all the way to the second trimester before the heartbeat stopped and I had to have a D&C *and* a later emergency surgery. The fourth was actually my first pregnancy, or the first I'm aware of: I went in for a surgical abortion when I was young and under no small amount of stress. They couldn't find a heartbeat, even though at that point in the pregnancy there should have been one. So that pregnancy probably would have ended as a "miscarriage," too.

† I'll go into more detail on this in the "Love" chapter, but for now I'll just note that human men and women are still actively involved in the mammalian sex war despite an evolutionary history that probably didn't have significant amounts of rape or male domination.

ever they could from the plant world's pharmaceuticals in order to control reproduction.

Plants are constantly at war with parasites, herbivores, and one another. As a result, numerous plants have evolved to produce chemical compounds that improve their chances to survive and thrive. These compounds directly affect the health of the creatures who eat plants. Most will learn to avoid ones with toxins. And many animals—including primates—also seem to seek out plants with compounds that help them improve their own health.

The field of research is fairly new, but primatologists have been able to find tantalizing evidence of self-medication. In one case, the medicine in question was the bitter pith and juice from shoots of the *Vernonia amygdalina* plant. Mahale chimps, sick with parasitic intestinal worms, spend up to eight minutes carefully peeling away the bark and outer layers of the shoots in order to get at the extra-bitter innards. They chew on the pith and suck out its juice. This isn't tasty. Nearby adult chimps who are not sick avoid the stuff. Primatologists sampled the poo from before and after this pith-eating behavior and found fewer parasite eggs in the post-medication poo. And it just so happens that local humans *also* had the habit of using this bitter pith in traditional medicine for treating intestinal parasites. As with humans, the chimps presumably learn to treat themselves this way from other chimps.

Similar sorts of self-medicating behaviors have been found throughout the primate world. From chimps and gorillas to baboons and macaques, nonhuman primates seem to have the habit of selecting plant foods with secondary compounds that can make them feel better.

And it also looks as if primates use plants to influence their fertility.

Phytoestrogens are compounds in plants that work, in animals' bodies, quite a lot like our own estrogens. Eating a lot of phytoestrogens can "trick" the body into functioning as if it were at a different stage of the menstrual cycle. A woman who eats an excessive amount of soybeans—full of phytoestrogens—can actually hamper her fertility; many fertility specialists now advise their patients to avoid soy if they're having difficulty getting pregnant.* It's also why a number of people are up in arms about whether the estrogen-like compounds contained in certain plastics are messing with our bodies' natural estrogen balance. But do other primates seek out these plants for the purpose of manipulating reproduction?

In Uganda, a group of red colobus monkeys eat the leaves of estrogenic plants seasonally; in a given week, the plants can constitute as much as a third of the animals' diet. As a result, their estradiol and cortisol levels rise. And as those hormone profiles shift, so, too, does their behavior, altering how aggressive the males are, how frequently they mate, and how much time they spend grooming one another. Basically, they have a lot more sex when they're eating lots of these leaves.

Chimps in the Sudan, meanwhile, have been seen eating leaves from the *Ziziphus* and *Combretum* species. This wouldn't seem too remarkable—chimps eat leaves all the time—except that humans who live in the same area use these plants to induce abortion. *Combretum* is also used in traditional medicine in Mali: if a woman has been suffering amenorrhea, she'll drink a potion of its dried flowers to bring on menstrual bleeding. If it were the case that selectively eating these leaves detrimentally influenced the chimp population, they would probably avoid them, much as they avoid other toxic plants. But because females—not males—are the ones who eat the leaves, and because the plants are known to have abortifacient properties, that leaves a rather tantalizing question: Are these chimps controlling their inter-birth spacing by selectively eating plants that limit their fertility?

Trying to guess an animal's intentions is always a tricky business. But given that today's primates seem to possess a range of knowledge about the plants in their local environment—what's safe, what's not safe, and what might be good when you're sick—

* Soy seems particularly useful during menopause, however, helping alleviate some of the nastier symptoms—think of it as a kind of hormonal protocol from the plant world, but with fewer side effects than similar pharmaceuticals. But, as always, talk to your doctor.

it's probable that early hominins did, too. Habilis was likely taking advantage of whatever she could to influence her own reproduction. Since she didn't have anything as reliable as the Bruce effect or a trapdoor vagina, she would have been driven toward *behavioral* adaptations to exercise her choice. She was social. She was a problem solver. She was a tool user. Faced with her own faulty reproductive system, she would have tackled the problem as only a hominin could: socially and cleverly, with whatever tools she could manage to invent.

Being a clever social primate was always a boon for our hominin Eves. But the brainier we became, and the more complex our societies, the easier it would have been to take these early foundations of gynecological knowledge and build on them. While Habilis probably had a decent amount to work with, later Eves had the sorts of minds that could really connect the dots.

LEAVING EDEN

To each Eve, her Eden. Habilis never left Africa.* Like most species on this Earth, she'd adapted her body and behavior to the particular world she lived in, and as that world changed, she went extinct. Call it Uriel's ecological sword. But her great-granddaughter, *Homo erectus*, was one of the most successful hominins that ever existed. What Habilis started, Erectus inherited.† She took it and literally ran with it—all the way to China.

Quite a bit taller than Habilis, *Homo erectus* males stood a full five feet ten—a good inch taller than the average height of today's American men. And the Erectus Eve wasn't much shorter.‡

* Or at least, most paleontologists are pretty sure she didn't. There just aren't that many hominin fossils.

† Scientists aren't sure yet if Erectus and Habilis split from a common Eve, if one is a descendant of the other, or if they interbred. The fossil record does show that they coexisted in Africa for half a million years and their territories overlapped. We know in the beginning they both used Oldowan tools and they'd inherited this technology from earlier australopithecines. Some even propose that Habilis herself was a kind of australopithecine and that her stone tools were a natural extension of the way other apes use rocks to break open hard nuts.

‡ One fossil site seemed to indicate males were quite a bit larger with a more

Her limbs were long and graceful, and her face flatter than Habilis's—a bit more like ours, part of a long evolutionary chain that produced the modern human face. Erectus's brain was also bigger than Habilis's. And you can trace the evidence of that brainpower in the fossil record: not only was Erectus a tool user, but also she was the first hominin to take down big game and the first to use fire. We've found charred remains in a cave near her bones from one million years ago. It's not clear if she made the fire or just used a forest fire opportunistically. But she definitely brought a cooked dinner into that cave.

Erectus improved on Habilis's tool tech. She invented the Acheulean tools: long, thin, elegant hand axes and choppers. You couldn't make them with just any stone, but had to scout out the sorts of rocks that would work. You had to plan ahead, shaping the stones just so, thinking about certain kinds of flakes and what they would become. If the Oldowan tools took a while to make, Acheulean tools took significantly longer, becoming the sorts of prized possessions you'd probably try to keep with you.

All of that means that while Habilis was smart and capable and social, Erectus was all of these things and more. And we know she could really travel, which means she was an adaptable problem solver, clever enough to take on new challenges. But that extra brain came at a cost, given that her pelvic opening was still narrow. In all likelihood, Erectus's pregnancies and births sucked *even more* than they did for Habilis, because she delivered infants with even bigger heads and shoulders. That means Erectus needed gynecology. She needed it badly. *Homo sapiens* would need it even more.

Despite making it out of Africa and colonizing a number of places, leaving fossils and her stone tools along the way, Erectus went extinct over time. Humanity's hominin Eves—from creatures like Erectus all the way up to ancient *Homo sapiens*—repeatedly tried to leave their Edens. Some might have speciated into new creatures, evolving in ways that left their old bodies and habits behind. But with the exception of *Homo sapiens*—who are

prominent browridge, but this is unclear. In general, the hominin evolutionary line has progressively less sexual dimorphism—the closer you get to *Homo sapiens*, the more similar the sexes' body size. More on that in the "Love" chapter.

not yet dead or speciated into something else—all of the other Eves are gone.

That shouldn't be surprising: creatures who aren't prolific reproducers hit environmental and competitive challenges and, lacking suitable work-arounds, they fail to adapt.

This is particularly true when you're dealing with migration. In order for a species to move to new environments and thrive, it needs to build what's called a minimum viable population (MVP) in that new location. This is a concept from ecological science: the minimum number of reproducing individuals that are needed to ensure a group's ongoing survival in any particular place. If your group has enough members to ensure both ongoing diversity and general reproductivity in your local environment, you've got a healthy chance of survival.

What migrating Eves needed to do, in other words, was to make *babies.* Nice, healthy, viable babies that could live long enough to make more babies of their own.

This wasn't exactly the hominins' strong suit. By the time Erectus came around, their placentas were greedy, their birth canals were a gauntlet, and their babies, once safely born, were highly dependent for years and years and years. Maybe that's why as few as 50 percent of human pregnancies actually produce a human baby. Maybe that's why a healthy woman having sex on the day she ovulates still has only a 9 percent chance of becoming pregnant.* If pregnancies, births, and child rearing are biologically expensive, then you'd expect the bodies that have to do all those things to evolve ways of ensuring that only the pregnancies with the very best chance of success will continue.

If those sorts of miserable success rates were true of our hominin Eves, too, that's probably why only a few hominin species ever managed to get out of Africa. It's also probably safe to assume that it's a large part of why all but one species died off.

* If at this point you're thinking to yourself, "Yes, but my cousin so-and-so gets pregnant when she even looks at a guy funny," you're not *wrong*, exactly. Some people are especially fertile. But statistics like these are about averages—not your baby factory of a cousin, but what *most* women's bodies tend to do. Most women will not become pregnant by having sex on the day they ovulate, though they've got a better chance than someone who doesn't have sex during their fertile window.

Consider the armadillo: one of the reasons the strange, semi-armored little mammal does so well in its many difficult environments is the simple fact that it's able to control *when* it's pregnant. In the low belly of the nine-banded armadillo, the embryo, semi-miraculously, is able to stop developing. It just floats around after it's fertilized and waits, sometimes as long as eight months, to implant in the womb. So if an armadillo happens to be crossing a large, inhospitable stretch of desert, her embryo will just . . . *chill.* When she gets to a place with more food and water and happily settles down, the embryo begins developing again.

The armadillo, unlike early hominin Eves, is good at migrating for precisely this reason. She can rapidly adapt her "birth spacing"—when she has her babies and how often—according to any given environment's challenges.* All human women have to work with is a possible miscarriage (risky in itself—a failed pregnancy in the second or third trimester can easily kill a woman or render her infertile). Thus, the only way we're able to manipulate our birth spacing with any reliability is by doing things that decrease or increase women's fertility, depending on which benefits us more. And they would have had to call on all the gynecological knowledge they had when they tried moving those bodies, long adapted to certain environments in Africa, all the way up to the ancient Levant.

No one knows why Erectus left home in the first place. There's a "pull" scenario, wherein green corridors opened up to the north due to rising humidity, creating little pockets of newly available, hominin-amenable territories that Erectus happily moved into. We're pretty sure that happened for some of the later *Homo sapiens* migrating out of southern Africa: a large lake transformed into extensive wetlands, stretching northeast and southwest. That happened about 100,000 to 130,000 years ago, after a group of hominins had been doing pretty well around the lake for about

* Her embryo also splits into four, producing four identical offspring attached via one placenta—very unusual among mammals, and another reason she's good at migrating. Armadillos can build up a minimal viable population in any environment fairly quickly. Why they don't suffer from inbreeding's normal challenges with all those identical offspring isn't clear.

70,000 years.* But if Erectus was "pulled" north out of Africa into newly welcoming territory, it wasn't all that long until those new territories experienced climate change, forcing her to adapt her strategies yet again. And if it were a "push" scenario instead—wherein a local environment changes enough that a group has no choice but to move†—being able to quickly adapt would be even more important.

For Erectus and the migrating hominins after her, some of those changing environments would be better served by births that coincided with a fruit and nut harvest, or a wave of migrating animals. Some environments would be barren, challenging—best to *widen* birth spacing to reduce the burden. Some environments would prove rich enough to support heavier reproduction, so she'd also need gynecology to survive all those pregnancies and nursing children.

If she migrated slowly enough, evolutionary processes would ostensibly take care of those adaptations. But taking direct control of your reproduction changes the game entirely. Instead of waiting millions of years for her buggy hominin uterus to catch up, Erectus could directly influence her reproductive outcomes in her own lifetime. Which she did, given that she managed to spread into a dizzying number of different ecosystems: not just the entire continent of Africa,‡ but out and across the Middle East, up through Europe, into central and southern Asia, and down to the Pacific Rim. She took over the world.

* There's evidence for this from both fossil records and pollen remains, showing the evidence of climate change, and from tracing humanity's mitochondrial DNA back to its collective origins (Chan et al., 2019).

† For example, what's happening right now in many of the low-lying islands of the world, where rising sea levels are forcing large numbers of people out. At the current rate, the Maldives will be completely submerged in thirty years (Storlazzi et al., 2018). In the world's largest delta, the Indian Sundarbans, as many as 4.5 million people will be displaced in the coming century. Seawater will mix with the delta's freshwater in ways that will make the region's agriculture untenable. These millions of people will be "pushed" to migrate, as many already are (Pakrashi, 2014). Whether the rest of India provides sufficient "pull" before then remains to be seen, but generally speaking, most models of climate change in the coming years show widespread human migration. Many of us simply won't be able to live in the places we do now.

‡ Which is *incredibly* big, and she crossed it on foot.

Meanwhile, back in Africa, another population of Erectus invented those Acheulean stone tools. With fossils, we can trace that evolution on a map: The first wave of Erectus out of Africa used Oldowan tools. Those fossils were found in southern Russia, in India, in China, in Java, often with Oldowan stone tools alongside the bones. But *later* Erectus fossils in Africa start being associated with the more advanced Acheulean tools. Once she had that upgrade, she took her new tech with her up through the Levant and beyond.

This is the first record we have of the hominin success story: the fact that our Eves were able to adapt to a wide variety of new environments. They did it with big brains. They did it with stone tools. When they improved on those tools, they took that knowledge with them. Eventually, they did the same with fire and cooked foods.

But none of that would have been possible without gynecology. In each new place, we probably barely made it to our MVP, and we certainly needed primitive gynecology to get there. By one recent calculation, the MVP for a reproductive group of humans isolated for 150 years would be fourteen thousand, with forty thousand being a much safer bet. Of that forty thousand, only about twenty-three thousand would be the "effective population"—that is, males and females reproducing with one another. The rest are folk outside the birthing years. The latest, best estimate for *Homo sapiens*' first foray into the Levant? A thousand to twenty-five hundred individuals. That's it. A couple of thousand, barely managing to reproduce.

There was, in other words, a succession of such events—time upon time, a too-small band of ancient hominins migrated, ceased reproducing with anyone but themselves, and did everything they could to survive and thrive and have yet more genetically similar offspring. This is why you and I are so closely related to one another no matter where on the planet we live. We should be more genetically diverse, but we're not.

It's not hard to imagine why that might be. Here's a more realistic Genesis: about sixty to a hundred thousand years ago, a population of ancient *Homo sapiens* finally reached critical mass in southern Africa.

A small group of them then migrated to eastern Africa. Ten thousand or so years later, they finally flourished enough to enable another band to migrate, moving into the ancient Middle East. From there, it took only about five thousand years for subsequent groups to move into Europe, into central and northern Asia, and finally, only fifteen thousand years ago, into North America. We know this because most people who are descended from this migration are so similar.

Each time a group produced enough of a population for smaller bands to break off and colonize nearby areas, the genetic diversity of the new group would have been reduced. That's because at *each new site* the group would be breeding largely with themselves. Once those founding mothers left southern Africa, subsequent generations produced more offspring from within a limited genetic pool. And that limited pool would then, yet again, have the same effect when *they* left, compounding inbreeding upon inbreeding in ways that would naturally outpace normal genetic drift. This is the best argument going for why humanity suffered a genetic bottleneck right around the time we left Africa.* This phenomenon is called a founder effect—something you can easily identify in a species' genetic history, when a migrating group is reproductively isolated and their offspring become less genetically diverse than you'd expect.

When we finally managed to populate the world with *Homo sapiens*, a time paleoanthropologists call the "Great Expansion," we were also, simultaneously, reducing our species' genetic diversity. In order to avoid becoming the eleven-toed wonder, doomed to extinction because of inbreeding, each band of migratory humans would have been under *even more* pressure to build up and sustain a minimum viable population in that new location.

Intellectually, that makes plenty of sense. The numbers and

* A "bottleneck" happens when there's some sort of pinch point in a species' genetic history: a time that somehow radically reduces genetic diversity thereafter. One scenario is mass death: reduce the number of mating pairs, say, from a massive volcanic winter, and suddenly the species will have less genetic diversity because the pool is simply much smaller. The other plausible scenario is a founder effect.

the timing of our genetic bottleneck work out, too—this model fits a good deal of current knowledge across scientific disciplines for what really happened to our ancestors when they left Africa. But for those of you who've had a uterus that bore children, let me make this a little more real for you: Each group of ancient settlers needed to do *better* than the replacement rate. To build and maintain an MVP, each breeding pair needs to make *at least* two more kids and those kids need to do the same when they get old enough. Ancient babies died a lot. Two wouldn't be nearly enough. And the majority of our Eves had barely the remotest shot of surviving, let alone living beyond their reproductive years. Most hominins—until recently, most *humans*—were lucky to reach age thirty-five. That means if our Eves survived childhood, they would have spent the next decade, or at most two, having children, breast-feeding, trying to keep everyone alive—or hell, at least enough to launch two kids into adulthood—and then kicked the bucket.*

Human offspring are hardly self-sufficient at age two, so that means any kids these Eves had at age thirty-three would have had an uphill battle to make it to puberty themselves. The *likeliest* reproductive success scenario involves clustering your births at the start of your reproductive years, leaving yourself time to help keep your offspring alive until they become teenagers.† You could

* In other words, Thomas Hobbes wasn't entirely wrong: most of our Eves' lives *were* brutish and short. Though this "natural condition of mankind" was less nasty than he'd imagined, because collaboration would have been absolutely vital for survival, particularly when you're pregnant or nursing for half your life.

† Much ado has been made of birth clustering in the paleoanthropology crowd—it's often credited for hominins' success at migration. While it's true that having a hand on the levers of reproduction greatly improves one's chances, what's rarely mentioned is that, for *Homo sapiens* at least, having too many pregnancies spaced too closely together—particularly if those pregnancies come to term—significantly increases the complication and mortality risk for both offspring and mother (Molitoris et al., 2019). This effect occurs in pregnancies *less than thirty-six months apart:* less, in other words, than the average inter-birth interval of most extant hunter-gatherer societies. That means it's not the case that simply boosting fertility in the short term will reliably produce a viable population if doing so mows down a bunch of females. So if you're a group of ancient Eves going for the birth-clustering option, a bunch of you and/or your offspring may well die because of it. That doesn't mean it *won't* work—depending how lucky you get, and the size

also go the chimp route, simply having one kid and raising that kid until it can roughly manage on its own. For chimps, that means having kids about every four to six years. Still, six-year-old human children aren't great at surviving without semi-constant attention. That's as true in a modern kindergarten classroom as it had to be in the wilds of the ancient world. Either way—clustering your kids in your late teens or spreading them out across your twenties and early thirties—you're going to need gynecological knowledge to help you and your kiddos make it. Some of that would have been calling on the skills of midwives. Yet more would involve social and medical practices, including pharmaceuticals, that regulate your fertility. No strategy would be perfect, but clearly the *worst* strategy would involve a reproductive free-for-all without shared knowledge (and shared child-care resources).

In other words, for each transition point in humanity's ancient migrations, you should expect to find a group of skinny, scrappy people just barely producing enough kids to replace themselves, finding ways around the inherent problems of inbreeding, and miraculously surviving. A huge portion of that survival would have been tied directly to gynecology.

"THEY MOSTLY COME AT NIGHT. MOSTLY . . ."

Female reproductive choice is an incredible biological tool set. And once it evolved into something as effective as human gynecology, women had their hands on the actual machinery of evolution, directly enhancing their species' fitness in their own lifetime. If you can manipulate your reproductive strategies to suit nearly any environment, that means, as a species, you're finally in charge of your own destiny. Our Eves used the gynecological tool set to overcome their greatest challenge: the wonkiness of their own poorly designed reproductive system. That's why you're able to do things like read a book about it now—this was hardly the given fate of our evolutionary line. But just as our Eves used gynecol-

---

of your group, you might just pull it off. Certainly some of our Eves did. But they probably lost a lot of sisters and daughters and mothers in the process.

ogy to survive and thrive in deep time, we can still use it today to overcome some of our species' biggest threats.

As one good example, think about infectious disease. We know that the placenta regulates a pregnant mother's immune system, as it does for most mammals. But it's especially true in the human body, where our extra-invasive placenta has to work extra hard to hold its ground. Evolving ways to make the maternal immune system look the other way makes perfect sense for the embryo, because in the trench warfare of maternal-fetal competition, you really do want to strip the enemy of its bigger guns as soon as possible. But as we examined in the "Womb" chapter, down-regulating an immune system also puts the mother's body at risk of infection. Those infections can be run-of-the-mill things, like yeast infections or simple head colds—the bane of a pregnant woman's existence—or they can be nastier episodes of the flu, or an outbreak of intestinal worms, or infectious diseases like dengue or Zika.

In 2016, women across the world became terrified of the Zika virus, a fairly benign infection spread by mosquitoes in hot, wet places. Most people who get Zika seem to have mild symptoms, so it wasn't exactly a world health priority until women in Brazil started giving birth to tiny-headed babies. Microcephaly—a rare developmental disorder that makes fetuses' skulls and brains fail to develop normally—can cripple a human being for life. Most people with it will die young. Before 2016, no one realized that getting bitten by a mosquito carrying the Zika virus when you're pregnant could mean your child would be born with a tiny head. Because of our female physiology, Zika in women might as well be a different disease.

You can say that about malaria, too. Pregnant women seem to attract twice as many malarial mosquitoes as nonpregnant women.* And once a woman is bitten, she faces severe consequences. In places where malaria is endemic, a full 25 percent

* Scientists don't know if it's because of increased respiration—pregnant women breathe more heavily and more often—or because of increased blood flow or higher body temperature or because some pregnant women might have higher levels of sucrose in their bloodstream. Mosquitoes respond to all these things when they hunt (Lindsay et al., 2000).

of all maternal deaths can be directly tied to malaria. Pregnant women are three times more likely to suffer a severe version of the disease, and nearly 50 percent of those women will die. If they don't die, they'll suffer ongoing complications from the disease, which may well kill them later.

But it's not just mom: a malarial mother's newborn is more likely to be born early and underweight. That's probably due, in part, to the fact that the mother's anemic—a side effect of fighting off malaria—and that malarial protozoans accumulate in the placenta. Infants and children, with their naive immune systems, are already more prone to complications from malaria, which means wherever malaria lives, a hell of a lot of newborns, infants, and young children die. Child mortality rates are directly tied to how often women are usually pregnant, and stats support that trend worldwide. The mechanisms are fairly obvious: not only does a woman ovulate more often when she spends less time pregnant and breast-feeding, but cultural and—presumably—biological drives also push women into becoming pregnant again after a child has died. That drives up maternal deaths, since human pregnancy is always risky, and inevitably affects the social status of women in those regions. Malaria, in other words, is a matter of women's rights worldwide and, because it specially affects women's bodies, quite a lot of malaria research and treatment should fall under the umbrella of gynecology.

But it doesn't, usually, because many biologists and medical professionals have difficulty reconciling the fact that sexed species produce two very different types of bodies. We're only just now starting to hear voices in the medical community calling for different treatment paths for the sexes. But even outside the clinic, knowing how malaria works in pregnant women can help men and children, too.

Given how much more likely pregnant women are to be bitten by malarial mosquitoes, taking special advantage of pregnant female bodies could be part of the protozoan's broader life-cycle strategy. We know that the protozoans accumulate in placental tissue. If sequestering in the human placenta allows them to escape detection for longer, that's a clear advantage—the sort evolution typically selects for. As in HIV, such "reservoirs" seem to help preserve pockets of infected blood cells in pregnant women even as the rest of their bodies purges the infection.

Researchers aren't sure how the protozoans "know" to hide out in the placenta, given how fiercely the placenta normally fights off everyday infections (lots of babies with naive immune systems would die otherwise).* But hiding there helps the protozoans avoid detection by doctors testing the woman's blood. Infected pregnant women regularly test negative for malarial infection and, as a result, don't receive treatment for it. When the protozoans reemerge, they find their way to the liver, reproduce, and start their life cycle all over again.

We don't know yet if viruses like Zika utilize similar strategies, though the virus *was* found in placental tissue from women who'd had miscarriages. And Zika infection in the first trimester does appear to be linked to a higher rate of miscarriage, as does malarial infection, and is likewise associated with fetal malformations of various sorts. Chasing these kinds of questions down is precisely what you should expect to see in the future of human gynecology, but also in the future of global health research. Maybe mosquito nets and pesticides aren't the only strategies we should be using to fight these diseases. Birth control should also be a frontline defense—not simply to protect women and children, but to protect entire local populations.

Think of it this way: America rid itself of malaria in the twentieth century by killing massive numbers of mosquitoes. That was partly achieved by spraying epic amounts of insecticide in and around American homes. But it was also done by controlling the environment: draining standing water, for example, and targeting areas where malarial mosquitoes were known to breed. It might seem straightforward to us now, but even imagining that strategy

* As with anything in biology, there are probably a few different mechanisms. One known path is that the infected cells make small proteins on their surface. These proteins "snag" the cells against the walls of smaller blood vessels, preventing the body from sluicing them down and away into the spleen, where they'll be detected and destroyed. That's a large part of how malaria does so much damage to the body's organs: it gums up the works in delicate blood vessels. Because placentas are especially rich with small blood vessels, that may also be part of how the protozoan ends up there.

required a paradigm shift: effective public health requires not simply quarantining and treating *patients* but being proactive by considering larger *environments* in which diseases go through their cycles. Thinking about malaria as a gynecological problem—not simply that women and fetuses are "vulnerable," in other words, but that human pregnancy might be an important feature of how the disease works in a larger mixed-sex population—requires a similar shift. It means we have to think about spaces in the human body as *environments.* As we discussed in the "Womb" chapter, maternal-fetal competition is centered on the local environment of the uterus, and that means the pregnant human uterus has unique features that infectious diseases can evolve to take advantage of. If something like malaria uses human placentas as reservoirs, hiding from the mother's immune system, what could we accomplish by offering women safe, healthy choices about their reproductive destinies? The stakes aren't small: we're talking about the suffering of millions of people, now and in the future. What happens when we give women the choice and tools to simply reduce the number of placentas per square mile?

WOMB TRIUMPHALISM

Instead of twisty trapdoor vaginas, we now have the Pill and the diaphragm. Instead of the Bruce effect, we have methotrexate and misoprostol. Instead of waiting for a less dangerous birth canal to evolve, we have midwives who help our newborns squeeze through the gauntlet and the miracle of modern C-sections.* When, in other species, physiological evolution would have created a newly evolved feature to enable female reproductive choice, hominins used *behavioral* innovations instead—some of them social and others involving new tools and pharmaceuticals. That control we have over the most powerful levers of our evolutionary fitness got

* Though much has been said about the "medicalization" of human birth, I have many friends, with many children, who might have died without cesareans. That mother and child are now so likely to survive that sort of emergency surgery—which was in no way assured for the vast majority of our species' history—is, yes, miraculous.

us to where we are today. It allowed the early human population to finally explode, expanding into nearly every ecological niche our ancestors stumbled upon. It also improved the survival rates of every pregnant female with a too-narrow pelvis and a greedy placenta.

What got us here is not tool triumphalism but *womb* triumphalism. Our species' success was, and still is, borne on the laboring bellies and backs of women who made difficult choices throughout their reproductive lives. The deep history of gynecology isn't just the story of how we found ways for women to suffer less; it's the story of why we are alive today at all.

So maybe we need a better narrative to describe humanity's "triumph." Our story doesn't begin with a weapon. It doesn't begin with a man. The symbols of our ultimate technological achievements shouldn't be the atom bomb, the internet, the Hoover Dam. Instead, they should be the Pill, the speculum, the diaphragm.

Okay, Kubrick, take two:

THUS SPOKE ZARATHUSTRA

A sallow dawn rises over the land. The camera pulls in. A small band of hominins, adult males and females and children, gather around a watering hole. Their bodies are lean. Their fur is long and black. But there is manna in the desert: between patches of tan rock and scree, there are berries and tubers, and the little flowers that come after rain.

One of the females is heavily pregnant. She crouches near the water, grimacing as she braces herself on her long, muscular arms. The males largely ignore her, eyes on a far ridge. She leans to drink and then waddles off. An older female, curious, follows her.

The two scramble over a hill, leaving the troop behind. The pregnant female stops in the shadow of a large boulder as water rushes down the fur of her legs, pooling in the tan dust below. In labor, she strains and rolls, and the older female stays close, watching. Trying to stay quiet, the pregnant female pant-grunts at her, a submissive *don't hurt me.* Shaking, she extends one hand palm up: *help*. The older one is confused at first, but then she comes closer, touching that outstretched hand: *safe*. She moves behind her and sits, grooming her fur.

When the delivery begins, the older female moves to crouch between the mother's legs and helps guide the infant out. She clears the mucus from its mouth and eyes and lays it on the mother's panting chest.

Then we see a fast montage of female reproductive choice: hominins having sex, eating strange plants, having babies, nursing, walking with their offspring on their hips over the ridge to a green horizon. And back to the newborn, who suckles at her mother's breast as the two females move together toward the troop. Near the watering hole, the mother lies down, exhausted. The older female picks up the newborn and raises it overhead. Its profile clear against the blue sky, the newborn transforms into a human baby in a woman's arms, the two in profile against the window of a spaceship. We see the thin, bright arc of the planet in the background, the curvature of Earth. In the woman's free hand, the camera zooms in on a pamphlet: *Planned Parenthood: The Best Care in Low Orbit*. And *The Blue Danube* begins.

Homo erectus

CHAPTER 6

BRAIN

The little girl was sliding back in her chair, sullenly refusing her milk, while her father frowned and her brother giggled and her mother said calmly, "She wants her cup of stars."

Indeed yes, Eleanor thought; indeed, so do I; a cup of stars, of course.

"Her little cup," the mother was explaining, smiling apologetically at the waitress, who was thunderstruck at the thought that the mill's good country milk was not rich enough for the little girl. "It has stars in the bottom, and she always drinks her milk from it at home. She calls it her cup of stars because she can see the stars while she drinks her milk." The waitress nodded, unconvinced, and the mother told the little girl, "You'll have your milk from your cup of stars tonight when we get home. But just for now, just to be a very good little girl, will you take a little milk from this glass?"

Don't do it, Eleanor told the little girl; insist on your cup of stars; once they have trapped you into being like everyone else you will never see your cup of stars again; don't do it; and the little girl glanced at her, and smiled a little subtle, dimpling, wholly comprehending smile, and shook her head stubbornly at the glass. Brave girl, Eleanor thought; wise, brave girl.

—SHIRLEY JACKSON, *THE HAUNTING OF HILL HOUSE*

SOUTHERN AFRICA, TWO MILLION YEARS AGO

The mother had dragged the body for half a mile. It wasn't too heavy: she'd already torn open its soft belly and eaten the liver, and the heart, and the stomach, too, full of the nuts and fruits her prey had been feasting on when she found it alone, crouched under a tree. She'd even broken into the rib cage to get at the lungs—spongy little air bags.

She wanted to eat the rest of her kill in a safe, quiet place, but getting it into her den was a challenge. Her own body, sleek and long, just fit through the crevice that led to the cave. She tried pulling the mangled corpse in by the neck, but its limbs kept getting tangled and wedged. So she dropped it at the cave mouth and slipped in herself, then turned around and reached out a paw to snag it. No luck. Finally, she turned the carcass around in the dust, broke its shoulder in her jaws, then tore the arm at the joint, folding it up toward the lolling head.

Problem solved.

It was dark and cool in the cave, the air filled with the high-pitched mewlings of her children. Rumbling contentedly to herself, the mother started chewing at the base of the creature's head, holding it down with a massive paw. These tasty little apes teetered around the world on two legs. To get at its brain—the most delicious part of the kill—she just had to sever the roped muscles of the neck, and the head would pop right off. When it was free, she punctured the skull with her incisors, like tapping a coconut, and salt, water, sugar, and little rivulets of oil poured into her waiting mouth.

Soon the wadis would dry and meat would be scarce for an endless season. She knew this because she remembered it. She knew this because the very cells in her body were programmed to *eat, eat, eat* while she could. As her mother had. And hers before. So she sucked up fatty chunks of the wrinkled brain from under its broken dura, never dreaming that the descendants of these delicious little apes, we of the skinny limbs and fat brains, would one day name her *Felidae*, keep her cousins as pets. Nor that those kittens would spend a good part of their lives begging for industrial scraps that slid out of cans with a wet, jellied *slop*.

When the mother was full, her belly stretched, the taste of oily brain juice on her tongue, her kittens waddled over to nurse. The milk would be rich today. And their growing bodies would busy themselves parceling out lipids as they slept: Some for the eyes. Some for the muscles. Some for their own growing brains.

YOUR MISSION, SHOULD YOU CHOOSE TO ACCEPT IT . . .

As our increasingly humanlike Eves roamed about on their two legs, populating new territories and manipulating their reproductive strategies to try to survive, their brains started getting bigger. It didn't happen all at once, but we know from looking at fossilized braincases that they eventually swelled to an improbable size for such gracile little apes. The prefrontal cortex, in particular, grew and grew and grew.

By analyzing the tools of Habilis and Erectus and the many tool-using hominins that came after them, we also know that alongside that brain growth the many Eves of the hominin tree were becoming more clever and more social, if that were even possible. Presumably, these changes helped our Eves improve their gynecological tool set. At some point, midwifery must have become the norm. At some point, too, local knowledge about the use of plants to manipulate one's fertility would have become the norm. Eventually, human *language* would be born—though our hominin brains were rather big for a rather long time before that would happen.

All this brain growth was costly: it's terribly metabolically expensive to grow and feed brain tissue, which is, ounce for ounce, the hungriest part of your body. It requires specialized lipids. It requires a ridiculous amount of sugar. And given hominins' deep history of being *prey species*, such a nice big brain was probably an extra incentive for our predators. Dessert, if you like.

So the question of why we bothered investing and reinvesting in such a trait, time and again, throughout that long, murky history of hominin evolution in Africa, doesn't have a straightforward answer. If you think that having a big brain is a great thing, look around: very few species in the world bother building such a buggy, hungry, fault-prone football of neurological tissue. If big brains are so obviously wonderful, don't you think everyone would be doing it?

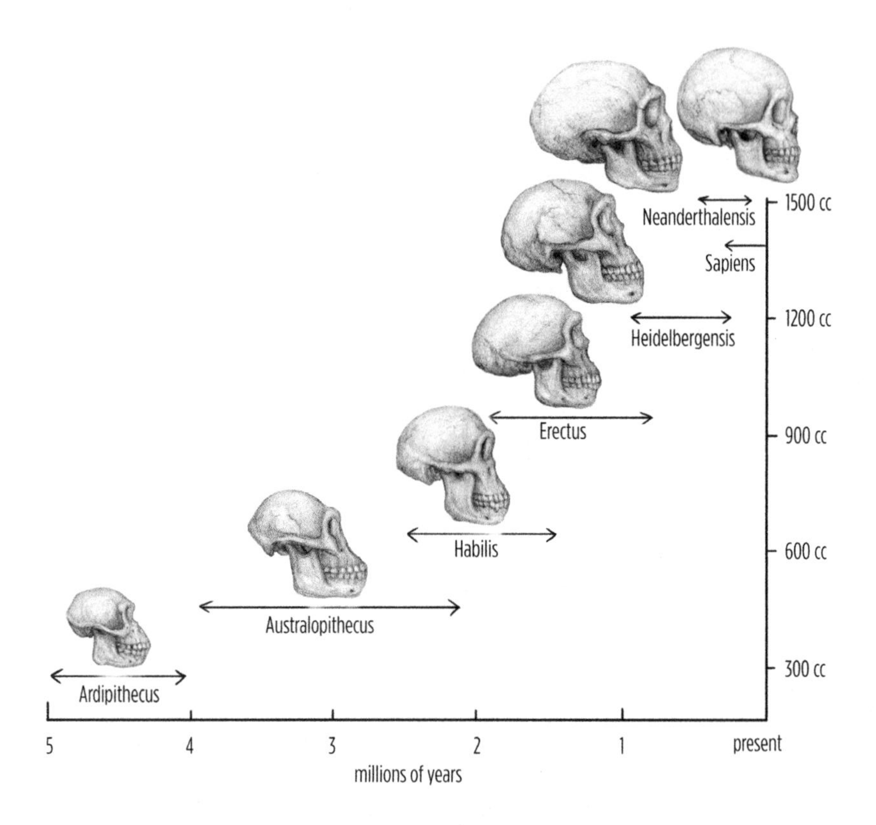

Brains and tools: a tale of hominin encephalization

So why did our Eves follow this path? We know they did: over time, in odd little bursts of rapid change, our hominin Eves' brains became more and more disproportionally *large* compared with the rest of their bodies. They finally got so big, in fact, that they had to build stronger clavicles to support neck muscles that could hold the silly thing up—which did a number on human childbirth, not to mention the fact that now our newborns can't hold up their own heads for *months*.*

The reason so many scientists have spent so much time

* As you might remember from prior chapters, giving birth to a big head is rather terrible, but flexible newborn skull plates do help out. Once the head makes its way into the birth canal, it's those wide *shoulders* that tend to get stuck.

thinking about this series of events, of course, is that the story of the human brain's evolution is the one most people think of as the story of when *they* became *us*—when our evolutionary Eves became something more like our human ancestors. We're terribly impressed with our brain. I'd argue we're in love with it. Which is to say, the human brain is in love with itself. If there's a single, physical trait that most scientists agree delineates humans from the other apes, it's our huge, lumpy, terribly intelligent brain.

Which is precisely why I didn't want to write this chapter.

We know so much about brains, every little special-topic journal brimming with new material, and at the same time so little. The entire field, precisely because it's so new and innately so interesting, provides a wealth of difficult ideas to chew on. I love that kind of literature, and I'm certainly not afraid of wading into the paleoanthropology minefield. The world of specialists who debate how hominin brains evolved and what they were good for—and, above all else, why they evolved at all—is wildly contentious. Because there are so few fossils, there are very few data points. *Nothing* is settled. This, too, I find incredibly fun.

The problem for me is writing about the human brain in *this* book—a book about the evolution of sex differences. My task, you see, is to wrestle with whether men's and women's brains are functionally different and, if they are, whether those differences are tied to something innate. Each part of that task is surrounded by a sociopolitical gender debate so dense it threatens to obscure the science. But there's just no getting around it. There are notable Eves of the hominin brain, and most scientists think of those ancestors as the beginning of our true "Humanity." What's more, there has been a flood of research into sex differences in the mammalian brain over the past two decades. Thousands upon thousands of scientific papers have come out on the subject, from big stuff, like social behavior, to little stuff, like cellular structure. Given that we're quite obviously *mammals*, it would be terribly odd to expect that none of that research would apply to us.

In fact, after spending years digging through the literature on the subject from dozens of different angles, I can actually report that the oddest thing about our species might be that the female human brain doesn't seem to be all that functionally different from the male. Adult human "female" brains are remarkably similar, in nearly every way one can measure, from cellular structures to outward function, to adult "male" brains. That's not true of rodents: male rodents have distinctly rodent-masculine brains, and the females have pretty obviously female brains; both are clearly about the same *size*, proportional to their bodies, but the way a female rodent's brain reacts to something like a particular pheromone is drastically different from what a male brain does. And those kinds of differences between male and female brains exist across the mammal kingdom. Given that the female mammalian body, particularly the placental sort,* has to be prepared for the high-stress, high-risk series of events we call motherhood, it wouldn't be surprising to find features in their brains that might prepare them for it.

For instance, the parts of the brain that have to do with anxiety (and, by extension, vigilance and its relation to learning) seem to have, in rodents at least, significant sex differences. We don't yet know if that's true of *all* mammals, but it's certainly tempting to imagine why it might be the case, and likewise tempting to imagine why males of most species seem more likely to exhibit risk-taking behavior and general aggression when compared with females. Usually scientists tell that story by talking about testosterone surges. But typical male mammals also have more androgen receptors in certain parts of their brain than females do—which is to say, it's not just the signal level but the density of receivers. And by comparison, male mice are a titch less good at learning from subtle negative stimuli than female mice. In other words—perhaps because her amygdala is differently wired into the rest of her brain, including her memory centers—the female rodent doesn't need quite as much of a shock to her paw to learn to avoid part of a cage in an experiment, whereas the males need a good strong electrical swat.†

* Who are the ones who do the lion's share of offspring investment by literally gestating the kids, and nursing them, and then taking care of them through early childhood while they're learning how to be whatever sort of Siberian tiger or pangolin they happen to be.

† Since there are also known sex differences in pain tolerance, this might also have

Whether that's a mammalian base trait—whether, for instance, it will help us understand why human women are so much more likely than men to be diagnosed with anxiety disorders—is yet to be determined. But for other sorts of things that model mammalian brains tend to do a bit differently between the sexes—differences in pattern-matching ability, or ability to track complex social signaling, or a host of other things—human brains keep coming up the same. So, again, the biggest question is why aren't most women's brains *more* functionally different from men's?

It's clear that some people don't realize this is the case. In fact, many believe that there is at least some truth to a number of the uglier stereotypes about women's brains: that women are innately less intelligent, that women are more emotionally fragile, that women are overall less *capable* of doing Man Things with our delicate "Female Brains." After all, the proof is in the pudding, right? Aren't women worse at math? At directions? Why are so few Nobel laureates in possession of two X chromosomes? Given that we're in *this* book, we also have to ask, If there are sex differences in human brains, how might that have played into how big hominin brains evolved in the first place? Did our Adams get smart while our Eves lagged behind, their intellects sapped by the rigors of childbearing?

Fair warning: if we're going to ask questions like these, we have to take every one of those famously sexist ideas seriously. After all, if the proposition is that the seat of the Self is sexed—not just gendered, but *sexed*—then there should be some scientific data to support it.

---

to do with females simply finding lower voltages more painful than the males. In one recent study, the pain tolerance—that is, the threshold at which they jumped away from the foot shock—had males jumping at roughly 0.11 mA and females jumping at 0.09 (Yokota et al., 2017). That might not sound like a lot, but remember that mice have sensitive little feet. We didn't know about this difference until relatively recently, given the history of biologists and behaviorists mostly studying males.

WHAT DOES BEING SMART ACTUALLY MEAN?

That big cat who ate our Eve in the safety of her ancient den was smart. Like most big cats and hyenas today, she lived alongside populations of humanlike creatures. She hunted intelligent prey. She was a good problem solver, as similar creatures still are. Tearing a shoulder to shimmy a body through a crack requires spatial intelligence, forethought, some autobiographical memory, and a good deal of creativity. Like most large mammalian mothers, she probably recognized and even cared about her children. She made decisions in anticipation of their future welfare. Some might even say she had a Self, in the deep sense of the word.

You can do a *lot*, in other words, without a human brain. You can be very smart and very social and solve complicated problems.

Having a big hominin brain didn't save our ancestors from becoming prey. It might have even made them a target, since those brains were delicious. On top of that, the average adult human's metabolism is massively higher than a chimp's in part because our brains are essentially supercomputers that run on fat and sugar. Feeding and maintaining these things is neither easy nor straightforward. Betting on a big brain, in evolutionary terms, is actually *not* a safe bet.

But it's clear that somewhere between Ardi and anatomically modern human beings—the entire hominin line, in other words, from Lucy the australopithecine to Habilis and Erectus and all the way forward to human beings—our Eves got brainier than the creatures that preyed on them. And brainier than the Eves that had come before. Because brain tissue is so expensive, most evolutionary biologists assume the hominins built bigger brains because we needed to, for some reason. Our ability to do all this human stuff, like math and engineering and language and complex social mapping, depends entirely on the kinds of brains our ancestors started making millions of years ago.

If womanhood is something the *brain* does, then it makes sense to assume that the evolution of our brains shaped womanhood. The best strategy, in fact, for an investigation like this might be to work backward from what we know about sex differences in *modern* human brains to see what that can tell us about our past.

So we might as well start with my *least* favorite question, since it seems to be everyone else's pressing concern: Did men evolve to be smarter than women?

IQ

When we call someone smart, we usually mean that the person is really good at a specific subset of brain activity. Though it's true that the human brain is important for what an Olympic pole-vaulter can do, we don't usually say that athlete is smart. We say "athletically gifted."

We make similar judgments about artistic talent, even though it's more obviously based in the brain than athletics. We also don't tend to call people who seem really good at complex *social* tasks, like getting others to feel at ease around them, smart. We call them "likable" or say they're a "people person" or even, if we notice that they use such skills to advance themselves, a "gifted politician."

We usually reserve "smart" for those who are good at things like problem solving. Smart brains are ones that can quickly assess problems and find creative solutions. Smart brains are good at remembering things and using those memories where appropriate. They're good at learning rule sets, understanding symbolism, tracking patterns.

There are a few different ways of testing a brain's ability to do these things, such as standardized aptitude tests tailored for infants and children. These measure how and when kids meet certain benchmarks: how quickly they can track familiar faces, at what age they learn how to speak in full sentences. School-age children are tested on what they know and are able to do at certain grade levels, not just in specific subjects, like history or science, but overall, like being able to read and understand complex passages from essays, or being able to use basic math to solve problems. And then there are tests that aim to measure general features of the brain itself—how *capable* it is of solving problems. That's what IQ tests are for. They're designed to test your intelligence quotient: how well and how quickly your brain can learn new things and solve problems.

The IQ scores of boys and girls up to age fifteen are about equal. But at puberty, boys start to have slightly higher mean IQ than girls, implying that grown men are naturally "smarter" than women. If that's true, then the "Female Brain" might really exist—or start to exist—somewhere around puberty.

To test that theory, we need to figure out what those test results mean. IQ tests are weird. If you're American or attended college in the United States, you might have taken the SAT.* IQ tests are similar, comprising short games or puzzles you have a limited amount of time to work out before moving on to the next one. For question 1, you might see something like an IKEA assembly diagram—some box you need to imagine folding in the right way. To get the correct answer, you need to be able to "see" the result in your mind. Or maybe you have to sort a bunch of letters or numbers in a certain order or do a bit of code breaking.

It may seem like a parlor game on steroids, but IQ scores are given a lot of weight. Numerous studies have found that your IQ is strongly correlated with what you'll be able to *achieve* in life. It predicts how far you'll go in school, your possible income range, how many children you're likely to have, and even your longevity. IQ scores also seem to be strongly heritable: identical twins separated by adoption tend to have similar IQ scores, while fraternal twins do not. People who aren't directly related to one another, but have very similar genes, also tend to have similar IQs. Right now, most researchers think IQ is anywhere between 50 and 80 percent heritable, and the latest research proposes that it's closer to 80.

That does imply that every human being is born with a set potential for intelligence, hardwired into our genes.

But the whole idea of IQ is controversial. For one thing, white Americans tend to have higher IQ scores, on average, than African Americans. But if you control for family *income*, most of those differences disappear.† Similar problems pop up in tests like the

* The results of those two tests tend to be correlated: if you took the test in the 1980s or 1990s, just divide your SAT score by 10, and you've got your likely IQ score, plus or minus 10.

† Taking a test in a non-native language is almost always a drag on your score—most Americans would not do well on an IQ test in French. Meanwhile, one recent

SAT, where your score determines whether you have a shot at a top U.S. school; the differences in results here also largely go away if you control for family income. That implies that the way test questions are asked gives people with certain backgrounds greater advantages. It also implies that the way children are raised shapes their cognitive development. It's extremely stressful to be poor. It might also be stressful to be a girl in a typical test-taking environment.*

But if you test a large enough group of African American people, the variations in their scores will be *greater* than the average differences between that group and a group of white Americans or Asian Americans. It's impossible, in other words, to draw any meaningful conclusions about a racial group's "smarts" based on IQ scores. The bell curve of IQ test results for any group of human beings tends to have a long tail in either direction. There's too much variation—and too much overlap—to be able to associate IQ meaningfully with race.

The same can be said for the differences between the scores of men and women. The average woman and the average man will both tuck themselves neatly under the big hump of that curve. Where you tend to find the most difference is at the tails. That's why the mean shifts for men—they have wider variability overall,

---

study from 2015 did show race outweighing both family income and parent education levels as a driver for SAT scores (Geiser, 2015). However, the data were drawn specifically from applications to the University of California from 1994 through 2011 and show racial influence *increasing*, not staying steady (ibid.). In that same period, high schools in the United States became increasingly segregated, with as many as one in fourteen schools having 99–100 percent nonwhite students (ibid.). As such, race and class are becoming entangled again in educational settings: more nonwhite SAT takers are going to exclusively nonwhite schools, regardless of family income, and these "American apartheid" schools are famously underfunded and underserved.

* IQ tests are something you usually take in adolescence or later, and growing up in a stressful, impoverished environment tends to do things to your body, including your brain. What's more, your IQ scores tend to vary over your lifetime—tests designed for five-year-olds show a lower degree of difference between poor people and wealthier people than tests designed for eleven-year-olds (von Stumm and Plomin, 2015). Maybe instead of thinking of that as a "failure to thrive" because of some innate predilection for stupidity, it'd be better to think about that as potential evidence of accumulated harm.

but this shows up in some areas more than others. For example, if you isolate those things we call *mathematical* ability, male test takers have far more variability than females, with more male geniuses on one end of the tail and more male confusion on the other.

MATH

So let's dig into the math question. At some point in your life, you probably had to take a math class. You might or might not have enjoyed it. You might or might not have thought of yourself as a "math person," but I'm sure you've been told that women aren't as good at math as men are. You've probably also been told that's *why* there are more men in scientific and technical careers—why there are so many men at Google, at Facebook, at NASA, why nearly every scientist character you've seen in movies is a scrawny, under-sunned male wearing glasses.*

But brains aren't born with numbers in them. There's no wet bit of tissue in a baby's head that codes 2 + 2. Brains *are* supercomputers, but they have only so much original code. Everything else has to be learned. Boys and girls are both perfectly capable of learning math, but sex differences *may* make brains in XY bodies better at learning some things over others.

That's why it's important to sort out what "math" means here. Basic math involves tasks like counting and adding. It also involves problem solving that transforms symbols into ideas your brain can work with. Math also asks you to reason spatially, "moving stuff around" in your mind. When your brain "does math," it's usually performing a host of different cognitive tasks.

* These are also popular roles for South and East Asian actors, because, yes, Hollywood casting is sexist *and* racist. It is true that there are proportionally more South and East Asian men in math-related fields in the United States compared with Latinos or African Americans. But those ratios don't hold up in other countries—nor are Asian countries the dominant producers of math-related human knowledge!—so it probably says more about local culture (and the history of U.S. immigration and labor policies) than the innate math aptitude of Asian men.

Men and boys tend to do better on tests that involve spatial reasoning. If you ask a boy and a girl to rotate an imaginary 3-D figure in their brains, boys tend to do it slightly better than girls will.* That basic skill might influence all sorts of things in our daily lives. Adult men and women have subtle differences in their ability to navigate spaces, for example. Men tend to memorize paths more abstractly, while women tend to use *visual* landmarks around the path to remember where to go. This seems to line up with other sex differences concerning remembering specific locations—women are generally better at that, which may be tied to that visual landmark trick, whereas men are generally better at navigating virtual 3-D spaces.

But when you change some key features of spatial tests, you get different results. For example, if the tests involve humanlike figures, women do just as well as their male peers. Say you give someone a little map with a path marked on it and ask the person to imagine walking on that path and to write *R* or *L* each time they have to turn right or left. A man tends to do a little better than a woman does on that one. But if you include a tiny picture of a person on every corner, women do just as well as men. So, women may tend, ever so slightly, to pay better attention to other humans than men do, and to remember social details better as well. But that difference shows up less below age five and more from puberty on, so it could just be that girls are socially *trained* to pay more attention to other humans than boys are.†

Either way, the design of certain IQ test questions seems to reward male brains. At this point, no one knows if that's because male brains *excel* at the problems IQ tests set, and typical female

* If you give the test taker unlimited time to answer, however, both girls and boys seem equally capable of getting the right answer, which seems to imply boys may largely be better at answering questions about rotating 3-D shapes *quickly*, which could be a matter of confidence in test-taking scenarios, actual raw ability, accurate judgment of what one knows, or something else entirely (Loring-Meier and Halpern, 1999; Robert and Chevrier, 2003; Peters, 2005; Voyer, 2011).

† Or it might just mean the experiment itself is really brittle: if you can change an outcome significantly with a minor revision to your test, then the original results might not be trustworthy. Maybe the reality of how the brain navigates an imaginary path is just too complex for that experiment to deal with.

brains need a leg up to match those skills, or whether the tests are simply designed poorly.

Let's go back to that variability thing. On many measures of quantitative and visuospatial ability, men and boys have more *spread* in their results. More high end, more low end. Female test takers are more clustered under the norm.

Meanwhile, women and girls reliably outperform their male counterparts on many tests that have to do with language. This is especially true when the tests involve writing. And while the majority of males do a bit more poorly on language overall, once again males are spread wider, with greater numbers at the low and high tails, and much wider variability even under the bell curve of what's "normal."

But it's not enough to conclude that "girls are good at words" and "boys are good at math." The thing is, good math skills often *require* good language skills. Scientists, engineers, and mathematicians need to be able to adequately communicate their work to other members of their field and, ideally, also to people *outside* their field in order to secure funding and support. They also need to be able to read and understand the work of their peers in order to build on that work and engage in the major debates in their disciplines.

Even middle-school math requires a decent level of language skills to succeed, given that a number of math problems require that you write out your answers in explanatory sentences. Boys, as a rule, do a bit more poorly on those questions than they might otherwise do, despite generally doing better on SAT word problems in the math section than girls do. And as always, the effect size here remains rather small.

In other words, the evidence that the "Female Brain" is less smart than the male brain after puberty starts to crumble whenever you put pressure on it. IQ tests might tease out *some* sort of significant difference, but the results only sort of correlate with what other research has shown about girls' and boys' cognitive aptitudes. Of all the mess of what we do and do not know about sex differences in intelligence, the spatial logic piece is most compelling. But for math skills overall, it's complicated by the language bit.

USE YOUR WORDS

So we'll set the math question aside, because we must—the evidence that the Female Brain is the tiniest bit less attuned to math is both compelling and terribly wobbly.* Maybe that tiny difference drives the sex gap in STEM fields, or maybe it doesn't; the differences in tested ability are much smaller than the differences in who gets the jobs.

But the language test results are fairly robust: girl children do better at language tests than boy children, and those differences are still present after puberty. So, is the biologically female brain more innately *verbal* than the male?

Across multiple cultures, people do seem to think women talk more than men. But there are very few scientific studies that measure how many words men and women use in a given day. What's more, though there are tons of studies about how many words men and women utter in *specific* situations, the scenarios presented to subjects in a lab aren't exactly drawn from real life. For example, women tend to speak less in professional meetings where men are present.† This is also true in classroom settings.

* When differences are that small, and assigned to such narrow features of the broad range of cognitive functionality we call "math," there simply isn't much to support a broad claim of obvious male advantage in "math." There *are* sex differences in spatial logic, particularly in mental rotation tasks, and there do seem to be sex differences in *strategy* for a broader range of such problems. But most human brains, regardless of their sex chromosome arrangement, still manage to produce remarkably similar outcomes when presented with these tasks.

† Tannen, 1990. In a handful of studies on small-group tasks, women do speak slightly more than men, but they spend more of their time *reacting* to others' statements, while men spend more of their vocal time on the task itself (ibid). So a woman is more likely to spend her time using phrases such as "I like your idea, but are you sure . . ." while a man is more likely to *skip* that kind of social verbiage and jump to his ideas about the task at hand. We're talking here about college-age people and older, so these differences are more likely to be driven by learned social norms rather than anything innate. It's also true that many studies like this fail to control for *power* relations. For example, the Schwartz lab finds that gender-

But because speech in many such places is controlled by formal constraints, like being called on by a teacher, the likelihood of your being called on is the biggest predictor of how many words you'll speak. As a rule, women and girls are called on less in business meetings and classroom settings, and as a result they speak less than men.

And yet we seem to believe the opposite. That belief is so deeply seated it defies reality: when listening to recorded conversations, we're usually pretty good at estimating how much of the total time each person speaks if both participants are the same sex. But when we listen to a conversation between a man and a woman, we usually think the woman talks more than she actually does—even if she's reading a script with the same number of words as she'd spoken opposite another female.

So adult women aren't any gabbier. But maybe the stereotype comes instead from watching little girls. Because language isn't something we're born able to do, our general facility is often anticipated by how quickly we learn it. For whatever reason, girls produce their first words and first sentences at a younger age than boys. In those crucial early years, girls also have larger vocabularies and use a wider range of sentences than boys the same age.* And those early advantages pay off: in a recent large-scale international assessment, girls consistently scored higher on verbal tests.

But just like math, not all language tests are created equal. For example, the SAT verbal test includes a number of verbal *analogy* questions, where you're trying to determine whether one word is similar to another. Unlike most types of language tasks, this requires the test taker to build a conceptual map of relationships between different things, and males scored *higher* on that test than females did.

---

related communication patterns are also reflected in groups in which women have less social power, and males who similarly lack that sort of power can also have these patterns, even within the intimate bounds of private relationships (Steen and Schwartz, 1995).

* This is particularly true for vocabulary acquisition in the first two years, though other studies show these verbal differences persisting. Though there's some controversy in the science about this, the general finding has held true in research from 1966 all the way through 2008 (Lutchmaya et al., 2002; Halpern et al., 2007).

So, when we say "girls are better at language," what we really mean is that girls score better in verbal *tests*, depending on what sort of verbal test is being given.

Women *are* generally better at reading and writing. This is true at every age of testing from age five on, and the gap tends to increase with age until puberty and stays relatively steady thereafter. Large data sets from the U.S. Department of Education support this, and these sorts of differences pan out internationally as well. Across both language and cultural barriers, girls tend to out-read boys early in life and continue that trend throughout their lifetime. Men make up only 20 percent of the people who buy and read novels. The numbers improve for history and other nonfiction, but overall book publishers throughout the Americas and western Europe are selling books to women.

There might be all sorts of cultural influences around those sales numbers, of course. But it's worth noting that reading is the interpretation of written language and writing is the production of written language—very different cognitive tasks. As a rule, boys aren't great at either, but their scores are *much* lower on writing than reading.

There are a few different possible reasons for this. First, reading itself is a deeply strange activity. You're asking a human brain to tune out nearly all sensory information from the outside world for a long stretch of time in order to focus on a small area of somewhat obscure black markings on a white background, carefully shifting the eyes across those markings in a given direction. And while the eyes are so carefully focused, the ears are supposed to ignore any sounds in the environment so that the mind, meanwhile, can discern those markings as bits of language and immediately *interpret* that language without any of the usual cues speakers give: no facial expressions, no hand gestures, no useful variation in pitch. . . . Reading is an extraordinarily difficult thing for a human brain to learn how to do. Our perceptual organs evolved for the explicit purpose of carefully tracking the world. Millions upon millions of years have trained the eyes and ears to pay attention to what's going on around us. Our Eves' very survival depended on it. Human language likewise evolved in primate brains with our primate sensory organs. Our brains prioritize sensory information as we process any given moment in our day-to-day lives. They also do that for language.

So, it's not at all surprising that our species didn't manage to invent writing until roughly four thousand years ago, nor is it surprising that most human beings weren't even remotely literate until only a few hundred years ago. People who have difficulty reading silently for long periods of time should be the norm among our species, not the exception.

And maybe they are. As it becomes more socially acceptable to be forthright about difficulties with reading, the number of children diagnosed with reading difficulties of one type or another has increased accordingly. Not all reading problems qualify as dyslexia, but dyslexia is also fairly common. The mind of a person with dyslexia may *flip* the order of words or letters as it tries to read, sometimes even turning them upside down. This makes it difficult for dyslexics to read as quickly or as accurately as other people. For reasons that are still unclear, boys are two to three times more likely to be dyslexic than girls. Furthermore, given that schools aren't great at identifying these issues—as of 2013, less than 20 percent of students that researchers identified as having reading impairment were categorized as "learning disabled" by their schools—boys are probably not receiving help with their reading problems as they move through the education system.

Does that mean we're failing our boys in school? Unfortunately, maybe. As with math, the gap in reading ability widens as boys get older. And unlike with math, the reading gap is pretty robust between the sexes: from infancy on, male children meet verbal benchmarks later than girls, so it could be language in general that has a sex bias, not just the cognitively odd task of reading.

And then there's writing. Many writing tasks involve rhetoric, so they also require a high degree of both logical reasoning and social awareness, given that any successful argument tends to involve a high degree of *anticipation* of your reader's needs. You need to quickly create a simulation of your reader in your own mind and then shape what you mean to say according to how you anticipate your words will affect that person. So, if the Female Brain does better than the male at writing tasks—at least, in a testing environment—it may not be that women are necessarily more innately "verbal" than men. Maybe women score better on writing tasks because, for one reason or another, their brains are good at anticipating what other people want.

Whatever's driving it, what *does* seem clear, just about any way you try to measure it, is that the very few functional differences in general intelligence between the sexes don't add up to all that much. In boyhood, male brains lag behind in verbal abilities a titch, though they tend to catch up well enough. In girlhood, girls' brains seem to be pretty good at test taking of all types, falling behind in math more obviously by adolescence, except for a very specific subsection dealing with imagined 3-D rotation and a few minor other spatial tasks—and even there, the differences barely reach statistical significance. But let's come back to what happens in puberty a bit later. First, let's head to another major category of functional differences in the human brain: mental health and recovery.

THE FRAGILE SEX

The Female Brain is supposed to be fragile, an idea that's been around for thousands of years. Women are considered depressive, moody, hysterical, and easily prone to mental breakdowns. As has often been pointed out, the word "hysterical" comes from the Greek word for uterus. Until a little over a century ago, otherwise intelligent Europeans believed the uterus drove women to huge, disruptive emotional outbursts. Originally, Europeans thought an angry, irritable uterus could actually move, floating upward past the stomach and the diaphragm and into the throat, to somehow choke a woman's brain.*

* Though the mobile uterus was disavowed, "hysteria" remained: As late as the 1920s, clitoral stimulation was considered the proper treatment for feminine hysteria. That meant doctors—typically male—were obliged to stimulate moody women to orgasm in clinical settings. Hilariously, most of the doctors seemed to find the task boring and tedious, which drove the invention of the electric vibrator in Paris in the late nineteenth century. Far from a sex toy, the vibrator was explicitly meant to provide "hysterical paroxysm" as a medical treatment not only for hysteria but for a range of problems a woman might suffer, including constipation and facial wrinkles (Maines, 1999).

Though now we understand that the uterus doesn't move around, some of these ideas are still with us. Women are supposed to be "moodier" around menstruation, for example. And that can be true. There really *is* such a thing as premenstrual syndrome, and one of the common symptoms is mood instability or, for the unluckiest of us, what amounts to short-term bouts of clinical depression. Not all women get it and not all of those who do get it with every period, nor do all women have brain-based symptoms. But fluctuating sex hormones do seem to have a direct effect on many women's brains. There are two well-documented times in our lives when this happens: just before and during menstruation, and during pregnancy.

But does that mean the Female Brain is more *unstable* and fragile than the male's?

Let's dig in. The most obvious place to start is depression. From puberty on, women are more likely to be diagnosed with major depressive disorder than men are. Some of that might be diagnosis bias: maybe women are more likely to seek psychological therapy or be sent by others. It's also possible that women's symptoms might "look" more like depression than men's, even if the underlying cause is similar. For example, in data from the United States, boys and men tend to act out when in psychological distress, whereas women and girls tend to turn *inward*. So among people with mental health issues, a woman might be stereotypically more prone to things like self-cutting or severe diet restriction or becoming socially withdrawn, whereas a man might do things like punch a wall.* No one knows whether those trends have to do with fundamental brain differences or social training.

To try to get away from the diagnosis problem, then, we're better off looking at the points in many women's lives when predictable volatility in sex hormones seems to align with common diagnoses of mental illness. When we *know* a woman's hor-

* Though it's true that women significantly outrank men in eating disorders, at least some of that can be attributed to social pressure on women's weight. But men and boys do have eating disorders, and these diagnoses have been on the rise in the last twenty years (Galmiche et al., 2019). Many cases involve men and boys who spend a lot of time on social media, which can have an especially damaging influence on adolescent self-image and self-worth (Gorrell and Murray, 2019).

mones differ strongly from her body's norm, is she more likely to be depressive or anxious?

In the first trimester of pregnancy, women tend to report more emotional variability than they would otherwise. This can happen even *before* they are aware that they're pregnant—sometimes it's the symptom that prompts a woman to go buy a pregnancy test. She'll find herself crying at emotional movies, laughing hysterically at something that's not actually *that* funny, feeling angrier than she's used to feeling at little annoying things. But for a smaller subset of pregnant women, the general moodiness tips over into something more serious. If something might normally make her sad for a little while, she might instead find herself spending an entire day, or many days, unable to leave the house because she's just too *sad* to deal with the demands of regular life. Everything seems to hurt. Everything looks as if the color's been drained out of it. Nothing makes her feel *good* or *happy* or *hopeful.* It's as if her brain had somehow been detached from its reward centers.

That's clinical depression. Not everyone will get it while pregnant, but women are more at risk, especially right after giving birth. Postpartum depression (PPD) strikes as many as one in every eight women worldwide. Women who get it usually report feeling as if they've crashed through the floor. Instead of bonding to their babies, they feel detached, adrift, anchorless in a world suddenly rendered in gray scale. What's worse, many of these women feel guilty for having these feelings. As if they were not good mothers. As if they were not good *women.* But they may be suffering because they're *especially* womanly: PPD may just be a matter of how some women's brains respond to the normal, female assault of wildly changing levels of sex hormones.

Women's estradiol and progesterone rise sharply as their ovaries and uterus work to maintain the pregnancy, causing all sorts of effects throughout the body. Right after giving birth, those hormones plummet to their pre-pregnancy levels, usually within the first twenty-four hours. That quick crash can have devastating effects on the brain. Estradiol has a direct relationship with serotonin, which is part of how it influences the dilation of blood vessels. But it also seems to greatly influence the brain's ability to access and maintain overall happiness. The world's most popular antidepressants work directly on the serotonin pathways, increasing the availability of serotonin. Imagine a brain that's gotten used to high levels of circulating serotonin for around nine months. Now cut it by as much as half in twenty-four hours or less and imagine what can happen.

Women who suffer from depression also report similar, if smaller, effects when estradiol and progesterone levels fluctuate around menstruation and also while taking certain kinds of birth control pills designed to mimic the hormone patterns of pregnancy. PMS can make certain women's brains feel more depressed than usual, especially if those brains are already predisposed to depressive patterns. Bipolar women will also sometimes report more manic and depressive swings around their periods.

Young girls *don't* seem to be more depressive than young boys, meanwhile. And postmenopausal women who were used to depressive episodes on the Pill, or around pregnancy and menstruation, sometimes say they feel "freed" from these things after their sex organs quiet down (but if they'd suffered from depression previously, they're six times more likely to be diagnosed during perimenopause and the menopausal transition, making their forties and early fifties a bit of a rough ride).*

All of this does seem to paint a picture of the Female Brain as prone to emotional fragility—that because we have menstrual cycles and give birth, we're doomed to suffer more debilitating sadness and general moodiness than men. You might then also assume that women would be more prone to the extreme sorts of emotional instability—things like bipolar mood disorders, or hypomania, or any of a host of extreme outbursts of emotion. Funny thing, though: we're not. Though women are about

* Standard treatments for these women usually involve hormone therapy, SSRIs, and/or talk therapy with a suitable counselor. Notably, women who are having more symptoms in addition to their depression can be given a combination of hormone therapy and SSRIs, and there's even evidence for hormone therapy being useful all on its own (Clayton, 2010). As with most things for trans bodies, research in this area simply doesn't exist for trans men and nonbinary folk who have ovaries and reach menopause, though it may prove a sensitive time and deserves more attention from clinicians and scientists alike. As always, if you're a person with ovaries of any age, do take your body and your feelings seriously and ask your doctor if something's troubling you.

12 percent more likely to receive treatment for mental illness, men and women are *equally* diagnosed with psychiatric illnesses.

We do tend to have a slightly different set of disorders than men do—for instance, being twice as likely to be diagnosed with depression. Men are slightly more likely to be diagnosed with schizophrenia, which seems strongly genetically driven, and they're also more likely to be diagnosed with any disorder whose main symptoms involve violence and/or inappropriate social outbursts. Men are also more likely to have debilitating drug and alcohol addictions, which *may* be tied to some sort of innately masculine obsessiveness or compulsiveness, while women are more likely to be diagnosed with anxiety and self-harming disorders. But men and women are equally likely to be diagnosed with OCD. It's a bit of a Venn diagram of differences and overlaps, but the overall rate of occurrence of mental illness is probably about equal in men and women.

Among bipolar people, women patients do seem to have more depressive episodes than men—in that case, the she-brain seems more "down," and the male brain more prone to hypomania—but in terms of overall moodiness the sexes come out equal here, despite all the hormonal wonkiness that comes with menstrual cycles. More women than men, in fact, seem to have a milder form of the disorder.*

Let's be clear: no scientist or clinician in the world has a complete picture of how the brain falls ill with something like depression. We know, for the most part, what heart failure looks like. But we have absolutely no idea how a brain becomes depressed. We know that some people seem to be genetically predisposed to depression, and we also know that hormones play a role and that environmental stress likewise makes brains more vulnerable. A brain that's processing the death of a parent, for example, is

* But when they *do* have the more severe form of the disorder, they also tend to have a more rapid cycling between mood swings—four or more per year—and the rapid-cycling type is unfortunately less responsive to pharmaceutical therapies (Erol et al., 2015). That could mean that male bipolar disorder is driven by different underlying functional mechanisms from women's. Or it could mean that the hormone balance in the typical female brain is somehow interfering with how certain drug therapies work in these brains.

far more likely to become clinically depressed than a brain that's watching a sad movie. But no one knows why. And if the Female Brain is more *depressive* than the male brain, it can't be said to be more fragile because of it.

So how should we understand fragility here? A biologist might say, "Well, what actually kills you?" We have some data on that. If there's any single marker of a human brain experiencing organ failure, surely it's a brain that's become so ill it's managed to convince itself that jumping off a bridge is the best solution for its trouble. Women commit suicide roughly three times less often than men. That's a massive difference. Some used to think it had to do with the success rate—that men who try to kill themselves succeed more often than women because they tend to use more obviously violent methods, like guns, while women are more likely to take pills, increasing the chance that someone might save them in time or the attempt will somehow fail. That does explain some of it, and women patients do report more frequent suicidal thoughts than men do, but that depends heavily on self-reporting—it may be that men think about it but don't come into the clinic, or when they do, they're less honest. Whatever's driving the difference, the end result is clear: men end their lives dramatically more often than women. It's not exactly the battle of the sexes you'd want to win, but in this arena men are far ahead.

There are a few different ways of interpreting this. More women than men suffer from clinical depression, but most depressed people are not suicidal.* It is, however, a dangerous

* As the psychiatric community is discovering, one need not be depressed to be suicidal—they just *tend* to go together. In fact, 54 percent of people who die by suicide did not have a diagnosable mental disorder, which may be because they failed to receive treatment and be diagnosed or because the onset was simply too rapid or unusual to catch (Stone et al., 2015). A brain that appears to be functioning quite well can still have suicidal ideation, even to the point of acting on it. Being suicidal can even be something that comes on quickly in an otherwise healthy mind, without the stereotypical symptoms. Some famous recent examples are the rare people who have bad reactions to certain drugs, who suddenly become suicidal without much or any history of depression. (Sleep aids are one of the categories where this problem is known.) People with bipolar disorder are even trickier for the psychiatric profession: after starting treatment, some may find themselves suicidal, despite not having had any of those feelings before medication. To want to *end* one's life isn't always tied to feeling a lack of joy or reward over time.

comorbidity: people who become suicidal after suffering from depression may be more likely to act on suicidal thoughts. Still, among those people, women are significantly less likely to attempt suicide than men.

Researchers usually attribute this disproportion to women having a more robust social support network—when you have a reliable "web" of connections to other people, that "web" may be a mental safety net. Sometimes even just being aware of the web may be enough: you can rely on other people, but people also rely on you. And there may be some sex differences there, too: if women feel more social responsibility to keep on living, even when their diseased brains would rather not, then maybe that helps catch them when they're falling. Despite our uniquely female postnatal depression, being a mother makes a depressive woman far less likely to feel suicidal, and suicidal mothers are less likely to try. Unfortunately, this isn't as strongly true for fathers, not because men care less about their children, but (in this model) maybe because they have a harder time understanding that they're *needed* than mothers do.*

Still, not all men are fathers and not all women are mothers, and the large difference in suicide rates between the sexes can't simply be attributed to sex-normed features of parenthood. And while women in some societies do seem to have more robust social networks than their male counterparts, it's not true that men have *no* intimate relationships in those cultures. In fact, while some of the outward features of that intimacy might look different between the sexes, the overall feeling of "closeness" seems to be about the same, particularly as applied to "best friends." That means we can't just boil the suicide question down to men feeling less *close* to other people, though what might feel "allowable" to express in the space of that intimacy—for example, admitting suicidal thoughts—might have strong gender norms.

Okay. So, if there is a Female Brain, it may be more prone

* If that's true, the most obvious root is a social norm that makes motherhood more immediately "important" than fatherhood and ties a woman's worth to her ability to care for children. Men are left with a model of fatherhood that doesn't seem all that vital. So that's a case of sexism screwing over both the oppressed and the oppressor. We'd *all* be better off valuing fathers more.

to depression and anxiety and certain kinds of self-harm, but it's far less vulnerable to catastrophic failures like suicide. With the exception of things like postnatal depression, the Female Brain doesn't seem to be more fragile. It might even be more robust: for instance, men are more likely to wind up in the ER from severe traumatic brain injuries, but women are more likely to recover from them.

What takes that male patient a year to start doing again—say, walking, or speaking, or being able to dress himself in the morning—might take a female only six or seven months. This is true even if she suffered the same kind of injury to the same place in the head: the same amount of force, the same type of impact. This isn't because women are better at taking a hit in general. It's just that a female-typical brain seems to be better at repairing itself or even at preventing certain kinds of damage in the first place.

The main problem with getting hit in the head really hard isn't the actual spot where the brain gets crushed or cut, but runaway inflammation. When any part of it is injured like that, the entire brain will swell. If you don't give the swollen brain tissue room to expand, it will crush itself against the skull.

Most of the damage from traumatic brain injuries is caused not by external force but by what nearby cells do in response to the damaged cells. Similarly, when you have a stroke, it's not just the little bits of tissue starved of blood downstream from the clot that die. Lesions can form around those dead cells, and it's very hard for an adult brain to repair that tissue. The best it can do is try to wall off the danger zone, rerouting signals where it can.

Male brains seem to suffer more extensive inflammation and lesions around injury sites than females' do. And this might be because progesterone and the estrogens—the classic female sex hormones—have a protective effect on brain tissue, dampening that inflammatory response. If you do terrible things to a rat's brain in a lab, and then immediately dose it with a combination of estrogens and progesterone, that brain will recover faster and more fully in *both* female and male rats. In fact, as I write, clinical trials for human beings are under way to establish whether doses of female sex hormones will help people suffering from a recent traumatic brain injury to stabilize and recover. If those trials pan out, ERs of the future will have a ready supply of female sex hormones to help heal their patients' brains.

How exactly these hormones work is still unclear. Estrogens do seem to stabilize the blood-brain barrier, for one thing, which might help stave off extra fluid rushing in, causing runaway inflammation. Progesterone, too, seems to play a role in tamping down inflammation, as well as helping cells to chomp down on free radicals and other problems with oxidation.

Still, some of women's improved prognosis isn't simply that a typical female brain might be a super regulator of inflammation. Female patients also seem to have higher self-awareness in terms of their limitations after injuries and illness, which might lead them to take fewer unnecessary risks after they leave the hospital. What's more, in one of the rare beneficial twists of sexism, a woman's friends and family may expect *less* of her after an illness or injury, since they believe her to be more fragile than a man. As a result, they may distribute her life responsibilities among themselves, giving her more time to heal, allowing her to go slowly, rather than dive headlong back into her previous life.

But something else in the very cells of that female brain might help. If you culture XY and XX neurons separately, with no exposure to sex hormones, they *still* behave a little bit differently. It mostly comes down to how they die.

Every cell in the body—neuron or no—has to deal with stress. Sometimes they recover from stress, and sometimes they "choose" to die instead. If you dose both dishes of neurons with things known to stress out or even kill them, XY neurons die faster and more often.* The main reason for this, as far as scientists can tell,

* What's more, they die in different ways. The XY cells usually die in a way that depends on a pathway responding to apoptosis-inducing factor, the primary signal a cell responds to when the local environment "calls" for cell death. XX cells, on the other hand, usually die in a way that depends on cytochrome c, which can be used to induce or prevent apoptosis. That seems to imply female cells are dying from a failure to *prevent* cell death, rather than a response to a signal to commit hara-kiri (Lang and McCullough, 2008).

is that male XY cells have more difficulty dealing with oxidative damage.*

Take Parkinson's disease. Men are significantly more likely to suffer from it than women are. When women do get it, their symptoms tend to be different. Men are more likely to develop the characteristic rigidity, while women, no surprise, are more likely to suffer from depression. Women are also more likely to get dyskinesia—the problem Parkinson's patients have with uncontrollable movement. Diseases of the nervous system are mysterious, and Parkinson's is no exception, but the fact that it strikes more men than women, and that women who do have it tend to exhibit a different pattern of symptoms and disease progression, probably means that some part of most female brains are wired differently. That difference could be in how cells respond to hormones, or it could even come down to how the cells themselves deal with certain kinds of stress.

So the Female Brain is, by these measures, differently fragile, not more fragile, than male brains, depending on which question you want to ask. Some of that has to do with sex hormones, and some of that has to do with deeply coded differences in how cells with a Y chromosome go about their business of living and dying.† Obviously, neither sex is particularly *bad* at these things, or we wouldn't have a human population that's roughly 50 percent male. We do, however, have some marked differences in how the sexes go about making *more* of themselves, which, as we saw in the "Tools" chapter, matters when it comes to expanding one's territory. And in the end, that might be why modern human brains are so functionally similar between the sexes: The evolution of the hominin line wasn't just about surviving in *one* place. It was about building a body and a set of behaviors that could work for *lots* of places.

* Specifically, XY cells are a bit crap at regulating the amount of glutathione inside their walls, which helps protect against oxidative damage (Tower et al., 2020). So, if a bunch of cells *around* a male-typical neuron start to go, the neuron is probably going to die, too.

† More on that in the "Menopause" chapter.

PROBLEM SOLVING FOR PROBLEM SOLVING

Each time the world changed, our Eves changed. They were among the lucky: their bodies managed to mutate, adapt, and survive. Their children and grandchildren used those adaptations—slowly, slowly—to outcompete their cousins. We didn't *suddenly* have milk. We didn't suddenly start walking, either.

Australopithecines like Lucy were probably better adapted for walking than Ardi's kind, but they, too, spent a lot of time in the trees—in fact, some paleo-scientists think Lucy died by *falling out* of a particularly high tree, plummeting more than thirty feet to the ground. According to that story, she tried to catch herself, but her arms and wrists snapped, and when she hit the ground, her pelvis shattered. The force of the fall also shoved the long bone of her right arm into her shoulder, where it broke in four places.* Human ER surgeons see similar injuries today, when car crash victims try to brace themselves by locking their arms against the dashboard.

Our Eves survived the asteroid. They survived the earth splitting into separate continents. They survived the move into the tree canopy. They survived having to come *down* out of the tree canopy when the East African plateau shoved upward, turning their former forests into a mosaic of rivers and grasslands and little woods. At each turn, their bodies changed to adapt to that new environment and slowly adjusted to the new normal.

The biggest trademark of the hominin line isn't our big, fancy brains. It's the fact that we use those brains to survive just about *anywhere*, at any temperature, in any environment: Desert. Grasslands. Forests. Even the Arctic. The fossils of our hominin ancestors can be found in wildly different places, spanning eastern Africa, the Middle East, the Mediterranean, Central, South, and East Asia. Our brains are a big part of how we did that. The pressure to be adaptable may even be the reason we have them in the first place.

* It's also possible Lucy's skeleton shows fracture patterns because fossilized bones tend to snap and crush over time as the earth shifts around them. As with most of this sort of work, you'd need a time machine to know for sure.

Just as our Eves didn't get milk all of a sudden, the hominins didn't get big brains suddenly, either. The size of hominin brains slowly increased over millions of years. And then, over a span of about 1.5 million years, the brains of a wide range of hominins started massively expanding. This is also when you see the first hominins migrate out of Africa. This is when you see that effort collapse, followed by a *second*, more successful migration.

For many of today's scientists, the reason our Eves got so much brainier through this stretch lies in climate change.

Animals are generally fine with very short periods of changed climate, like seasons—cold some of the year, hot for the rest. But let's say the lake your species uses as a food source *dries out* in less than ten thousand years. And let's say a few of you manage to adapt to that cooler, drier environment. And then the lake fills back up, and everything is hot and sticky and wet again. How many of you are going to survive that reversal?

Not that many. And it's the species that are *less* specifically adapted to an ecological niche who are the ones most likely to make it.

Once upon a time, an Eve of the modern hippopotamus liked her rivers a lot. She was *so* adapted to her rivers and lakes that when they dried out, she died. The modern hippopotamus, meanwhile, is a bit smaller, a bit more omnivorous, and can move over longer stretches of dry land. If her river changes, she probably won't die.

The same is true of the ancient baboon. A very long time ago, *Theropithecus oswaldi*—massive, baboon-like creatures that weighed more than 120 pounds—roamed ancient grasslands. They had big, horsey teeth, entirely adapted to eating grass. Then their grasslands dried out and they died. The last fossil we have of them is at least 600,000 years old. Their cousin, the ancestor of today's baboon, was a little smaller, more omnivorous, more adaptable. And also, judging by the brain size of the modern baboon, a bit smarter. It survived.

This rule holds for *all* Mammalia, actually: historically, being omnivorous is the best way to survive. In a recent study of fossilized teeth, it seems the mammals that had more diverse diets were the ones that survived a massive planetary die-off thirty million years ago.* The specialized animals went extinct. Nearly all mammals are descended from Eves that were lucky enough to have mouths and guts that could make do.

Homo sapiens hadn't arrived when those grasslands that were home to giant baboons dried out—that would take roughly another 400,000 years—but the hominin line was in full swing. We'd already been wandering around eastern Africa for a good 5 million years. Ardi came and went. Lucy, too. And all the hominins you've ever heard of—*Homo habilis*, *Homo erectus*, *Homo rudolfensis*, the usual suspects—had been wandering the earth, doing chimpy stuff, and generally surviving in a variety of habitats.

But as time passed, those habitats became ever more sharply varied. Scientists have established that in a few ways. First, by looking for things like fossilized pollens and vegetable matter, we can tell what sort of climate hosted those plants. That's how we know Ardi lived in a mixture of woodlands and grasslands and that Lucy and most of the australopithecines did, too.

Another way to know is by looking at what was happening in the oceans. Tiny creatures called forams live on the seafloor, as they've done for hundreds of millions of years, long before there were mammals or dinosaurs. When they die, they leave a useful layer of microscopic skeletons. In those skeletons, traces of stable oxygen are woven into the matrix of the fossilized bone. One type is more common when the world is warmer; another when it's cold. So, if you grind up a little pile of foram fossils, you can get a pretty good model of ancient weather.

About six to seven million years ago—when our Eves split from chimps—climate change sped up. The weather started swinging between wet and cool and hot and dry in just a few thousand years. There's a lake, then no lake. There's a forest, then a

* de Vries et al., 2021. This was driven by a global cooling that marked the boundary between the Eocene and the Oligocene (ibid.). It was especially bad in Africa, however, because massive volcanoes blew up in Ethiopia about three million years *after* the world had already started to cool. If you're keeping track, that lands about halfway between Purgi and Ardi and squarely in the part of the world where Ardi was eventually discovered.

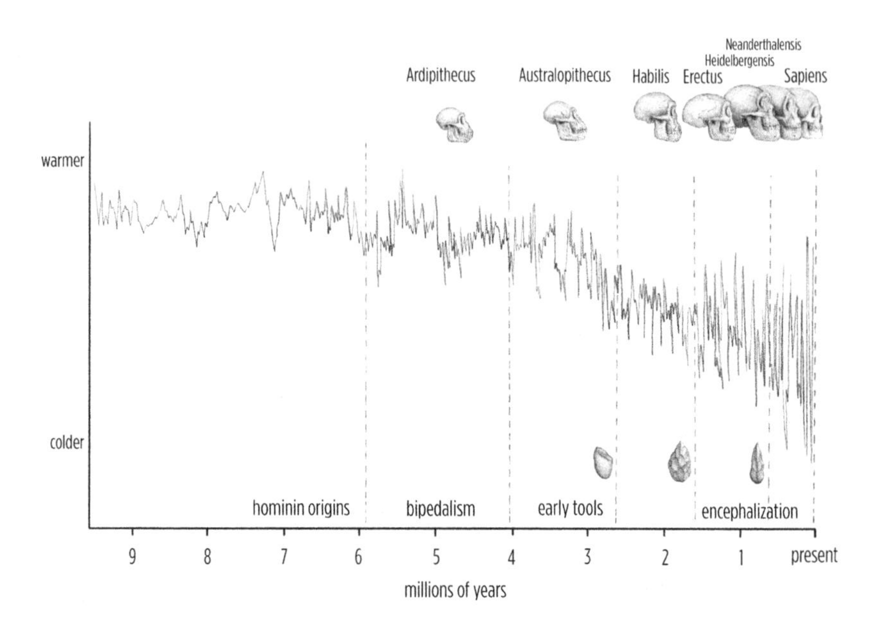

The variability selection hypothesis

grassland, then a desert, and back again to a forest. As a rule, simple mutations aren't going to be fast enough to adapt to a world that changes wildly every thousand generations.

But some species, instead of adapting to *specific* environments, evolve a set of traits and behaviors useful in many different environments. This is called "variability" selection. Being omnivorous is a good example—having a particular food disappear isn't going to kill you.

Or even better than being omnivorous, what if you found a variety of ways to make just about anything edible? That way, no matter *where* you went, you could make the local food work for you. Cooking does that. Pounding tough plants with rocks does that. Breaking open bones with sharp tools does it. Learning how to store and transport water helps, too. Those are behavioral changes. Software, not hardware.

But to run such software, you need a bigger computer. More powerful processors. Faster memory. A nimble set of algorithms.

To learn how to change your behavior to make any environment work for you, you need a supercomputer.

And that's what the human brain is: a supercomputer that runs on sugar.

HOW TO BUILD A SUPERCOMPUTER

Human brains are structurally a bit different from our ape cousins'. For instance, we have a massively expanded prefrontal cortex. How, precisely, this helps us be so "smart" is still rather mysterious, but given that it's the most obvious physical difference, how much more our brains can do compared with a chimp's, and how many things go wrong when we damage those areas of the human brain, it's clearly a major player in why we're so different.

But here's the funny thing: when human beings are born, overall, our brain size is roughly equal to a newborn chimp's. We're a lot *fatter* than chimp babies, to be sure, and we only get fatter from there, but our brains aren't so very different. What happens after we're born is the kicker: the biggest difference is what happened to our ancient ape brain when hominin evolution beefed up that frontal cortex and gave it a superlong *childhood.*

Chimps come out of the womb with brains significantly more developed than human babies': about 40 percent of their adult size for a chimp and a bit under 30 percent for a human. Some of that difference can be attributed to the fact we seem to be born roughly three months developmentally premature compared with other apes. But that isn't the whole explanation. Human babies also develop more slowly overall. Chimps are able to walk around by the time they are four weeks old. Though they'll keep developing for years, chimp brains are significantly further along than a human baby's by nine months. Human babies can't even crawl until six months, at the earliest (many need closer to ten), and they usually won't take their first upright steps until twelve to fourteen months. By the time they are two, their brains are still only about 80 percent of their adult size.

This is a huge part of why the newborn human skull is basically *soft*, with two gaps between the bone plates called fontanels. This seems, on the face of it, like a terrible idea: Why come into the world with two giant soft spots right over your *brain*? One good hit and you're done for. But that's just one of the developmental trade-offs that human evolution has made. In order to get a brain to grow to such an enormous size, we can't have bone blocking the way. But we also can't do what the chimps do and build brains to 40 percent of their adult size in the womb. If our bodies tried to do that, it'd kill both mother and fetus during delivery (or in a metabolic catastrophe well before then).

So that means somewhere deep in the hominin line—somewhere between Lucy and *Homo sapiens*—the hominin genome started messing with three things inside the womb and in early childhood: skull, brain, and fat.

Let's start with fat. Human fetuses build up their fat stores in the third trimester and continue building body fat throughout infancy and early childhood. Some of that is about hedging in case of a decrease in the mother's milk supply, but our kids *need* to hedge so much because our brains are so greedy. Since brain tissue is the most expensive bodily stuff to build, our kids have long evolved to dump every bit of fat they can into storage.

Also, human babies' metabolisms burn white-hot. Newborns drink 16 percent of their body weight in milk every day for the first six months of their life. To put that in perspective, an average 150-pound woman needs to eat and drink only about 5 percent of her body weight per day—a third of what newborns need. Babies put a massive portion of all that energy and fat and protein directly into building their oversize brains.

After our brains reach 80 percent of their adult size at age two, we take a much longer time to build the remaining 20 percent. Our brains aren't done internally organizing until somewhere in our early to mid-twenties. Probably the biggest innovation the hominin line came up with is the long childhood, which is precisely the reason we're as clever as we are—it's not the size, if you like, it's also how you build it.

The two basic tactics our bodies use here are bloom and prune.

First, there's hardware. As the brain grows bigger in those first two years of life, neuron stem cells seem to *migrate* from one portion of the brain to another, building out the frontal cortex massively and laying down highways between this "higher order" brain region and the areas that control movement and sensory information.

Some sex differences do seem to show up in this process. For example, as I've mentioned, girl babies babble and talk a little earlier than boys. They're able to maintain eye contact and point to things they want and generally communicate with their caretakers a little bit earlier, too. Even their fine motor skills tend to outpace the boys: girl babies are better at manipulating toys, eating with utensils, and (eventually) writing and drawing more clearly. Boy babies, meanwhile, tend to squirm and kick a little bit more than girls and to reach *physical* benchmarks involving large muscle groups a little earlier. But both girls and boys usually start walking at around the same age, so whatever boys were doing to build out the movement-related portions of their brains, girls manage to catch up in time for major locomotion.

No one knows why these developmental differences exist. One possibility is that boy babies are more likely to be born slightly prematurely—maybe because of some mysterious immunological conflict with the mother's body, or for some other reason—and even slightly premature babies usually take a little longer to catch up to their peers. But it could also be that the influence of sex hormones in the womb does something to how the brain builds a plan for itself. And it might have something to do with how the brain blooms and prunes. The human brain reaches peak synaptic density—that's when the most neurons are the *most* wired to other neurons—when we're around two.* Then the brain starts violently pruning itself back, like an overzealous master gardener. Glial cells move in and gobble up synapses. Inhibitory cells start damping signals in some pathways, effectively increasing the strength of signals traveling nearby paths, a bit like redirecting traffic. The brain of a standard toddler is effectively *rewiring* itself, dramati-

* That's part of why toddlers start seeming so much smarter all of a sudden. It's also part of why they throw so many tantrums: the emotional centers of the brain are more densely connected to every other part of the brain, the theory goes, and once you start a kind of experiential emotional "cascade," it's kind of hard to stop.

cally reshaping the material it just built. One theory, in fact, for the development of childhood autism has to do with this pruning process—some scientists think certain kinds of autistic brains over-prune or under-prune some regions, leaving others alone.

We don't know precisely when this modern pattern of brain development evolved, but we do know that ancient *Homo sapiens* were already on the path to an extended childhood, given how radically our life patterns differ from those of chimps and bonobos. Wild chimps enter puberty around seven years old, with females reaching reproductive maturity around ten and giving birth for the first time anywhere between ten and a half and fifteen; they are considered "subadults" until age thirteen or so. Males, meanwhile, begin ejaculating around age nine, but don't reach their full adult weight and physical maturity until age fifteen. Because social factors heavily influence the likelihood that any chimp male would be able to father a child (having *access* to fertile females kind of matters here), male chimps are likelier to be fully adult by the time they successfully pass on their genes.

We don't know for sure if Neanderthals, despite their bigger brains (massively larger than Erectus's and all the Eves before her, competing in size even with our own), also had these chimpy childhood patterns: maturing faster than we do (and possibly dying earlier, too). But if *Homo sapiens* did capitalize on childhood to the nth degree, that may be part of an explanation for why we managed to succeed where Neanderthals ultimately didn't.

Nowadays, human boys tend to catch up to girls on most cognitive things by preschool (ages four to five), but not all differences go away. As I've mentioned, girl students tend to receive higher grades in school, across all subjects, until puberty. Then it all goes to pot.

So why do teenage girls who *previously* outperformed their age-matched male peers start to fall behind?

MORE BLOOMING AND PRUNING

If you're on the hunt for the Female Brain, there's no way to ignore adolescence. After all, that's when most human bodies become sexually mature. Testosterone pumps out at a massive rate in male adolescence; the same is true for estradiol and the other estrogens in female teens. Both shifts in hormone profiles are known to influence brain development, so predictably teenagers experience significant brain changes. Having something tilt brain development in one direction or another is bound to influence its functionality.

One of the biggest things the human brain needs to do in a sexual body is carefully map out its shifting role in the local social environment. It's not just a desire to get laid; in most human societies, once children are reproductively mature, their responsibilities change, sometimes quite suddenly. As they move away from dependence on their parents, human beings in every known culture need to learn what "independence" means for their day-to-day lives. Human societies usually mark these transitions with formal "coming of age" rituals—some before the social life of the child markedly changes, like the Mexican quinceañera at age fifteen, and some closer to young adulthood, like the American tradition of getting ridiculously drunk on the night you turn twenty-one. There are graduations, religious ceremonies—for example, the bar and bat mitzvahs, the Catholic confirmation. Some of these rituals so symbolize a new identity that you actually change your name.

And then, of course, there's marriage, which for many cultures signifies the final border between adult and child.

That's a lot of cognitive work. But our brains' development patterns seem hardwired to handle it. Though all of this research is white-hot and new, a number of different mechanisms seem to be at work. Stem cells in the brain seem to migrate outward toward the frontal cortex, blooming in little clusters in these areas as the brain grows and reorganizes itself. The adolescent "bloom" isn't nearly as prolific as a two-year-old's, but more of a mini growth spurt, usually timed to the growth of our long bones. So just as a young man is groaning through the night because of the painful stretching of his ligaments and bones, his brain is also growing.

But it's also *changing*. There's a massive, secondary "pruning" process that occurs during puberty as we eliminate some of the synaptic connections we've built from toddlerhood through the preteen years. There's also a big insulating task going on, with key pathways getting extra myelinated (the fatty coating over nerve fibers), especially in the corpus callosum.

Girls tend to start this process between the ages of ten and twelve, and boys start later, typically between the ages of fifteen and twenty. While female and male brains prune themselves roughly the same amount, males prune later and *faster.* That might be one reason why schizophrenia hits boys so hard and so predictably in mid- to late adolescence, whereas female schizophrenics don't typically fall ill until their mid- to late twenties. These shifts are also tied to depression and pathological anxiety—"teenage angst" is a real thing in the brain. As all that pruning and myelinating tapers off, most brains adapt just fine. But maybe because of genetic vulnerability, or maybe because of environmental influence, some people's brains don't.

Whether during the toddler transition, the adolescent struggle, or any of those long years in between, what children's brains are doing the most is social learning: paying extremely careful attention to what others want, trying to predict those wants, and likewise trying to figure out fast and dirty ways of communicating their *own* wants to others.

Take coffee. Human toddlers don't know that it's a *bad* thing to harass their mother with endless requests before she's had her morning coffee. Older children have no problem learning that rule, and thousands of other social rules like it. It's not necessarily that older kids "care" more about their effect on others, but rather that they've managed to learn a set of parameters for dealing with their mother's cognitive state. Babies know when they're being paid attention to by eye contact and physical touch and are prone to cry if they don't have those things. But older kids learn to realize that they don't need to say "Mommy, look at this" more than once or twice to understand that she probably heard them and will eventually look at the thing. What's more, they've learned that she might get *irritated* if the pestering continues. That's a theory-of-mind task. And theory of mind—building a model of another's internal cognitive state, mapping out its potential desires, and communicating accordingly—is something human beings are extraordinarily good at.

Toddlers, for example, can sit at a table and point at something they want. Chimpanzees, however, seem to feel the need to get up, clamber over the table, gesture wildly, and continually look toward the thing they want and back toward their caretakers. Chimps communicate with a combination of brute physical gestures, vocalizations, and facial expressions. Most human kids, no matter how "hyperactive" they might be, seem to "get" that you can simply point at a thing, make sure someone else saw you point, and anticipate that the other person will understand what you mean. That means they're good—probably *innately* good—at quickly building shared social understanding.

Some of that may be a deep feature of our hominin line. But a good portion of it also just has to do with how human mothers and children interact, and how older humans interact in view of the child. For one thing, kids learn how to point in part because they *have* to, since unlike chimps they can't get about on their own until they're at least seven to twelve months old. If a human baby wants something—an object, or to go somewhere, or to *stop* being trapped in a high chair—they have to ask others for help.

Being handicapped in this way might be part of how baby brains become more human—they have no choice but to *ask* for stuff. They have no choice but to become better at communication, and specifically *referential* communication. Chimps don't have to do that for very long, because their bodies give them independence earlier than human babies get it.

It's a chicken-and-egg problem: Did humans evolve to have a needier first year of life because they grew brains that could accommodate that neediness? Or did we grow brains that were good at social communication because our relatively handicapped babies needed to figure out how to ask for things? We'll never know. There's also no reason it couldn't be both. How we build our supercomputers has a lot to do with childhood training, so any minor shift in the genome that influenced fetal and child development could potentially tweak our ancestors' brains. And once our climate became so very unstable—maybe somewhere around *Homo habilis*—the general trainability of our babies' brains would have made a huge difference in their ability to thrive. That general ability would serve such an important end that either sex would need it. In other words, maybe the reason human brains have so few sex differences in their overall functionality is that the need for that adaptability *overrides* many of the built-in sex differences left over from our mammalian heritage.

Our Eves' children needed to learn how to solve problems in their environment—not just specific problems, but *any* problems. Social interdependence is a very good hack for solving a host of problems because it builds a server bank, if you like, of supercomputers, instead of just stand-alone machines. To learn how to do that, you need to spend *years* carefully training your social brain.

Which may be the real takeaway for most questions about the Female Brain—not simply what it is, but *how we build it.*

MOM BRAIN

Brains don't simply arrive fully formed at birth. In fact, hardly *any* part of the body is near developmental completion when we take our first breath. That's normal—nearly all life on Earth has planned life phases. For animals, that's usually the egg, the embryo and fetus, newborn or neonate, juvenile, and reproductive adult. As we discussed earlier, the transitions between life phases are often dramatic and involve all sorts of bodily reordering. For something like a butterfly, that can mean entirely losing one's jaw. For human beings, the visible processes are usually a matter of stretching and lengthening and thickening, with—for females—some obvious breast budding during puberty. But deep inside the human brain, most of these phase transitions also involve a lot of that characteristic blooming and pruning and general violent reordering. That process is simply part of how we build our giant human brains and put them to work over our lifetime.

So on the hunt for the Female Brain, we're going to have to talk about human childhood. But first, we're going to have to talk about mothers—because clearly, while we build some portion of the human brain in the womb, there's a rather active human brain in the body that just so happens to *house* that womb, and likewise goes on to support that new baby brain once it's out of the womb.

And there's one part of the Female Brain I haven't really talked about yet: if what is unique to humanity is our brain, and further, what's *most* interesting and unique about our brain are the ways we seem to have evolved to take advantage of the normal physiological changes that come with these predetermined transitions in the human body's natural life cycle—newborn to juvenile, juvenile to adolescent, adolescent to adult—then we can't ignore the fact that some adult women do go on to have babies of their own.

The reason that matters for the evolution of human brains, of course, is that pregnant and breast-feeding women just so happen to have brains that are doing *very* similar things to what human brains do at other major transitions in our body's life cycle: they violently rearrange themselves. A pregnant woman's brain will, quite reliably, shrink in volume by as much as 5 percent during her third trimester, followed by a steady rebuilding during the first few months after giving birth.* Similar things seem to happen in other mammalian mothers, but it's particularly dramatic in human brains.

The pregnant brain doesn't shrink everywhere—the volume loss is most notable in areas of the brain strongly related to how we humans go about building emotional attachments, general learning, and memory. Some researchers suspected it was mostly fluid loss (brains don't have noticeably fewer *neurons* in late pregnancy, but lower general volume), but many now suspect it has to do with a large, quiet, and ultimately violent cutting away of synaptic connections: particularly in gray matter, and particularly in specific brain regions (though losses have been measured in many different areas).

Thus, human women might have evolved to be capable of an *extra* phase of brain development, of much the sort all humans go through when we're children: a deep pruning that precedes a massive period of social learning.

* From what I could find in the literature, it doesn't seem clear if this change occurs in *every* third trimester and postpartum period for human beings, or just the first—one presumes a shortage of cadavers and MRI scans of pregnant and postpartum brains to properly answer that question.

No male body will ever experience this phase of development.* No woman who lives a birth-free life will, either. This phase is unique to pregnant women who make it to the third trimester and then give birth. It's something the human brain does, presumably adaptively, to prepare for the intense phase of life that is to come: caring for an extraordinarily needy human newborn and then continuing to raise that child in deeply social settings for a very long time. Not unlike adolescence, however, those brain changes do seem to come with a short-term functional cost: problems with short-term memory, emotional regulation, sleep dysregulation (not simply from an uncomfortable body, but also from hormone levels rising and falling in the brain itself). It's kind of a mess inside a pregnant mother's third-trimester brain, and likewise in the early months after giving birth. But like adolescence, so long as we survive it, it thankfully ends, and our newly shaped brains are better able to handle the life that comes afterward.

I'm not implying here that women become what we're ultimately "meant to be" when we become mothers. That's wrong. Women who never give birth are perfectly prepared to continue through the rest of their adult lives as fully functional, fully productive members of human society. But for those who give birth, a human mother's brain, as exhausted and buggy as it feels to use, uniquely adapts to succeed at this extremely difficult task, mean-

* Some brain scans of new fathers do show some structural changes in regions similar to new mothers (Diaz-Rojas et al., 2021), but clearly no male goes through the same sort of preparation for these changes the way mothers in their third trimester and during early breast-feeding do. This doesn't mean men are innately less capable of the sorts of cognition required to be a good parent—I've met plenty of tremendous fathers—but it does mean most women's bodies evolved to have some arguably useful cognitive ways to prepare for new motherhood that seem to be, like puberty, triggered by dramatic hormonal changes. The study that investigated new fathers' brains notably concluded these fathers had anticipated their fatherhood for considerable periods of time, were largely housed in the same place as the pregnant woman involved, and were deeply invested in child care during the newborn's early life—which is to say, there are life history and cultural issues at play that may boost male brains' responsiveness to fatherhood, which may not prove true of the brains of all fathers. As with most things involving the human brain, social influences have cognitive outcomes.

ing it unlearns quite a lot of how we've gone about a day and learns new ways to do things.

We need to socially bond with our infants, because, let's face it, that's just about the only way we will reliably choose not to kill them.* We need to be able to recognize their needs and try to fill them and, above all else, learn to communicate with them as they radically fail to be able to talk for *years*. What few people in the scientific community have written about, however, is the social learning that the mother needs to do as she adapts to her new role as a mother in a *community*.

It's a common complaint: We tend to forget women when babies are around—all eyes go to the baby, whether they're social eyes in a living room or academic eyes falling on the idea of babies and women and how they might have evolved. But just as it's wrong for mothers to find themselves suddenly invisible behind their new children—as translucent as a thin curtain blowing in front of a window—it's also strange, scientifically speaking, to think that human motherhood is only about a mother's relation with her offspring. New human mothers, profoundly bereft of sleep and general wellness, recovering from a typical round of pelvic trauma and—especially if it's her first—the daily injuries of breast-feeding, need to learn *how to be* in their social networks. They need to learn how to ask for the things they need, even to realize what those things are. They need to reevaluate many of their relationships given their new life circumstances—which of the people around them will be most useful in child rearing? Who can be trusted with shared care of the baby? What new things will be expected of the mother, and what old things that *were* expected of her will change? Are there social norms to support all this? Are there ways around those norms when they're not working? *Whom do I trust? Whom do I lean on?* These adaptations would have been true for ancient human mothers, too. And as our ancient societies became increasingly interdependent, the more complex the social rules of motherhood would inevitably have become, too.

* When a small baby is screaming at 3:00 a.m. and there seems to be very little you can do to make it stop, loving that child helps tremendously.

What I'm saying, in other words, is that human women's brains seem to have evolved a process, unique to pregnant women and new mothers,* that helps them adapt to the deeply ancient, ever-challenging sociality that comes with human motherhood, and that this process is neurologically violent.

In that light, motherhood is not the *completion* of womanhood by any means. That is the last thing I mean to imply. But because motherhood has long required a uniquely challenging period of social learning for human women, it shouldn't be surprising to find that human women's brains may go through a unique phase of brain *development* to prepare them for those profound challenges.

Much like puberty, this phase seems triggered by a specific sequence of hormonal shifts in her body—in this case, the ones that naturally occur as she enters the third trimester of pregnancy and prepares for birth and breast-feeding. And though her brain will continue up that long on-ramp of social learning as she parents her child through many phases of growth, the most dramatic period of adaptation is likely during those first few, critical months of motherhood, which many call the fourth trimester, when her child is particularly needy, and motherhood, if it's her first, is particularly new.

So if what's unique about the evolution of human brains is fundamentally about our *childhoods*—that is, our extended period of social learning and the many things our brains do during those periods to optimize for living in deep webs of interconnected social groups—then maybe, when we think about human mothers and *their* brains, it's useful to ask whether similar processes might be at play to prepare women for especially challenging motherhoods. New mothers have to shove a lot of new information into their heads, and it looks as if we've evolved third trimesters that usefully make room.

And the timing of it all might matter: what isn't clear is whether this presumably adaptive feature initially arose in a way that overlapped with the standard sorts of brain development we now associate with adolescence. As I discussed in the "Womb"

* Not unique, necessarily, to mammals in general, but perhaps repurposed in our highly social, brainy, human-type lives.

chapter, most women in today's hunter-gatherer societies do not have their first periods until their mid- to late teens, and likewise don't go on to have their first child until a bit later than that. Whatever you might have heard about aristocratic marriages in Shakespeare's time, for most of human (and presumably earlier hominin) evolution, it simply wasn't the case that young teenage girls were ready to have babies. From what we can see in hunter-gatherer communities today, the beginning of ovulation happens in the late teens, which neatly aligns with the tail end of our long period of human juvenile social learning and brain development.* We don't know exactly when in the hominin line that would have occurred, but beginning the childbearing years later, when the "juvenile" period of brain development had largely taken place, certainly sounds like an evolutionarily sound strategy. There are obvious scenarios for how that would be adaptive, and of course, even if it weren't immediately beneficial, there are scenarios in which it could have been essentially harmless.† Either way, we carried the trait forward: for most of human history, human girls arrived at menarche when our bodies had enough gluteofemoral fat and bone growth (and a suitably low amount of daily stress) that becoming pregnant might not be too harmful, which also usefully aligned with a point in human-typical brain development that would allow for maternal brain changes to not overlap with earlier puberty.‡ And social norms frequently

* After all, if you're still in the thick of your juvenile period, why on earth would you go out and have a baby—few other species do something as silly as that. In fact, most social species' life cycles reflect both physical development and whatever social learning may be required for individuals to function as reproductive adults in their respective societies.

† That is, if it didn't significantly affect the likelihood that those girls' babies passed on their genes, then timing ovulation to occur later in brain development could be passed down without being directly "selected" for in the classic sense. Not all mutations are immediately beneficial or harmful. Some just happen to make their way into the gene pool because they essentially don't matter all that much for reproductive success. Whether those traits eventually become harmful or helpful is another matter.

‡ As mentioned previously, some girls are now beginning puberty as young as age eight, clearly before their bodies or brains are ready to become mothers! No one knows why this is the case, but researchers suspect rising obesity, genetic predisposition, and hormones (potentially estrogen-mimicking molecules in the environ-

reinforce a more appropriate date, whenever that physical ability arrives.*

Still, modern social norms, in the end, may actually be one of the biggest drivers for many of the stereotypes we have about the Female Brain. After all, while motherhood seems to be a major neurological metamorphosis in a female human brain's life cycle, quite a lot has already happened to that brain. For example, it's already gone through what we call, for lack of a better word, "girlhood." And until the world changes for the better, her daughters' brains will, too.

GIRLHOOD

There is a moment in every young girl's life when she realizes that she's being watched. That her body is a thing that's *seen*, and that men are the ones who are doing the seeing.

As a term, the "male gaze" means too many different things to be useful here. But this fundamental experience—this moment or loose assemblage of moments, somewhere between ages eight and fourteen, wherein a girl starts to *know* that being visibly female means being a thing that's seen differently—rings true for me. When I asked the women I know if they could remember it, the majority said yes, absolutely. Some had pitch-perfect memories of a specific event, usually on a sidewalk; others recounted a kind of creeping *feeling* that accumulated over time, a growing paranoia tightly wound in the warp and weft of their young Theory of Mind.

The members of older generations I talked to usually had

---

ment sending a false trigger to the ovaries, or some other unusual new factor) are producing an unusual combination that fast-forwards the onset of puberty (Winter, 2022). This is both new and potentially dangerous, and getting to the root of the problem is going to require both cutting-edge science and radically better networks of public health institutions and private doctors. Meanwhile, the best thing we can do to protect our girls from early puberty is a combination of healthy diet and reducing exposure to toxic chemicals, and meanwhile protecting them from adverse social reactions to their changing bodies.

* In some cultures, social norms allow a girl to become pregnant when she's clearly not ready. That never goes well, but more on that in the "Love" chapter.

more difficulty remembering a particular moment, though all agreed with the general principle. One of my professors at Columbia remembered reading James Watson's description of Rosalind Franklin, the woman whose work—nearly forgotten to history—was the basis of the double helix model of DNA. In his memoir, one of Watson's central complaints was that "Rosy" never prettied herself for the lab: he noted that she never wore lipstick to lighten her features and that "her dresses showed all the imagination of English blue-stocking adolescents." Somehow over two decades of scientific achievement, my professor had *forgotten* that lipstick was a thing. Reading Watson's idiotic account of the woman whose work enabled his own Nobel Prize was her reawakening to the reality of sexism.

My first awareness came at age eight in a rather clumsy scene on a Georgia sidewalk after I borrowed my mom's red high-heeled boots. I remember my reawakening more starkly. I was a PhD student sitting in the audience at a prominent scientific conference watching one of my mentors set up her projector before she was to speak. Behind me, I heard an older man say cattily to the person sitting next to him, "You know, a lot of older women are doing young hairstyles like that. I just don't know. I don't think it's appropriate." He seemed to be referring to my mentor's bangs. This is a woman who is a brilliant scientist and well known in her field. I sat in that little folding chair fuming. I wanted to turn around and make *him* look me in the eyes, maybe say something biting about his own hair and sagging jowls . . .

What is the cost of these moments? Individually, small: thinking about what I *should* say, but couldn't, interrupted my ability to properly focus on my mentor's talk. Cumulatively, these moments—as they often do for other women who do research—affect my ability to feel like a member of a scientific community that includes people like *that guy*. But the general cost of dealing with sexism as a woman—how these things accumulate, I mean, in the brain, over the course of one's lifetime—goes all the way down into some basic functional features. And that might finally give us a better definition of the Female Brain.

There are essentially two networks the brain uses to deal with challenges and threat. The first is the sympathetic-adrenal-medullary axis (SAM). We mostly use SAM in classic fight-or-flight moments, when things happen *fast.* Say you hear a tsunami alarm and realize you have to run. Your brain sends a signal to your adrenal medulla to pump epinephrine throughout your body. That's the same stuff ER doctors use to restart your heart after a heart attack. Epinephrine is what's going to let you run up the mountain to get away from the tsunami. It's what lets the gazelle run away from the lion.

The second network for stress is the hypothalamic-pituitary-adrenal axis (HPA). The HPA axis is what triggers the release of cortisol—the classic "stress molecule." You always have a little bit of cortisol in your body. If you need to be vigilant, cortisol is how your body is going to pull it off. But when cortisol levels are high for a long time, they disrupt the sleep cycle. They screw with digestion. They make short- and long-term memory a bit wonky. Cortisol suppresses the immune system. It hardens arteries. A little bit of stress is good; a lot of stress is famously bad.

The HPA axis is something our brains use in longer, grinding, "life stress" periods. Say your kid hasn't been doing well in school for the last couple of years. Say you know your company is going to close and you don't know what your next job will be. Or maybe you're a Black engineer at NASA. Or you're a woman in a position of power in a sexist culture.

"Stereotype threat" is real. Psychological research is pretty clear about this. If you tell a woman that girls are bad at math and then give her a math test, she's not going to do as well as a woman who wasn't exposed to that threat. This effect is astonishingly robust; it works at every age, in nearly every possible experimental scenario, and even when you're not testing women. When you tell male subjects that men aren't as good at interpreting emotion, they'll be worse at a test asking them to discern what facial expressions mean. If you tell Black subjects that Black people aren't good at engineering, they, too, will score lower in subject tests.

In people who encounter threat every day, the HPA axis is overactive. They're waking up with higher cortisol levels than people who aren't stressed in this way. After a certain amount of time, chronic stress causes knock-on effects in many different parts of the body. But in the brain, especially, you'll see that classic pattern: difficulty with memory access, generally slower processing, and higher distractibility.

You can see similar patterns in people who suffer from chronic pain or depression, and in refugees who've recently had to flee a conflict zone. Too much cortisol every morning. Too many random bursts of epinephrine. Too much, and too frequent, *vigilance.*

Then, if you experience enough low-grade stress over enough time, you'll tend to develop emotional and perceptive *detachment.* Such numbness is essentially what happens when the brain itself adapts to be less responsive to its own signals: cortisol has a lesser effect, and to get a boost, those brains require more epinephrine.

In universities, many professors and researchers work in fields that stereotypically aren't "for them." Women in STEM. African Americans in economics. In a number of different psychological studies, these individuals will often show a kind of "psychological disengagement" over time: feeling detached from their work and social interactions with colleagues, feeling less positively stimulated by their own research.

Human brains are long evolved to carefully track how each individual fits into a larger group. We each have specialized roles in our groups—roles that can shift, depending on circumstance. We spend *years* carefully learning how to successfully live inside our deeply social world. It's one of the most characteristic features of our species: that extended period of social learning. Our brains are built for it. Our species depends on it.

When you break a social rule, usually you suffer consequences. So, you learn to perform in ways that fit, and learn how to fake it a bit when you can't. You're not going to be conscious of most of these performances. That would take too much energy. You just know you're supposed to smile when someone else smiles. You don't usually need to think about that.

But what if you've learned you're sort of *always* supposed to smile? Even when it's not a direct, appropriate emotional response to someone else's smile? For example, what if you're a woman walking down a New York street and some guy on the sidewalk yells, "Hey, why aren't you smiling?"

That should count as a stressor.* It's a reprimand. It'll probably train you, consciously and unconsciously, to smile more.

But social monitoring also does *good* stuff. For example, being able to accurately spot opportunities to deepen social bonds with your friends and family and peers and colleagues is going to give you a more robust social support network. And women, as we've discussed, tend to have more robust social support networks than men.

So if women *are* more socially attuned than men, maybe it's because they *learned* to be: a matter of the sheer number of hours that women and girls feel obliged to devote to such skills. A kind of cognitive muscle memory. Do a thing enough, and you get good at it. Maybe teenage girls are in tune with the kinds of social threats that surround the idea of girls being good or bad at math, good or bad at school, good or bad at being competitive or ambitious, good or bad at being desirable. One way to cope with that threat is to obfuscate. Play dumb. Grown women do it all the time. Do we really think teenage girls don't? That they might choose to give up on their math homework sooner than boys because they believe they're not supposed to be any good at it? That they might choose to spend more energy on subjects that win them social praise instead of alienation and ridicule?

It's not right to say that stereotype threat makes teenage girls psychologically fragile, that they lack "grit." Sometimes, grit means faking your way through a minefield. So, in a world that punishes you for being smart, if you pretend to be less intelligent than you really are, does that mean you don't have grit? Or does it mean your mind is quickly, quietly, even unconsciously, learning the rules for how to survive?

In the end, if you want to find the Female Brain, it's not just about the hormones or the hippocampus. It's not about grades in

* The trendy term for this is "microaggression," but the outcome in any deeply social human brain is easy to name: it's stress. Like a fine-grit sandpaper, little bits of social stress can wear you down over time. The damage accumulates. One doesn't need to intend to cause another stress to do it; the smaller the act, the more likely that person didn't think about it in the slightest.

school—that's a symptom, not a cause. Really, until puberty hits, there's no reliable way to tell the difference between an XY brain being raised as a boy and an XX brain being raised as a girl. The same can be said of a young trans brain. You'll find some minor differences, maybe, between the amygdala and the hippocampus, and some structural differences in the olfactory bulb if you sluiced the whole thing down and had a computer count the number of cells. To find a model "Female" human brain, in the way most people mean it, you probably need to find an adult mind that's been convinced it's terrible at math, hyper social, sort of flighty, super moody, a bit fragile, and generally good at only a narrow range of things.

It takes a whole girlhood in a sexist environment to build a brain like that. You have to have gone through puberty as a female in that sexist world. You have to have *felt* that moment when walking down the street changed because men started to look at you differently. When your understanding of your life's possibilities began to shrink, and you felt powerless to stop it. You'd be hard-pressed to build that sort of mind in modern boyhood.*

As for XY babies who went through gender reassignment as newborns because of "ambiguous genitalia"? Let's say that until puberty those children had, for most intents and purposes, a stereotypical "Female Brain in Training." They had a girlhood. They were treated as girls. They were trained to be girls. Maybe, in some cases, they were more "tomboyish" than some girls, but some XX girls are tomboys, too. Maybe the XY girl wriggled a

* If you think this sounds suspiciously like Simone de Beauvoir, who held that "one is not born, but rather becomes, a woman," you're not wrong (de Beauvoir, 1949/2011). It's true many philosophers and feminist theorists would say that I'm guilty of "biologism," in that I don't believe *anything* we do is somehow exempt from the natural biological mechanisms that produce those behaviors. I do think our "meat space" is fundamentally what creates the human Mind and anything that mind might do. Complex systems naturally behave complexly, so being genderqueer is just as "natural" as being cisgender. And for me, given what we now know about human brains, the idea that "girlhood" (that is, childhood brain development as a female-identified person in a sexist society and the accumulated, influential, remembered experiences associated with those years) might be one of the driving features of the rather odd set of things that happen to so many adolescent girls' cognitive test scores is both true and ultimately freeing.

little bit more as a baby or enjoyed rough-and-tumble play. But some XX girls do, too.

And what about trans women? It's clear that trans women *are* women. Their brains create a gender identity because they're wired the way they are and went through developmental shifts the way they did. The vast majority of human brains naturally seem to create an understanding of themselves as somehow gendered. It's probably as instinctive and natural as a sex drive—older, in that case, than many of the other higher-order features of the human brain.* The trans experience of identifying as a gender is as authentic as anyone else's, and equally driven by ancient biology. Having a brain-based gender identity that doesn't neatly match a society's expectations for the rest of the body it's housed in doesn't make that identity less *real* than it would be in people who do "match." To put it in plainer terms, if your *brain* produces an experience of identifying as a woman, but your genitals happen to include a penis, does that mean your identity as a woman is less *real* than another's?

Absolutely not.

If having a physiological mechanism driving one or another trait is what makes that trait *real*, then having a brain do something is as obviously real as having a liver or a lung do something. It's true that no one knows what functional features of the brain make any given individual identify as a gender other than what that person was assigned at birth. But so what? We also don't know what makes a woman like me identify as a woman.†

* Gender fluidity and/or plurality also exists, and should likewise be welcomed, because clearly the best authority on one's own internal experience in constructing a Self and its gender identity is that Self. Most people who currently identify as gender fluid aren't especially *neutral* on the matter of gender—in fact, having strong feelings about this is usually what prompts that person to consider it in the first place. In other words, I suspect what's innate is a drive to construct *some* kind of gender identity in the deeply social lives of sexed hominins like *Homo sapiens*, and because we're the only primates who can talk, we're the only ones who can self-report our experience with depth and nuance.

† The fact that the fine nuance of biological complexity elides our simplistic understanding of the causality underlying gender identities is deeply important. For me, the scientific worldview is a reductionism of liberation: all atypical sexualities and gender identities are fundamentally "natural" because nothing a body does (including its associated mind, which is itself a product of the body) could ever be

I can say that it seems incredibly likely that such mechanisms would involve some or all the parts of the human brain that are known to intersect with general sociality, given that gender—as opposed to biological sex—is fundamentally a set of social behaviors tied to how one's self and one's body interact in a social environment. It's clear that there's nothing in one's DNA that codes for wearing a *dress*, in other words, but there might be things that "code" for being more likely to have positive feedback loops with social affirmation around gender presentation, or negative responses when one's internal sense of a gender identity doesn't seem to match with social expectation and/or when that person perceives negative social feedback. But given that plenty of cisgender folk like me—again, that's a contemporary term for people who are normally assigned one of two genders at birth and are generally content with that assignment for the rest of their lives, barring normal responses to living in a sexist and queer-phobic society*—have a range of comfort with their own gendered social experiences, and likewise have a huge range of how they process those responses and integrate them into an ongoing identity, what lies with one's genetic predisposition and what lies with one's social environment are not going to be easily parsed. And that's because the human brain is simply too social, too plastic, too malleable, too revisable to pin down like that.

As our world becomes less and less sexist, being trans will become less distressing for the people who experience it. If people

---

unnatural. Whether something is "immoral" is an entirely separate question, but as a humanist I'm simply never going to find non-harmful sexuality between two consenting adults immoral, nor would I ever consider myself a better authority on another person's gender identity than that person. What, a human brain did something unusual in its identity construction? That's not weird. A tenrec can have twenty-nine nipples. *That's* weird. Trans women are just women whose bodies are atypical. Tenrecs are the mammalian equivalent of "hold my beer."

* Being "content" with one's gender is a complex thing. I've never met a woman, cis or trans or otherwise, who 100 percent enjoys the experience of living as a woman or girl in a sexist society. . . . Sexism is real and it's awful and it's a part of the everyday lived experience of *all* women and girls. So you can't say all woman-identified folk are simply "content" with their gender, even if, from the outside, a woman's appearance seems perfectly acceptable under the umbrella of local gender expectations. One can be comfortable with one's gender identity and still be exhausted by the experience of living it.

of all genders are allowed to live however they like, and wear whatever they like, and talk however they talk, and take on jobs they find fulfilling, and do any of a host of things they might want to do, what difference would it make for a kid in a boy-typical body to feel she's better suited to living life as a girl? Why would she feel stressed about dressing differently if she were *always* allowed to wear whatever she wanted, even if her parents didn't know she was a "she" when she was born? And why on earth would it matter which toilet one uses if no one feels that the body is shameful—or more important, if no one assumes that seeing another's body would automatically make the viewer feel entitled to sex?

In that gender-egalitarian future, it's also safe to presume that the *stressor* side of stereotype threat will go down. Despite what you may feel about recent trends in the United States, that threat has been going down for more than two hundred years.* And so, because our girlhoods are different now, the average adult female's brain is also probably a bit different from what it was a hundred years ago. You wouldn't expect someone who'd been starved as a child to be six feet tall, even if her genes hold that potential. Neither should you expect a brain that's been effectively starved to reach 150 IQ, even if the genetic potential is there. Sometimes it'll happen. But it's harder. A Marie Curie in her day is actually more impressive than a Marie Curie today. It would have taken *that much more* to grow a brain in a female body that could fulfill Marie Curie's innate potential. It's easier for us to do that now. Not *easy*, but easier. And presuming the trend continues, it will only get easier going forward.

Not because girlhood is ever going to go away. Just because it'll suck less.

* Not under al-Shabaab, mind you. Not under ISIS. Not in the gaslit hellscape of Afghanistan. Not in the hidden enclaves of any misogynist religion. Not at the Supreme Court in Washington, D.C., either, or in any number of U.S. state legislatures. But for the rest of us, if you look at data from the past two hundred years, the trend is clear.

Homosapiens

CHAPTER 7

VOICE

History here is an oral tradition, legends passed from mouth to mouth, a communal myth created invariably at the base of the mango tree in the evening's profound darkness, in which only the trembling voices of old men resound, because the women and children are silent, raptly listening. That is why the evening hour is so important: it is the time when the community contemplates what it is and whence it came.

—RYSZARD KAPUSCINSKI, *THE SHADOW OF THE SUN*

So caught up was I in my assigned role as ingénue that I was perfectly willing to entertain the possibility. . . . He was already telling me about [it]—with that smug look I know so well in a man holding forth, eyes fixed on the fuzzy far horizon of his own authority.

—REBECCA SOLNIT, *MEN EXPLAIN THINGS TO ME*

VERMONT, TWENTY-FIRST CENTURY

Someone found him in a heap on the side of a country road, his motorcycle yards away. It was one of the worst traumas the community hospital had ever seen. The man was only forty-one. Underneath the mash of bone and flesh that used to be his face, he was struggling to breathe. The nurse tried to intubate him, but getting in through the nose was hopeless. When she attempted to slide the tube down his throat and into his airway, she hit swollen tissue. His heart was beating. His lungs and liver were fine. But if she didn't open his airway, he was still going to die.

What he needed was a cricothyrotomy—a "crike." By slicing a hole in his throat, they could bypass the swelling and feed fresh air to his lungs. The nurse couldn't do it, so she paged the surgeon on staff.

Cutting into a human throat is asking for trouble. The blood vessels that feed and drain the brain run through there, along with huge tangles of critical nerves. You also need to dodge the blood vessels and voice box. Cut in the wrong place, or in the wrong way, and you damage a patient for life, maybe render him mute. Or kill him. Most patients who have trouble breathing can be intubated. But most people don't usually fly off a motorcycle at speed and land directly on their face. The surgeon on call hadn't done a crike in twenty years.

Luckily, the community hospital was part of a new program that Vermont was trying out: telemedicine. The surgeon was able to hail a doctor who worked at a Level 1 trauma center at a faraway hospital and turned on the video camera, giving his expert colleague a live close-up of the patient's gory face and neck. The doctor on the screen agreed it had to be a crike, and it had to be now. Speaking slowly and clearly into his small microphone, he walked the surgeon through the procedure.

*First, find the Adam's apple on the patient's throat. Now feel for the next bump, down about an inch.** *Between the two is a membrane. That's your target.*

The patient wheezed, his lips turning blue.

The country surgeon, feeling as if he were in med school again, focused on the trauma doc's voice. His left finger found the spot. With his right hand, he brought the scalpel into position, gently touching the razor edge to the skin in the middle of the dying man's throat.

Vertical incision. One centimeter deep.

The skin gave way to the blade, uncovering a slick, fibrous membrane just underneath.

Now horizontal.

It was tough and took some pressure, but the doctor's blade sank through.

* That's the cricoid cartilage—connective tissue that protects the lower throat.

Now flip the scalpel. Push the handle in and twist it ninety degrees.

The membrane opened like a buttonhole in a jacket. Blood oozed around the metal of the scalpel handle and down the sides of the man's throat. The nurse was ready with the plastic tube, and the surgeon slipped it into the hole as he pulled the scalpel out.

The man on the table started breathing again, ragged at first, then slow and deep. On a monitor nearby, the numbers started going up: 60 percent oxygen, 70 percent, 80 percent, 85 percent.

They had no time to celebrate. Now the surgeon needed to relieve the pressure on the patient's swelling brain. He picked up the drill and bored a hole through the bone. It worked. When the man was stable, they transferred his broken body to the state's only trauma center, hours away in Burlington. The man would live.

ORDINARY MAGIC

It really is like a magic trick. Without moving anything, without building anything, with little more than a skittering of electricity along tiny threads spindling off the ends of cells, your brain tells your throat and mouth to make a sound. With just a few pulses of air, the sound jumps across space to someone else's ears, and in hardly any time at all—milliseconds—your idea arrives in that person's brain.

You didn't have to show her anything. You didn't have to pee on a lamppost or wave your hands. And yet you can deliver a dense package of information from an organ inside your body into *another person's body.*

No other animal in the world is able to do this. No dog can teach another dog how to do a crike by barking into a mic from hundreds of miles away. No chimpanzee can make that happen. No whale. *Homo sapiens* are the only animals, in the entire history of animals, that have managed this phenomenal trick.

We are the only talking ape.

We're so linguistic, in fact, that we've even managed to figure out ways to create language without any sound at all. Those among us who aren't able to hear, or hear less well than most people, can use their hands to make language. Just a few thousand years ago, we even figured out how to make marks to represent the words we make. That means brains can miraculously download ideas into other brains they'd never even met.

It might seem ridiculous for me to be making such a big deal of this. After all, to speak to another human being is such an ordinary, everyday thing. But it's not ordinary. Here on Earth, peeing is ordinary. Sweating is ordinary. Moving your body so that another member of your species can see what you're doing, and maybe even loosely understand what you want, is quite ordinary. So are most animals' vocalizations—they sing, they squawk, they bark and growl and hiss, conveying rudimentary "messages" that other animals can understand.

But those messages are usually as simple as a smoke alarm. And they produce simple, automatic responses that are hardwired from birth. Most animals emerge into the world ready to communicate with one another. Puppies already know how to "bow," hunching down on their forelimbs, to signal that they want to play. No one has to teach them this. Cuttlefish know how to change their color to say they're angry, rattlesnakes know how to shake their tails, and honeybees know how to perform their strange waggle dances to tell the rest of the hive where the flowers are.

No other animal has human grammar. They don't have *language.* They can't cook up complex ideas and dump them into each other's brains simply by swapping around the order of a few sounds. They can't teach someone how to open up a man's windpipe with a scalpel and insert a bit of tubing and then drill a hole in his skull to save his life.

Speaking to someone isn't ordinary at all.

And it's entirely unclear how, or when, our ancestors managed to pull it off. But every living human culture has language. We might have started talking as far back as 1.7 million years ago. Or as recently as 200,000 years. Some think it was only 50,000 years ago, which might as well be yesterday in our evolution.

There's no way to know for sure, but there are likelihoods—things that changed in our ancestors' bodies and behavior over time that made language more or less likely. When *Homo habilis* started making her stone tools, she probably wasn't speaking yet—the configuration of her throat and mouth and chest would have made it *very* hard to pull it off. Her immediate descendants probably weren't talking, either. Their throats were wrong. Their mouths were wrong. Their brainpans, too, didn't seem to have the classic shape that linguistic human brains do, with the right bulges in areas we now know are associated with language processing.

If that's accurate, then all those elaborate stone tools and early gynecology were learned and passed on through *direct observation* and super-simple gestures and sounds. Monkey saw, monkey did. Maybe they also had rudimentary sign language. We could have been using complex hand gestures long before we were able to make modern language with our vocal apparatus. Our cousins still do this: chimps utter gentle *ooo* sorts of sounds, combined with a hand outstretched, limp wrist, palm down, that translates roughly as "Hey, you're the boss, don't hurt me, I'm not a threat." But that's not the sort of thing that can teach a doctor how to do a crike.

For the vast majority of hominin history, we left little trace of our culture. So if we had language, we weren't doing a heck of a lot with it. The soonest hominins seemed to have a modern *vocal* apparatus—throat, jaw, and tongue in the right place—is only a few hundred thousand years ago. So that's the earliest we could have been physically capable of producing the complex vocal language we do today. Neanderthals, Heidelbergensis, *Homo sapiens*. Those three alone.

Once we had language, it would have quickly spread through the entire gene pool because it was so useful: suddenly you could problem solve en masse. No need to wait for innate behaviors to get encoded in DNA. You could hack your challenges in real time.

There is one point in human history, between fifty thousand and thirty thousand years ago, where innovation seemed to explode. Before then, we had relatively simple tools and very simple cultures. After that, we had rapidly diverse technologies. What's more, we had *symbolic* culture: Cave paintings. Symbolic carving. Burial cultures. We took our old stone tools and made *far* better ones. These innovations spread rapidly—up through the Mediterranean to Europe, back down through Africa, and fanning out into Asia and the far Pacific.

In other words, innovation spread at a pace that most scientists think must have *required* language. But we don't know how long we'd had it before this moment, and we don't know whether complexity in rudimentary language had somehow changed. But why did our Eves invent it in the first place?

Most stories about the origin of human language have been pretty male. Take a look at those cave paintings in Lascaux, in the Levant, and scattered through northern Africa: the smoky, rubbed-in lines of aurochs, deer, bison. What is humanity's earliest art all about? Hunting. These drawings are spare with details suggesting human sexual characteristics, but the assumption most make is that the hunters the cave artists depicted are male.

The majority of scientific stories about the evolution of human language fall in line: at each turn, human innovation has been driven by groups of men solving man-problems. One popular tale holds that language happened because we *became* hunters, forming large parties (of men) who needed to shout complex directions at one another across wide savannas.* But wolves are pretty fantastic hunters, do it in groups, come up with surprisingly complex plans for the hunt that depend on members performing diverse roles, and don't have a lick of language.

* We don't actually know whether most big-game hunters were men, of course. Among known hunter-gatherer groups, the gender roles vary, but men are strongly associated with big-game hunting. But ancient evidence from the Americas implies women were strongly and commonly involved in big-game hunting, which might not have been unusual for our preagricultural Eves (Haas et al., 2020). In today's better-known model, women often take on the more "traditional" role of gathering plant stuffs, processing foods that would otherwise be toxic, and hunting smaller and less dangerous game. In terms of how the sexes contribute to the group's total protein intake, however, it's a wash: even if the females gather only plant stuff, bugs, and small game, they're contributing just as many grams of protein to the group's total intake as the males do.

Also, most of humanity's ancestors weren't particularly adept hunters. If anything, we were scavengers and prey—the favorite snack of large hyenas and lions and pretty much anything else that managed to catch us. Many scientists think even *Homo erectus*, the likeliest candidate for big-game hunter among the more ancient hominins, still relied more on scavenging.

So a better theory might be that vocal language evolved among our more fearful ancestors, calling out to one another when they spotted a predator in their territory. Campbell's monkeys do that now—they have different alarm calls for eagles and big cats and can even convey which direction the threat is coming from. The "big cat" call causes them to scatter up into trees; the eagle call makes them duck. The warning calls are so flexible, in fact, that simply changing the order of the sounds seems to function as a kind of proto-grammar: *eagle up and west*, *cat down and east*.

Maybe males, with their larger, more muscular bodies and more powerful lungs, were the obvious choice for the job of warning the clan against such dangers—protecting those fragile females and vulnerable children. And once they had language, those male groups would have been supercharged. No more whining and gesturing. Now they could engage in all the complex problem solving and social interaction that human ancestors needed to do in order to compete, survive, and thrive. So maybe men were the drivers, and women were the gabby, backseat passengers—participants in the language game, not leaders.

Perhaps that's why in study after study human subjects like listening to male voices more than women's. Maybe that's also why men are so often political leaders, with their big, powerful voices that carry so well in large rooms. The great orators of history—Lincoln, Mandela, Atatürk, Churchill—are also male, nearly all of them more than six feet tall, with long, masculine throats and barrel chests, their voices as resonant as a drum.

I admit, giving men the credit for the most definitive human trait doesn't sit well with my modern feminist principles. But the history of humanity isn't kind or egalitarian. So, let's set aside how we want the world to be and take the idea seriously.

A TALE OF TWO CLINTONS

AND SO, MY FRIENDS, IT IS WITH HUMILITY . . . DETERMINATION . . . AND BOUNDLESS CONFIDENCE IN AMERICA'S PROMISE . . .

Philadelphia, 2016. Hillary Clinton was about to do something that no American woman had ever done. In a scene somewhere between a brutalist rally and a children's birthday party, thousands of cheering people had herded into the Democratic National Convention, surging between the folding chairs and cotton bunting and waving placards.

Twenty-four years earlier, Clinton had watched her husband, Bill, do the exact same thing: accept a major U.S. political party's nomination as its presidential candidate. She was ready. She was polished. She was arguably more prepared than any other candidate in the history of American politics. There was just one problem: Hillary's voice was failing.

It wasn't just because she was sleep deprived from excitement and relentless campaigning. It wasn't just that she was about to turn sixty-nine. No—millions of years of evolution had led up to this moment. She stood at the lectern, all eyes on her. There she was: the second of two Clintons, faced with performing the same feat of vocal prowess. And somewhere along the line, a series of unfortunate events made Hillary's voice *different* from Bill's. Different, because she was female.

PRESSURE

In essence, vocal speech is just an elaborate way of holding your breath. In the moment before Hillary tried to say, "I accept . . ." she needed to take in a sip of air that would last her to the end of that sentence. She wouldn't get to inhale again until her sentence was done.

This isn't nearly as straightforward as you might think. Our brains and diaphragms learn how to power our words with our breath when we're young. Babies can't do it. Toddlers are better, but still pretty lousy. Mature breath control, the sort adults use to talk every day, doesn't seem to kick in until age five.*

At any age, talking is hard work. That's because holding your breath gets in the way of delivering oxygen to your blood; the rest of your body blows through your reserves fast. Men have bigger lungs than women, which means they have more oxygen still circulating while they're talking. That's one reason the male Clinton found it easier to deliver his acceptance speech. He simply had more hot air to work with.

Not only is Hillary's body smaller overall than her husband's, but her lungs are *proportionally* smaller than his. Men have 10–12 percent more absolute lung volume per pound of body mass than women do, which means that at any given moment they should have more oxygen to yell out warnings about incoming tigers. More oxygen to make an escape. And more oxygen, presumably, to squeeze out really long sentences about the Democratic Party nomination without getting light-headed.

From the moment Bill Clinton was born in 1946, his alveoli—those little bubbles in the lungs where air exchange happens—multiplied a bit more quickly than Hillary's would after she was born the next year. As boys grow older, the comparative differences in lung growth only get bigger. When Bill hit puberty in the early 1960s, his chest expanded and deepened, forming that characteristic V shape, with the wide shoulders and straight waist. His throat also lengthened and thickened, muscles girding his widened jaw. His larynx dropped lower in his throat, forming the Adam's apple, and his cartilage and vocal cords thickened.

Teenage Hillary did a bit of the same, but to a much lesser degree. Her chest cavity got bigger, but not as big as Bill's. Her larynx dropped and her vocal cords thickened, but nowhere near as much as Bill's did. Just like Bill's, her lungs grew bigger to fuel

* This is why you may find musically gifted children showing their talents by way of an instrument before they are five—Mozart did that—whereas singers don't start until later. They don't have the voice control, and they don't have the lungs. Hand-eye coordination and pitch recognition start long before a child is capable of properly singing. My own son, now a toddler, spends half his day sing-yelling the alphabet, but his pitch and breath control? Not so good.

her growing body. But they stopped short of filling the space under her rib cage. And that's because women's ribs don't sit in our bodies the way men's do. Women's ribs pinch *inward* at the bottom, just a bit, which is a big part of why women's waists are narrower than men's.

Evolution endowed teenage Hillary with that female rib cage for a good reason: She needed room for future Chelseas. By the third trimester of a human pregnancy, the fetus is so large it pushes the other organs out of the way. The stomach and intestines are smooshed. The liver is crammed. Soon, it's pretty hard to take a full breath because all those displaced organs get shoved up against the woman's diaphragm. Over the course of her pregnancy, her tilted ribs shift to accommodate the new organ arrangement, pushing out toward the side walls of her torso. That's why heavily pregnant women look as if they have a wider back: those longer ribs are doing their best to help stabilize and shield all the organs shoved out of place by the growing uterus.

Neat trick. But not so great for all the time you spend *not* pregnant and could use more lung capacity. Like when you're addressing the nation in one of the biggest moments in American political history.

But that's not the only challenge Hillary faced. She also had to maintain even *pressure* in her lungs as she slowly deflated them to power her speech. Our lungs shouldn't be able to do this, really—the pressure should substantially decrease the longer we speak, like a balloon going slack as you let the air out of it. But because vocal speech requires that you finely control the distribution of that pressure, you effectively bounce it back and forth between the voice box and the lungs. If you didn't carefully control that moving pressure, you'd be liable to tear up tissue—the force of air in the human respiratory tract when we're speaking is remarkably high. If our human muscles and neuro-wiring didn't do what they do so remarkably well, each time we talked, we'd either bloody up our vocal cords (literally) or seriously damage our lungs.

Bill had an advantage here, too. Not only could he take deeper breaths with his larger lungs, but he had more muscle mass surrounding those lungs, allowing him to better control the release of that pressure over time. Recent research supports this: when we speak, women's brains send more frequent impulse signals to the diaphragm and "inspiratory" muscles than men's do. To put it simply, women ask them to work harder and more often, which requires more involved neurological control. It's possible this bias toward greater control makes us better at fine-grained differences in voice control (more on that in a moment), but in the nuts and bolts of making sure our lungs don't explode from pressure differentials, the male chest wall has an easier job.

As far as we know, we're one of the only mammals able to prolong and control our exhalations through multiple tiny, forceful bursts of air. Other primates don't do it. Not even the noisy ones. Those long, raucous calls of our noisiest cousins—the booms that howler monkeys do, the screams from vervet monkeys—are fueled by repeated, forceful inhales. No single monkey call approaches the length of a middling human sentence.

Dolphins and whales are able to hold their breath for a long time, and even pulse out streams of bubbles, but their primary communication is made of clicks, squeals, and sonar that don't particularly involve the lungs. On land, the only other species that seems to do what we do with our lungs are songbirds.

But birds don't produce sound the way we do. Much like their dinosaur ancestors, today's birds have nine different air sacs that function like bellows. They breathe into their air sacs and out through their lungs. That means they have way more oxygen available at any given moment than mammals do, so it's a lot easier for them to do ridiculously energetic things like *flying.*[*] And singing all day long. Singing is, in many respects, a fancy way of holding your breath, much like talking.

Hillary actually took *five* breaths to utter that crucial sentence: "[*inhale*] And so, my friends, it is with [*inhale*] humility, [*pause*]

* Bats manage to pull it off by having a much more efficient method of flying than birds or insects: their stretchy wing membranes and many-jointed wing bones let them make tiny, efficient adjustments to the shape of their wings as they fly (Tian et al., 2006). That's why they look "flappy" and erratic when they fly, but it's also why they're able to fly at all. If they couldn't, they'd either be dead on the ground, resort to simple gliding like flying squirrels, or somehow have to grow *much* larger lungs—mammals just aren't built to be hummingbirds.

determination, [*inhale*] and boundless confidence in America's promise [*inhale*] that I accept your nomination [*inhale*] for President of the United States!"

All those breaths allowed her to speak with more control and precision. They allowed her to pause for emotional emphasis—the ways in which public speech is both musical and rhetorical, the emotional import of "waiting a beat"—and they also gave her enough air pressure to increase the volume of her voice. But when she did, she sounded strained. That was one of the biggest criticisms she received on the campaign trail: "Hillary sounds like she's *yelling* all the time."* That's probably because she was.

While often sexist, the criticisms of Hillary's voice in 2016 weren't entirely off base. Despite the acrobatic breath skills that evolution endowed us with, women's voices regularly fail us. We strain our vocal cords more than men do. This is especially true of women who talk and sing for a living: teachers, professional speakers, actors, tour guides. If you're a woman who uses your voice professionally, you're more likely to see a doctor about your strained vocal cords than a man who does the same work. What's odd about this is that the female vocal instrument isn't inherently more *fragile* than a man's. We might even have some mechanical advantages—finer control, for example, over our respiratory muscles, faster responses in nerve pathways between the brain and mouth and throat. The problem is probably that women unconsciously train our voices to mimic men's, especially in the public, political, and business spheres.

Standing behind that lectern, Hillary spent a lot of energy just trying to be heard, even with a microphone to help. The acoustics of most classrooms and auditoriums accommodate male voices pretty well: so long as you can "project," people in the back can still hear you. (This is especially useful for male listeners, of course, who—as we learned in the "Perception" chapter—begin losing their ability to hear higher pitches in their early twenties. To reach the *men* in the back seats, you have to be both loud and

* In September 2016, a number of Republican pundits even took pains to point out how often she *coughed* during an interview, as if the merits of one's candidacy could be measured in quantified throat clearing.

precise.) But when you're a woman like Hillary, whose speaking voice is naturally higher pitched and a bit quieter than Bill's, "projecting like a man" is harder.

She's yelling, in other words, even when she's not trying to. By the time Hillary began her run for the presidential nomination, she'd been effectively yelling for decades—projecting her voice at certain registers to fill large rooms designed for men's voices, making herself heard above the din. And her throat isn't built for yelling—if anything, women's throats seem to be built for a lot of precise, close-range vocal communication. In that sense, Bill's throat and lungs are a little closer to the older primate model. Maybe even closer to the moment human language originally evolved.

NICE THROAT SAC, MAN

When you want to get louder, your spine sends a signal—in a tiny, unconscious pulse of electricity—to your diaphragm and intercostals: *More volume, now.* As a result, they release a little more pressure, letting the spring of your lungs snap the air back out, hitting your larynx and vocal cords with determined force. This movement is ancient: our earliest Eves learned to control air pressure to make their cries louder. But we're *all* quieter than we used to be, because hominins lost their throat sacs.

Like many primates, today's chimps, gorillas, and orangutans all have throat sacs. Or more specifically, "laryngeal diverticula"—big culs-de-sac of flesh coming off either side of the larynx that they can fill with air. In chimps, these sacs run down the entire length of the throat into the upper chest. In orangutan males, they form a huge network of inflatable balloons that rest luxuriously in a flap across the neck and chest. The balloons fill with air and resonate when a male calls, thrumming out a chesty *barooooom* through the forest. In this way, the throat sac helps him warn competing males when they might be headed in his direction. It also lets females know when a male is nearby.

Careful study of the fossils of hominin neck bones suggests we had throat sacs until very recently. Lucy and the australopithecines still had them. And it's easy to see their legacy in today's human throat, which has deep folds on either side of the larynx. If Bill Clinton were an australopithecine, those folds would have extended out into pouches. When he exhaled, his breath would have poured into those pouches and vibrated, making his voice louder and more resonant. When he inhaled, the pouches would empty into his lungs, a bit like the way a bird's do.

Female primates also have throat sacs, but they're typically smaller. In the males, they bloom during puberty as part of sexual development. So when our ancestors lost their throat sacs, the males probably suffered the bigger loss, whatever they used those throat sacs for—maybe claiming territory, maybe intimidating rivals, maybe being extra sexy for an ancient Hillary.

Imagine if hominins hadn't lost them. Picture the U.S. Senate in the 1990s, a younger Bill Clinton giving his yearly address, the mostly male Democratic senators majestically inflating their throat sacs and thrumming in approval at each dramatic pause. And across the aisle, the Republican senators inflating *their* throat sacs, too, booming out their competing calls. The roar would carry a full kilometer down Constitution Avenue, rippling lightly over the Reflecting Pool all the way to the obelisk. Tourists would line up to hear it on the National Mall—the creaky, deep *rumble* of democracy's dawn chorus, broken only by the alarmed chittering of birds.

Still, while a hefty throat sac lets you get loud, you can't be *precise.* That's not a problem if you're communicating with only a limited range of hoot-pants and alarm calls. But if you want to *talk*, booming through a throat sac just won't do.

We don't know if spoken language came before or after the loss of throat sacs. But we know speech benefited from their absence. By using computers to simulate a human voice with the ancient throat sacs still in place, researchers found that listeners had trouble discerning subtle differences between the speaker's vocal sounds.*

* Some unlucky human beings still end up with laryngeal pouches, typically as a result of vocal strain or smoking. These people sound windy and imprecise

Presumably, at least for the males, the gains *had* to have outweighed the losses. Language is a pretty big gain. Maybe one of the biggest. But something else might have pushed the change: reducing the risk of infection.

Infections of their laryngeal pouches are one of the leading challenges in keeping captive primates healthy. Many primate researchers used to strap macaques upright in a chair, which made the animals terribly prone to such infections. When you're a normal macaque going about your day, your head is usually tilted forward or even parallel to the ground. With your head secured upright, the contents of your sinuses will drip straight down into the opening to your throat sacs, which can then get infected.

So imagine our upright ancestors, with throats now directly underneath the back of their sinus cavity. Having a throat sac there might have been more of a vulnerability than it was before hominins began to walk on two legs. Maybe especially for the males. You're not going to be very good at making sexy, competitive calls if you're constantly coughing up phlegm.

Still, knowing that throat sacs are mostly a guy thing counts against the idea that male physiology lent itself best to the evolution of human language. If what we wanted for the development of speech was precision and comprehensibility, being able to boom through a throat sac wouldn't have been as beneficial as the smaller, up-close perks of a female vocal instrument.

PITCH

Deprived of a resonating throat sac, and the larger lungs of her male counterpart, Hillary Clinton had to rely on her diaphragm to do most of the work of making her louder. Puffs of air buzzed and thrummed against her vocal cords, bouncing against the walls of her throat, before launching out of her mouth and through the

---

when they talk, and their throats usually feel sore, with a visible bulge on one or both sides of the neck. Men are more prone to this, especially saxophone players. (Luckily, Bill doesn't play that often.)

microphone to the nineteen thousand delegates hanging on her every word at the Democratic National Convention.

But Hillary wanted more than to just accept the presidential nomination. She wanted to accept it with emphasis. She decided to go for a *crescendo.*

To do that, she had to reach for another deeply evolved vocal trait. Starting around the time of *Homo erectus*, the hominin larynx dropped lower in our throats, giving the tongue more room to do all the complex, twisty, acrobatic stuff we do to produce spoken language. A lower larynx also lets us better manipulate the pitch of our speech—a key feature of the modern human voice.*

In human infants, the larynx drops lower in the throat about three months after we're born, and drops again at puberty, most dramatically in boys. (Chimp newborns have that first drop, too, but they don't have the second.) As their testosterone levels jump, the larynx shifts down in the throats of boys and their vocal cords thicken and lengthen, somewhere between ages thirteen and sixteen. The transition is so dramatic that boy's brains often have a hard time adapting to their new instruments. That's why teenage boys' voices "squeak and creak" so much, jumping wildly between the old, higher registers and their new, lower ones. When girls go through puberty, their voices drop a little, too, but the male voice can drop by as much as an octave. Hillary's? She probably dropped by only a few eighths. That's all well and good, but there's an evolutionary factor here that makes the male the likelier beneficiary: human males are able to hit bass notes that would normally be made only by animals three times their size.

In many species, moving the larynx lower in the throat allows for deeper-voiced vocalizations. When a male red deer calls out during mating season, it actually *moves* its larynx all the way toward its breastbone, producing a deep, throaty, and frankly intimidat-

* It also, unfortunately, is one of the deadliest new features of human physiology: choking kills an American child every five days, with similar stats worldwide. Adults fare a bit better, but not nearly as well as you'd think—it's still the fourth leading cause of "unintentional injury death" (that is, when you die from injuries, but not because you or someone else meant for you to be harmed). Other animals don't choke as much as we do, because their throats are differently arranged.

ing sound. (It also pumps its penis up and down while it's making its calls—red deer aren't subtle.) Larger animals have longer vocal cords, so mimicking the sound of a larger animal by making your voice sound deeper than it would otherwise is a common evolutionary adaptation for species that aren't of a particularly intimidating size. For many mammalian species that do this, the male is the one benefiting from that deeper voice the most.

For today's men, lower voices seem to be perceived as more "dominant," whereas male voices that are somewhat higher pitched are perceived as more "likable." Pitch for women is trickier, largely because of cultural ways of thinking about women's voices in the public sphere. Lower-pitched women's voices are usually considered not "dominant" but deeply unlikable. Higher-pitched women's voices are more desirable and more likable. In modern Japan, for instance, young women famously speak to men in a higher pitch, saving their "normal," lower-pitched speaking voices for conversations with women. But in the United States, women usually use the lower end of their vocal range when they're trying to sound "sexy" (often amping the "breathy" qualities, too). It can be hard to untangle what parts of the human voice are cultural and what parts are evolution-driven, but women with naturally lower-pitched voices tend to have less estrogen in their systems overall. So, the desirability of higher-pitched female voices could simply be a matter of fertility signals.

Menstrual cycles also play a role. Just after ovulation, progesterone is high and estrogen is low. Then, just before menstruation, progesterone plummets and estrogen peaks, a fluctuation that can affect a woman's voice. No one's entirely sure why it happens, or why some women are affected and not others—like many features of women's lives, this is a new area of research. But hormones are the likely answer. The lining of a woman's larynx seems to change over the menstrual cycle. During the weeks leading up to ovulation, the lining proliferates and happily lubricates the vocal cords with watery mucus. At ovulation, both a woman's larynx and her vagina seem to hit "peak mucus": the cervix creates extra in order to help sperm swim up and find the egg, and the larynx's lining and vocal cords become plump and happy and flexible. Across the menstrual cycle, women often favor their own voices around ovulation. Singers can hit all the notes in their vocal range, lowest to highest, without any problem. Professional speakers report the least amount of hoarseness and strain.

And then, just as the lining of the uterus shifts and breaks down after ovulation, the epithelium that lines a woman's laryngeal folds also seems to change. Its mucus gets thicker, tackier, and drier, and the larynx can get irritated. Many professional singers find they can't hit their high notes or sing as loudly. Some will avoid recording or performing altogether for a good week out of every month, because their vocal cords are inflamed. Some professional opera singers deliberately go on the Pill, not just because they want control over their reproductive lives, but because it's not economically feasible to be on vacation thirteen weeks a year.

As with PMS, these changes are more dramatic in some women than others. Those with more bothersome symptoms of PMS may be more likely to have more noticeable changes in their vocal quality around menstruation.

Most women also notice changes in their voices at menopause. Many find their voices drop as much as an octave in their fifties and sixties. Aging does that to men, too; the larynx, so flexible when young, gets harder and stiffer. The vocal cords thicken and get less flexible, too. But for women, these changes can be dramatic. As estrogens drop with menopause, the entire vocal system can get a little out of whack.

Which brings us back to Hillary Clinton. After she spent decades trying to make her female voice louder and more deeply pitched to address crowded rooms, the hormones in her body shifted during menopause. If her larynx was anything like the typical postmenopausal woman's, it probably struggled to adapt to its new, lower-estrogen environment. The vocal cords and larynx walls probably grew inflamed even as her professional career demanded that she "project" her voice more and more often, in larger and larger rooms. So, it's not hard to imagine how she ended up sounding the way she did at the convention: a bit hoarse, lower pitched, struggling to maintain her resonant crescendo and—critically—still be understood while doing so. A vague shout isn't what she was after.

PRECISION

The strongest muscle in the human body, in terms of absolute pressure, is the masseter muscle of the jaw. The uterus is the strongest muscle in the body in terms of constricting pressure. But when it comes to muscles that have both strength *and* flexibility, the clear winner is the human tongue, which has to roll and push a bolus of mashed food from side to side around the mouth, getting the un-mashed bits better mashed before swallowing, all while dodging the powerful slice and crunch of the moving teeth. If you've ever accidentally bitten your tongue or your cheek, you know chewing isn't always straightforward. Having a strong and flexible tongue is important.

But if chimps are any example, our tongue is *far* more flexible than the tongues of our ancestors. Chimps can't force air through the mouth and teeth to make a high-pressured *ess* sound—they don't particularly hiss. Chimps are good at *ah* and *oo* and can even do a screechy long *ee*, but consonants aren't their thing. Even if a chimp *wanted* to say, "And so, my friends, it is with great humility," it'd be a bit of a disaster. Chimps are largely content with vowels, grunts, a few lip smacks, and the occasional well-placed raspberry.

The human tongue starts lower in the throat than the chimp's, anchored by the hyoid bone. That extra bit of leverage helps us do what we need to do. Also, a large hole in our jaw called the hypoglossal canal lets a fat trunk of nerves pass from our brains to our neck, jaw, and mouth. These nerves control the careful coordination of our larynx, throat muscles, jaw, and tongue in the act of speaking.

Australopithecines used to have their hyoid bones essentially where chimps do: right at the base of the tongue at the back of the mouth. X-rays of fossilized hominin head and neck bones have shown where and how different sorts of ligaments would have attached, which lets us get a sense of how the vocal instrument was arranged. It was only around the time of Neanderthals and Heidelbergensis—*very* recent hominins, the sort *Homo sapiens* had sex with—that the larynx and hyoid bone are as low in the throat as they are in modern humans. The lower position of the hyoid lets us anchor the muscle of the tongue more effectively, which then allows us to flatten, curve, touch the tip of the tongue behind or between the teeth, and so on.

But we're getting ahead of ourselves. Why did the tongue move farther down the throat in the first place? It doesn't follow that the tongue dropped *before* verbal speech, since it's part of what makes speech possible. The best argument going for the change in position is simply that we started walking upright. And as we did, our heads tilted on their axis, pushing the jaw farther back toward the throat and shrinking the horizontal space at the top of the airway. Human tongues are fairly large. As our faces became flatter, the tongue had to either dramatically shrink, loll out the side of our mouth, or move its base farther down our throat.* Whatever the shift precisely was, that change probably started *before* we were properly talking.

If you're noticing a trend here, you're right: when you look across the span of recent research on the evolution of language, the latest science is moving away from the "humans are just so special" angle toward something a bit simpler, a bit more accidental. A significant part of why ancient hominins were able to invent vocal language may be that our Eves evolved to walk upright. Balancing our skulls on the tip of an upright spine naturally changed the structure of our throats and mouths over time. Not all those changes were beneficial. Choking was a problem. Infected throat sacs, too. The loss of the throat sac might have led to males developing a deeper voice to compensate, but that's hardly a hero's story, and certainly not enough to support the idea that men are inherently better speakers than women.

While our instruments differ a bit, there's no overwhelming difference in the mechanics of how men's and women's vocal instruments are set up. Women have some very small speech advan-

* If you'd like to see this sort of process in action—particularly where it fails—look at the Pekingese dog. Many petite dogs, whose skulls were bred into evolutionary strangeness faster than other parts of their bodies, now have tongues that don't fit their mouths, so they loll out the sides. Thankfully, since their dog bodies aren't upright, this doesn't seem to make them any more prone to choking—but for ancient hominins this just wouldn't do. We kept our big tongues, which are great for talking, and anchored them in the upper throat.

tages in how our smaller tongues fit inside our slightly smaller mouths—we find it easier to pronounce consonants and the tricky transitions between sounds. Girls are less likely to develop lisps and other functional speech problems than boys are; they're also easier to understand at lower volumes, especially if they're talking at speed—so long as they're not talking to older men who may have trouble hearing the full range of women's voices. But all those advantages in precision weren't quite enough to aid Hillary Clinton in the most important speech of her life.

She began her crescendo just fine, but as she followed it to the climax, pushing her voice higher and louder, it finally cracked. She smiled through it, like the pro she is, and the giant room she was trying to project her female voice into still erupted in emotional frenzy, applauding and screaming. Job done. In the many videos of this event, you can watch everyone trying to process what just happened. Years later, it still feels a little unreal. Even Hillary paused—for about fifteen seconds, in fact—which probably gave her just enough time to rest and clear her throat before she spoke again. But to me, one figure stands out in the frame in this moment—someone rather important for our purposes here.

She stood offstage, a bit to the left, in a cherry-red dress. Her name is Chelsea. And though she is not the *reason* Hillary succeeded or failed in her bid for the presidency, she is most certainly the reason human beings continue to have language.

FROM ZERO TO A THOUSAND IN THREE YEARS

Exactly zero human babies are born with the ability to speak, but most are language ready. Our unique human genes have preprogrammed our brains to be capable, hungry even, for learning language. But learning to speak involves a lot of data. It involves a lot of rules. It requires an incredible amount of highly specific, lightning-fast problem solving. None of these things can be passed on in DNA.

To learn to speak, you need a human childhood. For language to evolve and be maintained in the way it has, ancient babies needed constant exposure to another language user while their brains were growing. For all of human prehistory, going back to the origins of language itself, human beings have learned how to speak primarily by interacting with their mothers.*

So the male narrative of the evolution of human language misses the point. Language isn't like opposable thumbs or flat faces—traits that evolution wrote into our genes. Our capacity for learning and innovating in language is innate, but nevertheless, for the largest gains in intergenerational communication to persist over time, each generation has to pass language on to the next with careful effort, interactive learning, and guided development.† Language, in other words, is something that mothers and their babies make together and is dependent on the relationship between them in those first critical three to five years of human life. A long, unbroken chain of mothers and offspring trying to communicate with each other—that's what's kept this language thing going from the beginning.‡ Though you have no memory of it now, you, too, experienced this language learning curve.

A newborn's ability to learn and use language is minimal. It takes a good six months to even remotely understand what the giant milk-beast is chirping at you, and another six or so months after that to manage to say your first word. Still, your brain is developing at a phenomenal rate. And though you can't speak yet,

* Don't worry, dads, you can do this job, too. But for huge stretches of human history, dads probably didn't. And most dads still don't. Sex-egalitarian societies are incredibly rare. Females have been the primary caretakers of our offspring for at least the last 200,000 years, if not the last 200 million. The mother-child dyad is the most common and most important communicative pairing of most mammalian species, and that's still true in the vast majority of *Homo sapiens.*

† It would be great if a bunch of kids could, just maybe, invent a new language whole cloth if they somehow missed out on a fully fluent dyad experience with mom. But how the heck would that help them communicate with prior generations or vice versa? How would *knowledge* persist without resorting to the old monkey see, monkey do?

‡ There are many different models of child rearing, including biological parents of all gender identities and all sorts of nonparent caretakers. None of them are more valuable than another. None of them are more innately destined for success or failure. But since *most* people first learn language in the context of a mother-child dyad, which would have likewise been true through our species' history, I use that model here.

or really understand, you do manage to communicate with your mother, mostly by crying.

In the first three months of your life, you quickly figure out the difference between human voices and nonhuman sounds, and you pay more attention to the human sounds (in part because they often come with food or the removal of that uncomfortable wetness that frequently envelops your bottom). You also start mimicking the musical qualities of the language around you, which is probably something you learned in the womb. For instance, newborn French babies cry in a rising melody, which happens to be the typical way French people speak, with their pitch tending to rise a bit at the ends of words or phrases. German newborns, meanwhile, cry with a pitch that falls *down*—a typical German speech pattern.

By the end of your first three months, you're much better at being alive. More than likely, your eyesight and hearing are fully functional. You're also able to communicate a wider variety of cries: some signaling wetness, some hunger, some oh-God-I'm-so-bored. Your mother has probably even learned how to give you what you want when you want it, for the most part.

When you're not directly asking her for things, you'll spend hours and hours a day babbling, testing out random strings of pitches and syllables. Simple sounds are easier at first, the *puhs* and *buhs* and *muhs* that don't involve the tongue.* Sometimes you babble to get attention. Sometimes you try to imitate the noises around you. Sometimes you just find it pleasant to hear a human voice, so you fill the air with your own. You babble when you're happy and you babble when you're upset. When your mother

* Babbling isn't just something *human* infants do. Juvenile songbirds chirp and whistle in randomized, repeating patterns much as human babies do (Lipkind et al., 2013). What's more, songbirds such as the Bewick's wren share a regular set of fifty gene mutations with human beings (Pfenning et al., 2014). As with most genetic research, we're not *entirely* sure what those fifty genes are doing, but they seem to be critical for vocal learning. They're more active in language regions of the brain. Even more tellingly, birds that don't need to learn complex songs don't have this set of genes. And neither do other primates. At least in terms of vocal learning, that may mean human beings are more similar to birds than to other primates. So maybe, instead of the *talking* ape, it might be better to call humans the singing ape.

smiles at you, you smile back and babble at her. She seems to like this. And you feel happy when she seems happy. And her milk tastes a little sweeter, too.*

Once you've been alive for at least six or seven months, you finally start to understand that the weird string of noises the people around you make are individual *words*. Or at least some of them are. A baby starts fresh, with no point of reference, and it takes a while to realize that *muh* isn't meaningful, whereas "muh-ther" is.

When babies babble, they are testing out their vocal apparatus to see what sounds they *can* make. They're also testing out their brains' language faculties—seeing whether people around them respond more to one sound or another. Imagine learning an instrument before you even have an idea what *music* is. You play a note or two, hear it, see if you like it, see if your audience likes it, and then you play more. Except that the instrument is located in your own chest, throat, and head. Meanwhile, your brain is rewiring itself with simple sorts of rules for communication by paying careful attention while your main caregiver talks to you.† To become truly fluent in your first language, your brain needs exposure as early as the first six to seven months of life. Babies who don't get such exposure, for one reason or another, struggle with things like syntax for the rest of their lives.‡ That's *really* young. You're not even crawling yet. Before you're even mobile, your brain is already figuring out the building blocks of language.

And if your life is anything like that of the majority of human

* Think back to the "Milk" chapter here: When mothers and babies are stressed, there's more protein and cortisol in the breast milk, whereas "happy" milk is comparatively higher in milk sugars. For human babies, making mothers happy is rewarding.

† And she needs to be in the room with you. Babies who watch educational video programs don't learn as well as babies who hear language spoken to them in person (Anderson and Pempek, 2005), though having another baby in the room when it's happening oddly seems to improve things (Lytle et al., 2018). As with most human learning, social interaction matters.

‡ Babies who are born totally deaf and without any sort of sign language at home also have problems with language learning. That's partly why a number of doctors are now recommending deaf babies be fitted with cochlear implants before that window closes, alongside sign language to reinforce that linguistic learning, because cochlear implants don't always work as well as one would hope (Wolbers and Holcomb, 2020).

beings alive in the last 200,000 years, your mother's is the main voice you hear. Hers is the main face you see. Without her, you wouldn't be able to survive, sure. But she's also most of your social life. If there's anyone in the world you need to figure out how to communicate with, it's her. You'd been preparing for this, after all, before you were even *you*—newborns recognize (and preferentially respond to) their mother's voice, which they've been listening to since they grew ears in the womb.*

Assuming you succeed at communicating your needs to your caregiver, and assuming you make it to your first birthday, you'll finally be able to produce your first word. Some babies—usually boys—take a little longer. But even before you can say them, you start recognizing words. You'll even respond to basic requests (when you're in the mood), like "stop that" or "come here." The language regions of your brain hit peak density around your third year, which is precisely when your vocabulary explodes. Before, you had only a few dozen words. Now you rapidly learn hundreds. *Thousands.* Your grammar, too, becomes more complex. Your sentences will leap from two or three words long to ten or more. By age three or four, you'll have a word for almost everything in your environment. And if you don't know the name of a thing? You'll name it yourself, toddling boldly through the world like Adam in the Garden, blathering out new names without a second thought. Best of all, your mother will know what you mean and rarely correct you.†

Both parties are motivated here. After all, if your mama takes

* Infants who are born fully deaf don't have this advantage, but they are known to preferentially respond to their mothers' faces soon after birth, as do most sighted infants (Field et al., 1984). People who are born deaf-blind lack both innate pathways to social bonding, which may—on its own—inhibit early language learning. But these children do find other ways to both bond with their caretakers and learn language, particularly with therapeutic assistance, and a new language called Protactile (a deaf-blind variation of ASL) may be especially promising for families of deaf-blind children (Leland, 2022).

† Though oft translated and debated, the oldest texts we have of the Genesis chapter of the Bible hold that God made stuff and "brought" it to Adam to "see what he would call [it]," and whatever he called it, that became the name (Genesis 2:19–20). It's hard not to read that model of the Hebrew God as a terribly patient parent doing whatever he can to please a toddler, indulging whatever silliness the creature declares to be true.

too long to give you what you want, you're liable to throw a fit. All those dense synaptic connections? Between ages two and four, you have a *very* difficult time sorting out all the strong emotions you're feeling. But if having that emotionally unstable toddler brain of yours *also* makes you better at learning language, the gains could outweigh the tantrums. Your growing brain is engaged in a very special sort of cognitive development—building a communication engine inside the narrow window when your brain is *just* plastic enough to be able to wire itself for the job.

Human brains do seem to have a cutoff for such wiring. If you learn a new language after puberty, you're never going to achieve true fluency. You'll be able to function. But unless you're a very rare bird, no American is going to speak French well enough to pass as a Parisian.* You can certainly brutalize an older brain into memorizing the new rules of grammar. But there's something about how the brain learns language when it's young that older brains just can't do. For fluency in a second language, the cutoff ranges anywhere between ages ten and seventeen, depending on whom you ask.

Which brings us back to mothers. Among songbirds, evolution has long since optimized parent-child interactions to take advantage of the critical window. During that window, zebra finch parents, for instance, communicate with their offspring in ways that seem particularly good at teaching them how to sing. After that window closes, the parents spend much less time fussing over the kids, who slowly gain their independence.

Because milk is part of how we make and grow babies, we mammals have a preestablished period of childhood when the mother has to closely interact with her offspring. If a mammal were to have a critical window for language learning, it would make sense for evolution to optimize for that while the child is still breast-feeding. Among modern-day hunter-gatherers, babies aren't completely weaned until somewhere between ages three

* During my brief time in Marseille, my exceptionally poor French sometimes let me "pass" as someone from Spain rather than from the United States. But that was only because I had the bad habit of rolling my *r*'s behind my teeth instead of at the back of the tongue. I'd studied French in high school under a perpetually disappointed nun.

and five—precisely the stretch where their brains reach peak synaptic density and when most children's vocabularies and grammatical sophistication explode.

You could call it coincidence. Or you could call it a useful optimization. If humans do have a critical window for language learning, it would be useful if it coincided with the time the child has regular, necessary, up-close interaction with an adult language user. Given how expensive brain tissue is to grow and utilize, it'd also be handy to have that window coincide with a time when the child's food supply is regular, easily supplemented, and dense with sugars and brain-friendly fatty acids.

So when we think about the evolution of human language—how it's actually passed on from generation to generation—it's useful to remember that what seems to be the most critical part of the so-called critical window happens while the child is spending regular portions of the day in its mother's arms. While there's more collective child-care happening among humans than among chimps or gorillas, most human infants and toddlers still spend most of their time in close contact with their mothers.

Mom, in other words, is at least half of *how language happens.* And she's not passive. Not at all. Human mothers have evolved to be language engines—prodigious users and *teachers* of language. This is especially true during the synaptic blooming of their children's brains. All the while, how mothers talk to their babies is so ingrained, and so clearly universal, that scientists have even come up with a name for it.

MOTHERESE

The first thing a mother does after she recovers from the exhaustion of birthing her baby is change the *music* of how she talks.*

* While human beings have a normal range of pitches, and those pitches do vary, they don't tend to vary that much. But very few people speak in a true monotone—doctors regard that as a classic sign of trauma, disease, or some underlying mental illness (for example, schizophrenia), and clinicians in ERs are trained to watch out for it during patient exams. But speaking in wildly *varying* pitches is also rare. It's not that we don't do it; we just don't do it with other adults.

I'll bet if you haven't spoken motherese yourself, you know what it sounds like.* So give it a try. First, say this phrase as you would to a friend or colleague: "Who's a good baby?"

Now say it as if you were talking to a baby. There you go: that's motherese.† The pitch goes up, we over-pronounce consonants and certain vowels (especially "oo"), often while we're exaggerating what our mouths do to make the sounds—pursing our lips more than usual or opening our mouths wider. We speed up or slow down syllables (the "cadence") in places we normally wouldn't. We tend to simplify the grammar, and we also *repeat* things more—from individual syllables, to words, to whole sentences. We don't, in other words, talk to kids the way we talk to adults. And the younger the baby, the more dramatic the difference in our speech.‡

Across most cultures, women are especially prone to using motherese, and we're also more likely to exaggerate pitches and shift the overall register up. We do it without even thinking about it. From Arabic to English, Korean to Marathi, Xhosa to Latvian, and back again, mothers talk to babies in essentially the same ways. If you play a tape of a woman talking to a baby in a language you don't understand, you probably can still tell she's talking to a baby.§

* Same goes for the hearing impaired: Parents who use sign language to communicate with their kids have their own version of motherese (Masataka, 1992). Instead of varying *pitch*, they tend to slow down, vary the intensity of gestures, use simplified grammar, and more greatly emphasize the individual parts of each sign and the breaks between signed words than they would with adults (ibid.).

† In the scientific literature, this is also named child-directed speech, child-directed communication, parentese, doggerel (when addressing pets), and so forth. Because I'm acknowledging the overwhelming dominance of the mother-infant dyad in early language learning, I'm just going with the simplest and most obvious name for the thing.

‡ Motherese also usually involves some combination of sound stretching, emphasizing consonant boundaries against vowels, and widely exaggerated facial expressions. We know this because it's been incredibly well studied since the 1980s, both in English and across widely varying language groups.

§ This is a robust result—a number of different studies have found this to be true. When it comes to motherese, most human beings don't need to know what someone's saying in order to know that someone's talking to a baby. The patterns may be innate: the features of child-directed speech and the songs we sing to babies are remarkably similar across large numbers of human cultures (Hilton et al., 2022; Cox et al., 2022).

Men do it, too, though a bit less and a bit differently. Motherese is so universal, in fact, that we do it not just with babies but also with our pets or to tease an adult we think is acting childish.*

All this is why so many scientists think motherese is something we evolved to do to help babies learn how to be functional human beings—or at least how to be members of a particular social group, since it turns out that motherese may not be limited to the human species.

Much like us, rhesus macaque mothers "speak" around their infants in a more musical, higher-pitched vocal pattern than they do when only adults are around, and it seems especially effective at getting the infant's attention. It's also useful when it comes to smoothing social interactions with other mothers. Squirrel monkeys also call to their babies with widely varying pitch and contour. Even dolphin mothers communicate with infants differently than they do with the rest of the pod, and they also give them distinctive "name" whistles that seem to last for their lifetime.

So is motherese just a successful way of getting a baby's *attention*?† Or in the human case, is it specifically adapted to teach babies how to talk?

Consider this: There you are, on your mother's lap, gurgling and babbling and listening to her speaking to you in motherese. Just outside your window, there's a bird's nest. In the nest are a couple of baby songbirds. They're very different creatures from you, and yet mother bird and baby bird are doing a lot of what you and your mom are doing.

Songbird babies "babble" a lot like human infants, producing spontaneous combinations of notes and volume. Like us, they'll

* Instead of "Who's a good baby?" say "Who's a good puppy?"

† Human babies like a range of more "dramatic" stimuli: bright colors, bold and distinctive shapes, exaggerated facial expressions, music with a lot of repetition and varying pitch, and simplified patterns. Subtlety isn't really an infant's *thing*. And because attention is strongly tied to memory, getting infants to pay more attention to you is certainly going to help them remember whatever you're trying to teach them. Some parts of motherese, in that case, might simply be a matter of boosting the *signal strength* of early language exposure. But most scientists who study motherese think it's more involved than that.

do it with their mom and dad, but they are also quite happy to do it on their own. Songbird parents also direct a more pitch-varied, exaggerated sort of song at hatchlings. Baby songbirds who don't get to hear any parent's song have a terrible time managing adult song later; ones who hear a motherese-style song seem to have a leg up over birds that only hear adults singing at one another. Bird babies who *directly* communicate with a parent who's singing in motherese do best of all.* But the effect is still there, even absent direct interaction. The sound itself has its own benefit.

Studies demonstrate that babies have language advantages when their mothers use motherese: Mandarin-speaking children, whose language depends on subtle pitch variations, are better at language tests when their mothers hyperarticulate lexical tones and divide their phonemes with more emphasis—a very common feature of motherese across languages. The most obvious reason motherese might help is its higher pitch, which is easier for baby ears to hear and understand. So shifting the register *up* a bit already gives a baby a hand. Like the Mandarin-speaking mothers, we exaggerate the phonemes—the *smallest* parts of human speech, like the "fuh" in "far" or the "ah" in Hillary Clinton's "accept"—to make them more distinguishable. Babies whose mothers put more emphasis on vowels tend to perform better on language tasks later. And the phonemes, meanwhile, might help us distinguish different words in a string. They also help us learn our mother tongue. Up to the first year of life, babies can distinguish between all sorts of different phonemes. But once they pass a year, they're only able to distinguish phonemes from their parents' native tongue. Chinese two-year-olds, for example, aren't very good at hearing the difference between *l* and *r*, because Mandarin Chinese doesn't distinguish between the two in the same way as English does.†

* Worth noting that most of these songs should probably be called fatherese, given that the songbirds studied are usually species that have elaborate *male* songs, particularly during mating season, and males in those species are also often known for being good caretakers of their hatchlings. Mammals are female heavy in caretaking largely because females are the ones who make milk; among non-mammals, there's a wide range of models for caretaking.

† English-speaking adults who didn't grow up with Mandarin are also famously

In the end, most professionals who study these things agree that motherese is useful. But is it necessary? And, more important for our purposes, are the distinctive features of motherese encoded in your genes? Is there an innate instinct to produce this kind of child-directed speech?

It's hard to say for certain. Since most of us are spoken to in motherese as we first learn language, it could be something passed down from generation to generation in an unbroken chain from the Eve of human language—not through genetics, but through the simple fact that it's an effective strategy for communicating with children. It's a thing you do because your mother did it and it *worked.* The typical range of pitches in motherese just so happens to closely correlate with a child's particular range of hearing, and if you're a caretaker, it's *always* beneficial to communicate in a way that's easy for your offspring to perceive. If you and your offspring live in a social group, vocalizing in a distinctive way is also useful: you want your kid to hear *you* best of all. It's also perfectly normal for a daughter to grow up to communicate with her kids the same way her mother did. We model ourselves after our parents. Humans do it. Rodents do it. Dolphins and songbirds probably do it, too.

Except. Children whose mothers emphasize vowels more—the way you do in motherese—arrive at language milestones faster than other kids. And they perform better at language tests, too. Children of parents who *don't* use any sort of motherese lag behind. Among tonal language users, mothers who more greatly emphasize phonemes when talking to their kids end up with kids who learn the language faster and with more accuracy than mothers who don't. So even if you don't *need* motherese to learn language, it does, in many cases, seem to give kids a leg up.

And when it comes to evolution, getting a leg up is everything.

---

terrible at pronouncing Mandarin words correctly. In a tonal language, slightly altering the pitch of a syllable or word can entirely change which *word* the speaker is using. As much as 70 percent of the world's languages are tonal—from East Asia to Africa and even South America. European and Central Asian languages don't have this feature.

THE STORY OF STORY

Despite the weirdness of our vocal instrument, how hard it is to learn how to play, or the *years* we spend blathering nonsense before we're remotely fluent, it's still extremely rare for a human being to be nonverbal. It's such a universal ability, in fact, that some scientists think we are born with a kind of "language instinct": a hardwired drive to both learn and develop language, enabled by unique features of our oddly evolved brains. For example, deaf schoolchildren have been famously known to develop their own sign language in social groups, even if they haven't been taught sign language at home.* But those deaf kids *did* have important and healthy communicative dyads with their caretakers during the critical periods of their early childhoods and had already developed home signs for things they wanted: water, milk, food, bathroom, and so forth. While they didn't learn a complex grammar the way a child might learn from a fluent speaker, they did have the basics of language: they knew what words were, for example, having cracked that code as they developed their home signs.

Other cases of kids being isolated from language haven't exactly gone well. In nearly every instance, they never develop real linguistic fluency.† There seems to be something about forming those critical relationships with other communication partners—first in infancy, then throughout toddlerhood, especially, and on through childhood—that really *matters* for developing the sort of fluency we associate with human language.

Which is to say, maybe the story of language is a lot like the story of human brain evolution in general: It's not necessarily that we are able to learn patterns, rules, how to map social environments, and how to anticipate our communicative partners'

* In the psychology of linguistics, this is a pretty famous case: it's basically considered cognitive development 101.

† Sadly, these are usually children who were severely abused and neglected and isolated and/or completely abandoned—hence the extremely rare circumstance wherein they failed to learn language. Some of them were also suspected to have learning disabilities or other cognitive mishaps *on top* of all that abuse. What is obvious is that the caretaker-child dyad is so vitally important in human childhood that in nearly every case where it is damaged, bad things happen.

desires, among other complex things, or that we innately seek out certain types of learning, or even that we have a childhood. All of those matter, of course. But lots of mammals have these things, especially hyper-social apes. Rather, what's unique about us is that we have a long childhood full of those drives and capabilities, with extended and unique bursts of brain development usefully timed to stages where we need to learn really hard, complex stuff in order to be able to function in our highly social societies.* So in essence the story of language may be about windows of brain plasticity: times in our young lives when our minds can still build those critical pathways, which just so happen to be *perfectly* timed to coincide with breast-feeding and motherese.

But it's not the words that are important, particularly. The real payoff is grammar—the very stuff of human thought.

Grammar feels so natural to us that we take it for granted: we just *know* how to divide the world into "agents" that can take "actions" and, by taking those actions, cause predictable effects. That's what nouns and verbs really represent: the lion (an agent) waits in the grass (action); the goat (another agent) walks by; the goat doesn't see the lion; the lion catches supper. Most intelligent mammals can suss out some of why things happen and change their behavior accordingly.

* We still don't have a solid grasp on the exact mechanisms that underlie our unique abilities. But the state of knowledge is inching forward. For example, a mutation heralded as the "language gene"—*FOXP2*—seems to be more about pattern complexity and learning than language per se (Schreiweis et al., 2014). You can dump its analogue into a mouse, and he'll make more complex, chirpy sounds—but more interestingly, in his juvenile period and throughout his life in the lab, he'll also *learn* faster. Mice with this mutation are better at switching from step-by-step to repetitive learning (ibid.). For example, maybe when they go into mazes, turning right takes them to where the food is. If that's true often enough, they'll still turn right even when other features of the maze change. That's actually similar to how human children learn language: after exhaustive exposure, we switch over from step-by-step learning to derived rules, and then creatively innovate on those basic logic patterns. Humans with differing *FOXP2* mutations tend to have a range of language and cognitive issues, and while no one knows precisely what *FOXP2* does in the brain, it does seem to be strongly related to plasticity in language-related brain areas (ibid.). It's also involved in the fetal lung and gut, by the way, so as with anything in the body, assume some evolutionary repurposing and multitasking.

But when you are able to *talk* about a chain of events, the very language you speak can change your cognition. For example, just by changing the tense, you start to understand *time* and your place in it. You know that things happened in the *past*, and you understand that there's a nearly unlimited amount of past, which means there's a future in which all sorts of things could happen. You can talk about, and think about, things that *could* happen in that future. Things like sunrises and earthquakes and perfectly brewed coffee. Things like *Star Trek* and bachelorette parties and a cure for cancer.

Language is an infinitely flexible framework for cognition. That's what grammar does. *That's* what your mother worked hard to help you learn. Yes, Faulkner was able to write a single, grammatically correct sentence that contained 1,292 words, but that was just an artist at play. The point, really, is that the endless flexibility of human grammar lets us express an *infinite* number of ideas with a finite vocabulary.* With grammar, you don't need a word for everything you'll ever see or hear or want or do. Without grammar, you'd need millions of unique words.

Evolution doesn't like waste. It doesn't allow you the brain space for billions of word combinations, but it does allow you the ability to learn and create flexible rule sets that let you solve just about any problem. Your brain has evolved that ability to learn and create grammar. We're the only species on the planet who's *ever* managed to do that.†

With human grammar, we can make anything behave like an agent: a shoe can want; an eyelash can whisper. Likewise, we can turn anything into an action: we can *table* a discussion; we can shoulder the blame. We can make subtle combinations of ideas to get at something more nuanced. We can create what-if scenarios. We can treat the impossible *as* possible.

* The formal term is "recursive."

† Us, and *maybe* certain monkeys. Campbell's monkey "language" involves a total of four distinct vocalizations and an extremely simple grammar (Ouattara et al., 2009). Still, that discovery rattled linguists, because we assumed grammar was the real line between us and them, and it was shocking to realize that another species had even rudimentary grammar. Chimpanzees and gorillas can be taught some sign language, but it's vocab. Grammar, fluent syntax, never stick.

That's where it gets really crazy. As I've written, packs of wolves can form complex hunting parties. Without any language at all, they still manage to learn some basic "rules" of the hunt and improvise thereupon. But they can't *plan* a hunt the way we can. And they can't imagine anything like a unicorn. The impossible stays impossible for the non-lingual mind. Wolves will never dream of where they come from or wonder what they're supposed to feel when they watch a rabbit die. They'll never look up at the sky and create stories about the stars, never build a rocket ship, never make plans to go to Mars.

Everything humans care about is possible *because we have language*. The human mind is made for language, yes. But it's also made *of* language. The same sorts of logic paths that rule language, that combine known things into new ideas, that puzzle out the code of others' communication into knowable thoughts and desires, also write stories and build meaning and tease out the finest, strangest features of the universe. They make us what we are.

That's why grammar is one of the most important things your mother ever helped you learn. You picked up the salient features of motherese ambiently, and will mimic its music to your own children, should you have them, and thereby aid their own language learning. But the moment you learned *grammar* might well have built the most human part of your brain. Once you've managed to learn grammar, someone can teach you how to perform an emergency crike. You can also invent the crike and teach whole generations how to do it. But the coolest thing you can do, really, is invent *civilization*.

THE FIRST HUMAN

I haven't forgotten. I know we haven't talked about the Eve of the human voice yet. That's because of all the Eves in this book she is the hardest to trace.

She's also the most important. She is nothing short of the Eve of Humanity.

We can't point to an Eve of *communication* any more than we could have picked an Eve of *vision* or an Eve of *reproduction*—these are fundamental features of what it means to be a living organism. But we can find an Eve along the evolutionary line who seems, in some deep sense, the one that best represents a trait that's become *more* human than it was before. The arrival of human language left no fossils, no cache of sharpened stones, but we can assume that this Eve had a fully modern voice instrument, which lands her neatly among Neanderthal and *sapiens.* She was probably an anatomically modern human, a very recent ancestor. And she had language.

But are we "human" at the very start of language?

I don't think so. I strongly suspect human language came about in fits and starts, along a very long stretch of evolutionary time, not unlike the evolution of the hominin brain itself. Our Eves, no doubt, had all sorts of complex, social communication before they had recursive grammar. How else could they have survived so long? How else could they have become competent midwives?

But even that wouldn't have been enough. Even once they had grammar, our Eves probably still weren't *human* the way you and I are, because they simply didn't think about the world the way we do. There's something deeper at play here. So I think there was one moment in the evolution of human language that marked a dividing line: before it we were not yet human, but after it we were.

It was probably the smallest thing, neither heroic nor grand. More than likely, it was the intimate moment, probably late in the evening, in the low blue quiet before dreaming, when a single human being told the very first story.

I doubt it was told to a group. If anything, it probably took shape between two people who already spent most of their time trying to talk to each other: a fussy child who needed to sleep and a mother who needed to sleep even more.

So picture a mind that has language but has never yet told or heard a *story.* Brief, self-serving lies, yes. Exaggerations, sure. These are phenomena we find in other animals, too—deception is ancient. But no story. No religion. No morality tales. No afterlife. No gods. No fables. No legends. No origin stories. No just-so stories. No stories at all. The mind that existed as an intelligent, creative, fully cognizant human being *before* the beginning of nearly everything we mark as human culture was a truly alien mind.

So, I pick her. The Eve of the most important feature of the human voice had a mind that must have been profoundly different from human minds today. And that mind must have, at some point, in some deeply ordinary circumstance, invented the world's first story.

I won't give her a name. She was probably *Homo sapiens*, though anatomically she could easily have been *Homo neanderthalensis*. Both had modern vocal instruments. Both had that characteristic swollen bulge on the left side of the brainpan that we assume signals language, both had a widened hypoglossal nerve canal, both had the hyoid bone and trachea in the right spot.

But the timing makes *Homo sapiens* more likely. Somewhere between thirty thousand and fifty thousand years ago, human culture exploded. We went from using the same, relatively simple tools to a cultural revolution, not only advancing our tools, but massively increasing the amount of art we made, burial rituals, obvious jewelry . . . Symbolism was suddenly *everywhere*. Before this revolution, there was lots of the same for a very long time. After, there was Humanity everywhere you looked. Africa, the Middle East, southern Europe, Central and South Asia, China . . .

The change happened *so* quickly it's a little suspicious, frankly—the sort of shift that gives rise to the theory that visiting aliens made us smart, the sort of rapid, inexplicable change that kept Kubrick in business. Ten or twenty thousand years, max. Boom, all of humanity adopted complex symbolic culture. All of us. Everywhere. Again, most think it's the sort of speed that can happen only with language. Where genetic changes are slow, language-fueled behavioral changes can spread like wildfire. I suspect this is what happens when an intelligent species already capable of language suddenly gets symbolic narrative.

And who else to tell the first story but a mother to her child? After all, while men and women were (and are) equally adept at language, female bodies are slightly better at up-close communication with fine detail. Most adults use the music and style of motherese to aid language learning in children, but women do seem slightly more likely to use it and slightly more adept at it, at least in terms of pitch manipulation and adapting and responding to the unique sensory array of human infants. But a better reason, I suspect, is that of all the instances of communication between two people, that coupling of mother and child is the most common—she will talk more to her young child in its early life than nearly any other person. Of the many communicative scenarios involved, quite a lot of them would have to do with the child being fussy, and the mother needing to find a way to soothe the child, and if not soothe, then at least instruct and hopefully *amuse*.

Whether one is talking about historical or present-day parents, trying to distract or instruct or amuse a child with a story is a common go-to.

But what was that first story about? After all, *story* is as much its about-ness as its structure—not all tellings of events are "story." I could tell you what happened today, but it would just be an uninteresting string of facts. Urgency doesn't cut it, either: even Campbell's monkeys can tell you an eagle is in the sky. No monkey is going to tell you about the eagles in Tolkien.

But let's say it *was* a just-so story—an imaginative explaining of some feature of the world. Why snakes have no legs. What happens when we die.

That still wouldn't have been all it was about. Most modern-day just-so stories have to do with some moral quality—some set of social rules that the characters (and audience) need to abide by or there will be consequences. They're typically about love, or familial loyalty, or adherence to a social hierarchy.

Yet none of those themes would have been part of the first story, because little of our familiar social hierarchy would have existed. There were leaders or alphas, but nothing at all like a lord or a king. There would have been plenty of love and sex, too, but nothing like "marriage."

Instead, maybe it would have been simpler. There is one abiding theme that's stayed with humanity since the very beginning: hunger.

If the story of our ancestors is about anything, it's about survival. Hunger, and migration—the unyielding force of Death, driving us ever forward and out, into the gray line of a long horizon. That is where we came from. It drives us even now.

Homo sapiens

CHAPTER 8

MENOPAUSE

And yet, and yet... Denying temporal succession, denying the self, denying the astronomical universe, are apparent desperations and secret consolations. Our destiny (as contrasted with the hell of Swedenborg and the hell of Tibetan mythology) is not frightful by being unreal; it is frightful because it is irreversible and iron-clad. Time is the substance I am made of. Time is a river which sweeps me along, but I am the river; it is a tiger which destroys me, but I am the tiger; it is a fire which consumes me, but I am the fire. The world, unfortunately, is real; I, unfortunately, am Borges.

—JORGE LUIS BORGES, *LABYRINTHS*

Damn, I got out of hand!

—BORGES'S MOTHER, ON HER NINETY-EIGHTH BIRTHDAY

JERICHO, EIGHT THOUSAND FIVE HUNDRED YEARS AGO

Another dawn. The old woman woke to birdsong, thin streaks of early light drifting down over the mats on the floor. She rolled onto her side. First her eyes sought her sister, whose face was still smoothed in sleep. Then she heard the gentle whimpering of her granddaughter. The girl was long pregnant, her belly fat and swollen and hanging low, like an old fig. So the old woman struggled to her feet and made her way to the girl's mat, ignoring the way her hip and hands throbbed in the mornings. No time for an old body's complaints. She crouched next to the girl, brushed a sweaty bit of hair from her cheek, and laid one hand on her stomach, feeling the womb move in a strong contraction. Her granddaughter reached for her other hand and held tight.

The baby was coming, for sure. The girl's mother had died the previous year, lost to a flood, so helping this baby come into the world—the fourth generation of her family, a rare thing to be alive to see—was the old woman's job. She woke her sister and sent for freshwater.

They labored all morning: the girl cursing and crying, the old woman and her sister doing what they could to ease the pain. The village shaman came in unbidden, and she shooed him out the door—chanting and burning herbs wouldn't help here. The father poked his head in, too, so she sent him to fetch more water. Everyone seemed to want a job. But she was the oldest person in the village—people did what she said now.

By the time the sun was high and hot outside the hut, the old woman knew something was wrong. Squatting down between her granddaughter's knees, she saw it: a bloody little *foot*, its toes curled, wrapped in a flap of tissue. This baby was trying to enter the world the wrong way around.

She'd seen this twice before. When she was just a girl, her aunt had a baby who came feetfirst. It killed her. The second time, a woman turned the child in the womb—just reached her arm up there and turned the baby around, pressing on the stomach with the other hand to help. That time the baby lived, but the mother didn't.

The old woman sucked air through her teeth. She was long past having babies of her own, but she'd survived her births and witnessed many others. She had to try. She eased the girl back and propped up her hips with a thick bolt of hides. Then she washed both arms up to the elbows in the water basket, took a deep breath, and plunged her left hand into the girl's body.

THE MYSTERY

At a certain point—usually in her forties—a woman's menstrual cycle starts getting a little odd. At first, her periods might become heavier and more frequent. She might start feeling unusually warm during the night. Whatever patterns she used to have with PMS (headaches, moodiness, bloating) will shift a bit. She might even start to get arthritis, tied to the changes in her hormone levels. This is called perimenopause. It can last a couple of years, or as long as ten.

Then she'll enter menopause proper. That's usually when the worst symptoms happen. Because her estrogen and progesterone levels are dropping, and can vary wildly, she can suffer from headaches, mood swings, hot flashes, digestive quirks, vaginal dryness, sore breasts, dry mouth (or excess salivation), weight gain, fat redistribution from the butt to the stomach, and new and exciting hair growth on the limbs, upper lip, chin, and nipple areas. Hearing about menopause is one thing. Watching your own mother or aunt sweat their way through it is another. But actually feeling your own body change in these ways can be hard to wrap your head around.

Unless you're an endocrinologist, you probably don't know that the ovaries are an important part of the endocrine system—that there's a kind of three-way hotline between most women's reproductive organs, her body fat, and the pituitary gland at the base of her brain, constantly regulating the shifting balance of her sex hormones. And while these hormones have obvious roles to play in sex and reproduction, they also have important functions in the digestive, circulatory, and neurological systems. There's no part of the human body that sex hormones don't touch. That's why a woman can experience all of these seemingly disconnected symptoms during menopause.

Take hot flashes: More than 60 percent of menopausal women get them. They happen when fluctuating hormones trick your hypothalamus into thinking the temperature in the room has risen. It then sends the signal to dilate blood vessels near the surface of your skin, so the blood your brain *thinks* is too warm will pump through them and cool down.* Your face and your neck

* This is especially true where you have a lot of blood vessels close to the skin: the face and neck, the hands, the lower back, the feet, the underarms, the crotch. You don't have as many blood vessels close to the skin across your stomach, or down in

will feel as if they were burning; you'll sweat and your heart rate will rise; you might even want to take off some of those layers people advise women of a certain age to wear. Since levels of sex hormones naturally fluctuate over the course of the day, typically dropping to their lowest levels in the evenings, that's the most likely time for women to get hot flashes—until the menopausal body adjusts to lower levels of estrogen and they pass.

Other menopause symptoms follow similar principles. Lowered estrogen levels can make the vaginal walls thin and dry. Maintaining an active sex life may help with that, but menopause can also be tricky for libido—in some women, it intensifies; in others, it falls off.

Sex hormones also help your bones hold on to their calcium, possibly because the greedy placentas I wrote about in the "Womb" chapter have a sneaky way of trying to leach the calcium from your bones. Estrogen and progesterone seem to protect women's bones against the worst of it. Once menopause lowers those levels, a woman's body can start to lose calcium, which is why older women, especially, are more prone to osteoporosis.

Thankfully, menopause doesn't last forever. Each system in the body has been trained, since puberty, to respond to a certain pattern of sex hormones. So each system needs to relearn how to respond to a very different pattern. It's not an endless penance for having once been fertile, but a *transition.* The sign that this transition is over is simple: she stops getting her period. The uterus goes quiet. And so do the ovaries.

Once a middle-aged woman hasn't had a period in more than twelve months, she's not called menopausal anymore, but postmenopausal—and that's the phase she's in for the rest of her life. These days, most women will live a full third of their lives with no possibility of pregnancy. No more periods, no more babies. To many who have passed through this portal, that seems perfectly normal—even a relief, given that they don't have to worry about

---

your calves, so that's not where you'll sweat. But your upper lip? Your forehead? Tons of blood vessels and sweat glands there. These are also the places you'll sweat when you're nervous. Similar mechanisms.

birth control or tampons or menstrual cramping anymore (just more brittle bones, and being more prone to heart attacks).

But for scientists who study evolution, it's really, really odd. Evolution works by passing genes down through the generations. Thus, the more fertile offspring you have, the more likely it is that your particular genes will live on. In evolutionary terms, anything that reduces your chances of passing on your genes is a huge price to pay. Baby making should be the top priority, one that species generally only sacrifice in order to help the babies they already have. Most animals keep reproducing until they die. That's true of primates. That's true of birds and lizards and fish. That's even true of most *insects*. With the exception of orcas, no other species does what we do.

That's why human menopause is one of the biggest mysteries in modern biology, right up there with why we die. We know the general path of things. We've learned quite a lot about the mechanisms of aging—how tissue wears down, how cells commit suicide—but not why. In principle, any given cell should continue to reproduce itself forever. Given the right environment, enough food, enough oxygen, and someplace to get rid of metabolic waste, all cell lines should be immortal. But they're not. Tissue wears down. Cells kill themselves off. Parts of the body that did the same job for years seem to decide, after crossing some invisible line, that they're done, thank you very much.

And for whatever reason, a woman's ovaries give up the ghost a lot faster than the rest of her. We stop having children, but we keep on living. It's as if one part of our bodies were aging *a lot* faster than the rest.

Figuring out why that is may tell us a lot about how and why human beings die (and why some of us die so much sooner than others).

THE GRANDMOTHER HYPOTHESIS

In an otherwise healthy female body, why would you cut off the chance of having another child?

Until very recently, the scientific consensus was that humans have menopause because we're social. While making babies remains the general priority, the idea was that we made this sacrifice in order to protect our siblings, nieces, and nephews—our kin. Think of it this way: if your efforts boost the chances that your genes get passed on, even through a relative, then evolution will favor such efforts, including the sorts of bodies (and genetic underpinnings) producing them. The scientists' own grandmothers, for example, had cared for them, tended to their every bump and bruise, and cooked dinner when their own mothers were busy. Useful, right? These were the beginnings of the grandmother hypothesis.

So, what if ancient humans needed grandmothers to stop being fertile in order to succeed? What if, as humans became increasingly social, with increasingly specialized roles in society, new mothers needed more help taking care of their needy, vulnerable offspring? If the child's father or grandfathers couldn't (or wouldn't) do it, maybe the grandmothers could—but only if they weren't busy with babies of their own.

Though each scientist tells the story a little differently, the grandmother hypothesis usually holds that human beings evolved a kind of switch—a mechanism that shut down the ovaries, allowing grandmothers to stop making babies themselves and take care of their grandchildren instead. Scientists pointed to models for this sort of arrangement in other animals. Ants, for example, have a whole class of asexual workers who don't reproduce. Technically, the workers are female, though they develop in a way that makes them infertile. The colony's queen becomes huge and capable of laying eggs, while the workers stay small and strong, their ovaries stunted.* Worker ants forgo their own drive to reproduce in favor of assisting the colony.

So maybe, the theory goes, ancient human women evolved to support that kind of society: males doing whatever they're doing, young mothers tending to their children, and a significantly large,

* The males tend to live shorter lives, briefly fertilizing the queen and, when useful, defending the colony. But mostly, male ants are a sperm delivery system.

eusocial "grandmother" class assisting in child rearing. Assuming their granddaughters would benefit by such an arrangement, a "menopause gene"* would spread quickly through the population. Over time, it would be so useful that every girl would be born with the genetic code that switched off her ovaries by age fifty.

It's a nice story. I, too, would like there to be a specific, beneficial, evolutionary history for my grandmothers. They were lovely women. One was into embroidery. The other died when I was very young, but I still remember she had a large, apple-shaped cookie jar stuffed with Milano cookies. I remember the red of the apple, the generous curve of its shape. I remember her thin, knobby hands lifting off the lid. I want the idea that human evolution leads inexorably to my grandmother's cookie jar to be true.

But the grandmother hypothesis has problems. The idea of an "off switch" is the biggest one.

WHERE, EXACTLY, IS THAT SWITCH?

Here's a modern love story for you: A friend of mine recently asked if I would be willing to donate my eggs. He and his wife, both professors at Harvard, wanted to have a child. But like many accomplished women with challenging careers, my friend's wife was already in her early forties before she could seriously consider getting pregnant, and, as it turned out, she didn't have a healthy egg left. As far as I'm concerned, this is about the most flattering thing a person could ask you: "Say, friend, would you mind giving us your gametes? We're hoping there's a remote chance that our child could end up like you." I said yes.

There were hoops to jump through, including a rather extensive health questionnaire, involving information on all the possible genetic issues that may run in my family (which is extensive: Irish Catholics—in New York, no less—so my mother has eight siblings, most of them with multiple offspring). Because I was already

* Or rather, a suite of mutations—I don't think anyone assumes a single gene would drive something so complex.

in my early thirties, I also had to prove that my egg reserve was still robust. It was, happily enough, but the fact that it might *not* have been is one of the main reasons the grandmother hypothesis might be wrong. The IVF clinic needed to check my egg reserve because it turns out that there is no date-specific switch that triggers menopause. Rather, our ovaries just slowly run out of eggs. We actually start losing egg follicles—those little, fluid-filled sacs in the ovaries that harbor our eggs until they properly develop—before we're even *born*. If we do have an innate ovarian expiration date, it must be set in the womb.

Call it the "empty basket" theory. While men keep making new sperm until they die, a woman is born with all the eggs she'll ever have. Or rather, all the egg *follicles.*[*] Each month, as she moves through her ovulatory cycle, the pituitary gland cooks up a batch of follicle-stimulating hormones. In response, her ovaries begin "ripening" a handful of egg follicles. Typically, only one of these will go on to become a fully mature egg and make its way down the fallopian tube. It's a kind of in-house competition. Only the best survive.

This is presumably what happened to my friend's wife. Like nearly every woman on the planet, she was born with roughly a million immature egg follicles. But every year since, thousands of her follicles died off and were reabsorbed by her body. By the time she became a teenager, she had only about 300,000 to 400,000 follicles left. From then on, she lost about a thousand of them every month. If she started ovulating at age thirteen, she was destined to run out of eggs somewhere in her early forties. Which is precisely when most women stop being able to get pregnant without medical assistance. My friend had been on the Pill for many years, which you might assume would have saved some of her eggs. But no—delaying the ovulation process by going on the Pill doesn't save your eggs. In fact, every year of being on high-dose hormonal birth control seems to move the start of menopause up by about a

* Recent research indicates there may be stem cells in ovaries that do, indeed, regenerate immature egg cells, but the research is controversial (Grieve et al., 2015). And the result—steady egg loss over time, with a fairly predictable pattern—still holds true.

month.* That's because the loss of egg follicles isn't triggered by ovulation. Instead, ovulation *saves* about 20 follicles a month from early death, of which usually only 1 will go on to become a mature egg and find its way down the fallopian tube. But for those 20 that are saved, 980 die off.

Some women lose a few more egg follicles a month than the average, and some women lose fewer. And for whatever reason, some women in their thirties and forties retain more higher-quality eggs, while others seem to have more "bad" eggs left: eggs with more chromosomal malfunctions, eggs with buggy mitochondria, or eggs that are just, for whatever reason, no longer up to the task. But we don't have a clue as to why our bodies have evolved to discard so many eggs in the first place.

I did worry if donating my eggs to my friends would threaten my own chances of having babies later. Happily, no—women who donate eggs don't seem to have any lessened chance of becoming pregnant themselves, despite the invasive way clinic professionals extract the ripe eggs.† But no one could say whether donating eggs would make me go into menopause sooner than I would otherwise. (The data did suggest that I wouldn't.) Still, why *do* we burn through so many follicles every month? Why not lose a hundred instead of a thousand? How does the body know which eggs to save? Do good eggs become damaged over time, or are there only ever about four hundred good follicles out of the million we're born with?

In other words, are most of a woman's eggs *duds*?

For nearly half a century, the scientific community figured that mammalian eggs may have an expiration date. That would help explain human menopause at least a little: maybe it helps prevent genetic disorders. My friend's body might have discarded so many of her egg follicles before she reached her forties because the eggs had major flaws in their genetic blueprints, such as more "double strand breaks" in their DNA. There may be something wrong

* Luckily, today's standard lower-dose birth control doesn't make menopause start any sooner. It's also far less likely to cause cardiovascular problems—as usual with the body, less severe interventions produce fewer side effects.

† A long needle, guided by ultrasound through the vaginal walls, punctures the thin bubble atop the follicle in which each egg develops and sucks up the egg.

with the thousand eggs that most women get rid of every month, probably a result of the fact that eggs are just so much harder to make than sperm, so there's more opportunity for screwups.

While half of your DNA came from your dad and half from your mom, most of your mitochondria and cytoplasm came from your mother.* Sperm are basically an information delivery system that dumps the father's DNA into the egg, whereas eggs have to provide all the construction materials to build that embryo. And that's the major reason eggs are about four thousand times larger than sperm: they're not just half a set of blueprints; they're half a set of blueprints plus the entire factory.

Given that sperm don't require that much material, testicles don't have to work that hard or long to make their gametes.† Ovaries, on the other hand, have to exert more effort, over a lot more time, to help an egg mature—remember, the human fetus builds its egg follicles while still in the womb.

The longer a cell lives, the more chances it has to be damaged by accumulating waste and free radicals. There are mechanisms in place to repair damage, but those mechanisms get less reliable over time. It's also true that older eggs are more likely to have genetic problems of the sort that can lead to Down syndrome.‡

* We used to think *all* of a baby's mitochondria and cytoplasm came from its mother, but recent studies have shown that sometimes sperm manage to get some of its material into the egg (Luo, 2013). It seems to be a kind of breakthrough process, however, with the sperm mitochondrial DNA mostly eaten, jettisoned, or drowned out by the egg's cellular engines after fertilization (Al Rawi et al., 2011; Luo, 2013).

† Just about two and a half months, if you're counting. But if you're a person with sperm and you really, really want to impregnate your (willing) female partner with the best you can muster, it's better to have been living a healthy lifestyle for many years before the fact.

‡ Though the risk remains low—a forty-year-old mother has only a one-in-seventy-five chance (Cuckle et al., 1987). That's up from one in fourteen hundred in her twenties, but it's still long odds (ibid.). The age of the father is also a problem: older fathers increase the risk of chromosome malfunction, but until recently very few studies bothered to factor that in. From what we know now, having babies with a man over age forty means your child's risk of autism goes up, and schizophrenia, and Down syndrome (Callaway, 2012). Each year, it seems, there's a new study admitting that it's not all the mother's fault after all. So maybe it's better to think of it this way: stuff goes wrong with aging sperm, too. But there's just more that can go wrong with eggs because eggs are made of so much more stuff.

For the same reason, older women have more early miscarriages. So maybe ancient humanlike bodies somehow anticipated those problems, discarding all those egg follicles to avoid giving birth to disabled babies.

Since most mammals don't live as long as we do, maybe they don't have to deal with genetic damage to old eggs. There are some outliers, though, and they kind of punch a hole in that theory. Elephants give birth into their sixties, without any increase in genetic mishaps. Some whales do, too. Even chimps can give birth in their sixties, though it's rare and seems to happen only in captivity—in the wild, most chimps die before age thirty-five. Among the rare mammals who regularly live as long as we do, females usually keep reproducing late in life.* Generally speaking, all of these geriatric mothers produce perfectly healthy babies. That means aging mammalian eggs can't be the only reason human beings have menopause. If other mammals can keep giving birth late in life, why can't we?

The answer may lie in deep code: something about how our primate ovaries are "programmed" to function that's fundamental to our overall body plan and might be too costly to change. But since we don't know exactly why other mammals are able to give birth so late without problems, all we've established is that there's nothing about being a *mammal* that excludes old moms. That means that either human menopause is a really surprising change in the deep code of primate reproduction or it's actually a totally normal side effect of preexisting code that somehow proved too hard to significantly tweak in the long grind of our Eves' evolution.

It's been a very long time since we've been close cousins with elephants or whales. Though they don't live anywhere near as long as we do, maybe a good place to look would be closer on

* Ellis et al., 2018. Of those we're able to easily study, that is—the arctic bowhead whale seems to live two hundred years or longer, but we don't know enough about their sex lives to establish whether older females are commonly giving birth at two hundred. We only learned they live as long as they do because we've found nineteenth-century harpoons in their sides. It's very hard to study the longevity of whales that live in deep, cold water, particularly when most scientists are only professionally active for forty-odd years.

the family tree, at the ovaries of other great apes, and how *they* go about getting older.

SEXY GRANNIES

Once upon a time, our apelike Eves had massive labia. When their bodies were ovulating, those labia would swell into giant cushions of blood and other fluid to handily advertise that they were fertile. Chimpanzees and bonobos still have them. Our more distant cousins, the orangutans and gorillas, and other primates have them, too. Some are more dramatic than others, but it's a pretty common primate trait: when a female is in her fertile phase, her genital area swells and fills with blood, becoming warm and red and, for interested males, pretty darn inviting.

Scientists figure that when hominins began to walk on two legs, there wasn't room in their upright pelvises for giant genital displays.* These flaps shrank, but even now a woman's labia may swell just a bit when she's ovulating. The inner labia—nearly diaphanous flaps that nestle around the clitoris and its hood—can turn a bit darker with blood whenever we're particularly turned on, and in a more pronounced way around ovulation. As women age, the inner labia tend to stay darker, a leftover from lifelong cycles of fertility.† When a woman goes through menopause, her outer labia may shrink a bit—just another part of our menopausal

* Some think swollen primate genitals help encourage paternal care (Nunn, 1999; Alberts and Fitzpatrick, 2012). Others think that "hiding" our fertility might have had its own benefits in terms of female sexual choice—for instance, if males don't know when you're fertile, they can never be sure when having sex with you will actually produce babies. This can reduce the pressure on a female during her actual fertile periods, leaving her with fewer guys to fend off and potentially increasing her chances of selecting the males she prefers. It could also benefit the female in terms of paternal uncertainty, because sexual swellings in primates do align with other measures that influence paternal uncertainty (Nunn, 1999), though any *conscious* register might require more brainpower than early hominins had to work with: *Did I or didn't I have sex with Lucy when she had big labia? Let's see, how many months has it been . . . oh, right. I'm an australopithecine. I don't do math.*

† It can also trigger more melanin production in the skin down there—many parts of our skin can change their color patterns a bit as we age, and the genitals are no exception.

fat redistribution—even as her inner labia stay the same size or lengthen.

This happens to chimps, too—which is one of the central ways we finally figured out whether chimps have menopause.

It seems that, like us, most chimpanzees stop ovulating around age fifty. Or rather, their reproductive organs "senesce," the formal term for aging. Their ovaries get old, and so do their genital swellings. Quite unlike us, however, a fifty-year-old chimp is *very* old. Her teeth and fur are starting to fall out. Her joints are creaky and brittle. She's lost muscle tone. Even in captivity, where chimps live longer than in the wild, they usually die in their fifties or sixties. In other words, perhaps chimps don't have menopause the way we do because they *die too young*.

But, as opposed to human cultural norms, the older the chimp, the sexier the boys find her. The hottest gal on the block is already a grandmother. Maybe even a great-grandmother. She's got graying fur. She may even have a cataract or two. But the males can't get enough of her, and the younger females don't stand a chance. Primatologists aren't exactly sure why this is the case, but they agree that in general chimp grannies are very sexy.

When human women start looking older, it often does mean they're becoming less fertile. So in evolutionary terms, it makes sense that men may find them less sexually attractive.* But in chimpanzees, gray hair doesn't necessarily indicate a chimp's ovaries aren't working anymore, because visible signs of aging arrive *earlier* in a chimp's reproductive years than they do in a human female's. In fact, for female chimps, looking older can signal that she's the bearer of high-quality DNA. She might also have a pretty good standing in local society, since it's harder to live a long life as a social outcast. Put the two together, and it's quite a package.

But come back to that number: age fifty. If chimps manage to live that long, many of them seem to stop ovulating, just as we do. Other primates follow similar patterns. Looking across the primate reproductive plan, it seems primate ovaries age at similar

* Or at least they *say* they're less attracted; the proliferation of "MILF" and "granny" pornography on the internet may reveal a different reality. As with chimps and bonobos, human sexuality isn't purely tied to reproduction.

rates. If that were true, then from baboon to gibbon, chimp to human, each of us would lose roughly the same percentage of egg follicles each cycle, and our reproduction would follow the same slope of decline over time.*

In other words, the deep structure of primate ovaries might be fundamentally geared for a life span of about fifty years. We *can* live longer, but we won't be as good at making babies, and the rest of our bodies are also shutting down. If that's the case, then it would seem that the thing that changed in our Eves might not have been in their ovaries. Instead, women somehow delayed aging in the *rest* of their bodies, and human ovaries haven't had a chance to catch up yet.

But that still doesn't quite answer the deeper question: Why? Why did we need a bunch of older ladies in the first place? If it wasn't simply being alive without their own newborns, what else was there in being *old* that was useful?

BACK TO JERICHO

The girl's womb heaved and clenched. The old woman had to be careful. If she tore something, her granddaughter would bleed to death. And the child would probably die, too. The cervix was wide—that was good—and the girl's hips felt loose on either side. There was the foot, but she felt just one of them. If only one leg came down . . .

* There are outliers, of course—chimps who give birth after age fifty, for example—but that's true for humans, too. The majority of chimps will not successfully give birth in their fifties and sixties, and barring interventions like IVF, ovarian tissue transplants, or whole-uterus transplants, neither will the majority of women. As we've seen in the "Tools" chapter, our species tends to technologically intervene in our baby-making capacities—such interventions may even be distinctive of humanity, and something we should consider fundamental to our success—but that doesn't mean our bodies *evolved* on a longer timeline to reflect that. Certainly, having more women survive the reproductive process makes it likelier for women to manage to survive to old age, but that alone didn't produce human menopause. In other words, while IVF is a natural extension of gynecology, it's not true that human ovaries will suddenly change their baseline primate blueprints because our social environments can now support older pregnant ladies. As usual, cultural innovations far outpace genetic mutations.

Time was rushing by, the girl's life on the line, so she did the first thing she thought of—she pushed the foot back up into the womb. The baby's knee tucked up near the chest. With two fingers, she felt for the baby's slippery bottom, talking softly to calm her granddaughter down. She was delirious with pain.

As quickly as she could, the old woman pushed hard on the girl's open legs, and she heard one of the femurs slip out of its hip joint with a great wet pop. The child came quickly after that—butt first, arms tucked tightly around the chest. A boy. *Figures.* The old woman placed him on his mother's stomach, and they both rubbed the newborn's back. He wasn't blue. He wasn't crying either, but they could see him breathing. He'd live.

She wasn't sure about her granddaughter. The girl was pale and sweaty, her legs soaked in blood. The old woman's sister reached over to try to tug on the umbilical cord, but the old woman moved her hand away. It was better to let the placenta come out on its own. They'd tugged on one of her aunt's umbilical cords once, and a great rush of blood followed.

The next hour or two was critical. If the girl survived it, the old woman would deal with her injured hip. She told her sister to keep the gawkers out of the hut. Nothing to do now but wait.

WISE GRANNIES

The old woman of Jericho I've been imagining is actually two Eves in one: the Eve of human menopause, and also the Eve of the *elderly*—meant to represent one of the first women to live into old age with other old women around her.

For most of human history, elderly people were like unicorns. Maybe you'd know one of them. At the most two. Maybe you'd only see the stark white of an old woman's hair from a distance. Or maybe she was your grandmother. Maybe she fed you bits of meat. Maybe she shared food with your mother. But for the most part, people simply didn't *survive* long enough to become truly elderly.

Ten thousand years ago, when human agriculture really took off, our ancestors had collaborative lifestyles, medicine, and a full million years of gynecological behavior to call on to help women survive. Our Eve of menopause had to be the Eve of the elderly, too: not a *rare* woman who'd lived a third of her life past her ovarian stop date, but a woman who did that and lived among other women who had done that, too. In other words, while the mechanisms of menopause are physiological, being a "menopausal" species may be a deeply social phenomenon—you need to have *most* females routinely surviving to sixty and beyond, living a third of their lives after their reproductive years. Because evolution takes a phenomenally long time to standardize changes in a species' body plan, it couldn't be a one-off. Culture changes quickly. Physiology, as a rule, does not.

Though their lives are in many ways as "modern" as the rest of ours, we can look for some clues in well-studied hunter-gatherer populations. Among today's San hunter-gatherers, 50 percent of all children die before age fifteen, and the average life expectancy is forty-eight. Among the 10 percent of San people who manage to live to sixty, a majority are women (women outlive men everywhere, but the gap is more pronounced among the San). So do the San have menopause? The answer is yes, despite all that mortality.

But our ancient ancestors probably didn't have a body plan ready-made for menopause. From what we've seen in the fossils, it was incredibly rare, for a very long time, for hominins to live past their thirties. Even anatomically modern *Homo sapiens* didn't seem to in the very beginning. In fact, the reason I've chosen a woman living in Jericho as my Eve here is that many paleoanthropologists think that before the rise of agriculture, human beings didn't regularly live to sixty. That was only about twelve thousand years ago. Women's bodies might have been set for menopause before then, but maybe our lifestyles didn't support that potential until later. Until we know more about the genetic underpinnings of aging, we're not going to be able to backdate with much accuracy—we have to keep relying on what we find in ancient bones.

Still, if we limit ourselves to saying human menopause started when there were *societies* of the elderly, then it's possible even twelve thousand years is too early. Creating and maintaining a regular class of postmenopausal grandmas might not have been possible before the rise of more densely populated agricultural towns. And grandmas—or rather, the elderly, most of whom were women—would have been particularly useful for the rise of agricultural society.

Consider the killer whale: Transient orca pods are the *only* nonhuman social mammals that have verifiable menopause. They're hard to study, of course, because they're killer whales and the ocean is massive. But from what we've been able to determine, like human women, these females live a full third of their adult lives after they've stopped having children. The society is matriarchal. The sons stay with their mothers their entire lives. If their mothers die, the surviving sons don't fare as well. They don't have as many children. They don't retain status in the pod. The success of their lives, in other words, depends on their mother. They inherit her social status, and they receive daily perks accordingly, ranging from food rights to which females they get to have sex with, and when, and how often.

But a grandmother orca's duties don't involve spending a lot of time taking care of the grandbabies. That means orcas don't fit the grandmother hypothesis. From what the research has shown, postmenopausal orcas don't spend more time caring for their grandkids or other young offspring after they stop giving birth. They also don't spend more time defending the kids from outside threats, nor do they spend extra time gathering food for the family to eat. The fact that they stop having babies of their own doesn't seem to be in the service of the cetacean equivalent of free child care.

What the grandmothers *are* responsible for is teaching the pod in times of crisis. When food is scarce, the grandmothers are the ones who lead the way to places that are more likely to have good food. Once the pod arrives, the grandmothers are more likely to be the ones to demonstrate how to get that food, should there be particular challenges. For instance, creating bow waves to wash seals off ice floes and herding fish.

What grandmothers do, in other words, is *remember.*

Living a really long time as a social mammal is good for two things: reinforcing the social status of adult children, and ensuring the well-being of the group overall in a crisis by remembering how to survive in a world that changes over time.*

Maybe, instead of the grandmother hypothesis, we should think about two things: Postmenopausal grandmothers may help their children to maintain their social status and resources over time (call it the mother hypothesis). And maybe grandmothers are also helpful because they're really good at remembering things. Old people can be valuable because they're *wise.*

We need to look past our own grandmothers' fondness for cookie jars and think about what ancient humanity really needed from its old people—like the wisdom that is asked of this chapter's Eve, the old woman in Jericho.

It's not hard to find her counterparts in grandmothers today. For example, consider an Afghani woman named Abedo. Like many women from her part of the world, she was widowed when her husband was killed in battle. I first read about her in a small article by a young war correspondent after my brother had been an embedded reporter there; as time passed, I dug deeper. Abedo was the wife of a member of the mujahedin in Afghanistan in the 1970s—a situation that was hardly unique. But when she learned that he wouldn't be coming home, rather than flee with her children like the other refugees, she decided to fight. She started dressing as a man, which seemed the only way possible to do what she believed was God's will, and she came to lead many mujahedin during the war with the Soviets.

In 1989, the Russians finally withdrew like a glacier, leaving

* To be clear, just because you're old enough to have memories that younger people don't have doesn't mean you're always going to make the right decisions. For instance, shoving a foot back up into a woman's laboring uterus in a small, dirty, dimly lit hut is a *terrible* idea. Never do that! But not tugging on an umbilical cord right after a woman's given birth? That's a good one. And there are stories about doctors in the field finding nearly acrobatic ways to widen the birth canal, which can run the risk of dislocated joints (remember, she might be extra-flexible with all that relaxin on board). Dislocating a hip is not recommended, and certainly not standard practice, but under the right circumstances—especially if you don't have the tools to perform a C-section safely—who knows? It may help. That's probably the better way to imagine the ancient benefits of having older people around. They're not superhumanly wise elders but regular people, making a mix of good and bad decisions based on prior experience, whose overall effect helps the population rather than hinders it.

the land scraped flat by the rollers of war. For a time, Abedo managed to settle down to a more "normal" life back in her village. She even opened a shop, selling goods to people she'd fought with. Her children grew. Though it certainly wasn't normal for an Afghan woman to live the way she did, she maintained her independence and was well respected by her neighbors. Twenty years came and went. Her children had children. Poppies bloomed in the river valley, pink and white.

Then, after another war burned half the cities down, the Taliban started interfering with her business. They told her not to sell to the U.S.-backed government. The government, meanwhile, told her not to sell to the Taliban. She refused to take a side. She'd probably still be living her ordinary village life if the Taliban hadn't decided to set fire to her shop. After that, with the blessing of the American-backed government, she recruited ten young men for her own paramilitary troop. When I started researching this chapter, she'd survived, a cross between a wizened grandmother and a commandant, continuing to defend the daily life and well-being of her village with well-oiled guns. Given her extensive experience as both a fighter and a military commander, that U.S.-backed government had consulted her for security intelligence and strategy in the region. "Modern-day youngsters in the police and army don't have experience," she said to a journalist, "and it's easy for them to get killed in combat because they don't know how to fight."

No one I was able to contact knew if she'd survived the disastrous American retreat from Afghanistan in August 2021, nor if she even lived long enough to see it happen. One presumes the new Taliban government wouldn't see her as an ally. But at least we know that for a surprisingly long time Abedo was alive because she knew how to fight. She was also alive because human women usually live longer than men. And like many older women, she still had her wits about her, which helped keep the men fighting under her alive, too. She taught them because she remembered how war *works* in her river valley. She led them because she knew how, and they followed because they knew she knew.

Maybe Abedo is an unusual model for the evolution of menopause, given that modern Afghanistan is obviously not the same as ancient Jericho. But she's a woman who survived long enough in difficult circumstances to be able to offer important knowledge and leadership in a social group. Rather than thinking of menopause as a thing we evolved to provide extra child care, we should think about what it really means to be old enough to remember events that neither your children nor your grandchildren have experienced. Imagine someone like the old woman of Jericho seeing crops destroyed by a flood, something that hadn't happened in twenty years. Her kids wouldn't know what to do or how to survive. But *she* might.

And when you have a complex social group doing something as hard as figuring out how to live on cereals you grow yourself—and sharing and storing food at a scale no human had ever done before—maybe you need old people to pull it off. If that's true, once agriculture was invented, there should have been a kind of aging-agriculture feedback loop, each benefiting the other.

Remember, the start of agriculture was a bumpy ride. Stationary living brought the challenges of seasonal famines, waste-borne diseases, and nutritional deficiencies from a less diverse diet. And not all foods—not even foods we cultivated—were easy to eat. Eating cereals and tubers isn't like eating figs off a tree. You need to know how to prepare them to make sure they won't kill you. Many of today's domesticated foods are modifications of plants that, in the wild, could make you very, very sick indeed. For instance, manioc root, widely used today in South American and African cuisines, requires soaking, boiling, and pounding to remove the poisonous alkaloids from the raw tuber. Even the lowly potato needs particular knowledge. If potatoes are exposed to light for too long, they turn green and, if you eat too many green potatoes, you can become terribly ill; green potatoes contain solanine, a chemical that essentially prompts cells to kill themselves. Nausea, diarrhea, and vomiting are the milder side effects. The nightmares are also survivable. You'll have a harder time getting past the hallucinations, paralysis, hypothermia, and death. Freezing to death on a hot afternoon because you've eaten too many green potatoes isn't a good day for the advent of agriculture. And heaven help you if you eat the leaves, stems, or shoots.

The *reason* quite a lot of agricultural plants have dangerous side effects if not properly processed is that plants, like animals, often defend themselves, usually with chemicals. Plants that had already evolved with certain pesticides and other self-defense measures built in would have been great for planting in ancient gardens: they'd usefully resist beetles and other bugs that might eat them before humans got a chance to. In other words, you are more likely to be poisoned when you eat plants than when you eat a diet of meat.* The shared social knowledge of hunter-gatherers helped our ancestors navigate that dangerous poison-filled plant world alongside their meat-eating habits. But agriculture required knowing not only which plants to eat and which to avoid but also how to plant and grow the right ones, how to store and process those foods in ways that wouldn't make them *become* toxic over time, and of course how much of one thing or another is okay to eat, after which it becomes drastically *not* okay anymore. That requires far more social knowledge than our ancestors' prior lifestyle. It requires a lot of collaboration. And before the advent of written language, it might have required a certain density of old people like our Eve. People who have experienced a lot and learned from it.

In ancient Jericho, you'd need someone who remembered how the old woman's brother froze to death on a hot afternoon after eating the wrong thing. Someone to teach the community how to sow lentils and peas and emmer wheat, how to boil the bitter vetch to drain out the nasty compounds, what kinds of seeds to plant near one another to help keep away the pests and enrich the soil.

Once agriculture took root in human culture, there were plenty of advantages to having old people around. But outside genetics, extending life span still requires essentially the same things today: food, medicine, social stability, and a decent crisis plan. Agricultural societies can provide the first three. And old people were

* You might become *infected* from meat eating, which could certainly be deadly if those animals were carrying bugs that could infect your body, too. But cooking and salt are both pretty great ways to avoid that scenario, so long as you aren't eating meat that's been left to rot. The central reason *old* meat is so dangerous is that not only have bacteria had a long time to reproduce and munch on the decaying tissue but they've also had time to put out a lot of dangerous toxins—toxins that simply washing and cooking the meat may not solve.

useful for the fourth—what to do when a flood washed out your crops, what to do when there hasn't been enough rain, what to do when a conflict arises with a neighboring group, what to do when in-group conflict threatens the community's welfare overall. They were elders.

Before we could write stuff down, it was especially important to have someone in the group who could remember earlier crises. It's usually not hard to find someone who can remember a difficult thing that happened ten years ago. It's much harder to find someone who remembers a difficult thing that happened forty years ago, or how, precisely, the community managed to find a workaround. Oral history provides only so much after the storyteller dies. Living long enough to see a rare crisis happen again is the most reliable way to know whether a piece of knowledge is something the entire group should learn.

Today's hunter-gatherers don't have different patterns of menopause from urban people, so it can't be the case that inventing agriculture drastically changed our genes. And in fact, whatever genetic shifts might have happened to help extend our life span probably happened long before the Eve of Old Women.* The reason agriculture matters for menopause is that it was a critical moment in human history: We were trying to do something really, really hard. It often made us sick. It required whole new ways of living. Having elders who remembered what had worked and what hadn't would have been really useful. Such elders would have benefited hunter-gatherer societies, too, but maybe sustainable agricultural societies made societies of the elderly simply more common.

I think that's a simpler answer to the mystery of menopause. Rather than the grandmother hypothesis, which proposes a complex model of human eusociality requiring radical changes to both

* One study on fossilized humans and Neanderthals potentially pushes this date as far back as thirty thousand years (Trinkaus, 2011). But the work is controversial, and it might still have been very rare for more than one or two females in any social group to live longer. One way of thinking about this is that the hominin genome might have produced some mutations that allowed females to live longer as far back as thirty thousand years ago, but it might have taken something like the rise of agriculture to reliably see larger societies of elderly menopausal women.

our genetic programming and our social lives, let's consider the alternative: Maybe we didn't evolve to have menopause. Maybe it wasn't selected for. Maybe, instead, it was a natural side effect of our extending life spans. In principle, bodies do just about everything they can to avoid death. So, it's not hard to imagine evolution selecting for traits that helped us dodge the grave. But in social species, it can also be useful to have the elderly around. That can put further pressure on selecting genes that extend life span and, in women, lead to menopause.

In that way, the selection of this chapter's Eve is about finding a good-use case: new farming communities needed the memories of the elderly. It's not that farming made us better equipped to support our grandmothers—at least not in the early days—but rather that we *needed* them more than ever. The real start of menopause is when enough women survived into old age that a girl could *expect* to become a grandmother herself one day. The Eve of human menopause is really the first woman who lived among a group of other old women. We're looking for the first ancient knitting circle—except they probably weren't doing a whole lot of knitting. They were probably leaders. A council of elders. Our Eve wasn't the helpful grandmother, necessarily. She was the *wise* grandmother.

Thus, the point of menopause isn't that we stop ovulating. It's that we keep *living* past our predicted—and biologically tuned—expiration date. We made it normal to grow old. That means what's interesting about menopause may not be menopause at all, but how human beings manage to stave off death. And by human beings, I mean women.

Throughout the world, women are simply better at not dying than men are. So long as we manage to survive the ridiculous death ride our reproductive system takes us on, we usually live longer, healthier lives than men do. And that fundamental difference becomes only more obvious the older we are. In the United States, the average woman will outlive the average man by only about five to seven years. But that's talking averages of the whole population. When you control for *age* cohorts, the gap widens dramatically. With each passing decade, more and more of the men in a cohort start to die, while fewer of the women do.

Centenarians used to be unicorns. Now the United States has more than fifty-three thousand of them. Canada has nearly eleven thousand. Japan has more than eighty thousand. Italy, nineteen thousand. The U.K., just over fifteen thousand. And by and large, they're not men.

More than 80 percent of today's centenarians are female.

SUPER GRANDMAS

Of the three people alive today who have verifiably managed to live to age 115, all are women. The longest-lived person in the world, a Frenchwoman named Jeanne, lived to 122 years and 164 days before dying quietly in 1997. The oldest man was Japanese and died in 2013 at 116. But exceedingly few men make it past 100, because men's bodies age faster and more problematically than women's. The thing all of these incredibly old people have in common is that they live essentially free from old-age diseases until just before they die. No cancer, no heart stuff, no dementia riddling their brains, lungs clear, no diabetes, and no gut problems. What's remarkable about them, in other words, isn't simply the number of their years but how very few of those years they spent detrimentally *aging*.

We don't really know how female bodies do it. For decades, scientists wrote off the difference in longevity as a matter of lifestyle: Men are more subject to violence, to accidents, to trauma. Some said that maybe male bodies are more stressed because they have to work all day outside the home. Maybe men do more taxing, dangerous, heart-pounding jobs, which wear down their bodies at a faster rate.* Maybe it's the red meat. Maybe it's the commute. Maybe it's the cigarettes and booze.

* Actually, the findings are mixed. On the one hand, physically demanding jobs make men 18 percent more likely to die sooner than the average man (Coenen et al., 2018). But other studies show working physical jobs promotes a longer life span than desk work (Dalene et al., 2021), and most believe that activity is simply better for the human body than being sedentary. It's generally true that continuing to be physically active in your later years—even if it's just pottering in the garden—tends to make you live longer.

But even if you take two perfectly healthy people, one man and one woman, with similar amounts of stress, and similar types of nutrition, and similar sorts of jobs and habits, the woman is more likely to outlive the man. How and why that happens is a mystery, but the fact itself is no longer controversial. And it's true among our ape cousins, too: among both wild and captive chimpanzees, gorillas, orangutans, and even gibbons, females usually outlive the males.

That's why, from a genetic perspective, we probably shouldn't think of human menopause as the result of evolution selecting *for* nonreproductive elderly females. Rather, whatever helps female bodies live on may simply benefit male bodies less, and losing more males may not cost primate societies that much. It sounds harsh, I know, but it's true: From a scientific perspective, males don't really need to live as long as females to perpetuate the species. That's especially true of mammals. As biologists are fond of saying, male mammals are "cheap." By which they mean easily replaced.

So long as a human male makes it to adulthood, it takes him only two to three months to successfully pass on his genes, and the bulk of that time is spent making new sperm in his testicles. Once the sperm are built, it takes only sixty seconds to ejaculate them. Women, meanwhile, need a minimum of twenty-one months to pass on their DNA: twelve months for the egg follicle to fully mature and another nine months to gestate the baby. And then there's breast-feeding. Most of the hard work of reproduction and early caretaking is done by female bodies. That's why losing a *female* is usually a great loss for a species' evolutionary fitness. Losing a male? Well, there are more where he came from.

Since there's simply more *pressure* on the mammalian genome to preserve the life of the female, maybe, over time, certain mechanisms have evolved that protect against the bad stuff in the female body's aging process. Again, living longer than men is really about *not dying*. There are age-related markers all mammals have as they get older, like changes in body fat and arthritis and muscle loss. There are things that happen to the skin, which women's magazines are all too ready to recommend some expensive serum to counteract. But you can live a really long time with loose skin on your knees. The wrinkles under your eyes won't kill you. *Survival* is the real game. So, let's talk about what actually kills you.

First of all, death is what happens when your brain dies. What *usually* kills your brain is organ failure: your heart, your lungs, your kidneys, your liver, shutting down in a cascade. The blood reaching your brain isn't properly filtered. Not enough oxygen, too much CO_2, too many toxins. Or not enough blood reaches your brain. Maybe a clot plugs up the works and the cells in your brain start to die. You'll usually lose consciousness before this happens. Eventually, the lights go out.

Unlike children in many hunter-gatherer societies, most of today's industrialized human beings survive childhood. When we don't die of something stupid, like preventable infections or violence or accidents, we usually die because we get old. But "getting old" isn't exactly what kills us. It's the big three: cancer, cardiovascular disease, and lung disease. These are the killers we're running from. And, as they get older, female bodies are just better at outrunning them.

Really, the only thing male bodies have going for them in this race seems to be social. Historically, we've paid more attention to male bodies—how they thrive, how they die—so modern medicine (and popular knowledge) give men a leg up here. Cardiovascular disease kills men significantly sooner than women, but because women's heart attacks can present with slightly different symptoms, most people in today's industrialized countries are on the lookout for what *male* bodies do when their hearts are seizing up: clutching their chests, burning pain through the arm or jaw, a feeling of a crushing weight bearing down, and so forth. Women, on the other hand, commonly say they feel as if they were having a particularly bad or weird bout of acid reflux, maybe with a side order of anxiety and dizziness. Some get that classic feeling of a weight on their chests, but many don't. As a result, more women currently die of heart attacks than should be the case, not because they get them more, but because they don't take their symptoms seriously enough or they simply don't know what they're supposed to be watching for. There are many campaigns to change social awareness around these issues, particularly in the United States and western Europe, which may eventually shift the stats somewhat. But the result will only reinforce the existing norm: fewer women will die of cardiac events because they'll recognize their symptoms and go to hospitals sooner than they might otherwise, and the doctors there will treat them with the appropriate level of care. In other words, even *fewer* women will die from heart problems than already do. The longevity gap between women and men will simply widen.

The simple fact is that the male cardiovascular system seems to wear out faster than a typical woman's. There's more stiffening in the arterial walls. There tends to be more cholesterol buildup along those walls, too, which may represent higher degrees of inflammation. And these changes start very, very young—possibly in the womb. The male cardiovascular system is more prone to higher blood pressure from an early age. This may be why young men who received some of the COVID vaccines in 2021 were more at risk of myocarditis and pericarditis after their shots—inflammation of the sac around the heart or of the lining of the heart. But, of course, men and boys who fell ill with COVID-19 were *also* more likely to suffer cardiovascular problems such as these, and likewise were significantly more likely to die during the pandemic than women were. While most people thought of COVID-19 as a lung disease, many now think it would be better modeled as a cardiovascular disease, given that thousands of tiny blood clots clog up the lungs, each of which triggers yet more local inflammation and cell death, resulting in a particularly horrific bloody cascade toward lung failure.

Lung disease is another one of the Big Three that kills more men than women. Lungs, a bit like the brain, are incredibly foldy, containing a surface area equivalent to half a basketball court. And the immune system regulating all that body-world interaction is highly influenced by the body's sex—whether it's a postpuberty balance of sex hormones responding to signals on the fly, or deep regulatory stuff tied to individual cells' chromosomal makeup, sex matters for the immune system, and lungs are no different. So while blood clots were likely a factor for COVID-19's lung devastation, the simple fact of the masculine immune system can't have helped all those poor men who caught the virus and were unlucky enough to suffer a cascade of inflammation that ran unchecked in their lungs. Despite women having *smaller* lungs—and presumably, therefore, greater vulnerability to lung damage—female patients generally fared better.

So long as they weren't pregnant, that is. Pregnant women fell to the disease in droves. At first, they weren't sure; early in the pandemic the data were all over the place, and women aren't *constantly* pregnant, so naturally there were fewer pregnant patients to include in the data set. But as time wore on, it became clearer: pregnant women were more susceptible than most people their age to the deadlier forms of COVID-19. And there were probably two reasons for it: first, much like what happens when they catch the flu, pregnant women's bodies' screwy immune systems can underreact to initial infection and overreact to ongoing infection, making them both more likely to *get* the flu and then more likely to have deadly immune *reactions* when the flu invades the lungs and the immune system kicks in, creating deadly cascades of inflammatory signals. Second, pregnant women's lungs are always a bit compromised in the third trimester, meaning that things like the flu—and COVID-19—may have their deadliest consequences.

And there's the kicker: when lungs kill women, they usually do it during one of two times in their lives, either when they're pregnant, with lungs both squished and massively taxed by the swelling uterus and its associated placenta, or when they're postmenopausal and their hormone profile has changed. Still, while it's not great when your grandmother gets the flu, she's less likely to develop a severe lung infection from it, and her overall prognosis will probably be better than your grandfather's. When our lungs get older, it's just better if they're female.* The only way that's a problem, really, is that women who are diagnosed with lung disease are less likely to be given aggressive treatments for it than men are, which may actually dampen their chances of recovering. If women's lung disease were treated equally to men's, the stats could swing even more in their favor.

* So long as they're not smokers. Female human lungs seem to respond especially badly to tobacco smoke exposure (Langhammer et al., 2003).

As for cancer, outside genetics, many different lifestyle choices influence one's overall cancer risk: eating charred and fatty foods, sugar consumption, exposure to toxic chemicals, alcohol, failing to get enough exercise, *stress* . . . Simply knocking back one alcoholic drink a day raises an American woman's risk of *breast* cancer by 14 percent. But, in general, more men get cancer, they get it younger, and they are more likely to die of it. One in two men worldwide will suffer from some form of cancer before they die. For women, it's one in three. That's especially significant, given that aging, all on its own, is a cancer risk, precisely because the various things our bodies do to regulate our cells' ongoing division become less reliable as we age. We're rather good at trimming a few hairy snips of malformed genetic code in our youth (or dive-bombing wayward cells when they fail to self-regulate) and less able as our immune systems get older. Cancer occurring in one's youth is strongly tied to having a Y chromosome, and cancer in old age is slightly less so. Worldwide, in any given year, for every four boys under age fourteen who are diagnosed with cancer, only three girls will be;* men in their seventies (should they survive that long) are only slightly more likely to receive such a diagnosis than women of the same age.

One central reason most researchers think that may be the case is that the Y chromosome is tiny compared with the X chromosome: The X carries about eight hundred genes, while the Y only carries about a hundred to two hundred, leaving large portions of the X un-partnered in the male cell.† The reason that

* There's some problem with diagnoses and sexism here: the ratio is closer in wealthy countries, while developing countries often show a wider gap. The assumption is that boy children may be more likely to be taken to the doctor when they fall ill, where the cancer is diagnosed. That's likely a factor, but it doesn't explain everything, and it certainly doesn't explain the difference in survivability: boy children diagnosed with cancer are significantly more likely to die of it, compared with girls diagnosed with the same sorts of cancers (Dorak and Karpuzoglu, 2012). If you're a child who develops cancer before puberty, you're usually better off being biologically female.

† Cheap joke, but instructive: it's not the size that matters, but what you do with it—simply having more genes doesn't necessarily make what's on that chromosome more fundamentally *important* (the *SRY* gene on the Y chromosome has

matters, of course, is that in the womb female embryos shut off or "inactivate" one of their two X chromosomes, presumably so they don't double code for things and gum up the works. Thus, while each cell line *carries* two X chromosomes in a female-typical body, each living cell is normally activating genes on only one of those two chromosomes.

Presuming she's born, for the rest of that female's life, each cell somehow remembers which X was initially shut down in the womb, and instructs all subsequent cells in that line—each time the cell and its progeny divide, for the body's *entire* lifetime—to keep shutting down that particular X. This turns out to be true *except* for about fifty of those eight hundred genes, as researchers just discovered in 2017; some of those genes seem particularly important for cellular DNA self-regulation and metabolism—precisely the things a cancer cell tends to screw up, both when tumors form and in determining how fast they'll grow, reproduce, and eventually metastasize.

So if a male has some screwy genes on his X chromosome, his dinky little Y chromosome isn't going to be able to keep the dampers on potential tumors the way having two X chromosomes would. This problem is so characteristic of men with certain kinds of cancers, in fact, that researchers decided to call the still-active X genes EXITS: escape from X-inactivation tumor suppressors. Across twenty-one different sorts of cancers, five of these EXITS genes were more frequently mutated in men than in women. Being *male*, in other words, was very much a part of what was killing them. And presumably, as we slowly figure out how to develop treatments for those sorts of cancers, we may well be trying to make those cancer-ridden male bodies essentially more female.

But at this point, from what animal studies have shown, there's honestly only so much medicine can do once a body has already gone down a male-typical path. Unless our understanding of the

---

a huge effect throughout the body). But it is true that when it comes to certain genetic mishaps, males may have problems with their comparatively tiny sex chromosome.

biology of sex changes drastically—and more important, our ability to *intervene* in that biology—women will continue to outlive men by many years.

LIVING WITH THE DEAD IN JERICHO

The ancient people of Jericho buried their dead under their houses. We know this because we found their bones, thousands of years later, after digging up the foundations from the packed earth that covered them. We found their skulls, some decorated with plaster, some with cowrie shells where the eyes had been. We found stone carvings of women. We presumed "ancestor worship." We helplessly called it religion.

We know they lived with their dead. We don't know how they did it. We don't know if they called out to them with quiet prayers while they cooked on their hearths, or if they thought of the dead under their houses as they ground the dried barley. As they braided their daughters' hair. As they gave birth, blood soaking into the earthen floors. We don't know what they thought of their lives, living so close to the dead, every day—every day, the dead under their houses.

We know there was a nearby spring, which is the reason they built the city there. We know the wadis flooded, which is why they built a wall around the city. We found the wall. We found the foundations of their houses. We held their cowrie-eyed skulls in our hands.

During the great wars of the twentieth century, Americans and Europeans wrote a lot of pop songs. They were usually about love. But love in absence: boyfriends and husbands leaving town, girls waiting on letters home. The whole idea of a home front was female: Women sowing their victory gardens in a time of rationing. Women packing bombs to be shipped thousands of miles away. Women stitching parachutes in a factory in the hopes they would catch the bodies as they fell.

To be a woman in those war years often meant you were a person who loved someone who wasn't there.*

It's a very old story: Penelope waiting for Odysseus to come home. There are versions in Sumerian, in Akkadian, in the little cuneiform arrowheads that line ancient clay tablets. Even the story of Inanna, the Sumerian goddess of love and war, resolves with her mourning the death of her beloved Dumuzi.

But it's not just wars that take men from us. Their bodies betray us, too. Women stand in a field of accumulating absences. Holes open in the earth. Caesuras.

I have a brother I love more than just about anyone in the world. But he's five years older than I am. Neither of us smokes. Neither of us uses any drugs to speak of. Though we didn't grow up with much money, we live pretty well now. We have good health care and eat good food. Our cities don't have much pollution. I'm a bit fatter than he is and a bit less healthy, and he certainly exercises more than I do.

I understand, painfully, that he's probably going to die before me. I might live as many as ten years without him.

Statistically speaking, that's the number I'm working with. It's not for sure, but it's likely. Five years for the sex difference, and then the five years he's older than me. Ten years.

I haven't wrapped my head around how I'm going to handle that.

That's the real story about menopause. It's not the night sweats. It's not the dry vagina. It's not really about menopause at all. It's that we outlive the men we love. We outlive our brothers and husbands and lovers and friends. We have to live on, all of us, and watch them go.

* This was true of women in Japan, and China, and India, and the Pacific Rim, and parts of Africa, too, where many men were pressed into war—I speak only of "Western" women here because those are the songs I know.

Homo sapiens

CHAPTER 9

LOVE

> And a human being whose life is nurtured in an advantage which has accrued from the disadvantage of other human beings, and who prefers that this should remain as it is, is a human being by definition only, having much more in common with the bedbug, the tapeworm, the cancer, and the scavengers of the deep sea.
>
> —JAMES AGEE, *COTTON TENANTS*

> The man has a theory.
> The woman has hipbones.
> Here comes Death.
>
> —ANNE CARSON, *DECREATION*

You do a lot of math when you're broke. Rent, gas, the credit card tango . . . Algebra drifts through the mind like the chorus of an old song you don't even notice you're still humming: *If I drive only twenty-six miles a day, this tank of gas should last until Tuesday.* So it was as I drove on that Indiana highway at the turn of the millennium, numbers buzzing behind the visual data: skinny trees, seamed concrete, square buildings, giant signs for Brake Depot and Jesus Saves and Midnight Runners XXX. I remember rain on the windshield. My red Nissan had a break in the door seal. Water leaked onto my shoulder. I crossed the city lines and looked for my exit.

The ad in the paper said they were hiring someone to answer phones. I'd wanted to get a job at Lilly Pharmaceutical—$12 an hour—but they hired only college graduates, and I was still short one semester. So. I was only twenty, but I'd already had plenty of jobs: model, pharmacy clerk, caterer, pastry chef, transcriber. I even made a little money as a guinea pig for a research hospital.* But stitching it all together wasn't working anymore. I'd been a phone girl before. I could do that job.

I looked away from the road to change the channel on the radio. I still had a bandage on my arm from a blood draw at the lab. One of the doctors had wanted to use me for another study, this time on diabetes. I would be in the control group, since I didn't have the disease, but the study would still involve piercing a major artery in my groin and ran the risk of extensive blood loss, difficulty walking, and/or large clots that could—and the paperwork assured me this was rare—cause a blockage in the heart or a stroke. It paid $1,000. I'd declined. A stroke was worth at least ten grand, in my mind.

Not far from the highway, I pulled into a nondescript industrial park and scanned a series of gray doors for the address. A few cars around. Not many. Before we shipped much of it abroad, American telemarketing usually happened in these out-of-the-way zones, strip malls of temporary industry. Low rent. Clean. Relative anonymity.

Maybe I was young, or just exceptionally stupid. But it was a good ten minutes into the interview before I realized I was applying to answer phones at an escort agency.

I still remember the fabric on the armchair I sat in—nubby, tweed—as the madam explained I just needed to "sound friendly" when the johns called. But she didn't call them johns. For $8 an hour, thirty-five hours a week, I'd deal with the company calendar, connecting "service providers" to "clients," arranging "drivers." Two hundred eighty dollars a week was good money. Nearly twice what I'd earn in a kitchen. I smiled. She showed me around the call center. Standard-issue cubicles and headsets.

As we were shaking hands and exchanging numbers, the

* I had to take Saint-John's-wort for a month and save all my pee. Five hundred dollars. I carried a jug around with me.

madam stopped and said, "You know, I'm sure you'd do a good job on the phones, but I think you should be one of our girls."

That'd pay $200 an hour.

There are things you can't unlearn: when I was twenty years old, I learned that the most money I could make, of *anything* I could possibly do, was putting my vagina up for rent.*

I didn't take the job, in the end. But I came damn close to it. I remember thinking, Is there anything, really anything, *immoral* in the sale of a body? Is it really all that different from dating someone who buys you dinner? Takes you on vacation? What about the lab—weren't they buying my plasma? Buying my time, my daily habits? Didn't I smile at professors because I felt I had to? Wasn't my mother raised to "marry well"?

What parts of the body, exactly, are we allowed to sell? If not the genitals, then the mouth? Can we make the body smile, make it say things, put food in it or not, put a fist in it or not, smooth the edges of the voice, drop the pitch, change the rhythm—let them hear it, but not see it; let them see it, but not touch it; let them touch it, but not own it, drag their fingers along it, the way you'd run your hand over the hood of a car.

I wasn't the sort to lie, or somehow I told myself I wasn't—not when it mattered—so I told my boyfriend I was thinking about it. Bless his heart, he was incredibly clear: he said if I took the job, he'd break up with me.†

I'd like to say some feminist revelation shone down on me in that moment—bell hooks in a halo and robe—but it didn't.

I just loved him. I loved him and I was terrified he wouldn't love me anymore.

* Or more accurately, letting a middle-aged woman run a series of Airbnb-style sublets of my vag, all catering to well-moneyed clientele, each event buoyed by a rotating crew of temps: Drivers. Waxers. Understudies. A guy to run the website and four girls to run the phones. What should we call that—the sharing economy?

† He didn't offer to help with the rent, mind you, nor offer the use of his own apartment, where he lived alone with twelve guitars, a water bed, and an old poster of Tori Amos.

So I didn't even call the madam—I just ghosted. And then I landed a scholarship that took me to England and then an MFA and finally the PhD at Columbia. Fancy. I even got a stipend and discounted rent in Manhattan. And I went to a lot of parties with wealthy men, some of whom—and I can tell you this is true—brought call girls with them. Not always. Not usually. But sometimes, yes.

WOMEN IN LOVE

I am not the Eve of human love. That isn't why I told you this story. There probably isn't an Eve of love, really. But I am an Eve, as are you, just like every single living human today. We are the drivers of our species' tomorrows. We are all writing the future of humanity through the choices we make, day to day, in these bodies we inhabit, in the children we have or help raise and protect, in the societies we push against and collaborate with and innovate on. We live, at all times, both in the present and in the long rivers of evolutionary time. So these lives we're living are all the lives of an Eve. These hours. These small things. My memory of rain leaking through a car door. Wherever it was you woke up this morning. How you drew the first, conscious breaths of your day.

But we've come to the end of a book like this, and there's really just one thing left. There's something distinctive about our species today—often left out of biology textbooks, discussed largely in graduate seminars and science-interested forums. It's the unusual way we love one another: our distinctive, complex, often bizarre and overpowering love bonds, and the way we're able to *extend* those loving bonds to people we're not related to. Though many other species have sex the way we do, make children a bit as we do, arrange lifelong mates or date around or build a home and cheat on a spouse, help a good friend and mourn them when they're gone, the unique ways that human beings go about loving each other over the course of our lives are things both biologists find curious and most people think deeply define us as human.

And it just so happens that idea of human love is woven into the warp and weft of how scientists and historians alike tend to think about human women. Some of that has to do with mating strategies. Some of it has to do with how we associate the idea of women with the idea of raising children. More of it probably has to do with sexism. But I can tell you that from the very first day I arrived at Columbia to start my PhD—flush with that small stipend, the barking hounds of financial debt put back in their kennels for a while, the memory of the madam who was almost mine fading like some old tintype—my mentors in both the sciences and the humanities, no matter how feminist and smart and well meaning, were basically telling me the same two stories about women:

The first was what I just told you, that the thing which makes us *most* human is our ability to love. To truly love someone. And while they weren't always talking about heterosexuals, nor even romantic or sexual love necessarily, they were, by and large, thinking about it. And they were most certainly thinking about women's role. It was, as academics like to say, the "dominant frame."

The second: that the history of women is a history of prostitution—the "world's oldest profession"—and that the evolutionary origins of human marriage can be found in that first moment when some ancient ape traded meat for sex.

I'd prefer to think that neither is perfectly true. "Most" is often code for "best." Is loving a man actually the *best* thing a woman can do? As for the second, I'd greatly prefer that the story of womanhood not be summed up as elaborate whoring.

But just as we've done with other unpalatable ideas, we need to explore these two threads. How *did* human beings evolve to love one another, and what role did women have in that evolution? Is it prostitutes all the way down? Was this world of "love" bonds always male dominated, as it is today? Is love the defining characteristic that makes us human?

Every human culture is steadfast in feeling that their particular way of dealing with love and sex is right while others are wrong. Many liberal scholars draw on written history, noting how patriarchal many of the world's major cultures have been. They point to Solomon and his many wives and say polygyny (one man, many women) must have been the way our ancestors used to do it.

Others talk about sexual jealousy—how common it is, how apparently innate—and say monogamy is the way we evolved.

Evolutionary biologists, meanwhile, tend to look to our fellow mammals for answers. Some look at chimps, with all their bullying and promiscuity. Others look to gorillas and other animals that have harems, with one dominant male and a gaggle of females, to make the case for polygyny. Thinking back to how early hominins migrated out of Africa, a few even draw on wolves, where packs are usually led by two parents, a male and a female, with all the children following in social dominance. Maybe *that's* what ancient humans looked like: patriarchal, monogamous family bands, traveling the savanna, with fathers at the head and daughters marrying out into other families.

In other words, when it comes to love and sex and whatever is most "natural" for us, no one agrees. Not the scientists, not the ethicists, not even the religious people. Most theories point to patriarchies of one sort or another, but before the invention of the written word the evidence for each case is not nearly sufficient.

To dig for the real story, you need something older: the human body itself.

WRITTEN ON THE BODY

For all the storied wisdom of King Solomon, the man lived at most three thousand years ago, his body and its songs made of a clay already long evolved.

If our ancestors were mostly polygynous—like gorillas and King Solomon, with one dominant male mating with many females—then our bodies should tell that story. If we were promiscuous, like our closest primate cousins—with everyone pretty much having sex with anyone they wanted to—we'd have traces of that history written on our bodies instead.

Because male mammals are usually the ones who compete for sex with females, male bodies are often the best place to look for telltale signs of mating strategies. Among our fellow primates, there are two physical traits usually tied to polygyny: teeth and body weight. The males have big canines—the eyeteeth, or "fangs"—and their bodies are significantly larger and heavier than the females'. This is as true of baboons as it is of gorillas. Male chimps and bonobos, meanwhile, are also bigger than the females, though the size difference is less significant. And while their canines are smaller than those of gorillas or baboons, they're still far more intimidating than any hominid's. No one in their right mind would want to be on the bad side of a full-grown male chimp—that's two hundred pounds of muscle and pointy-toothed rage.

Aside from shredding food, big canines are mostly for threat displays. Males threaten other males when they're competing for females. They also bare their fangs to compete for social dominance. So most scientists think these teeth are the way they are because that species has a lot of male-to-male competition for females. This seems to be as true for modern-day mammals as it was for our pre-mammalian ancestors: fossils going back 300 million years *also* have these sexy "show teeth," better designed for flashing a lusty (competitive) smile than for eating.*

Male primates usually have these huge, scary, pointy-toothed bodies precisely because it's better *not* to fight. Better to make a lot of noise. Beat your chest. Yell a bit. Flash your face-weapons. Looking scary is generally good enough.

In the biology of sex differences, this is a general principle: the harder it is for males to get a chance to reproduce, the harder they compete with one another for a chance to have sex.† Developing bigger, intimidating bodies with bigger, more intimidating teeth is a proven strategy for winning those competitions, ideally without having to lose an ear for it.

So, are humans more like the promiscuous chimps? Or the harem-style gorillas?

Let's start with weight class: human males are only 15 percent heavier than females on average. By way of comparison, adult male chimps are 21 percent heavier than females, male bonobos

* These fossils are the therapsids: those lizard-like creatures that came before mammals, the ones that eventually produced Morgie.

† Unless you're talking about anglerfish, but come on, those weirdos—however delightful—aren't exactly a good model for "normal."

are 23 percent heavier, and silverback gorillas are a whopping 54 percent heavier. Mandrill males, who don't live with the troop and show up only when the females are fertile, are nearly 163 percent heavier.

In other words, despite whatever you might have seen in bodybuilding competitions, human women aren't that much smaller than men.

But it wasn't always that way. Looking back in the primate fossil line, males were usually significantly bigger than females—it's one of the ways paleontologists can tell the bones apart when they don't have a fossilized pelvis. By the time hominins arrived, however, the males were getting smaller and the females were getting larger. This is fairly recent news: a paper in 2003 finally determined that male and female *Australopithecus* had about the same body size ratio as modern humans. That is, females such as Lucy were only about 15 percent smaller than the males.

And the males were already losing their big canines. If you line up the hominin skulls over time, the male canines keep getting smaller and smaller, until finally the biggest male canine you'll find is the sort you now see in the grins of men like Tom Cruise: a bit longer, a bit pointier, but not very different from a woman's. Tooth size seems to be modulated by a mutation on the Y chromosome, and human men still tend to have larger teeth. But the show canines are mostly gone.

So, if our ancestors did have harems, they were probably very distant ancestors. Maybe even further back than when we split from the chimps and bonobos. That means Solomon and his wives, and any other harems you've heard of, represent a very recent innovation in our sex lives. The trend, if anything, is convergence: men's bodies getting lighter and less intimidating, and women's getting bigger.

But what about promiscuity? Were ancient hominins having tons of sex with one another, like the chimps and bonobos? And if we were promiscuous, why didn't our bodies settle somewhere closer to the chimps, whose males still have nasty-looking teeth?

Looking at the fossils, it's hard to say. For one thing, teeth are also what we use to *eat*, and many early hominins were in the habit of eating starchy tubers, nuts, even tree bark and the occasional grasses—hard stuff to chew. (We weren't regular meat eaters until much later in our evolutionary history.) Have you ever broken a tooth on something? Imagine breaking your big, fancy, show canines on a hard nut and dying of a tooth infection. In the long run, that's not going to work for preserving long-tooth genes.

It's possible our teeth evolved to be good at heavy, regular *grinding*, rather than slicing. Likewise, if food was especially scarce, smaller bodies with bigger fat stores made more sense, rather than large bodies with a lot of bone and muscle. Though our bodies do tell a story about our hominin ancestors reducing male competition and aggression, some other factors could have pushed those features, too.

Like the testicles.

Promiscuous primates have gigantic balls. This is a fairly universal trait—chimps have them, and baboons, and so do bonobos. That's because in promiscuous societies, females have sex with more than one guy, so the sperm of individual males have to compete with one another. If you want your sperm to win out, you basically have to blitzkrieg the female's cervix with huge numbers. To make huge numbers, you need huge testicles.

Gorillas? Tiny little balls. Peanuts.* But silverback gorillas don't have to worry that much about other males having sex with their harem. What's more, the females aren't in estrus for very long—only two to three days a cycle, compared with chimps' ten to fourteen days—which means male gorillas don't have to make as many sperm. So, if you don't *need* as many, why waste all that energy on growing big balls?

In primates, testicle size is so deeply linked to male competition that sometimes their size will even change depending on the social status of their owners. When mandrills compete with one another for dominance, the winner's balls increase markedly in

* Testicle size is also tied to the size of the female's reproductive tract—that is, how far the sperm have to swim to find the egg. Longer swim, greater losses, and so a greater number of sperm needed. But even controlling for that factor, the tininess of gorillas' balls is still remarkable.

size, and his face markings become more colorful. Losers' testicles gradually shrink over a few years of defeats, and their faces become less colorful.

Human males, as a rule, have medium-size balls. A bit like Goldilocks: not too big, not too small. Since there's currently no way to determine how big ancient hominin testicles were, we don't know if modern men's testicles are bigger, smaller, or about the same size as they used to be. But given the changes we can establish, it's not hard to imagine that our forefathers' balls were quite a bit larger than testicles are now. Regardless of how they got that way, having medium testicles *now* does imply that our ancestors weren't especially promiscuous, or at least not as much as the chimps.

And there's another count against promiscuity hidden in our bodies. Producing more sperm, via larger testicles, isn't the only thing competitive males do. When male mammals want to make sure the females they're having sex with will have their babies and not some other male's, they sometimes produce a clumpy, sticky seminal fluid that "plugs" or blocks the female's cervix against later intruders. Among primates at least, the more promiscuous the species, the thicker this seminal plug. Chimps have the thickest of them all: inside the female's vagina, the fluid in the male chimp's semen turns into a four-inch-long piece of clear rubbery spunk. Primatologists know this because they've watched such plugs *fall out* of a female's vagina, usually when they're dislodged by another male's penis. Many scientists gather these from the forest floor like prized gems.

Human semen also thickens, but not as much as a chimp's does. And it's only thick and sticky at first, liquefying about fifteen to twenty minutes after the man ejaculates. Still, it's not hard to imagine that it might stick to a female cervix and block any other semen from getting through. Except that human females produce a *lot* of cervical mucus when they're fertile, good for getting sperm up and through the cervix, should a woman so choose, and also really good at *flushing out* excess material from the vagina during that period. When in contact with a woman's fertile cervical fluid, human semen can liquefy more quickly than it does in air.

And then there's the fact that we walk upright. A goodly portion of a man's partially dissolved semen plug could *fall out* not long after a woman stands up. No need for another man's penis to dislodge it. Which means a woman's vagina is pretty much good to go for a male competitor within a handful of minutes. Thus, unless our female ancestors were in the habit of lying on their backs for hours after sex while they were ovulating, it's unlikely that modern human semen evolved to block other men's sperm.

Medium balls, runny sperm, short teeth, smaller bodies—that doesn't sound like King Solomon to me. Doesn't sound like King Chimp, either. If ancient hominins had a lot of male-to-male competition going on, our bodies are pretty good at hiding the story.

But there is one other way an ancient hominin male could have tried for reproductive success: he could always try raping his way to fatherhood.

This is one of the more taboo subjects in the science of human sexuality—whether human males evolved to be prolific rapists. It's not hard to see why we'd ask the question: right now, everywhere in the world, men raping women is common. It's especially prevalent in times of war and violent social conflict—the Congo is rampant with rape; ISIS uses it as a primary weapon; and after Russia invaded Ukraine, reports immediately began to emerge accusing Russian soldiers of committing rape as a war crime.

All rape is horrific. Sadly, it's not unique to our species, either. So does the human body tell a story of a rape-filled evolution? Instead of Solomon, should we look to Zeus?

Better to look at our closest relatives. Given that there's very little rape in chimp society, there was probably even less among ancient hominins. For one thing, it was dangerous: a fully grown female hominin could beat the bejesus out of anyone who tried, and so could a female chimp. Though chimp males can be absolute jerks to females in their troop, they rarely engage in violent forced copulation. That's likewise true among bonobos, baboons, mandrills, and even gorillas. Aggression, coercion, general harassment, yes, but rape is incredibly rare.

In fact, when it comes to sex, chimp males are typically more cajoling, solicitous, even friendly. Or they employ tactics remarkably similar to what human domestic abusers do. Male chimps will physically and vocally harass females, often in an attempt to socially isolate them, stress them out, and wear them down. They put their aggressive male bodies in between the female they're targeting and the rest of the troop. They do their best to prevent the female from associating with other males—and if they do observe the female hanging out with other males, they're more likely to backhand her later. Primatologists called this "mate guarding," and it does seem to give male chimps a reproductive advantage: while dominant males still get the best chance to pass on their genes, less dominant males who mate guard have a better chance than the guys who *don't* beat up females regularly.

But remember chimps are, by and large, a male-dominated society. Bonobos are female dominated. When a male bonobo tries to backhand a female, he incurs not just her wrath. All the female members of the troop are liable to descend on him. Bonobo females have an incredibly tight-knit, interdependent social web, and they use that web to defend one another from any male who gets out of line. They might even chase a too-aggressive male out of the troop entirely. So bonobos don't do a lot of mate guarding.

We don't know if human ancestors were more like the chimp or the bonobo. Genetically, we're equally related to both. We do know that sometimes male domestic abusers are also rapists. But not always. And while we don't seem to have any reliable data on whether abusive human men have more offspring than non-abusers, it does seem to be the case that males with significantly less money and social status are more likely to be violently abusive toward their partners than men who don't have those problems.*

* (Flynn and Graham, 2010). To be clear, human domestic abuse and rape are present in every social class. Physical abuse, however—that is, reports of physical abuse and subsequent arrests, which is where most studies on the matter draw their data—is more likely in places where people live below the poverty line (Bonomi et al., 2014). In the United States, Canada, the U.K., and many countries in Europe, intimate partner violence and murder disproportionately affect poor people and people of color, with men far more likely to be the perpetrators than the victims (Stockman et al., 2015). Nonheterosexual and trans people of both genders also disproportionately suffer from domestic violence and rape, but when you control for race and income, some of that difference may fall away (Rothman et al., 2011; Flores et al., 2021). It is, in other words, terribly expensive, in every sense of the word, to be someone who exists on the margins of society. And those costs extend even into the supposed safety of one's home.

So maybe human men *have* evolved to use violent mate guarding as a reproductive strategy. Or at least, as primates who are incredibly similar to chimps, maybe our bodies and our brains were abuse ready: given a social scenario in which less dominant males have the option of using mate guarding as a strategy, it wasn't that much of a stretch for our ancestors to start doing it. Maybe that's part of why some men still do it today.

It's a sobering thought.* But it still doesn't quite answer the rape question. Human abusers, while predominantly male, aren't always rapists, just as rapists—also usually male—aren't always domestic abusers.† There is one thing both categories of men have in common, however: extremely boring penises.

Assume that male bodies "want" to pass on their genes. Assume that female bodies want to as well. Let's also assume that male bodies want the *best* females, and female bodies likewise want the *best* males. But the game isn't equal. Not at all. Though females can technically "rape" males, they can't exactly rape males in a way that would force them to father their children.‡ That's because male bodies don't actually contribute that much to reproduction—as a rule, males have testicles, but they don't have a womb. Females usually do. So if males can somehow manage to force their sperm into a female's reproductive tract, they get a chance to pass on

* At the very least, it should challenge what we mean when we use the term "rape culture." What if "rape culture" is, at its core, something deeply influenced by class conflict and male competition? What if one of the best ways to combat "rape culture" is actually *economic*?

† Rape is so common that the numbers nearly follow the general population; you're more likely to rape someone when you earn less money, yes, but not *much* more likely. Unless you're literally in a war zone in the Congo, the person most likely to rape you is someone you already have an intimate relationship with (BJS, 2017). Not a stranger, but your actual boyfriend or husband or some other member of your family or group of friends.

‡ In fantastically rare cases, human females can, but at that point we're talking about tricky, nuanced, and very modern ideas about consent. Yes, a modern human woman could force a male to have sex with her against his will and thereby impregnate her (which is the only sort of rape we're concerned with here, because we're talking about rape that perpetuates a rapists' genes: forced sodomy and other horrors don't count). Yes, that rare case would count as rape. But no, that's not something ancient hominins would have been able to pull off, nor any other animal on the planet—at least not in any way that would cohere with a modern definition of rape.

their genes. It's a reasonable ploy. But given enough time—the sort of evolutionary scale that allows some genes to be favored in response to environmental pressure—the female body is likely to produce counterploys. So if ancient hominins were particularly rapey, it's reasonable to think some trace of that history would be written on our bodies.

Think of it as a weirdly sexy Cold War. Rape is common across the animal kingdom. But species that commonly use rape as a reproductive strategy are often the ones with the more elaborate penises, such as the mallard duck's curlicue. That's because the vaginas they're raping have their own agenda, or at least the genes that produced those vaginas do, which generally "aim" to be passed on in the most competitive way possible. But human vaginas are only a tiny bit foldy. For the most part, it's a straight path to the cervix. The human penis is likewise straightforward: long, medium girth, no bells or whistles to speak of. It doesn't corkscrew. It doesn't knot. It doesn't have *any* obvious structural weapons. Hell, it doesn't even have a baculum—that little bone that other animals use to prop up their erections. That means a man who regularly tries to forcefully shove his turgid weapon into an area of a woman's body inconveniently housed between two muscular, flailing limbs—not to mention the proximity of her very real and very strong pubic bone—is likely to break the thing.* And human penises do break, even when they're not trying to rape. Left untreated, the injured member is far less likely to be able to transfer sperm to any female in the future.

So if the human penis and vagina evolved in a rape-fueled competition, our current anatomy doesn't betray that history. If anything, our bodies seem to reveal a lot of consensual sex without very much violent male competition, and maybe even a continually *reduced* competition over time, with our older ancestors being more competitive, and our more recent ancestors getting less and less so.

There are two features of our sex organs that support this idea. Chimp penises don't have any obvious head, but they do

* Chimps and gorillas still have a baculum, though it's rather small. Ours disappeared somewhere along the hominin line.

have penile spines. Human penises, meanwhile, have that classic arrowhead design, and the shaft is completely smooth. Both traits might have come about because our ancestors changed the way they went about mating.

Let's start with the head. The chimp penis—which isn't very long, by the way—is thicker at the base and narrower at the head, forming a kind of elongated wedge shape. The human penis has the glans—that fluted arrowhead—which is typically thicker at the flared end than the shaft.

There is one good reason chimps *don't* have a wide head and narrow shaft—it would be really lousy at dislodging another guy's seminal plug.* If you want to try to dislodge a rubber lump from inside a capped tube, your best bet is a narrow wedge: something that manages to get inside the vagina to the side of a competitor's plug, and then, by thrusting quickly, help draw it out. Trying this with a wide penis head would just drive the silly thing in there deeper, potentially bruising your own penis in the process.

One theory about the shape of the human penis goes as follows: so long as a competitor's sperm isn't too thick, maybe the human penis head is good at *scooping it out.* One lab even created an artificial vagina and a range of artificial penises. Then they filled the vagina with a semen-like runny oatmeal puree. (No, really.) The penises with the most humanlike head shapes—that slightly fluted glans, thicker than the shaft—were the best at clearing the vagina of the most pseudo-semen. Thus, the paper concludes, the human penis evolved its particular shape in order to help men with sperm competition.

It might have gone that way, sure. But like many such papers, the authors ignored one key feature: the incredible unlikelihood of an ancient woman lying on her back being continually inseminated at a rate that exceeded three different men per hour. Remember, human semen thins out after twenty minutes, gradu-

* The arrowhead shape can also make some men's penises get a bit trapped inside their own foreskin when they get erections. It can be very painful, and obviously isn't all that conducive to genetic selection. But luckily the most severe cases are rare, obviously solved with circumcision, and modern medicine can safely take care of that problem *without* circumcision should the man prefer. Chimps also have foreskins, but their narrow head means they don't really have this problem.

ally dribbling out of the vagina like any other fluid. What's more, being upright means we have *even more* seminal fallout than our primate cousins.

So what else could the glans be good for? How about sucking out fertile cervical mucus. Though the human penis doesn't fit so perfectly into the vaginal cavity that it makes a total vacuum seal, most *do* produce a slight sucking force during thrusting. So each time the penis draws out, a small amount of suction draws away material from the upper vagina down the fluted rim of the penis head and along the shaft.

The acidic vaginal environment is actually toxic to human sperm. The pH is too high. Sperm die inside the vagina rather quickly. Fertile cervical mucus, however, is the right pH for sperm. It also has a useful structure that helps sperm swim through the cervix toward the uterus and fallopian tubes. So the more fertile mucus that surrounds the upper vagina and cervix when a man ejaculates, the better chance his sperm have to *get the hell out of there* through the cervix before the pH of the vagina kill them all. The very best position to be in, actually, is to ejaculate as close to the cervix as possible, with a lot of fertile mucus *around and behind* the sperm at the moment of ejaculation, shielding that tiny bubble of semen for a few crucial minutes while they swim desperately toward their north star.

A penis shape that was particularly good at getting them as close to the cervix as possible, with any potential barriers cleared and "docking station" usefully sealed, would be a benefit.

To do that, however, you can't just thrust once or twice with your fancy new arrowhead penis. It's gonna take more work. And we do work: human males take, on average, more than *four times* as long to ejaculate during sex as chimpanzees. And part of how we manage to pull that off, sadly, may be that the human penis became less sensitive.

Chimp penises, like many other primates', have little spines on them made out of keratin, the same material as your hair or fingernails. These mammalian penile spines come in all shapes and sizes. In cats, they're proper spikes. In chimps, however, they're more like nubby polka dots. The bigger they are, and the more a male has of them, the faster he's able to ejaculate when he has sex.

That might be because nerves in the chimp's penis are responsive to signals from those little bumps. In other words, they likely feel really, really good when they're rubbed.

Of course, having a sensitive penis is a reward in itself. When sex feels good, you're more motivated to have it—enabling your genes to get out there and carry on. Also, in the competitive chimp world, there isn't a ton of time for long bouts of sex. With all the males competing with one another for access to females' sex organs, if you're going to get something done, you're probably going to want to get it done quickly. Having a sensitive penis could be one way to do that.

Having sex feel *less* good would have to be especially rewarding, from an evolutionary perspective, in order for that trait to survive. If our ancestral penile spines really did feel good, there'd probably need to be a decent reason why the hominin line lost them. By comparing the genomes of chimps, humans, and Neanderthals, geneticists are just starting to suss out which parts of our DNA have been deleted over the course of our evolution. Thus far, it looks like about 510 deletions. One—an androgen receptor that triggers certain kinds of development in male bodies—is probably responsible for the loss of our penile spines. And men lost them fairly recently: somewhere after we split from the chimp line, but before we split into ancient humans and Neanderthals 700,000 years ago. We also lost the genetic sequence that gave us sensitive facial whiskers, which might also be part of how we lost the sensitive penile spines.

Either way, they're gone, and they're not coming back. We most certainly lost them by genetic accident, but given how important sex is for evolution, it's possible their loss had some sort of advantage. Maybe longer bouts of sex led to greater male-female pair-bonding. Or it helped suck out more fertile mucus to protect the guy's sperm. Nobody's really sure.

But for whatever reason, it wasn't as big a deal for males to take longer to ejaculate, which does imply there was less *immediate* threat from other males.

In a rape-driven reproductive environment, you'd expect all sorts of signs of violent male competition, both in men's bodies and in women's reproductive organs. Human beings don't have those signs. There's no evidence of invisible mechanisms, either: whether a man forces a woman or they have consensual sex, so long as she's in her fertile period, he still has a roughly one-in-four chance of producing a child. Among mallard ducks, a raping drake has only a 2 percent chance of having offspring—far less than if the female were willing. A proliferation of duck rape has been around long enough, in other words, that the female's body evolved to compensate—not true for hominins. So, no matter the prevalence of rape in modern times, our human ancestors probably weren't very rapey, they probably didn't have a lot of violent competition for mates, and they were only about as promiscuous as you'd expect from a medium-balled primate.

BABY KILLERS

If the fossils (and our current physiology) tell one story, it seems to be this: over time, hominin males competed less and less with each other for mates. But why? What drove all those changes in teeth and body size, penis shape and semen behavior?

Monogamy. The most popular argument in the scientific literature is that ancient humans started being monogamous and didn't have to compete as much for mates. If each male had a good chance of having *exclusive* access to a female, then more "little guy" genes would start to show up in the gene pool. Since having a smaller body size and smaller canines is less expensive than having a big body and big teeth, eventually the smaller version would win. Genes don't just influence behavior; behavior can change the likelihood a gene will be passed on.

What we see in the fossil record, in other words, may be the beginnings of the nuclear family: one husband, one wife, an appropriate-for-the-circumstances number of kids. A male doesn't have to compete with other males, because, more than likely, there'll be a female out there to have sex with him, and no one else, and give birth to and raise all his offspring. In such a society, after the males tacitly agree with one another not to steal each other's wives, they grow smaller. The females, meanwhile, spend a few million years getting a bit bigger, a bit taller, in part because they're eating well off their mate's contributions. And all the while, our big, vulnerable babies manage to survive to adulthood because their mother has a husband to help look out for them (and her).

It sounds like a good deal for a female. In exchange for being sexually exclusive, she has a husband who'll help her feed the family and help defend it from predators. Because her hominin children are so helpless, she needs all the assistance she can get. And the bigger her brain gets, and the greedier her placenta, the harder it is to be pregnant and give birth, making females that much more in need of assistance. Over time, all those big-headed babies need longer and longer breast-feeding, creating even more strain on the mother's body, and an even bigger need for food, which means she needs that much more help from her mate. So why not offer him exclusive sexual access? That way, he knows her kids are *his* kids—he couldn't know otherwise—and he'll feel that much more obliged.

That's science's way of saying the history of human womanhood is a history of whoring, of trading sex for protection and food. The end.*

It does fit neatly with the fossil record. It also helps explain why human sexual culture is so very different from that of our primate peers. There's just one problem: monogamy wasn't such a sweet deal for female hominins. As with other apes, our ancestral promiscuity wasn't just a pleasurable habit. It was a strategy—a necessary one. See, primate males aren't just a danger to one another. They're incredibly dangerous for *babies.*

In *all* of our closest primate cousins—chimps, bonobos, and even orangutans—promiscuity has a clear purpose for the female. She's not just getting her rocks off. She's making sure that no local male knows who the father is. In biology, this is called "paternal uncertainty." When researchers talk about the evolution of

* Very few papers about the evolution of monogamy bother to talk about whether the male was being sexually exclusive, too. Of course, if you're talking about the evolution of human *harems,* the male body would presumably look very different—that's the gorilla model, or worse: orangutan face flanges—so one presumes there's a more even distribution of male and female fidelity in the rise of human monogamy.

human monogamy, they usually talk about what the female gains by letting males be certain who's fathered all the children. But they rarely talk about how dangerous that is for the females and their young children.

While chimp males rarely kill chimp babies in their own troop, when they war with other troops, males regularly kill their enemies' babies, since babies produced by the enemy males' sperm are no benefit to them. They also have a habit of raping—or at least violently coercing—their female enemies, presumably to both enforce their dominance and potentially father new kids.* Thus, many argue, the main thing keeping chimp males from killing babies in their own troop is that they're never really *sure* the kid isn't theirs.

That's not true in harem-based societies. Among mountain gorillas, over *20 percent* of child deaths happen at the hands of an adult male—gorillas have harems, so paternity is more certain.†

That's the big problem with the monogamy argument.

Picture a group of ancient hominins. Really ancient ones, maybe even before *Australopithecus.* There they are, having sex, making babies. They're probably being as promiscuous as chimps, and the fathers aren't sure who their children are. Then picture a female deciding to be sexually exclusive with a male in exchange for food. That guy better be *huge.* Because now he doesn't just have to guard his mate. He has to make sure his kid doesn't get slaughtered by a rival, because all the other males in the troop know that the kid is *his.* Not theirs.

* It can be hard for primatologists in the field to draw a line. Typically, the warring females seem strongly *not into it* and protest greatly, but "consent" is a very human idea and good scientists prefer to avoid anthropomorphizing.

† (Robbins et al., 2013). To be clear, chimp males *do* kill babies in their group sometimes. In many cases, the attacker goes on to eat the child; chimps don't waste meat. So it's possible the males do it when there's a problem with the local food system. Or, since chimps are very aggressive animals, sometimes aggression turns on your own kind, even to your own genetic detriment. Chimp societies are also political: if a female pisses you off because she's friends with one of your enemies, you might punish her—and your rival—by eating her baby. Or maybe you just want her to be fertile again sooner than she would be otherwise. Mental illness is also a possibility. Who knows? We're talking about chimps. As with humans, questions about their behavior don't always have obvious answers.

In other words, when it comes to physiology, if there had been early hominin monogamy—pre-language, pre-culture—it should have turned these hominins into gorillas. Because every single one of our male ancestors had the obvious potential to be a rampant baby killer.

That means cooperative culture *had* to come before monogamy started. You had to have other cultural checks in place before measures to create paternal certainty made sense. You had to have bands of ancient hominins who were interdependent and had created clear and dire consequences for any behavior that threatened children.

What you basically needed was a matriarchy.

MAKE LOVE, NOT WAR

Since ancient human beings were, above all, *primates*, let's take a look at three well-studied primates that live in matriarchies right now: the olive baboon, the gelada, and the bonobo. Our ancestors could have been a bit like them.

For baboons, geladas, and bonobos, living in matriarchies doesn't mean they've reversed "male" and "female" roles. The girls aren't bigger than the guys. Girls aren't violently competing for male attention, either. Females are still the ones who have to invest more in reproduction, and as a result males still compete for them. So, their bodies look like typical primate bodies, as did our oldest ancestors'.

But in these societies, alpha females decide where the group will go for the day. Resources are divvied up in a way that tends to benefit the girls. If society is in flux, females are the ones who determine how things will pan out. If group members have conflicts, dominant females intervene to help one side win over the other. Females dominate the doling out of social acceptance and rejection—which is to say, social *credit*—not males. Daughters inherit their social rank from their mothers, and everyone else sort of scrambles.

Being in a matriarchal primate society is a bit like spending your entire life in a high school where the popular girls rule. It's

Mean Girls. The top girls form complex alliances that reinforce their own power and keep "lesser" girls in check. When the group as a whole decides to do something, everyone looks to the top girls for guidance. The most popular girls also tend to get the attention of the most desirable guys, while the lower-ranked guys do everything they can to raise their status. Sometimes they try to get in with the "friends of friends"—the lower-ranked females who are allowed to hang out with the popular girls. Sometimes they try to make friends with the higher-ranked males, a kind of being cool by proxy. Sometimes the non-popular guys "settle" for lower-ranked females, figuring it's better than being alone.

Except in bonobo society everyone's having a lot of sex. Males with males, males with females, females with females, even juveniles with adults. No one would recommend such a thing for a moral human society, but if you're a bonobo, it's how you solve problems. It's how you pass the time. It's just one of the everyday things you *do*, really, when you're not searching for food or grooming one another.

Needless to say, bonobo fathers have *no idea* who their children are. Paternal uncertainty is a given. And, as in the rest of the primate world, mothers are the ones who mostly take care of the babies. But unlike in chimp troops, all bonobo females look out for the babies. They form deeply bonded female coalitions, and heaven help any male who gets on their bad side. A lot of these "sisterhoods" consist of bonobos who aren't directly related to each other. That's because, like chimps, bonobo females leave the troop they're born into once they're sexually mature. They need to find a new group and quickly make friends with the females there—ideally, the highest-ranking one they can ally with, but just about anyone will do at first. The daughters of the highest-ranking females inherit their mothers' social status and are basically princesses, until they have to leave the troop to find a new home. These girls get groomed all the time. The males want their attention. When it's time to eat, they'll usually get some of the best food.

Chimps are matrilineal, too, but the females don't receive as many perks and the guys are still in charge. Just because you *inherit* through your mother doesn't mean females have all the social power in a matrilineal society.*

If our human ancestors *were* matriarchal, why would monogamy even start? What reason would males have to collaborate with the females on child rearing and food sourcing? Why wouldn't they just lie around all day, eating ancient bonbons? Is monogamous prostitution really the *only* way to get men off the couch?

One bizarre, if enticing, alternative theory: in a matriarchy, babies make good buffers for aggression.

Take savannah baboons. They're highly social, highly intelligent, and highly adaptive. They're matriarchal, so the females are in charge, though they don't use sex to resolve conflicts the way bonobos do. No, they fight—violently. And unlike the bonobos, daughters *stay* with their mothers their whole lives. It's the males who leave. So that means female social ranks are more stable in comparison to patrilocal societies, and male social ranks are in constant flux. Being a dominant male doesn't confer as many advantages in this model, since subordinate males also get a chance to mate with a high-ranking female if she so chooses. And she does choose: there's no rape in baboon society. Social manipulation, sure. Even some violent coercion. But no forced copulation. And there is infanticide, quite often from males, in contrast to bonobos—babies are certainly under threat here. But coalitions of males and/or females can deter quite a bit, and in large mixed-sex groups, killing a breast-feeding infant isn't as safe a bet for passing on your own genes.

So, what's an ambitious male to do?

Turns out, males form relationships with *babies*. Primatolo-

* For a human example, look at Judaism. In order to be "officially" Jewish, one has to be born from a Jewish womb. Having a Jewish father isn't enough—though if one's mother is Jewish and the father isn't, the child will still be considered Jewish according to most authorities in the Jewish community. Nowadays this doesn't matter as much—even Israel's Law of Return allows international Jewish people to have Israeli citizenship if their father or grandfather was Jewish—but once upon a time, excepting horrors brought by *non-Jewish* communities, being born to a Jewish mother afforded a kid all sorts of perks. And that tradition came out of a famously patriarchal culture.

gists have seen this many times in the field: Say a male is fighting with another male. The females largely ignore the conflict, so long as it doesn't bother them or their children. But then one of the combatants goes off and picks up a baby, who blithely clings to his chest hair or his back. Then he goes over to the male he was having the fight with. If the baby likes the male it's clinging to, the kid will scream at his opponent if he acts aggressively. So the other male either backs off or is mobbed by friends of the mother, spurred on by the baby's cries. It's so effective, in fact, that some males simply *carry a baby around* as a kind of adorable bodyguard, preventing fights before they start. A male will bring a female he's friends with to a fight, too, using her as a buffer. This can work but may not be as effective as grabbing a kid. Attacking a female isn't as taboo in baboon society as it is among the bonobos.

So picture a world in which early hominins are matriarchal like bonobos and baboons. Imagine two males fighting. One of them is already friendly with a local kid. The other isn't. What happens if that hominin uses a kid as a buffer? Having an in with the females and offspring in a matrilineal group is greatly beneficial for males. The greater the affiliation, the greater the benefit for the male and his offspring.

Geladas have similar social features.* They also live in complex matriarchies, subdivided into harems and the occasional all-male band.† But in an interesting twist, sometimes a gelada harem can have *two* regular male members, one dominant and one not. Only the dominant male gets to have sex. The other, typically younger, will help raise the offspring after they're six months old or so, which could eventually give the secondary male a "leg up" on his future prospects. Hanging around helping with the kids also gives him more chances to have sneaky sex while the dominant male is

* You've met them before in this book: our abortive monkey friends in Ethiopia. Gelada females help approve and arrange male coups. If they're early in their pregnancies from the ousted king at the time, they usually have miscarriages.

† Geladas are remarkably similar to baboons, even though they technically occupy a separate branch on the evolutionary tree. The primary way they differ, besides living only in Ethiopia and nowhere else, is that they eat a heck of a lot of grass. Olive baboons are more omnivorous and can live just about anywhere.

away—call it the monkey world's gender-flipped version of having an affair with the nanny.

Similar things can happen in baboon society—the friendlier you are with a female's offspring, the more likely you'll get to have sex with her, whether or not you're a dominant male.

Still, maybe all these males helping out with the kids is hard to imagine. It probably feels strange to think about, in part because we're so used to stories about human men being aggressive toward women and children. Women do commit domestic violence and men can be the victims. However, in the United States and the U.K., men are more likely to be abusers and much more likely to be *frequent* abusers, overwhelmingly of women. Likewise, women are far more likely to be murdered by their male partners and ex-partners than men are by their female partners—that's a big part of why human men are thought to be more aggressive and violent than women. It's hard to imagine a hominin past in which this weren't the case.*

Those ancient males were probably violent and aggressive, too. It's just that cooperative and affiliative behaviors might have rewarded them with more success than violence and aggression, especially in a matriarchal society. The helpful guys got laid. A lot. That means males who had friendly relationships with females and offspring were more likely to pass on their genes. So they

* It's actually slightly more likely for *mothers*, not fathers, to commit infanticide (Friedman et al., 2005). It's hard to know exactly how to interpret that, but it might come down to the sheer number of hours mothers tend to spend with infants: if that sort of malice and mental illness were evenly spread between parents, then statistically speaking such women would have more occasion. For neonaticide—that is, parents killing their newborns within twenty-four hours of the birth—the available data lean strongly toward the postpartum mother (ibid.), but those numbers are largely drawn from studies conducted on Western societies rather recently, where doing such a thing is both illegal and socially unacceptable. As many scholars have said before, the history of humanity seems to be one in which infanticide was incredibly common (Hausfater and Hrdy, 2017). As horrific as that is to the modern mind, from a biological perspective ancient females choosing to kill their offspring is a very different thing from living under the threat of having any local *male* kill their offspring without their consent. What I'm describing here is ongoing threat from ancient males, not what a postpartum female might do on her own.

got an advantage by being aggressive with other males, but they also got an advantage by changing out those behaviors with the females who were really running the place. This would be even more true if their societies were both matriarchal *and* matrilocal, meaning females stay put and males are the ones who "marry out" by relocating when they come of age, like the olive baboons. Having an in with the females in power in your new social circles would be that much more important.

But of course, that's not at all what we think of when we think of modern or historical human societies, is it? While there's *some* known history of matriarchies among human societies, the dominant model now seems to be *patriarchies.*[*] And not just patriarchies, but patrilocal, patrilineal patriarchies, with sons inheriting status and resources from their fathers, and many societies even having those sons stay "local" in the same family for their entire lives.[†] Male human society can be incredibly stable that way, with a respect for brotherhood that is both deeply meaningful and power reinforcing.

Sisterhood, meanwhile, is kind of in shambles nowadays. We're nowhere near in charge. Compared with primate matriarchies, our female bonds are weak. In most cases, we can't even rely on kinship to keep female coalitions intact: In the majority of historical human cultures, new brides tended to move to our husbands' family group—even to the point of changing our names. And if we inherited anything, which was not at all certain, we inherited primarily from our fathers.

What I'm saying, in other words, is that at some point in hominin history human society must have flipped on its head to make it the way it is now. What we do now is *not at all* what other primates do. Other primates can be patrilocal, but they're never patrilineal—outside harems, how would men even know whom

* Here I mean in the biological sense, and not as in "the Patriarchy."

† That locality can mean lots of different things: living in your dad's house, working in your dad's business, using his connections to advance your early career. "Inheritance" and "locality" of the sort we mean here have many manifestations in modern human society, but they're not that hard to trace.

they fathered?* And other primates are never truly monogamous. Males are essentially never limited to just one female, and unless they're in a harem, females are rarely limited to just one male.

So how on earth did we get from free-love matriarchies to male-dominated monogamy?

THE DEVIL'S BARGAIN

The transition wouldn't have been sudden. You can't just switch to a monogamous patriarchy on a random Tuesday afternoon. But you could start small, with ancient hominin males edging in on female power. There are a few different ways that could have happened.

One scenario: Deep in the past, somewhere in East Africa, adult hominin males find it useful to make friends with females and their babies. Like today's olive baboon and gelada males, they especially like making friends with *high-ranking* females. So they start helping out with child care. Trading food for social favors. Grooming. Getting in on the power coalition.

We don't know if those males are living in the same group as their fathers, as bonobos do, or if they've joined another social group, like baboons and geladas. In either case, they're still dangerous. They're still male primates, so they're still potential baby killers. The females know this on some level. But, thankfully, a violent sisterhood helps keep that aggression in check. Eventually, it becomes normal for the top females to have close male friends. Those friends get a lot of sex as a result, and many other social perks. The less friendly males don't.

But these aren't ordinary primates. They're hominins. And things are changing in their bodies. Over huge amounts of time, giving birth gets more difficult and dangerous for the females.

* There have been some recent tantalizing data that suggest some male chimps, in certain circumstances, may know and are more likely to treat those offspring preferentially, but maybe they're getting lucky by simply playing it safe with the offspring of a female they've had a lot of sex with (Murray et al., 2016).

They start collaborating with each other to try to survive and to take care of the children. Their favorite males help out even *more* with the kids. So those guys get laid even more and pass on their helpful, collaborative, Nice Guy genes.

But it's not as if the Nice Guys stopped being primates. It's still potentially dangerous to let them know that the kids aren't theirs. Meanwhile, if pregnancies and births and early child care are getting more dangerous, that also means being super promiscuous is more dangerous. It's useful to have more control over how often you're pregnant. And STIs are always a potential problem.

In that environment, what if some of the females started making bargains with the friendliest males? In exchange for certainty over which kids are his, would a male offer protection from other males and competitive females?

Believe it or not, this is the kind of bargain primatologists are starting to find among today's chimps: females who spend more time with friendly males are less likely to lose their offspring to infanticide. That's probably because some chimp infanticide is committed by other *females*. In a violent, male-dominant society like the chimpanzee's, it's useful to find a balance between the threat of other murderous females and the threat of murderous males. Mind you, no chimp is monogamous. Spending a bunch of time with one male while nursing doesn't make a chimp female more likely to mate with that guy the next time around (which, for chimps, means about four to six years later). While a new baby is still vulnerable, it helps to have some extra muscle around, but paternal uncertainty is still valuable in chimp society.

So let's think about ancient hominins and their devil's bargain. In promiscuous matriarchies, females would already have more power than female chimps do, so maybe the kids wouldn't need a lot of protection at first. If a male were to start misbehaving, all hell would rain down on him—female coalitions don't allow aggression toward offspring. Maybe a few well-positioned males start participating in violent retribution against such transgressors. Maybe they start acting a bit more like *thugs*, even beating up their female allies' enemies. Sometimes the enemies' kids get caught in the cross fire. And if those chimpy enforcer Adams keep getting more exclusive sex despite their bad behavior, and they're still nice and helpful when it comes to their own kids, it might encourage other Eves to strike similar bargains.* At any given time, in any given generation, no one realizes it's happening. But slowly and surely, females are giving up paternal uncertainty. Those new behaviors, and whatever genetic underpinnings that better allow them, are getting favored because they *work*, allowing more babies to survive to reproduce themselves.

But this shift had to give a real advantage for these sorts of changes to stick. Remember, getting rid of paternal uncertainty is still a dangerous bargain. And it also opens the door to sons inheriting status from their fathers. In species like ours, males usually have to compete for rank. That's true in every single social primate species—except our own. Among our primate cousins, you can be born a princess but *never* a prince. You have to fight for that.†

Once our ancestors had princes, dominant males gained much more power. The ability to inherit social status bred tighter male coalitions. And finally, the small difference in body size between males and females could have started to have more of an effect. It's one thing for a group of females to gang together and beat up a single annoying male. It's another for a group of *males* to gang together and beat up a female.‡

* You can see hints of something like this in olive baboons. Usually, males move around between troops, especially if the ratio of males to females shifts and there are too many males in one troop. But sometimes, if a male has managed to have a lot of sex in a group and has fathered lots of babies, he'll stick around longer (Alberts and Altmann, 1995). Even if a bunch of young, studly males come along. He may not be getting as much sex as he was before, but something about his just being there seems to benefit his immature offspring. Maybe he's helping defend the young from misbehaving males. Maybe he's helping keep the female coalition together. No one really knows. But if it weren't beneficial for him to stay there, he probably wouldn't do it.

† Hyena males seem to inherit their mother's social networks and some of the status perks, but mostly if she's high ranking; because males don't tend to stay local, the effect is far greater for the daughters (Ilany et al., 2021). There's only one mammal we know of where males inherit their mother's social rank for life: transient orcas. Sons stay with their mothers their entire lives, inheriting *their mothers'* rank. They're matriarchal, too, and the only other species proven to have menopause.

‡ Male coalitions among bonobos aren't nearly as tightly knit as they are among chimps. And because they're matriarchal, most bonobo males aren't going to risk losing female favor by coming to the defense of their male friend.

Again, in this scenario, it doesn't happen all at once, but more like a slow tide: Males get more power. Brotherhood gets stronger. Groups of males start coming together to resist the Mean Girls. Some males even begin mate guarding, like the chimps. Maybe whole groups of males start mate guarding females. But I'm not convinced the story is as simple as this sounds: females falling victim to male power in ancient human history. Instead, I think females were probably *instrumental* in the shift to patriarchies.

The devil's bargain wasn't just a deal women made with men; it was a deal they made with other women.

THE SWITCHBOARD

Ever called a woman a home wrecker? Or even thought it? Ever been *mad* at a woman—maybe even a woman you've never met—because you heard she had an affair with a married man? Ever found yourself angrier at the woman than the man, even though he was the one who was married and "ruining" his marriage?

Yep. Me too.

It's an incredibly common reaction. As a rule, North American and European women are far more strict about women following sex rules than they are with male rule breakers. When men step out of line, women get mad. When *women* step out of line, other women get furious. Men in these parts of the world follow similar patterns, but they're usually not as judgmental about female misbehavior as women are. Sure, men will throw around the word "slut." But research confirms that women use that word about as often as men do.

And while most of this kind of research comes from Western countries, similar rules hold for the Middle East and Japan. Women are sexist. We think sexist stuff about other women. We do sexist things in the world. We create sexist rules and strongly reinforce them. So, the question is this: What possible motivation could we have to maintain a sexist culture that mostly disadvantages women?

I propose that women are sexist because we essentially *evolved* to be. It isn't Stockholm syndrome—we're not just internalizing sexism. It's not some cynical power grab, either. Most women are not looking for ways to succeed by crawling over the bodies of other women.

No—sexism is one of the ways our ancestors solved our hardest problem, which, as I've already discussed at great length, is that we categorically suck at making babies.

I think of sexism and gynecology as two sides of the same coin: they're two behavioral strategies our species employed—and still employs—to try to jury-rig a glitchy system. If pregnancies are dangerous and babies are needy, you need work-arounds. For example, birth spacing to control how often the girls in your troop are pregnant. Gynecology gives you tools for birth control and abortion. But you can also create cultural rules around when and where the males get access to female bodies, and then create punishments for those who break the rules.

That's the core of what sexism is: a massive set of rules that work to control reproduction. The aspects shift from place to place, but every single human culture has rules about what women should wear, where they can go and in what circumstances, whom they should talk to and when, and most certainly when and how and with whom they should have sex. Each rule tweaks access to a woman's body, shaping the parameters of her reproductive life. Having a rule keeping women out of the workplace is, at its root, about controlling when and where and in what context women can be in public spaces. It influences male access to women's bodies. To women's time. It influences how many hours women are supposed to spend on child care. In other words, it's about sex.

Men also have sex rules, but they aren't nearly as numerous or as strictly enforced as the rules for women. From a scientific perspective, the reason for that is simple: we're mammals. Our babies get made in wombs, and females are the ones with the wombs. Since the male's role in human reproduction is relatively small, controlling access to male bodies isn't as crucial. Human beings care a *lot* about sex rules, but especially when it comes to women.

So how did that happen?

There are no specific genes for individual sexist beliefs. There's nothing written in your DNA that makes you approve or disapprove of the length of a woman's skirt. But you *are* wired to care about sex. And you are wired to care about social norms. And the consequence of how much you care about sex and social norms is a massive rule book that mostly applies to women, built up over more than a hundred thousand generations.

No one ever sat down and signed a contract agreeing to a monogamous, sexist patriarchy. Lucy didn't know how to read or write, after all, and we didn't even have language for a long time after her. But men's bodies were already shrinking by the time Lucy came along. That probably means violent male competition was waning. So maybe by Lucy's time, we were already starting to move away from promiscuous matriarchies toward monogamy. Eventually, we built patriarchies. And there's a good chance sexism was built into those changes from the beginning. Not all human cultures ended up this way. Even in written history, there are accounts of more egalitarian and even matriarchal cultures. But the majority we've ever heard of are patriarchal and largely monogamous.*

So yes, at some point our Eves traded sex for food and protection and assistance in child care, and yes, it's quite possible that it got started inside ancient primate matriarchies, with males trying to get in on female power. And over time, sex rules became a part of how human beings built modern human culture. Maintaining those rules helped us take control of our reproductive systems, but the rules also destroyed the legacy of the matriarchies. Modern female coalitions are scattered, vulnerable, brittle.†

* The *more* patriarchal and sexist a culture, the more likely you'll see polygyny and harems in its history. But even within harem-having cultures, monogamy was the more likely arrangement for individual families. That's as true at the height of the Islamic empire as it was in Solomon's day. After all, having multiple wives was expensive.

† If human women had coalitions like the bonobos, every single member of ISIS would have been slaughtered ages ago. Every single human trafficker who tried to pimp out little girls. Thailand. The Marshall Islands. Armies of women, bristling with weapons, would have flushed Boko Haram out of their fetid little forest holes the very *hour* after they'd dared kidnap the Chibok girls. There'd be no limpid talk of "cultural difference" in a world of true female coalition—anything that threatened the well-being of women and their daughters would be quickly snuffed out. Primate matriarchies don't equivocate. Mean girls are mean to *each other*, but they don't tolerate a lot of BS from males. That isn't a society I want to live in, by the

But today, no one is really aware that they traded anything, or that we're continuing to re-up this contract with each generation. That's because the way human behavior produces human culture isn't straightforward. What we call culture is an emergent property of a huge, complex system: individuals making decisions, most often unconsciously, that collectively and over many thousands of years become ingrained in local identity.

Picture a switchboard. It's got all sorts of knobs and levers. Twirl one knob, and women are allowed to show their knees, so hemlines rise. Pull a lever, and parents have more control over their daughter's choice of mates, and you get things like arranged marriages. Other knobs affect breast-feeding. Other levers, women's paid work. There are thousands upon thousands of controls on the board, each manipulating some feature of local human culture, from the mundane to the profound. Not all the controls have to do with women's bodies—that's just a large subset. Another subset has to do with food, another with property. And like any massive switchboard, there's plenty of overlap and redundancy, with some controls having knock-on effects on other controls.

So the reason we want to shame women who have affairs with married men isn't simply that we've "internalized" male dominance. Frankly, that gives men too much credit and women too little. Every human being is an active agent in the generation and maintenance of his or her culture, and by extension, what it means to hold that cultural identity. When a woman has an affair with a married man in a society that has strong rules around monogamy, that woman's behavior is a violation of a number of different cultural standards.

Those standards do a lot of heavy lifting. From a biologist's perspective, primate cultural rules can reduce competition, resolve conflicts, and ensure lower-ranked members still get enough food. But the standards that control *sex* are some of the hardest settings to change because sex controls have a lot of evolutionary power. In our deep past, getting those settings "right" in any

---

way—primate matriarchies are violent. If humanity lived in that sort of society, the only thing you'd find of Boko Haram would be blood on the leaves and scattered teeth.

cultural group's particular environment could mean the difference between survival and annihilation.

Evolution doesn't care about suffering.* Human rights are irrelevant to genes flowing down through time. Evolution doesn't care if Hillary Clinton or Elizabeth Warren or Donald Trump becomes president.† Evolution doesn't even care about terrorist regimes like ISIS. If cultures that have overtly sexist settings on their switchboards produce more babies, and those babies survive—and those trends persist for many thousands of years, outcompeting cultures with different settings on their switchboards—then, in evolutionary terms, the sexist strategy was successful.

As each culture's circumstances change over time, sex rules also change. Human beings are tremendously adaptable. We've *evolved* to be adaptable. And our behavioral innovations are adaptable, too. If there were only one set of sex rules that universally had good outcomes, we'd *all* have the same sex rules. But we don't. So we keep tweaking the settings. In fact, it's one of the first things any human culture looks to in times of cultural flux. In such times, humans tend to become strict about enforcing their particular set of sex rules, sometimes on totally new populations. The first thing ISIS does when it takes over a town? It forces locals to serve as religious police and sends them on patrol to make sure women are covering their bodies when men are around. The Taliban did this. The *mutaween*, too. And when France gets especially nervous about its Muslim population, the government reestablishes the country's "Frenchness" by making rules against women wearing hijabs on the beach.

But it's not just a modern thing, and if we pull the camera back a bit, it's easy to see that it really has nothing at all to do with

* Technically, evolution doesn't "care" about anything. It's a system of cascading events in biological systems measured over massive amounts of time. The point is that the things *we* care about, as sentient social primates, don't often have a lot to do with evolutionary fitness. *Eventually*, ISIS-style settings will clearly lose, because they involve a lot of murder, near inbreeding, and child rape, which all winnow the gene pool and invite tremendous aggression from competing groups in their territory.

† Barring nuclear outcomes, that is.

Islam. European colonists made a big fuss about "covering up" the bodies of Native American women. Aztecs spread their own sex standards to the people they conquered, too. So did China. Japan. The U.S.S.R. Throughout human history, when cultures with different sex rules come in contact with one another, some rules get abandoned and others get violently enforced.

A good chunk of what right-wing French folk are saying about hijabs is old-fashioned bigotry. But cultural differences around women tend to be flash points. I think that's because our sex rules have been vitally important in the evolution of our species. That's why we care about them, and that's why we keep *tinkering* with them. We're not just selecting for specific rules; in effect, we're selecting for the urge to have sex rules at all.

It's very, very hard to stop. But it looks as if we might have to. At this point, sexism is killing us.

HEALTHY, WEALTHY, AND WISE

For now, let's set aside the very good moral arguments as to why cultures that are less sexist improve the lives of women and girls (and everyone else in that culture). Instead, let's investigate whether sexism is still doing the job it evolved to do. Does sexism help us the way it used to?

Our ancestors' birth control was only so good. Our midwifery could save only so many lives. Our abortions used to be really dangerous.* We still *needed* sexism to get where we needed to go when it came to survival. Over the millennia, gynecology slowly advanced as cultures constantly tweaked the switchboard to create better reproductive outcomes. The right number of babies,

* Today they're not. People who say differently know almost nothing about science, medicine, or women's bodies. So long as an abortion is performed by a well-trained, licensed medical professional in an appropriate setting, it's both safe and far less likely to create any long-term complications than a human pregnancy left to run its course. What is comparatively dangerous, in other words, is *pregnancy and birth*, not legal abortion. The same can't be said for illegal abortions, many of which are not provided by medical professionals in any sort of appropriate setting.

at the right time, raised in a way that worked with that group's resources. Birth control and midwifery did some of the job. Sexism did the rest.

But what happens when sexism turns into a runaway train? What happens when a culture's sex rules start to *reduce* the overall health, fertility, and competitive viability of a population?

Here's what a biologist would say: if a set of behaviors that *used* to be beneficial starts to make a group less "fit," then it's just a matter of time until these behaviors change. If the behaviors are in any way encoded in the genome, then we're talking about an evolutionary amount of time. But eventually those behaviors will be weeded out, either through cultural change within the group in question or through the die-off of that subpopulation. If the behaviors are global to the species—meaning, if everyone's doing them—the same thing should happen, except with more dire consequences. Either the behaviors change, or the entire species goes extinct.

Human beings are no exception here. The only difference is we have the cognitive capacity to recognize when something like that is happening. At this point, sexism in a wide array of different cultures is starting to hurt our species as a whole. To paraphrase one famous American, modern sexism is making us less healthy, wealthy, and wise.*

LESS HEALTHY

You'd think at the very least sexist rules would keep sexually active people healthy. Paradoxically, in the modern world they tend to have the opposite effect, accelerating the spread of sexually transmitted infections and unplanned pregnancies and reducing access

* Famous, and famously sexist, by the by: Ben Franklin also wrote that having a mistress in her mid- to late thirties is just as good as having a younger one because, so long as you cover the top half of her body with a basket, her aged genitals are indistinguishable. He also liked that the old mistress would be "so Grateful!" He'd also fathered a child by a mistress when he was twenty-four, and his common-law wife raised the kid (Franklin, 1745/1961; Isaacson, 2004).

to maternal health care. Sexism is making us sick. All of us: men *and* women.

Female chastity is a common sex rule. In most cultures, "good" women aren't supposed to have multiple sexual partners in their lives. Many Western parents still think that encouraging their daughters to be more chaste will protect their health in the long run. That seems reasonable. In principle, it should at least reduce sexually transmitted infections. Parasites, viruses, and bacteria have fewer chances of spreading if you lower the number of sexual partners an individual has. A chastity rule should produce cultures with much less gonorrhea, syphilis, HIV, chlamydia, herpes, and genital warts. From a biological perspective, that sounds like quite a boon: all of these STIs can screw with your fertility, and thereby your evolutionary fitness. Except it doesn't quite work out, because most women do not stick to sex with only one man—not now, and not historically.* But even more important, men don't have sex with one woman. If anything, men in contemporary cultures with "chaste women" rules are *encouraged* to have multiple sexual partners over their life span. In many of today's cultures, having a long and rich sexual history is a measure of successful manliness.

That leaves us with a double standard: women aren't supposed to have sex until they're in a monogamous relationship with one guy, ideally for life. Men, meanwhile, are supposed to have many sexual partners in order to achieve manliness. In biology, that

* Even in medieval and premodern Europe, where as much as 14 percent of the female population was celibate thanks to financial concerns and the influence of the Christian churches, the average man *still* likely had three or more sexual partners over the course of his life—often via prostitutes or domestic servants (if he had the money), who were often de facto sexual slaves of the men associated with households they worked for (Fauve-Chamoux, 2001; Dennison & Ogilvie, 2014; Karras, 2012). Many "celibate" clerics likewise dallied with sex workers and/or domestic servants, despite the risks to their livelihoods and social status (Ingram, 1990). If anything, the clearest benefits of Christian sex rules went to the church itself—without legitimate children who could claim inheritance from its clergy, the Church remained the uncontested owner of all its property generation after generation. The reason the Catholic Church remains so fantastically wealthy is not that little plate they pass around on Sundays. It's the legacy of an absolutely massive real estate portfolio held by the same institution for centuries.

sounds like a classic case of female reproductive choice butting up against the sex drive of sperm-spreading males. The only problem is, human women aren't generally free to exercise their reproductive choice. Too many cultural influences are at play.

Not to mention the simple fact that the math is, of course, impossible. What you end up with is a large group of both men and women with roughly the same number of partners and a minority of more promiscuous people at the far end of the curve. There simply isn't any group of extremely promiscuous women "filling the needs" of wannabe promiscuous men;* nor is it true that the average man has more sexual partners than the average woman.†

As you'd expect, historically it's been the most promiscuous among us who drive a lot of STIs. And the taboo reflects that idea of the "filthy whore" and the "nasty slut." Because that taboo falls squarely on female shoulders, cultures that play this game are setting themselves up for failure: the more male promiscuity is encouraged, and the more stringently you enforce chastity for women, the fewer checks there are on the spread of disease. This is where the evolution of human gynecology *should* come to the

* There actually aren't that many female prostitutes at work these days. By the most generous estimates, sex workers constitute only 0.6 percent of the U.S. population, and in places where prostitution is legal and regulated, sex workers tend to be *more* vigilant and consistent when it comes to safe sex (Platt et al., 2018). According to the latest statistics, you're actually less likely to contract an STI from a Nevada prostitute in a well-regulated brothel than you are having sex with an average young woman in San Antonio (Rodriguez-Hart et al., 2012; CDC, 2022). Note that I'm talking about research into known prostitution here that may well be coerced but is not overtly forced, unlike the global horror that is human trafficking.

† In the mid-twentieth century, American men famously reported three times as many sexual partners as American women (Kinsey et al., 1948). That shrank to merely twice as many in the 1990s, though it's unclear whether that was due to the average woman having more sex, the average man having less sex, both sexes being more honest, or some combination of the three (Wiederman, 1997). It's mathematically impossible for the average man to have three times as many sexual partners as the average woman. The only thing you can actually learn from these surveys is that among people who lie about these things, sexual mores tend to skew the data in one direction or the other: men say they've had more partners, and women say fewer, with the odd exception of New Zealand, where women report more partners than men—which is also impossible, given that Kiwi women outnumber Kiwi men by seventy thousand (Durex, 2007). Perhaps they're importing partners?

rescue. For example, since the mid-twentieth century, you'd think condoms would have solved the STI problem. Generally, they are the most effective strategy, so long as men wear them. Each time. Consistently.

Which they don't, particularly in cultures where male promiscuity is tied to the idea of one's overall manliness. Consistent condom use is remarkably low in places famous for their machismo, from Brazil to Texas, South Korea to South Africa. It seems everywhere men are expected to be promiscuously "manly," they also tend to fail to sheathe their manhood. According to one recent study, Latino men in the Miami area use spot judgments to decide whether to wear a condom, largely based on a feeling that the particular woman is "clean" or "dirty." (Their judgment is not, by the way, particularly astute in this regard.)

Research shows that STIs go down consistently in places where everyone is taught to use condoms and condoms are cheaply available.* But if *other* sexist notions remain in place—for example, that women shouldn't have many partners and men should—parasites and bacteria get a boon. In the United States, many promiscuous people are now more careful when it comes to practicing safe sex. But meanwhile, because of the *assumption* of safety among the less promiscuous—the idea that because they're more exclusive with their sexual partners, they're immune from risk—the less promiscuous are becoming significant drivers of disease. They're not using condoms, because they think they're safe.

The effect quickly snowballs: one partner acquires an infection from a prior partner, passes it on to their next averagely promiscuous partner, and that partner passes it on down the chain to subsequent partners, all of them neglecting to practice safe sex because they assume they're having sex with less promiscuous people.

If the less promiscuous people in question are women, then they would be more likely to acquire a range of STIs than the men. That's because the penis and the vagina are what they are: an ejaculator and a receptacle.† It's also because mucous membranes

* (Dodge et al., 2009). The actual teaching part is important. Leaving condoms out in a bowl next to an instructive banana doesn't help anyone.

† It's not hard to see the outcome: a full 75 percent of young African people who

are more vulnerable to infection than outer skin, and women's vaginas are lined with those membranes, while men just risk the tiny lining of the urethra.*

So basically, it's relatively *chaste*, modest, and serially monogamous women who are driving massive outbreaks of syphilis, herpes, gonorrhea, and chlamydia in places with cultures that promote female chastity and masculine promiscuity. The Centers for Disease Control has been tracking them throughout the United States: Minnesota hit a record high for STIs in 2014. Montana more than doubled the rate of gonorrhea transmission from 2013 to 2014. Louisiana, Mississippi, Georgia, and Texas lead the charge in syphilis, chlamydia, and gonorrhea—all states with some of the highest social emphasis on the importance of female chastity (and predictably, some of the lowest public funding for sex education and prophylactics).† Syphilis rates tripled

have HIV are women and girls (UNAIDS, 2004). It's not because they've failed to be relatively chaste but because their male partners aren't wearing condoms. Sex with *other* orifices has its own dangers, of course, and a goodly amount of sex happens with two partners of the same gender. Anal sex is particularly vulnerable to disease transmission, because the rectum didn't have the same evolutionary pressures as a sexual orifice and therefore the tissue there is more fragile. But because the majority of human sexual intercourse involves a penis and a vagina, the majority of STIs involve the combination of those two. That's why I'm being heteronormative here: we're talking about *huge* numbers, statistics involving large populations. I'm also talking about normative sex rules that regulate male and female heterosexual behavior, so queer populations—already taboo—work a bit differently under wider social sex rules. That said, male homosexual behavior is still influenced by local notions of manliness and promiscuity, which can likewise drive STI transmission in those groups.

* Men do acquire STIs on the outer skin of their genitals, and men of every sexual orientation may also participate in anal sex, which has its own risks. But when simply comparing a male urethra and a vaginal cavity, it's clear who's more vulnerable.

† Having no idea how to properly use a condom, or even that it's useful to do so consistently, is the most obvious driver for the spread of STIs in these communities. But given that cultural emphasis on chastity is a huge driver for the defunding of science-based sex education, it's not hard to tie the two as more than mere correlation. Presumably there could be a world in which both real, evidence-based sex ed and a strong cultural emphasis on chastity would peaceably coexist. A better bet is simply funding real sex ed and letting the cultural chips fall where they may. I don't think any teenager has ever been inspired to get laid *more* after learning about what gonorrhea actually does to the body. And fewer STIs means better fertility in the long run, so at least biologists would call that kind of policy evolutionarily successful.

between 2012 and 2014 in Louisiana, a state where more than 60 percent of the population regularly attends religious services.

Though the irony is a bit thick, at least sexually transmitted infections are significantly lower now than they were a hundred years ago. Latex condoms actually exist. But from an evolutionary perspective, it's not just the infection load that's the problem: it's the fact that STIs screw with female fertility.

Chlamydia and gonorrhea are tricky little bugs. The majority of chlamydia infections don't actually produce noticeable symptoms; while the infection is setting into a woman's cervix, she probably has no idea. The male partner who gave it to her also probably had no idea, because it's even less likely to produce symptoms in the male body than in the female. So on it goes, quietly irritating the tissue of the cervix, causing low-grade inflammation. That inflammation can then spread upward into the uterus and fallopian tubes, where it can cause something called PID—pelvic inflammatory disease, in which the female sex organs continually go through cycles of damaging inflammation. Untreated gonorrhea infections can do the same.

Sometimes PID is hideously painful. Sometimes, mysteriously, it causes few noticeable symptoms at all, remaining "subclinical" until the woman tries to become pregnant. And fails. Or worse, she becomes pregnant, and because her fallopian tubes are scarred from years of an undiagnosed chlamydia infection, the pregnancy is ectopic.* If the mother manages to survive the ectopic pregnancy—achievable only through modern gynecological intervention—then one of her tubes is likely damaged beyond repair. If infection managed to ruin both of her tubes, then only if she can *afford* a few wildly expensive rounds of in vitro fertilization can she pass on her genes.

If not, evolution takes another female out of the gene pool.†

* Roughly one in fifty American pregnancies are ectopic. In the U.K., the estimate is one in ninety (Cantwell et al., 2011), but it's unclear if the difference is due to differences in measurement strategies or actual difference in prevalence. It is true that chlamydia and gonorrhea don't cause *all* cases of ectopic pregnancy, but they're leading suspects.

† Being infected while pregnant also tends to mean you'll have a preterm birth, which is risky for the kid, and that baby may be born with eye problems that can

Sexism used to keep a check on things like this. Creating a taboo around female promiscuity worked well enough when human populations were *small*. But because our global population is so much larger, and transportation technology is so much better than it was two thousand years ago, infections spread fast. Every year, roughly sixty-two million people are infected with gonorrhea. It's burning across America like a brush fire, frying the fallopian tubes in its path.

Some think gonorrhea has been around since the time of the Old Testament, so clearly we haven't managed to out-evolve it yet. Luckily, human behavior can outrun it. We can, in fact, use condoms. We can, in fact, reduce antibiotic use to help curtail the spread of antibiotic resistance. Recently, a vaccine for chlamydia is looking promising, so we could even try creating herd immunity long before our genes would manage to produce it. That would require cultural agreement, of course, and speaking as an American in COVID times, I know that getting to herd immunity through vaccines is no small task. But it's certainly worth a try.

And, of course, there's always the less appealing option: we could have less sex. But I'm afraid abstinence isn't likely to win out. Historically, it never has. And at this point, rules to reinforce women's chastity tend to screw over women's fertility and the population's overall health more often than they help.

There are, of course, more extreme examples of ways that sexism damages our health. And I don't just mean things like female genital mutilation in parts of Africa and the Middle East. We're talking about outcomes that undermine the reason we adopted sexism in the first place. Outcomes like the death of the mother or child. While impaired reproduction has an obvious effect on long-term evolutionary fitness, know what's even more devastating and faster acting? Killing the mother.

For most of human history, most girls didn't reach sexual maturity until age 16 or 17. That's still true of today's well-studied hunter-gatherer groups. Among the !Kung people, the average age a girl gets her first period is 16.6. Among the Agta Negrito girls of the

---

lead to blindness. There are many ways in which STIs can reduce a population's evolutionary fitness.

Philippines, it's 17.1. In both of those groups, the average age of first birth is 19 to 20—two to three years after most girls' first period.

So why would *any* human culture marry off girls younger than eighteen? Even more inexplicably, why do some cultures marry off girls at *eight*?

A woman who gives birth at eighteen has got a pretty good chance of surviving, anywhere in the world—not only surviving, but having a healthy pregnancy with a healthy baby *and* going on to give birth to more children after that. That's even after accounting for humanity's lousy reproductive system. But if she's under fifteen, her chances of survival drop drastically. Under thirteen, the chances of survival are even lower. Maternal age is the single most predictive factor for whether a girl is likely to die simply because she became pregnant. Reducing the number of girls married before they are eighteen by even 10 percent can reduce a country's maternal mortality by 70 percent.

Thus, the sexist cultures that produce child marriage—places like Niger, Chad, Bangladesh, and Nepal*—are also the ones that kill the most girls, if for no other reason than that they force girls to marry and have sex with older men before their bodies are developed enough to be able to survive it. If they do survive, their reproductive fitness is grossly compromised. Girls married before they enter puberty often suffer from infections and trauma to the pelvis, sometimes to the point of prolapse, caused by performing their "marital duties" with genitals that aren't developed enough to handle it.

Obviously, this isn't sustainable on an evolutionary scale: no behavioral group that deliberately injures young females would be

* In Nepal, the government is committed to changing this and has made it illegal for anyone to marry before age twenty. Punishment involves both fines and a prison sentence. Yet somehow, 37 percent of Nepalese girls are still married before age eighteen (UNICEF, 2022). Niger seems to be barely trying: three out of four girls marry there before eighteen. In some areas, nearly 90 percent of girls marry when they're children (ibid.). In terms of sheer numbers, India is by far the worst offender, with 15.5 million girls married off as children (ibid.). But they're also one of the best improved, with rates declining from 50 percent of girls to 27 percent just in the last decade (ibid.). That their numbers are still high is due to their large population, but how rapidly they've shifted the bar when it comes to child marriage also shows how effective a concerted effort can be.

able to survive and thrive in the long run. That such practices are seen as "ancient" is only evidence of humanity's myopia. Sure, once upon a time, child marriage was also fairly normal in places such as China and Europe, but we're talking only a few hundred years ago, and since then it's fallen out of favor. Ancient Greece aimed closer to sixteen, as did ancient China, while the age of marriage in ancient Rome ranged from fourteen to twenty. What's more, in Rome the younger brides were often *wealthy* and married off as a matter of political exchange; the plebes generally married in their late teens or early twenties. The same was true of China and Greece.

It's safe to say that for most of our species' history, girls were not being raped into pregnancy at age eleven. We'd never have made it this far if that were the case. In the mammalian game, you can always make more boys. The loss of a healthy, young female is incredibly expensive.

But it's not just these dramatic cases of sexism that are holding our species back. Child marriage is egregious, but people in the Americas, in Europe, and in prosperous Asia can say, "We don't have that here."* Not in the sorts of places that have more money. The places where people read books like this one.

* Forty-eight of fifty states in the United States do allow child marriage with the "permission" of the parents—a legally sanctioned form of child abuse (Ochieng, 2020). Unfortunately, the United States allows parents to do all sorts of things to their children, usually under the mantle of "religion" or "cultural preference." For example, in twenty-one of our fifty states, it is legal to force one's daughter—no matter her age—to go through with a pregnancy when she clearly doesn't want to or, even worse, is simply too young to be able to understand the physical and existential consequences of doing so (AGI, 2023). If you're eleven years old and your parents tell you to give birth to a baby because they have a preestablished cultural belief, are you *really* going to be able to say no? And if you do, will you be able to run away and cross state lines and somehow get yourself an abortion within a time frame that allows the procedure to be simple and safe? No adult will be legally allowed to help you do so. Besides those twenty-one miserable places to be a girl, another sixteen states require the parents be notified about such a procedure, which is wonderful if you happen to live in an abusive household (often the case for a pregnant eleven-year-old). You may be able to petition a judge to get around them—if you have the resources and chutzpah to pull *that* off—but you'll have no guarantee that the judge will agree. The judge option exists only because the U.S. Supreme Court demanded a judicial bypass be provided, and even that might go away now that *Roe v. Wade* is gone. Meanwhile, no requirements exist for keeping track of how often those judicial bypasses are successful, nor that they be equally

So why, exactly, is the maternal death rate going up in the United States?

In the last ten years, pregnant women and new mothers are dying in the United States more than they used to. That's a direct reversal of the general trends of the last two centuries; normally, rich places have *fewer* dead mothers every year, not more. But a hot combination of racism, sexism, ableism, reduced public support for female health, and the crippling of science-based sex education has finally made it more dangerous for American women to be pregnant than it used to be.* Americans are boldly leading the charge back into some kind of dark age for women, but similar trends are popping up in parts of Europe. Though European maternal deaths are still going down, the rate at which they're dropping is starting to slow, especially among the less wealthy. So what's going on?

Part of it is obesity. While every pregnancy has risks, pregnancy is statistically more dangerous for obese women than for non-obese women. There are a number of things that can go wrong, medically speaking, and many of those are tied to a range of common comorbidities for obesity. While no one knows if obesity *causes* these things directly, or vice versa, it's true that obese bodies tend to have cardiovascular systems that show more strain and damage, body-wide inflammation, joint problems, and issues associated with poor sleep, such as sleep apnea,† which is to say, people who are obese frequently have a number of things going on that are generally hard on a body. It's also hard on a body to be pregnant—even the healthiest woman can be laid low by a pregnancy—so combining the two is obviously a difficult thing for that body to do. It's also true that not all doctors are prop-

---

accessible to all strata of society, nor are any protections offered to adults who may choose to help young girls in need when that help may be against the law. To put it bluntly, the United States simply doesn't care enough about girls to protect their rights over the beliefs of their parents. If it did, laws like these wouldn't exist.

* The maternal death rates are especially stark among African American women; some of that difference may go away when you control for income (the U.S. system is crappy to poor people, and systemic racism traps many people of color in the lower class), but not all of it (Hoyert, 2022).

† That's a condition where you actually stop breathing for brief periods while asleep. Getting enough oxygen is rather important for one's health.

erly trained to care for the unique needs of obese patients during pregnancy, and because of social shame associated with obesity, patients may struggle to form productive relationships with their doctors.* As for why obesity is on the rise, reduced quality of food is affecting poor people everywhere, as it always has, but the rise of cheap *sugary* foods and drinks is strongly tied to the rise in maternal obesity among poorer populations in Europe and the United States.

But it's not all down to the rise in obesity. Perversely, modern sexism directly inhibits the advance of gynecology. As much as sexist cultures seem to want women to be pregnant more often, they also have a habit of reducing the health care available to pregnant women. Where are most pregnant American women dying? Poor communities, yes, but particularly poor communities in Texas, the American South, and Minnesota. These are all places where women's access to health care and health education has been dramatically reduced in recent years through anti-abortion campaigns, abstinence-only education policies, and a simultaneous series of cuts to publicly funded health clinics. As a result, women are getting pregnant more often in these areas, but they're also getting more STIs, having more complications with their pregnancies, receiving less prenatal care, and typically having more difficult births. After those difficult births, they also tend to leave the hospital sooner than they should—driven in part by a lack of money. Going home too soon further increases their risk for postpartum hemorrhage and other complications. The state of women's health in these communities, in other words, is starting to look the way it did fifty years ago.

Presumably every species wants the healthiest mothers and offspring possible, within the resource limits of its particular environment. Allowing maternal mortality to go *up*? In evolution, that makes no sense at all. If the mother dies because of some local

* Far more has been written on these issues than I could sum up in a footnote. But in general, I think it's safe to say that *all* women need healthier relationships with medical professionals, and issues like gender and weight and race only compound these problems for patients and doctors alike. If we're going to fix the deep problems in women's health care today, women need to trust the science more, and scientists and clinicians need to trust women more.

antiabortion policy, that means she never gets to have more children. If she dies because she didn't have access to good health care and family planning, she doesn't get to have more children. This is the opposite of optimizing for the greatest number of healthy babies.

It's the biological equivalent of cutting off your nose to spite your face.

LESS WEALTHY

As an American, I can easily tell you how *expensive* poor health can be. But it's not just a matter of whether one has a nationalized health-care plan: poor health is terribly expensive for communities over *generations*, not simply by passing on debt, but by crippling the income potential of any given family as it wrestles with the poor health of its members. What choices do you make, after all, if you need to care for a sick parent? What if you're widowed? What if you're the main earner of a family, but you lose potential working time to deal with your own health? How well can you tend to your children's care if your own body is failing you? What will that do to your children's life paths?

The moral imperative here is clear. Sexism's cost to global health is immense, and that cost is both metaphoric and literal. But again, let's take a more biological approach to the question: What does it mean to reduce the wealth potential of a community in *evolutionary* terms?

Human wealth is one of the easiest predictive measures for a child's eventual success. How much money a child's parents have access to shapes not only how much wealth that child is likely to have as an adult but also how likely that child is to reach adulthood with fertility left intact.

It just so happens that the easiest, cheapest, and most reliable way to increase a community's wealth is to invest in its women and girls. As counterintuitive as it may seem, financially supporting females usually makes the entire community richer—richer, even, than giving the same amount of money to males in the same community.

There are a few ways of measuring this. Let's start with financial control and independence. In many of today's more overtly sexist cultures, men have full legal control over the financial resources of their families.* Women and girls do not have a say in where the money goes, even if their labor is the primary source of that family's income. But institute a policy that lets *females* have control over their own money and the results can be dramatic.

In a wide variety of studies, covering cultures ranging from rural America to urban India, women are more likely to allocate financial resources in a way that directly affects the welfare of their immediate households and local community. When given the opportunity, women are more likely to spend a family's money on food and clothing and health care and children's education. Men, meanwhile, are more likely to spend it on entertainment and on weapons and—if we're talking global trends—on gambling or the local equivalents.† Worldwide, girls and women spend up to 90 percent of their earned income on their families. Men and boys spend only 30–40 percent. When women in India were given the opportunity to participate in local governments as ministers and officers, those governments more greatly invested in things like public services and infrastructure, from waste management to potable water and railways—things that, as it turned out, seemed to *matter* more to female politicians.

It's not that male politicians don't care about community concerns and infrastructure. They just seem to care about them less—or at least, if they have those concerns, they act on them less. Similar trends can be seen in the voting habits of women in the United States and Europe. As troubling as it sounds, the data exist: when you leave men in charge, roads and bridges and dams

* Including American culture, until very recently—women weren't legally allowed to inherit before the late nineteenth century (Knaplund, 2008). There were various systems of dowries and gifts in place, such that girls could take some of their family's money with them into a marriage—where it would promptly become the legal property of the husband. Among the moneyed, becoming a *widow* was the surest path a woman could follow into financial independence. For the poor, such a fate was often devastating.

† We're talking about large-scale statistics here, not individuals. The ever-so-male father of my children has no interest in gambling.

are effectively left to rot. When women are empowered in local governance, for whatever reason, they are more likely to vote for local infrastructure (and health services and local, directly impactful public spending) than male politicians, and in Europe, they're even likely to improve government transparency.

Obviously, these data aren't talking about the Margaret Thatchers of the world. After all, most women aren't Margaret Thatcher: most women don't have lives with that kind of social power. So what's driving these numbers?

Some think these inclinations may be tied to the fact that women do most of the child rearing, and that keeps their focus on local concerns, but the real truth is we don't actually know what's driving these differences. Still, even without fully understanding the mechanism, we can say that you don't have to care about women's "rights" in order to find good reasons to financially empower women. You can just look at known outcomes. Maybe you can just care about the bottom line of your economy. Many well-regarded economists have written extensively about this: give women more money and give them the power to make decisions about where to spend it, and their communities will generally become more economically productive.* Whole programs at the UN, World Bank, and IMF are based on this premise. The president of the World Bank and the managing director of the IMF have both given speeches specifically on this topic in the last decade.† If you

* As with all things in science, it does matter how you measure it—for instance, while we have the data on recent programs that have purposefully invested in women in developing countries and can see the shorter-term local effect, it's much harder to extract correlation versus causation in places like the United States and Western Europe, where the economies have been relatively sex egalitarian for a while. It's possible the female-investment strategy would have the most effect in regions where sex equity is more rare. But because these interventions are usually only a few decades old, the field probably needs more time and more data to better interpret these trends.

† If you're worried the World Bank and the IMF are bastions of liberalism, in 2015 the McKinsey Global Institute came to essentially the same conclusion: improving women's labor equality could add as much as $12 trillion to the annual global GDP by 2025 (Woetzel et al., 2015). McKinsey. You know, the consulting guys. They're capitalism experts—more knowledgeable, one imagines, about working economies than your average academic economist. And they're paid through the nose for it. MGI is their research arm. In 2018, MGI got more specific: in Asian

want to invest in a community, a good bet is simply investing in its women. But it's not just investing in *grown* women that matters. You can also boost your bottom line by investing in girls' education.

Currently, everywhere in the world, men earn more per hour of work than women do, across nearly every industry you can name. It's also true that formal education reliably increases a person's eventual wages. But investing in girls has an even more dramatic effect on earning potential, both for those girls and for their local community. For every additional year you educate a girl, her average lifetime wages increase by 18 percent. For boys, it's only 14 percent. Part of that comes from the fact that in many countries women are far less likely to be educated, so educated females are dramatically more competitive in the job market. But that doesn't account for all of it. A big factor is simply that educated women have fewer children.

The World Bank estimates that for every four years of education, a woman's fertility is reduced by about one birth per mother. Let's put that in the simplest terms possible: four years of school equals one less baby.

The reason why the Indian state of Kerala's fertility rate is 1.9 per couple,* whereas the state of Bihar's is more than 4, is

---

Pacific countries, their models predicted a 12 percent increase in the region's GDP by advancing women's equality (Woetzel et al., 2018). Admittedly, their models are all supply-side—they acknowledge as much—which means accompanying job growth and educational expansion would be needed to prepare those economies to absorb all the increased female labor. It is, in other words, a best-case scenario. But still.

* By most accounts, the reproductive replacement rate in economically stable, non-warring countries is 2.1—that allows for random non-childbearing folk and those who die off early. (In countries in crisis it can be as high as 3.4.) (Espenshade et al., 2003.) But because India, like many places in the world, has massive internal migration, Kerala is in no danger of running into problems with an oversize aging population. And if India ever manages to reach a reproductive rate like Kerala's countrywide? Well, immigration and international work programs are always an option. Germany's been doing that for years. Most of the histrionics about German women not having enough babies are driven by cultural anxiety, not a looming financial crisis, despite Germany's many elderly. They've had Turkish, Bosnian, Russian, all *sorts* of other people come work in their country for decades. And their GDP? Their ability to care for their elderly? Yup, just fine. One of the strongest in Europe, in fact. Nearly all of the doom-and-gloom projections

probably the simple fact that more of Kerala's women are educated, whereas half of Bihar's are not. Even though Kerala lies in India's traditionally underserved south, Kerala's doing well right now. Though much of the local economy is still tied to tourism—a known threat to long-term economic stability—international companies are starting to set up shop. Google's opened an office there; other tech companies are following suit. Local wages are rising. While the rest of India's economically depressed south is lagging, Kerala is marching forward, expanding its average income, and nurturing hundreds of new tech and scientific start-ups, including a prominent biotech company founded by a local Keralan woman.*

This extends to other countries, too: the greater the number of girls who go to secondary school, the higher that country's per capita income growth. Some of that can be attributed to a general cultural value for education and intellectualism. As our world economies become more tech- and science-driven, having a better-educated population is going to help create the kind of workforce that tends to do well today. But even in agricultural communities, educating girls boosts the local economy. And one of the ways that works, again, may be in reducing the total number of babies being born each year.

A smaller number of children being born means a community has more of its wealth to dedicate to each one. When you have fewer mouths to feed, there's more to go around. Health costs go down. Costs for education go down as well. So you also have more money available for things like infrastructure, economic development, or any of a million things that money can buy to help build

---

around birthrates ignore immigration and foreign worker programs. Most of the news you've heard on the subject is driven by *identity* fears, in other words—not that an aging population couldn't be supported, but that Other folk might have to come in and work. They also tend to ignore the potential effect of technological advancement, wherein tech makes individual workers more productive, but that's a much longer discussion.

* It's worth noting that before the colonial era, Kerala was traditionally a *matriarchal* society. As recently as the turn of the twentieth century, properties were inherited along a matriarchal line, women were allowed to have multiple *husbands*, and women were frequently in positions of power in their local communities (Jeffrey, 2004).

a community's long-term economic stability. And hey, if local women aren't spending all their time being pregnant and disease-ridden, maybe they'll even take on jobs in governance and, you know, promote spending on local infrastructure.

We don't need to care about these issues just because it's good to care about others' pain. We should also care because it's good for our own security: terrorism and violent unrest are usually bred in places with a lot of economic and social instability. Make those places safer, and you make *all* of us safer. That means we get to spend less time and money and general anxiety on massive military projects and more on our grander goals. After all, we want to do things like fix the climate crisis, build sentient AI, extend the human life span, cure cancer. Most of all, we really don't want to go extinct before we get the chance to do any of those things.

There are many different ways to haul ourselves into whatever shiny future we prefer. But one thing's clear: to pull it off, we need as many of us as possible to be really, really *smart.*

LESS WISE

Being smart matters. It's not just that it helps you make "wise" decisions; it helps you make decisions in the first place. Your ability to solve problems, your ability to form deep relationships with other people, your ability to contribute to your community, your ability to keep your kids safe—everything you might want to do with your human brain is shaped by how smart it is.

But again, let's take a biological view here: how smart you are affects how likely you are to stay alive. If you have an IQ even fifteen points higher than the average when you're eleven years old, you'll have a 21 percent higher chance of surviving into your seventies. That's a bigger boost to longevity than just about anything you can think of—bigger than what's provided by your level of wealth and your access to medicine combined.

As I discussed in the "Brain" chapter, your IQ is influenced by your genes, but being "smart" isn't something you're just born with. "Smartness" is something that brains actively *do.* It's also strongly shaped by how your brain developed in the womb, in childhood, and even through the sorts of things you ask it to do when you're an adult. Sexism can compromise the cognitive development of children in both genders. In other words, sexism makes *everyone* less smart.

You might think I'm about to talk about education again. But let's start with something even more basic: food. The human brain is literally built out of food. All the sugar, protein, and fat a fetus uses to build its nascent brain come directly from the mother's body. So, what happens when you starve women and girls? Their future fetuses and nursing children starve, too.

In many Indian states, it's normal for young women and new brides to eat last. In Maharashtra, for example, the cultural rule is that guests eat first, followed by the oldest men, then the younger men, then the older women, then the children. In traditional families, a younger woman eats only once *everyone else* has been fed. That rule doesn't change if she's pregnant.

More than 90 percent of adolescent Indian girls are anemic. More than 42 percent of all Indian mothers are underweight. And that's not just down to poverty; only about 16.5 percent of sub-Saharan mothers are underweight. Even worse, the average woman in India weighs *less* in her third trimester than most sub-Saharan African women do when they *first* become pregnant.* Malnutrition is deadly and dangerous all the time, but especially when you're pregnant. If the mother and child both manage to survive, the newborn usually arrives too soon, too tiny, and too fragile. Many die within weeks of being born. Those who don't usually face severe health problems throughout their lives, including problems with cognitive development.

It's true that pregnant women in India's rural areas are more vulnerable to these problems. But rural Indians make up 68 percent of the country's population.† The majority of the world's second most populous nation live in areas where there often isn't enough food and, as a rule, pregnant women eat last.

* Some of that is because Indian women are short, but most sub-Saharan African women aren't much taller. It's really because, on average, they're skinny and anemic. When you control for everything else, the main reason that's happening is cultural.

† To put that in perspective, only 19 percent of Americans live in rural areas.

Sexism, in other words, is starving India from the inside out. At the same time, the country is investing a lot of its resources in a bid to become one of the biggest technology centers of the world. You need a lot of good brains to be a tech giant. Well-fed brains. To build them, you're going to need pregnant women to jump the line at dinner.

Now, I'm hardly the sort of person who wants to think of women as simply baby factories. But as a species, let's say all of us want to get smarter. That's what it takes to cure cancer. To solve the climate crisis. How do we do that? For a start, we might want to acknowledge that human brains are something that are made primarily out of women's bodies: first in their wombs, and then from their breast milk, and then from the quality of interactions mothers have with their children. So if you want the best possible chance to make a lot of kids with high IQs, you want healthy women who are *fed* well, and have been fed well, consistently, for at least two decades before they become pregnant. You want them to have had a rich and well-supported childhood education. And you want them to be well cared for throughout their reproductive lives, with readily available education about nutrition and healthy habits and newborn caretaking. You want them to have community resources available when they get sick and when their kids get sick. And, because STIs have such a proven effect on reproductive health, you want them to have ready access to prophylactics and good sex ed.

It's not nearly enough to say that simply being *wealthy* makes you more likely to give birth to and raise higher-IQ babies. Babies born to wealthy families do tend to have fewer obstacles in their way as their brains grow and learn things. But even born-wealthy babies are still subject to many of the obstacles that sexism produces.

For instance, it's becoming fashionable among Western upper-class women to aim for bodies with very low amounts of body fat. Thanks to the rise of so-called baby bod celebrity photos and endless how-to articles in the popular press, it's gotten to the point that such women expect to be thin *even when they're pregnant.* And if a woman does gain body fat during her pregnancy, she's expected to return to her pre-pregnancy weight as soon as possible after giving birth. This isn't what doctors are recommending to their patients. It's what the media is telling women. It's what women are telling other women. It's clear that a high-status pregnant body is thin, and a high-status breast-feeding body is thin, and everyone—high and low, rich and middle class—is scrambling to catch up.

Some amount of diet awareness is good to prevent pregnancy-related obesity and gestational diabetes. But when it comes to babies, maternal dieting is generally terrible—and that much more terrible if the mother doesn't have a lot of excess fat to start with.

As we learned in the "Brain" chapter, brain tissue uses the most energy, pound for pound, of any tissue in the human body. And it's fairly fragile stuff. When you starve it, the effects are drastic. If you've ever been "hangry," you know how food influences something as simple as your mood. If you've ever dieted for a while, you've probably also experienced the classic dieter's "brain fog," where everything seems to move a little more slowly. Conversations are murky. Problems can seem impenetrable.

And that's a too-hungry brain that's *already* built. For a fetus, and the child who comes after, malnutrition is an undeniable force—destructive, long lasting, and in some cases irreversible. Poor nutrition in early childhood is famously linked to lower IQ, even when you control for the mother's IQ. Behavioral outcomes suffer, too. Malnourished babies tend to become adolescents who have difficulty with self-control, long-term planning, violent impulses, and other social aggression. Malnourished mothers are far more likely to have malnourished babies and are also likely to give birth to underweight newborns—another factor in lowered IQ and stunted cognitive development—and/or they give birth before their due date, yet another proven factor that compromises baby brains. These kids meet their cognitive benchmarks later in life and tend to score lower overall in math, in spatial reasoning, in language. Every way you test it, in other words, screwing around with women's food and reproductive health tends to make everyone in the local culture a bit less intelligent. Not because we're genetically predetermined to be that way, but because we're *starved* into it.

So that's the first way sexism makes us less wise—across the span of human cultures, time and again, sexism puts us in danger of starving the very brains we build in the womb and early childhood. If you want a culture to produce smart children, then you have to take care of maternal and childhood nutrition.

But feeding a growing brain isn't the only thing that influences its potential. There's also the matter of how it learns. We know that brains assemble themselves as they grow: building crucial networks, learning social norms, paving shortcuts for language and math and problem solving of all sorts. When a growing human brain is *neglected*, it's probably not going to reach its full intellectual potential. Over time, a brain can easily learn that it doesn't need to be "smart"—or worse, that it isn't "supposed" to be smart—and to some degree will build itself accordingly.

So let's turn back to the costs of a sexist girlhood. Females are very close to half of all the world's newborns. But girl brains are far less likely to be formally educated. And when they are, they're far less likely to continue that education past the age of ten. When they do manage to go to school, their education is frequently cut short by early marriage, or a parent's decision that a daughter's education is less important than educating a son. While this is clearly a sexist choice, it's not an illogical one, given that formal education is not free in the majority of the world and poor families have to choose which child to invest in. If a daughter's education doesn't seem to have obvious returns, it's only logical that girls would be the ones to be pulled out of school. It's incredibly hard to persuade parents faced with such a choice that investing in girls' education will make *everyone* in the community smarter and wealthier someday. Many of those parents are dealing with more immediate problems.

Still, the difficulties these families face don't make it any less true: wide disparities in childhood education between boys and girls cripple the future workforce. More studies than I could possibly list support this idea. Not only is half the population significantly less educated than it should be in places like Niger or Mali, but because future *mothers* are neglected when it comes to education, those mothers' ability to fully support their future children's education is also compromised.

Let's switch gears for a minute. This isn't just about having communities in places such as Niger and Mali and rural India "catch up" to more egalitarian societies. There's also a ton of evidence that more sex-egalitarian education tends to be associated with the golden ages of human civilizations in our past. Our societies seem to be at our *best*, in other words, when we're educating girls.

One well-studied example is the history of Islam in the Middle East, Africa, and Europe. By many measures, medieval Islamic societies were more gender equal than the Arab world today. In fact, the Prophet Muhammad's first wife, Khadija—famously his most beloved—was older than him, twice widowed, already had children, and was a widely respected businesswoman when he met her.* In the twelfth century, the Islamic philosopher Ibn Rushd (Averroes) wrote that women should be considered equal to men in all respects, including education and opportunities for employment.

Remember, this is the Middle Ages. At the time, Islam wasn't just more egalitarian than European societies. It was also more *intellectually* productive. Because Muslims believed reading the Quran was vital for the soul, these societies expected all children, male and female, to be literate and well educated—not just in the Quran, but in a range of topics they found valuable: visual arts, mathematics, the sciences, even music. Public education was both available and well funded. Public schooling didn't take hold among the Christians of Europe and North America until the Industrial Revolution. If you were a child born between 1100 and 1400, you definitely wanted to be born in an Islamic society, whether you were male or female.

The payoffs were enormous. Islam's golden age produced algebra, chemistry, the magnetic compass, better modes of navi-

* According to the Quran, the Prophet Muhammad met her while she was his *employer*, and it was her idea to propose marriage, not his. He also refused to take on a second wife while she was still alive, quite contrary to local custom for any man who could afford more than one wife—and he could, largely because of *her* wealth and business connections, which were instrumental in the early spread of Islam. To put it in modern terms, Khadija wasn't just Muhammad's wife. She was Islam's angel investor.

gation, and all sorts of advancements in medicine and biology. While Europe was busy telling itself the plague was caused by an evil fog, Islamic doctors had already figured out that copper and silver instruments were best for surgery (the metals are antimicrobial). Philosophy also flourished, with new ideas about humane government and social interdependence, many of which directly influenced the rise of the European Enlightenment. The golden age of Islam, in other words, produced one of the most intellectual, egalitarian, cosmopolitan, and profoundly influential societies of its time. And women were right there at the fore, contributing to its success.*

This isn't to say the only reason civilizations falter is that sexism rears its ugly head. Many factors contributed to the decline of Islamic nations, colonialism not least of them. And money is certainly a factor in whether a civilization is likely to be intellectually productive. (Golden ages are called "golden" for a reason.) But as of 1989, many Arab nations had become incredibly wealthy and yet managed to produce only 4 frequently cited scientific papers. The United States, by contrast, produced 10,481. Why? For one, they'd systematically cut off education for half their population. Roughly sixty-five million adult Arab people are illiterate right now, of which two-thirds are women.† Many of those women live in wealthy countries, such as Iran and Saudi Arabia. Places that, once upon a time, shone with the brightest lights in human intellectual progress. But we'll never know which of these women

* The slow decline of that civilization also happened to start when Islam absorbed Byzantium and became more influenced by *Western* thought, including the increased seclusion of women and girls, so popular in Persia, and the de-emphasis of the importance of education and "worldliness" of anyone who happened to be female (Ahmed, 1986).

† Hammoud, 2006. Don't blame these nations exclusively: two-thirds of *all* illiterate adults in the world are women, according to a UN study from 2015 (United Nations, 2015). It's true that sub-Saharan Africa and parts of the Middle East are driving the numbers, given how many literate women live in other parts of the world. The data here on the Arab regions come from data and reports from 2002 and 2006 (Hammoud, 2006). It's worth noting that the illiteracy rates for women aged fifteen to twenty-four in Jordan and Bahrain are almost nonexistent—this is, in many ways, a generational problem (ibid.).

could have been a modern Khadija. We'll never meet their Marie Curie, their Ada Lovelace. Whatever contributions those women and girls might have made have been sacrificed to the symbolic function of their modesty. Unless, of course, they escape these more restrictive communities and get the support they need elsewhere—but what if they can't afford to?

Where women are undereducated, entire societies eventually go fallow. If history proves right, neglecting girls' education is a sign of a civilization's decline. You can neglect half the brains in your community for only so long.

ASTEROIDS AND ASSHOLES

So we evolved to be sexist. Maybe we're all tweaking and defending the base settings on our cultural switchboards because, once upon a time, sex rules helped us overcome our lousy reproductive systems. If that's true, maybe it really is too much to ask Americans to stop caring about whether a celebrity "stole" her husband from another woman. We can demand, out of a sense of fairness, that the standards become more equal. We can deliberately change American sex rules. But we can't ask people not to care about that change. Sex rules are built into our cultural identity. Those rules used to help us survive.

In part, that's because sharing and enforcing sex rules isn't just about making us more competitive baby makers. It's also useful to be the same *sort* of sexist as the people around you. Sharing cultural rules helps trick the human brain into thinking your neighbor is your sister.

Call it a primate hack. Social primates are pretty good at extending "kinship" behavior, which is why it's possible to have a group of 150 baboons in a troop, or 100 bonobos, or 800 geladas, even though many members of the group may not be immediate kin. That's also why it's possible to have such a thing as a human nation. The fact that humanity could even *conceive* of something like the "United Nations" is precisely because we're social primates. Human beings, much like the bonobos, have a long evolutionary history of finding hacks to make their brains *care* about people who aren't relatives. It's one of the coolest things we're able to do.

So it's not quite accurate to say that loving another person is the *best* thing that human beings do. Maybe it's how we're able to love our not-sisters in the way we love our sisters. That might be our best thing. The urge to protect others' children, because most of us have an urge to protect children in general. The ability to recognize our common humanity and value it. That's the best human thing—it's the way we took "primate" and made it better.

One of the ways humans make this happen is by telling each other stories about ourselves—stories that create odd ideas like "I am a citizen." Buttressing those stories is our shared switchboard of cultural norms: the things cultures do in common that help everyone signal to one another, "We belong here." Generally speaking, the more common the switchboard, the stronger a local culture becomes. That's a lot of what sociologists mean when they talk about "social cohesion"—it's what happens when all of the common features of the switchboard, and all of the common stories, together build this crazy human thing we call cultural identity. It's the main reason we don't dissolve into mutually warring family clans—we usefully trick ourselves into thinking people who aren't related to us are actually our kin.

It's not *just* humanity's crappy reproductive system driving sexism, in other words. It's also our deep social drive.

It's hard to pit two of your most valuable and unique behaviors against each other. Though its evolutionary roots run deep, gynecology is uniquely human. So is our kinship behavior. Shared social rules are one of the main ways cultures build extendable identity. And sharing sex rules—not just being sexist, but being sexist in the same *way* as other members of our cultural group—is one of the big ways we reinforce group membership. We like the feeling of being with people who "share our values."

Conservative American Christians, for instance, use their sex rules to help signal to one another that they all believe the same things about the world, that they all belong—whether that means acting in a way that's less welcoming of their non-Christian neighbors or extending their group membership to include Christians in distinctly non-Anglo parts of the world. Sex rules can also be a way into a group you wouldn't otherwise be a part of: Promoting gay marriage has found a foothold in some Christian communities that would never condone a promiscuous homosexual love life. "After all," many Christians say, "they're being monogamous and raising babies. We feel strongly about that. Maybe we can bend this one rule—against sex with the same gender—and include them."

Getting rid of sexism is hard. Maybe even impossible. But we have to try because, frankly, it just doesn't work anymore. Or at least not how it used to.

While sexism continues to serve as a force for local social cohesion, it also drives social *fission* between different cultural groups.* Most of us don't live in small cities anymore. Human culture has become global. Conflicts in other parts of the globe have *far* greater costs than they used to. When you rip a hijab off a woman's head somewhere in France, the story immediately fuels massive amounts of rage in the Middle East—rage that drives extremists' agendas. When ISIS rapes little girls under the false mantle of religion, the rest of the world becomes outraged—as well we should. Yet we don't get *nearly* outraged enough when countries deny contraceptive services to their female citizens. Not even when that denial keeps those women poor, which fuels social unrest, which leaves entire populations vulnerable.

The history of feminism—which is to say, the history of tension between individual female reproductive choice and collective strategies for reproduction—is certainly as old as our species. Feminism is, at a minimum, 300,000 years old. But we're just now arriving at an understanding of the true history of our species—finally starting to piece together what it really means to be "human," what it means to be a "woman," what the history of ourselves really involves, across a time span that's far longer than

* Signaling us versus them, especially. Do you or do you not wear the hijab in Uzbekistan? Will you or won't you let your immigrant parents arrange your marriage in the United States?

our mythic origin stories allowed for or even could have imagined. Armed with this understanding, we now get to decide, as a species, how we want to proceed. We get to *choose* how to balance individual reproductive choice with collective reproduction.

As with all things, we're probably going to head in a thousand different directions at once. That's fine. No human culture is any less evolved than another—by definition, every human being alive today is equally modern. And, in essence, every culture is a kind of *experiment* to figure out what works for us, in our given environment, for our particular needs. Some of these experiments work. Most don't.

We could use some guidelines. For example, while eradicating sexism seems pretty impossible, we *can* become more deliberate about the choices we make around sex rules. We can actively choose to create social institutions that combat the negative effects of sexism. We can reinforce the need to be more egalitarian. And above all, we can choose to support and defend the advance of gynecology.

Because while innovations on human culture are created by randomness—environmental pressures, local mutations, individual decisions that get adopted or nixed—it's not actually true that human cultures develop in entirely random directions. For example, once your culture's gynecological knowledge and traditions reach a certain level of effectiveness, they rapidly outstrip sexism's usefulness. And when there's finally *enough* gynecology, like safe contraception and abortion and proper prenatal and postnatal care, but there's still a lot of sexism, being sexist can even *undermine* gynecology. Screwing up women and children's health inevitably hinders the population doing it.

Time and time again, throughout history, sexism wanes and gynecology rises again. Despite today's sexist backlashes, I still think we're moving irresistibly toward our species' collective future: one of true egalitarianism between the sexes, supported by better and better gynecological medicine. We're taking control of our reproductive systems. We're *deciding* how we want to become pregnant, and when, and with whom, and we're going to have a more even distribution between the sexes when it comes to child care. It's not that men will start breast-feeding, but the sheer number of hours and labor and money required to raise children is going to be spread more evenly across the population.

We're escaping our evolutionary destiny, in other words. And we're doing it by being *human:* being smart, collaborative problem solvers who tell each other stories and revise those stories to make better ones.

But such progress (for lack of a better word) is always fragile, and right now two fundamental things are standing in the way: asteroids and assholes.

I mean that both literally and metaphorically: if something like an asteroid hits, and it's big enough, it could wipe out our entire species before we ever stop being sexist. Similar things have happened before. In fact, over and over again in human history, massive events that killed off huge numbers of human beings dramatically altered human history. Something knocked out a good chunk of the world's hominins about eighty thousand years ago—could have been a super volcano, could have been climate change in general, could have been a particularly bad cluster of things—so we had to leave Africa *twice* to become a global species. Massive global cooling wasn't all that great for human advance in northern Asia, either—we got farming in the Middle East, not Moscow. Fast-forward to the Middle Ages, when the Black Death killed a *third* of the population of Europe. Some say that's why Europe went through the Dark Ages, while the Islamic empires managed to flourish. Even the Spanish flu in 1918 had a sinister legacy: though it might have helped tip the war in the Allies' favor—imported from Kansas, of all places, it rapidly infected German troops on the other side of the trenches—it *also* left Germany that much more devastated after the war, ripe for the rise of resentful populism and, eventually, fascism.

Granted, massive death isn't always *all* bad: some say Europeans developed the premodern middle class because so many poor people died during the Black Death, upending the social structures that had reinforced feudalism. The result: the Enlightenment, the Reformation, the rise of the premodern. Similar things have been said about second-wave feminism in the United States—that if it weren't for the radical absence of young men during World War II, American women might have taken quite a lot longer to get used to the idea that working outside the home was an acceptable and useful thing to do.

But deliberately killing huge portions of the population, besides being immoral, doesn't necessarily produce greater freedom and more egalitarian societies in the long run. What's more, we can't very well control for the size of asteroids. There are random, catastrophic events in human history, and there will inevitably be *more* random, catastrophic events in our human future.

So that's our asteroid problem:* massive, outside events we can't control that can wipe us out or set us back hundreds of years, if not millennia. Cultures tend to respond to stress and threat with stronger adherence to local cultural identity, a kind of behavioral banding together to ride out a storm. Sexist rules are a part of every culture's local identity. Shifting *back* to older settings—things that feel safer, maybe, or more "proven," or at the very least more familiar than more recent tweaks—is a more likely outcome of asteroids than female freedom.

But it's not just unforeseen large-scale disasters we have to deal with. There's also the quintessential asshole. And not just the Hitlers, or Pol Pots, or Assads, or even the less overtly murderous types, like the Trumps. There's the consistent problem of *everyday* assholes. Enough of them, at the right time, under the right conditions, can have extraordinary influence over a civilization's progress. In India, like in so many places in the world, there's widespread corruption; many national governments run on the principles of organized crime. It keeps large parts of their populations poor. It degrades public confidence in the criminal justice system. And for the most part, it's not just a matter of higher-up government employees being on the take. (America has those, too.) It's a matter of whether you need to bribe your mailman. Or the guy who's in charge of local sewage. Or your neighbor. Or the police on a highway on your way to work. Or the guy who makes sure your area even *has* a functioning, regularly maintained high-

* Many of these are called black swan events, but not all our metaphorical and literal asteroids are actually so unpredictable, nor are they all so sudden. For instance, if we don't get a handle on climate change stat, it will—very predictably—destroy much of what we currently understand as "modern human life" on this planet.

way. The big assholes do enormous damage, but it's the little assholes that chip away at every citizen's confidence that they can *rely* on other people to do what needs doing to make a country work.

Though some of these examples shift when you're talking about places like America, that sort of thing happens here, too.* And when you don't feel that you can rely on big institutions, you fall back on your immediate family. Your friends. Your village. And, yes, the features of your local identity that keep those groups strongly bonded together. Including your local sex rules.

The more clannish you become, the more *local* and rigid your identity gets, the more you fall back on short-term planning, the more corruption spreads, the more institutions *break down*—weakened by a lack of funding and a lack of public confidence—the more vulnerable you are to the big assholes. The world-changing assholes. The demagogues. The autocrats.

The monsters.

Monsters don't have a very good record of bolstering human progress. Monsters who are given real social power usually set us back, not just through death and destruction and widespread despair, not just by bringing out the *worst* in human nature, but because recovering from monsters after they die is really hard. Cambodia still hasn't recovered from Pol Pot. Khomeini set back Iran in many ways—not just for women. Assad is going to die one day, more than likely safe and warm in a bed, tucked into sheets with a really high thread count. But Aleppo? All of Syria? They won't recover in our lifetime. They may not recover at all. Because all these beautiful institutions we build are fragile. Unless we work, together, *collectively*, to reinforce them, we'll lose them to any given asteroid or asshole.

So really, when I think how to answer that question at the start of this chapter, it seems to me that loving someone isn't the

* Many analysts believe the rise of right-wing extremist groups in the United States isn't simply a pushback against the success of liberal social inclusion but in fact a symptom of the deepening crisis borne by consistent disappointment with local governance. The causes are deep and wide, but some are fairly obvious: if you don't believe contacting city officials will actually result in ever repairing the pothole in your neighborhood, and you *know* those same officials repair the roads in front of their own houses, your trust in democracy is inevitably going to falter.

best thing a woman can do. The best thing any human being can do requires all of our uniquely human traits: an amalgam of our extended kinship behavior, narrative building, and problem solving. The best thing we do is create *institutions* to support and protect those fragile extended bonds. And those institutions, like them or not, are precisely what allow us to overcome our less desirable behaviors: territoriality, sexism, competition for dominance. They are the way we push beyond the limitations of our bodies' evolution. They are the means by which we become truly free.

I don't know if I could explain any of this to the madam who tried to pimp me. I wouldn't even know how to find her now, though she's probably alive, still running the same business in that little industrial park on the outskirts of town. She's probably a fairly intelligent person. It's not *easy* to run an illegal brothel with high-end clientele. If I ever do see her again, and try to tell her any of this, it's not that she *couldn't* understand. I just don't know if she'd care.

She might care about Assad—I think her family was from that part of the world. They still had a house, she told me, on a little Mediterranean island somewhere. I think she told me that because that's how they recruit their workers: they make girls believe their lives could be beautiful, if only . . . For American girls, the idea of a house on a little island is beautiful. It's so far away. It's sunlit. It's warm. It's the opposite of your life in an industrial park with gray doors you thought you were knocking on to get a job answering the phone.

But could I make her see that what she did when she tried to buy my body has a 200-*million*-year evolutionary history? That the moment she met me is part of that same melody—a weird little trill that goes all the way back to the dinosaurs—but also that it's not the *only* story of womanhood? That women used to be matriarchs. That our ancient grandmothers were a huge part of how we invented human culture. That women's mouths are the root of human language. How could I tell her that the very shriveled breasts she'd tucked into her old, stretched-out midwestern bra are part of how mammals took over the earth, the reason we have immune systems that can survive pandemics, the *reason* most of the world she's ever managed to see looks the way it does?

I wish I could tell her that it wasn't always like this. That a woman's world is bigger than the equation she'd figured out running her brothel. And older, and weirder, and more beautiful. I don't think I'd try to stop her. I wouldn't try to tell her she's not supposed to do what she does. But I think I would tell her to donate part of her money to women's health clinics. To children's hospitals. To research. To whatever will make the world easier for women and girls. And I wish I could tell her, as I will tell my own children someday, that every power men have ever had over women is something we *gave* them. We just forgot.

We forgot we can stop.

ACKNOWLEDGMENTS

This book wouldn't have been possible without the love and support and astonishing patience of my friends and family and professional communities. Among the many who have my eternal gratitude: My husband, Kayur, who supported me for years during my PhD and book writing. My brother, John, who repeatedly hauled me up from the depths in those initial drafts of the chapters. My editors, Andrew and Anne—whose work here can only be called heroic—who also usefully reminded me that not all my jokes can land. My enormously talented agent, Elyse, who has held my hand through nearly a decade of my life now. My advisors and mentors at Columbia, who somehow believed I'd still finish my dissertation despite getting a book contract at the same time I passed my quals, and continued to believe as the years stretched on, and continued to believe as even I lost faith, right up to the moment of my defense, and even said very nice things that I surely didn't quite deserve. My children, who are somehow still alive and even manage to like me, some of the time. But most importantly, every single scientist whose work is represented in these pages—their labs, their toil, their lost sleep, their grant proposals, their endless reanalysis of data, their internecine squabbles and tenure gauntlets and conference awkwardness and submissions and revisions and little victories and years and years and years of stubborn resistance to the perfectly reasonable urge to give up . . . we owe them so much. We would know essentially nothing about the biology of sex differences without their efforts. The tide is changing by the sheer force of their will.

NOTES

INTRODUCTION

3 "We did this": Rich, 1978.

3 a gruesome scene: Scott, 2012.

3 evidence that they may affect the sexes differently: Eid, Gobinath, and Galea, 2019; Sramek, Murphy, and Cutler, 2016; LeGates, Kvarta, and Thompson, 2019.

3 Prescriptions for pain medications: Mogil, 2020. Note that although the underlying mechanisms driving these differences are likely complex, ranging from how we process these drugs in the digestive system and the liver to how our nervous system responds, it may also be tied to sex differences in how our nerves process pain in general (Ray et al., 2022).

3 Women are more likely to die of heart attacks: Shehab et al., 2013; Shaw et al., 2008. Some of the disparity here can be attributed to age—women generally present in the clinic at older ages than men with these issues, and being old is itself a risk—but rates are increasing in younger women, too (Mozaffarian et al., 2015).

4 women and their doctors alike: McSweeney et al., 2016. For at least some good news on these issues as well as remaining problem areas, see the 2010 report from the U.S. Institute of Medicine Committee on Women's Health Research.

4 Anesthetics in surgery: Buchanan, Myles, and Cicuttini, 2009.

4 treatments for Alzheimer's: Ferretti et al., 2018.

4 public education curricula: van Hek, Buchmann, and Kraaykamp, 2019.

4 just the same as men's: For a recent review of sex and gender influences on health, which (rare for such efforts) actually considers *both* biological sex differences and important issues around the social construct of gender, see Mauvais-Jarvis et al., 2020.

4 more than 79 percent of the animal studies: Mogil and Chanda, 2005; Beery and Zucker, 2011.

5 what gets studied in the lab: Beery and Zucker, 2011; Hayden, 2010; Wald and Wu, 2010. Even the esteemed Francis Collins said as much in

2014, in a rather public effort to try to turn the tide, but sadly the number on the dial hasn't moved nearly enough (Clayton and Collins, 2014).

6 considered difficult and expensive: Prendergast, Onishi, and Zucker, 2014.

6 for mouse and human alike: There are simply too many papers to cite regarding the depth of sex differences in our mammalian body plan, but, for one recent example, Oliva finds sex differences in cellular gene expression present in numerous different tissue types in the human body (Oliva et al., 2020).

7 NIH clinical drug trials: Mazure and Jones, 2015; Heinrich, 2000.

7 move from clinical trial to market: English, Lebovitz, and Griffin, 2010.

8 psychotropic drugs than men: Simoni-Wastila, 2000; Serdarevic, Striley, and Cottler, 2017; Darnall, Stacey, and Choi, 2012; Herzog et al., 2019.

8 "jimmied" for a female patient: One important caveat on the pain med claim: the FDA reports that women are more likely to be given pain meds and at higher doses and for a longer duration than men. However, it may take longer for them to get it; post-surgery, men are more likely to be prescribed pain meds, and women are more likely to be given sedatives and/or Tylenol (Hoffman and Tarzian, 2001; Miaskowski, 1997).

8 for example, OxyContin: More specifically, they failed to test for sex differences in reproductive-age women, and they failed to properly design an experiment to test for sex differences at any age. We do know they conducted one clinical trial on about 130 mostly elderly patients with osteoarthritis, 76 percent of whom were female—perhaps not surprising, given the older (and thereby more likely to be female) population with this disease. The degree to which the pain was reduced was somewhat better than placebo over the fourteen-day trial, with the 20 mg dose better than the 10 mg dose. The degree of effect was not analyzed for sex differences, though the adverse drug effect was analyzed for sex differences, showing nothing significant. (Note how much larger the female signal would have been in the data here: with 76 percent of a meager 130+ mostly elderly people, it's going to be hard to tease out what the average male signal might be at varying ages.) Also worth noting: nearly a third (32 percent) of subjects who received the 20 mg dose dropped out due to side effects. But the study nonetheless recommended the 20 mg dose as the most effective one.

The reason we know any of this, of course, is that legal documents made the clinical trial information public. We do not have access to the clinical trial data, nor the earlier rodent studies, in the development of the original drug (oxycodone) that OxyContin is made from, but it's safe to assume, given that it was so long ago, that female patients of reproductive age were unlikely to be a significant portion of that study population. As of 2022, anyone in the public can access information on the Oxycontin clinical trial here: www.documentcloud.org/documents/6562785-21-Purdue-Docs-1-20-to-29.html.

8 released in 1996: The context of the OxyContin study is also worth noting. The company very specifically wanted to get this drug out into the non-cancer population and make it a common prescription. So the fact that the study is specifically studying an older chronic pain population who specifically did not have cancer isn't just a matter of cleaning up the science for confounds; it's literally whom they wanted to sell this stuff to (Chakradhar and Ross, 2019).

8 endometriosis and uterine-related pain: Freire et al., 2017; Lamvu et al., 2019.

8 put on methadone: Jones et al., 2010.

8 born addicted to opiates had tripled: Ko et al., 2016.

9 That number is still on the rise: Hirai et al., 2021; Mossabeb and Sowti, 2021.

9 many mothers didn't realize: Patrick et al., 2020.

9 wake up faster than men: Gan et al., 1999; Mencke et al., 2000; Kreuer et al., 2003; Sarton et al., 2000; Buchanan, Myles, and Cicuttini, 2009.

10 from the same ancient organ: Parra-Peralbo et al., 2021.

10 an article about liposuction: Kolata, 2011. Kolata was reporting on Hernandez et al., 2011. While the *Times*'s illustration mostly demonstrates larger arms, the study itself showed "redistribution" to the abdomen with only some redistribution to the upper arms, while the thighs remained smaller. (The text is more representative than the illustration, all credit to Kolata there!) All thirty-two subjects of the study were female and roughly in their mid-thirties—suitably premenopausal, in other words, but not post-reproductive in the classic sense. The study doesn't mention whether these women had children already, nor how long it had been since those potential births, nor whether they were going to have any afterward. One assumes, at least, that they weren't pregnant or breastfeeding during the period of the study, because surely that would have ruled them out in terms of subjects' risk profiles for surgery—not to mention massive metabolic confounds for interpreting any data.

11 chock-full of unusual lipids: Phinney et al., 1994.

11 most of them from our diet: Cunnane and Crawford, 2003, 2014. Cunnane is one figure in an ongoing scientific debate about whether our brainier Eves could have gotten enough of these lipids from terrestrial foods or aquatic sources (Carlson does a fun rebuttal in Carlson and Kingston, 2007). Cunnane is convinced our ancestors got brainy along the shore, but which particular shoreline might matter (Joordens et al., 2014); the fossil record shows brainy Eves in both near-water and more landlocked terrain; and it can be hard to tell in the fossil record whether things like catfish and turtles were part of the diet, because not all water

food leaves an easy trace in the rock with evidence of butchering (Braun et al., 2010). I admit I care less about the aquatic story than I do about our presently fat behinds; wherever they originally evolved, the special features of these depots in the modern human body make a compelling case for leaving them well enough alone.

11 female gluteofemoral fat resists: Rebuffé-Scrive et al., 1985; Rebuffé-Scrive, 1987; Guo, Johnson, and Jensen, 1997; Karastergiou et al., 2012; White and Tchoukalova, 2014.

11 building blocks for human babies' big brains: Lassek and Gaulin, 2008; Haggarty, 2004.

To be clear, no one would say this is *all* women's butt fat is for. For instance, mice show signs of prediabetes when you liposuction their rumps, which may imply some lower-body fat depots in mammals initially evolved as a way to buffer the body's general homeostasis against variable diets, with rare-lipid functionality as a later feature (Cox-York et al., 2015). The notion that different fat depots function differently is hardly controversial, with evidence even at the level of gene expression (Rehrer et al., 2012).

12 when she'll get her first period: Fredriks et al., 2005; Walker et al., 2006; Lassek and Gaulin, 2007. But there's an important caveat here: lots of things can trigger menarche, not just butt fat. For example, even outside potential genetic quirks—women with early periods tend to have children with early periods—continual exposure to things that tend to speed up growth in children, like slightly raised levels of cortisol, may be a contributing factor all on its own. Take Freedman et al.'s 2002 study of U.S. girls with a cohort divided by racial background—Black girls tend to get their periods sooner than white girls in the United States, which remains statistically significant even when controlling for height and weight and body fat percentage. Considering the well-known differences in cardiovascular disease among these populations later in life, which is sometimes attributed to the "wear and tear" pattern associated with continual stress, the early onset of menarche among Black American girls may be yet another physiological cost that comes with exposure to the stress of lifelong racism.

12 190,000 women undergo liposuction: American Society of Plastic Surgeons, 2021.

12 violent disruption of women's tissue: Seretis et al., 2015.

14 all three share so many properties: Parra-Peralbo et al., 2021.

20 "any large-scale picturing of women": Leibovitz and Sontag, 2000.

21 The Eves: All of the foundational papers on the Eves, whether exemplar or presumed true ancestors, will be cited in the individual chapters they're associated with. But this overview of the Eves is greatly in debt to both Dr. Advait Jukar, who did the enormous job of helping me find and select appropriate species for each trait and their associated ancient

environments, and the fantastic work of the Smithsonian's Human Origins Initiative. One thing that's particularly wonderful in the Smithsonian's work is how they offer simple lists of what's known and what's still unknown about our hominin Eves. You can see these for yourself online at humanorigins.si.edu.

CHAPTER 1 MILK

27 "No sooner had the notion": Rimbaud, 2011. "After the Flood," from 1886, was inspired by Genesis 9. I'm translating here from the original French, with a heavy debt to Clive Scott (Scott, 2006) and John Ashbery.

27 *Got Milk?:* The Got Milk campaign was massively successful, in no small part due to the extensive (and exclusive) photography of celebrities with milk mustaches taken by Annie Leibovitz. Her then-partner, Susan Sontag, even tagged along for one session so she could "meet" Kermit the Frog (Hogya and Taibi, 2002, in Daddona, 2018). The campaign was originally conceived in 1993 by an advertising firm working for the California Milk Processor Board, then overtaken by the Milk Processor Education Program, then very clearly in possession of the American public's collective mind in the 1990s (Daddona, 2018).

27 no bigger than a human thumb: Slater, 2013. Body masses provided in the supplement. For a nicely written overview of many features that I include in the Morgie scene, see Brusatte and Luo, 2016.

27 We call her Morgie: Kermack, Mussett, and Rigney, 1973; Kielan-Jaworowska, Cifelli, and Luo, 2005. Some of the nicknames I use for the Eves in this book are common in the paleo community already—the Smithsonian, for example, has called *Morganucodon oehleri* "Morgie" in its Behring Hall of Mammals, as does the National Museum of Wales in Cardiff—but some are done to personalize them. *M. watsonii* is the initial discovery in Wales in 1947, but the entire genus of Morganucodon is what I'm using for "Morgie."

27 all the little clocks in her clockwork: Liu et al., 1997; Gerkema et al., 2013; Borges et al., 2018; Morin and Allen, 2006.

28 heard them before she saw them: Grothe and Pecka, 2014.

28 sweet crunch of its chitinous body: Gill et al., 2014.

28 crawled down her tunnel like a lizard: See Luo, 2007, for a well-written overview of the increasing evidence for diversity in early mammaliaforms, including traits like burrowing. Most assume, however, that Morgie had burrows and, alongside other early mammal types, a sprawling pelvis.

29 Jurassic beasts: Carrano and Sampson, 2004.

29 early Eves evolved underfoot: Luo, 2007; Gill et al., 2014. In fact, there were all sorts of mammal-like beasties of varying size, even in the Jurassic and prior; some of the earliest direct evidence we have for mammalian fur comes from what appears to be a medium-bodied, otter-like creature

(Ji et al., 2006). But the idea that evolving "underfoot" (that is, in niches good for relatively small bodies) was a useful strategy for early mammals like Morgie is common thought in evolutionary biology.

29 Her fossils have been found: Kielan-Jaworowska, Cifelli, and Luo, 2005.

29 inordinate fondness for beetles: Gould, 1992. This quotation is often misattributed to Charles Darwin, who likely felt similarly (with perhaps greater belief in a Christian God), but has never been confirmed to have said as much. As for Haldane, the exact phrase may be apocryphal, but his friend Kenneth Kermack provided the following, which seems appropriate to include here: "[Haldane actually said,] 'God has an inordinate fondness for stars and beetles.' . . . Haldane was making a theological point: God is most likely to take trouble over reproducing his own image, and his 400,000 attempts at the perfect beetle contrast with his slipshod creation of man. When we meet the Almighty face to face he will resemble a beetle (or a star) and not Dr. Carey [the archbishop of Canterbury]" (Gould, 1993). Kermack and his wife were also instrumental in advancing our knowledge of Morgie.

30 Morgie nursed her young: Benoit, Manger, and Rubidge, 2016.

30 newborns are 75 percent: Shubin, 2013. Actually, the range seems to be 73 to 78 percent, depending which pediatrician you ask—my sister-in-law is a hospitalist pediatrician, so I had occasion to—but 75 percent seemed a suitable compromise. The reason newborns are so structurally wet is largely that their limbs are so very short and flimsy. Despite being born nearly as fat as a baby seal, the average human newborn's body is mostly chubby torso and big, fat head. At the tissue level, human lungs are roughly 83 percent water, while muscles and kidneys are about 79 percent, and the brain is about 73 percent (Mitchell et al., 1945). The reason most adult humans are only 60 percent water is that our childhoods and puberty built a lot of new bone, muscle, and fat. The average adult's legs, for example, from ankle to outer hip bone, make up nearly half of that person's height—another one of those things beginning art students tend to get wrong before they take a figure class. More on what we get wrong when our eyes look at bodies in the "Perception" chapter.

30 carefully ordered molecules of water: Khesbak et al., 2011.

31 breast milk is almost 90 percent water: Boquien, 2018. Note, however, that primate milk is especially watery, likely because we keep our babies so close for so long: primate mothers tend to have frequent nursing sessions "on demand," and the babies have lengthy juvenile periods. Each species' milk is tailored both to the baby's developmental plan and to the mother's caretaking pattern. If human milk weren't so watery, the mother's body would be rapidly stripped trying to keep up with our thirsty babies (Hinde and Milligan, 2011).

32 Morgie laid eggs: Hopson, 1973.

32 hers were soft and leathery: Stewart, 1997.

32 calcium-rich diets before reproducing: Larison, 2001. Not all such creatures live in places where enough calcium is readily available, however, and may store excess calcium in their leg bones, which is then depleted after egg laying (ibid.). This idea probably sounds familiar to readers who've been pregnant: there's long-standing evidence that human pregnancy leaches calcium from the mother's bones (Kovacs, 2001). Making babies, in other words, however one happens to initially house them, utilizes *every* part of the maternal body, not simply her reproductive organs. This is true across the animal kingdom.

32 Chickens in industrial egg-laying farms: Janson et al., 2001. Janson's study usefully qualifies this fact with the possibility that differing genetic lines can produce different outcomes, but did establish that at least in terms of frequency and consistency of egg laying, there's a significant correlation. Because chickens in industrial farms are manipulated to produce more eggs, more often than they would in wild-type conditions, the birds' long-evolved mechanisms for compensating for the calcium costs of egg production aren't quite up to the task.

33 dense in antimicrobial material: Oftedal, 2012; Griffiths, 1978. Mucus in general has ancient origins in our long war with microbes; there are good reasons our guts are lined with the stuff, and our respiratory passages, and a healthy wallop for our egg and birth canals, too (Bakshani et al., 2018).

33 a serious evolutionary boost: Oftedal, 2012; McClellan, Miller, and Hartmann, 2008.

34 Most mammals have this pattern: Hinde and Milligan, 2011.

34 Each of those fat globules: Harrison, 2004.

34 dense with immunoglobulins: Kunz et al., 1999.

34 what they called beestings: Colostrum was largely considered bad for children for centuries—an idea unfortunately owed to Aristotle (Yalom, 1997).

34 the first European textbook for pediatrics: Prühlen, 2007. Text was originally published in 1473. Quotation taken from the translation in Ruhräh, 1925.

34 geared toward a newborn's development: Hinde and Milligan, 2011. In humans, the act of breast-feeding within an infant's first hour of life is a powerful predictor of that child's risk of death (Boccolini et al., 2013). That's one of the reasons modern maternity wards now promote immediate breast-feeding after birth.

35 reliable laxative: Kunz et al., 1999.

35 Toll-like receptors learn: Carr et al., 2021.

36 preemie babies in the NICU: Underwood, 2013.

36 deter dangerous pathogens from adhering: Coppa et al., 2006; Kunz et al., 2000; Morrow et al., 2004. It's now widely accepted that milk oligosaccharides are "for" our commensal bacteria, in that they consume them, and also usefully work "against" our bacterial enemies in various ways,

which naturally creates a less competitive environment for bacteria we've evolved to host (Marcobal et al., 2010).

36 The 6′-sialyllactose that colostrum: Though its presence continues significantly beyond the colostrum stage, and how much the baby gets seems correlated with its cognitive development by eighteen months of age (Oliveros et al., 2021). No one knows precisely how or why—unlike other oligosaccharides, the metabolites of 6′-SL (specifically, sialic acid) seem to reach infants' brains, while others may work via the gut-brain axis or via the vagus nerve (ibid.). Other studies have demonstrated a connection between infants' gut microbiome and their cognitive development, which remains true to varying degrees in adulthood (particularly when associated with anxiety) (Foster and McVey Neufeld, 2013). NeuAc, the predominant form of sialic acid in humans, was approved as a food additive in the United States and China as of 2015, with the EU following in 2017, but its production remains incredibly inefficient (Zhang et al., 2019). The main thing to know here is that the human brain, at any stage, is deeply sensitive to our relationship with the local environment, and one of the most obvious sites of interaction lies in the digestive tract, constantly mediated by our many bacteria.

37 third-largest solid component of milk: Coppa et al., 1999.

37 make the whole system work: Probiotic bacteria have *also* been found in human breast milk, and likewise are found to be beneficial (Lara-Villoslada, 2007). Perhaps, then, one could think of some of those early colonizers of infant guts as coming with their own wagon of food and supplies, with some of them coming from the breast, some from the microbiome of the birth canal (Shao et al., 2019), and some from the placenta (Stinson et al., 2019, though controversial—see de Goffau, 2019). More on our birth-related microbes in the "Womb" chapter.

37 help these little patients gain weight: This is particularly true for preemie babies infected with dangerous gut bacteria, which is naturally a risk for those patients for a variety of reasons (Mowitz, Dukhovny, and Zupancic, 2018). For a well-written overview of the economic side of this topic, including the typical cost of Prolacta's supplement, see Pollack, 2015.

37 trying to create human-type oligosaccharides: Pollack, 2015.

38 Scientists are feverishly trying: Palsson et al., 2020; Xiao et al., 2018; Maessen et al., 2020. For fructose-type oligosaccharides (obviously easier to acquire than human milk derivatives) and applications for Crohn's, see Lindsay et al., 2006.

38 buy human breast milk on the black market: Easter and Freedman, 2020.

38 far less protein than cow's milk: Boquien, 2018. In fact, it seems to have some of the lowest protein content of all mammalian milk; for example, rat milk has roughly ten times as much protein, while human milk has significantly more cholesterol and LC-PUFAs than most mamma-

lian milk (ibid.). This is sensible: human infants have a much slower growth trajectory. And of the proteins our milk does have, as compared with macaque milk, they seem largely geared toward gut, immune, and brain development, which aligns pretty well with the human pattern for somatic growth; we simply have more development to do in these areas while we're breast-feeding, and our species' milk is tailored appropriately (Beck et al., 2015). So if human bodybuilders are hoping to grow particularly fat and brainy, with guts specially trained to drink *more* breast milk for a long time, they're certainly welcome to drink human milk. But given that developmental pathways are *timed*, it's doubtful any good would come of it.

39 giving birth isn't just when *you* reproduce: For more on the increasingly porous boundaries of the individual organism, I strongly recommend two books: Richard Dawkins, *The Extended Phenotype*—easily his most important work, despite how cranky he gets later in his career about overextensions of the phenotype—and Ed Yong, *I Contain Multitudes*, which is both wonderfully written and frankly just fun.

39 famously prone to *C. difficile* infections: Cammarota, Ianiro, and Gasbarrini, 2014.

40 You can see that responsiveness: Milligan and Bazinet, 2008.

40 human milk has the most: Newburg et al., 1999; Tao et al., 2011; Urashima et al., 2001; Urashima et al., 2012.

42 "mammary hairs": Oftedal, 2002.

42 "co-produced biological product": This phrase is drawn from Dr. Katie Hinde, a primate milk expert and director of the Comparative Lactation Lab at Arizona State University (though I originally met her at Harvard). I'm greatly indebted to Dr. Hinde's academic and public-facing work throughout this chapter.

43 Prolactin stimulates milk production: World Health Organization, 2009.

43 regulating salt balance: Dobolyi et al., 2020.

43 the more prolactin you have: Brody and Krüger, 2006. Unfortunately, it seems the relative increase is far greater after (heterosexual) intercourse than masturbation, which is great if you actually *have* an orgasm during sex, but for many women I'm sorry to say that is frequently not the case (Shirazi et al., 2018).

44 "Oxytocin makes you love your baby": Drewett, Bowen-Jones, and Dogterom, 1982.

44 "Oxytocin makes you love your man": Schneiderman, 2012. Higher levels of oxytocin in the initial stages of a relationship are predictive of the longevity of your relationship, assuming you're a college-age heterosexual person. Given that oxytocin is related to pair-bonding in other species, this isn't that much of a stretch. Causality is the question: Is it because you've got other mechanisms driving your attachment, and therefore you

produce more oxytocin, or is your feeling of attachment something that comes about as a result of having more oxytocin on board? One presumes feedback loops, but one little peptide does not make you fall in love.

44 "Monogamous men make more oxytocin": Scheele et al., 2012. Or rather, men who self-identify as being in a monogamous relationship with a woman tend to stand farther away from non-partner women when they shoot oxytocin spray up their noses. The lab interpreted this as meaning the men reduced their signal of "availability" and potential sexual interest for the non-partner women when their noses were full of oxytocin. In prairie voles—a largely monogamous species—oxytocin does promote pair-bonding among females, while a different molecule does the job in males (Insel et al., 2010). Perhaps the most intriguing theory for oxytocin as a behavioral modifier is the notion that it better *coordinates* the many different motor and sensory patterns needed for reproduction. Or at least that's true in the nematode—surely the most ancient user of oxytocin studied in the lab. Male *C. elegans* are worse at looking for, recognizing, and having sex with potential mates when they don't have the right receptors for their version of oxytocin (Garrison et al., 2012).

44 higher levels of oxytocin: Goodson et al., 2009; De Dreu et al., 2010.

44 hardly the angel of our better nature: De Dreu et al., 2010; Insel, 2010.

45 nursing mother might also experience: This is part of why breast-feeding is encouraged to cement mother-child bonding (and why nursing problems are considered a threat to that bonding process); another part, of course, is simply sexism—loads of mothers bond with their babies just fine, thank you, even if the breast-feeding thing doesn't work for that particular set of bodies. But it's true that the hormonal boost naturally involved in the nursing process can help it along. So long as the baby isn't unwittingly counteracting the process by mangling the poor nipple to a bloody mess (pain signals tend to ring a bit louder in the brain than oxytocin), nursing does help mammalian mothers feel bonded to their offspring, and vice versa.

45 Post-orgasm men and women: Cera et al., 2021. Please note that people do *all kinds* of things during and after orgasms, including laughing outright, so whatever normative claims are made about oxytocin and orgasms, one should assume it's happening in a wide mix of physiological and psychological states (Reinert and Simon, 2017).

46 Lactation scientists call this the "upsuck": Wilde, Prentice, and Peaker, 1995.

47 produce agents to fight the pathogen: Riskin et al., 2012.

47 nursing a child who's stressed: Gardner et al., 2017; Hinde et al., 2014.

47 can also work as an analgesic: Gray et al., 2002; Harrison et al., 2016. Note that this is a complex effect. Most studies are looking at infants breast-feeding, not simply drinking milk, so there are loads of confounds: skin-to-skin contact, pair-bonding with mother, olfaction, temperature,

sound of the mother's voice. Nevertheless, this is a widely accepted phenomenon: nursing makes babies feel less pain.

48 produces an analgesic effect: Drewnowski et al., 1992; Lewkowski et al., 2003. Note that the physiological pathways involved can backfire. For instance, there's some evidence that chronic stress greatly influences food type choice, and certain food types (namely, high fat and high carbohydrate), while down-regulating some of the physiological features of stress, can also create feedback loops that encourage those food choices as the body begins to rely on the food effects to feel less physical and psychological pain (Dallman et al., 2003).

48 the side of our face that is more expressive: Forrester et al., 2019.

48 other primates do this, too: Tomaszycki et al., 1998; Boulinguez-Ambroise, 2022. Interestingly, whether an infant is cradled on the left side also influences how likely that kid will grow up to be left-handed or ambidextrous—or at least that's true of baboons (ibid.).

48 This is true across many human cultures: Harris, 2010.

48 right hemisphere of the adult brain: Though it may depend on which type of emotional processing you're talking about, with cross-hemisphere involvement partially depending on valence (Killgore and Yurgelun-Todd, 2007). In other words, we use the right hemisphere quite a bit to process emotional faces, with some boost presumably given by cradling on the left (the right brain processes the left visual field thanks to the optic chiasm), but there's more going on than meets the eye (ahem).

49 baby personalities that are less risk seeking: Hinde et al., 2014.

49 Babies with low-cortisol milk: Ibid.

50 Higher-cortisol milk also tends to be protein heavy: Ibid.

50 lace a mother rat's drinking water: Casolini et al., 1997.

51 more likely to rate her child as "fearful" or timid: Glynn et al., 2007.

51 bottle-feeding their babies didn't describe them as fearful: Ibid.

51 mildly challenging environments inoculate: Crofton, Zhang, and Green, 2015.

51 her kids might likewise be more fearful: Hinde et al., 2014.

51 this may be a basic human behavior: Using your spit to clean a kid's pacifier is also common, and might even help lower the child's potential for allergies (Hesselmar, 2013). Having done this, however, I can also anecdotally report that this is a surefire way to make yourself sick with whatever the child has acquired from day care. Maintaining a unidirectional orientation of spit, barring whatever happens at the nipple, is better for the mother's health.

52 health of the large intestine's microbiome: Vitetta, Chen, and Clark, 2019.

52 There's a group of people: Hewlett, 1991.

53 historically known to have happened: Not all of these cases are good, of course. Male victims of concentration camps during World War II were

known to lactate after their rescue, presumably because starvation screws with the entire body, from glands to liver, and the glands recover faster than the liver; some men with advanced liver disease also begin to produce milk (Greenblatt, 1972; Diamond, 1995).

53 Trans women who want to breast-feed: Reisman and Goldstein, 2018; Wamboldt, 2021. Because of the obvious challenges of conducting a proper clinical trial for these issues, what exists in the scientific literature is a handful of case studies. But having personally interviewed a lactation consultant for trans women in Seattle, I've been told that there are far more cases in the field than there are reported in journals, and among clinicians who work with these populations, the Newman-Goldfarb protocol is the go-to, just as it would be for cisgender women who adopt and wish to breast-feed their new baby. What isn't clear is whether there are any unique risks for the trans population in adopting this treatment. More research is needed.

55 trans women taking heavy doses: de Blok et al., 2017. Importantly, this seems to happen primarily in the first six months of hormone therapy, and development is modest, typically resulting in breasts a bit smaller than a AAA cup size (ibid.). This is part of why "top surgery" is so frequently something trans women will choose, and shouldn't be seen as fundamentally any different from when a cisgender woman with smaller breasts might choose to have breast augmentation surgery. However, for both populations, such surgeries do not come without risk, including an increased chance of breast cancer (FDA, 2022).

55 breast augmentation surgery: American Society of Plastic Surgeons, 2021. The number of procedures actually lowered for the first time in twenty years in 2020, but this was likely due to the pandemic rather than any global shift in breast opinions.

55 Breasts are typically smaller: Slight asymmetries are common in the animal kingdom—in humans, for example, one eye is typically "higher" than another on the face—but radical asymmetries are not. Significant breast asymmetry in humans is actually associated with higher risk of breast cancer (Scutt, Lancaster, and Manning, 2006), which may point to deeper problems in tissue development in those bodies. As always, if something concerns you, talk to your doctor.

56 Venus of Willendorf: Weber et al., 2022.

56 small-breasted women regularly give birth: While larger breasts can store more milk in between breast-feeding, the vast majority of milk is made on demand; thus, large breasts might allow a nursing woman to go for longer in between feeds, but this would put her at greater risk of mastitis (Daly and Hartmann, 1995). What's more, the size of a nursing woman's breasts isn't stable; they usually shrink six months postpartum (Kent et al., 1999). The authors attribute this to tissue redistribution in the breast and greater efficiency in what remains. This period is also associ-

ated with the introduction of solid foods, however, so the six-month cliff may also simply be a matter of needing to make less milk now that the child begins supplementing its feeds.

56 hip-to-waist ratio is a better predictor: Singh et al., 2010.

57 estrogen-heavy phenotype: Jasieńska et al., 2004.

57 easier for our flat-faced babies to suckle: Bentley, 2001.

57 a two-legged problem: LeBlanc and Barnes, 1974.

58 long, deflated balloons: This is common knowledge. But if you want *documented* evidence, just take a look at any old *National Geographic* magazine with pictures of topless, multiparous women over the age of, say, thirty-five.

58 average vagina is only three to four inches: Lloyd et al., 2005.

58 the *average* erect human penis: Veale et al., 2015.

58 heterosexual women rate pictures of men: Mautz et al., 2013.

58 The "galloping theory" of scrotal evolution: Chance, 1996.

59 Leonardo da Vinci, a careful anatomist: Keele and Roberts, 1983; Di Stefano, Ghilardi, and Morini, 2017.

59 the idea of the *vasa menstrualis:* Galen is arguably the most to blame here. He discusses the transformation of menstrual blood into milk extensively in his work, even going so far as to say, "This is the reason why the female cannot menstruate properly and give suck at the same time; for one part is always dried up when the blood turns toward the other" (Galen, 170).

59 Sometimes, those changes are so dramatic: Kuhn, 1970. For a useful clarification of the history of Kuhn's thought and contemporary applications, see Parker, 2018.

60 our understanding of what the human body: This is well trod in the history of science. One barrier, in the case of Robert Koch's postulates, was that viral infections weren't something that could be cultured and observed with the technology available at the time (Brock, 1988).

60 complex systems are going to form: See anything the Santa Fe Institute is doing in systems biology, for example.

60 the discovery of agriculture: This is the dominant story in paleoarchaeology, at any rate, though it's possible social innovations came about independently (Emberling, 2003). As with all things in human history, the agricultural revolution came in fits and starts, with "domestication" arriving on a very long on-ramp before true agricultural societies took form (Fuller, 2019). And beyond the typical story of cereal crops, our human Eves also returned to their deep relationship with trees, which further drove growth of urban centers, because the long-term investment of orchard keeping required more complex urban social structures (and more obvious permanence) than the short-term gains that come from emmer wheat (Fuller and Stevens, 2019). More on the difficulties of becoming agricultural in the "Menopause" chapter.

62 blunt the growth of its population: Infection being the most obvious

case here, though presumably ancient societies had similar difference of effect between social classes. Throughout history, the poor are radically more devastated than the wealthy, a pattern that continues today with COVID-19 (Wade, 2020).

62 big cities did come about: Interestingly, in societies where agriculture was widely adopted, evidence of a fertility increase shows up in the historical data (Bocquet-Appel, 2011). Or at least an increase in children's bones found in graveyards. The usual assumption is caloric: having to work less for the same amount of food, female bodies in such societies might have been more able to reproduce. So long as they didn't poison themselves with undercooked tubers, that is.

62 African Ju/'hoansi hunter-gatherer tribes: Konner and Worthman, 1980.

62 four to five children over their lifetime: Ibid. Among the Hadza, it's four to six (Blurton Jones, 2016; Marlowe, 2010).

62 Women who do not nurse: Howie and McNeilly, 1982.

63 breast-feeding is Nature's Pill: This is well known to those in the medical field and biologists alike, but for specific mechanisms that might underlie this phenomenon among today's hunter-gatherer societies, see Konner and Worthman, 1980.

63 fewer offspring at a time: Jones, 2011.

63 nursing more than two children: Macy et al., 1930.

64 No wonder regulations for wet-nursing: Hammurabi, 2250 BCE.

64 many infants farmed out to the countryside died: Fildes, 1988.

64 French continued to employ wet nurses: Ibid.

64 African American women regularly nursed: West and Knight, 2017. This practice of obliged wet-nursing extended well beyond American borders, of course, and likely into antiquity. For a comparative view of the naturally traumatic effects of these practices in America and Brazil, see Wood, 2013.

65 Babylon had wet nurses: Gruber, 1989.

65 Mohenjo-daro, 50,000: Clark, 2013.

65 Columna Lactaria: Fildes, 1986.

66 the Sumerian gods were lazy: Though many versions of the Sumerian flood myth exist (Spar, 2009), the one I'm working with most here is the myth of Atrahasis, composed in Akkadian, produced roughly during the reign of Hammurabi's great-grandson, Ammi-Saduqa, around 1640 BCE. That text highlights how the gods created humans so they wouldn't have to work so much and how, when the cities were overpopulated and became noisy, the gods were annoyed, with the invention of death (and rules for which women could have sex, in which contexts, and with whom) as a useful population control after the waters receded (Dalley, 1991).

68 a full 30 percent of *all* cancers in women: American Cancer Society, 2020.

68 Breast cancer deaths: Siegel et al., 2022.

68 There's still a one-in-eight chance: American Cancer Society, 2020.

CHAPTER 2 WOMB

71 "Then the second angel": Translated from the Greek, presumably itself translated from either Aramaic or Hebrew. Many biblical scholars think that the book of Revelation is a deeply coded political document and that it's best understood within its historical context (Pagels, 2012).

71 The ash fell like snow: Bardeen et al., 2017; Vellekoop et al., 2014.

71 creatures who were small enough: Robertson et al., 2004.

71 ones who could live off the dead did well: Lowery et al., 2018; Robertson et al., 2004.

71 insects, like manna from heaven: Donovan et al., 2016, 2018.

71 Most think it was an asteroid: Though from where has been under debate, which has lent credence to the comet camp of late. A handful of prominent astrophysicists have been leaning away from a more local origin (that is, the asteroid belt between Mars and Jupiter) toward the Oort cloud, where any of a host of things could have hurtled the rock toward us, from Jupiter acting as a massive attractor for "sun grazers" to a plane of dark energy sheering a comet off course (Siraj and Loeb, 2021; Randall, 2015).

72 more than one billion Hiroshimas: Gulick et al., 2019.

72 soused with iridium: Schulte et al., 2010.

72 the world caught fire: Kring and Durda, 2002; Robertson et al., 2004; Bardeen et al., 2017.

72 pulsing heat over the course of many days: Robertson et al., 2004.

73 very small number of unrelated fish and lizards: Farmer, 2020. Technically, viviparity has arisen more than 150 times in the history of life on Earth (Blackburn, 2015), most of them squamate reptiles (some lizards and snakes). That sounds impressive, but the range of time involved for these uterine eurekas spans nearly 400 million years.

74 your warm-blooded body: Though there are a number of cold-blooded animals who give birth to live young (certain sharks, for example), live birth is mostly attached to creatures who have endothermy, and the control of developmental temperature may be one of the bigger drivers for the evolution of live birth in general (Farmer, 2020). One thing about those sharks, though: they may swim in warmer waters when they're pregnant (ibid.).

75 only about three inches long: Lloyd et al., 2005.

77 Roughly 200 million years ago: O'Leary, 2013. Timeline depends on whom you ask. Luo et al. (2011) are interested in about 160 million years ago and are also fond of the idea that tree dwelling might have been a useful ecological edge for early placentals, thereby "keeping clear of the dinosaurs below."

77 watch wallabies practice climbing: Drews et al., 2013. Please note that you can actually *watch this for yourself* in ultrasound video published online at youtu.be/Cig3ojSwoZY.

84 nothing life-threatening in a proto-penis: Nor are such bodies unusual in human history (Reis, 2009). Of course, this statement holds only so long as the patient lives in a place where it's unacceptable to violently punish one's child or wife for violating gender expectations. So-called honor killings, which are obviously gender based, have yet to disappear from the human world (Kulczycki and Windle, 2011), and the gender gap in countries like India and China strongly suggests gender-selective abortions, infanticide, and/or human trafficking (Hesketh et al., 2011), and that's happening to people with bodies that already *meet* gender expectations. Meanwhile, in the United States, surgical "correction" for atypical genitalia became popular in the 1960s and has only recently fallen out of favor, despite the widespread negative consequences of forcing gender assignment on these "intersex" infants (Dreger, 1998; Reis, 2009).

84 the impact killed off more of the marsupials' ancestors: Luo, 2007; Luo et al., 2011.

85 one in ten women suffer from urinary incontinence: Norton and Brubaker, 2006. This is a low estimate; some studies put it as high as 40 percent, depending how long it's been since the birth. Women who have C-sections can have incontinence, too (all pregnancies can damage one's pelvic floor and, to put it simply, rearrange things down there), but vaginal delivery is a strong, independent risk factor. I'm afraid the urethra isn't the only space that can be damaged down there, either: mothers who suffer damage to the supporting structures of their anal sphincters (something that can happen with third- and fourth-degree perineal tears, which occur in roughly 6 percent of vaginal births for first-time moms) report issues with fecal incontinence as much as twenty years later, and the impact appears to be cumulative, with more than one such birthing injury nearly doubling the risk for long-term problems (Jha and Parker, 2016; Nilsson et al., 2022).

87 the vaginal opening surgically closed: The procedure is called colpocleisis and is reserved for severe cases.

87 only 25 percent of women: Mahar et al., 2020. Note that I particularly mean cisgender, heterosexual, penis-involved sex—about which nearly all studies on the subject have been conducted, which is obviously problematic in this regard. Women self-reporting orgasm during sex often exclude receiving oral sex, which is more likely to produce an orgasm for many women (ibid.).

87 If you use a No. 4 camel-hair paintbrush: Parada et al., 2010; Parada et al., 2011.

87 much as it seems to be for human women: Jannini et al., 2009; Kruger et al., 2012.

88 "cloacal kiss": Herrera et al., 2013.

88 all amniotes have descended: Sanger et al., 2015.

88 useful to get rid of the penis: Herrera et al., 2013.

89 a small support bone: Though the baculum might have originally evolved not simply to support an erect penis but to stimulate the female: male rats with a bigger baculum have more success impregnating females, so long as they thrust a bunch (André et al., 2022). Because female rats are essentially in control of sexual events (they're the ones who solicit), one presumes they like this stimulation (Parada et al., 2010). Human women are quite varied in their preferences for vaginal stimulation, so the lack of a baculum in the human penis doesn't seem to be an issue in that regard.

89 typical two and a half hours of mating: These are measures from observation in the wild. In captive breeding environments, mating can take as little as half an hour, but other impediments, like uterine cysts or other reproductive woes, also take their toll (Nicholls, 2012). It certainly doesn't help that some species seem to ovulate in response to mating with a male, leading many conservation programs to rely on IVF (Foose and Wiese, 2006).

90 zoos have a difficult time impregnating them: Felshman and Schaffer, 1998; Foose and Wiese, 2006. What's more, because only a smaller portion of captive rhinos successfully reproduce, it's messing with species' genetic diversity (Edwards et al., 2015).

90 rigged up a mechanical uterus: Partridge et al., 2017.

92 genes that code for necessary proteins: Brawand et al., 2008.

93 the ancient placenta probably evolved: Though rudimentary on-ramps to the structure might have come about earlier and true lineage diverges much later. For a good overview of how incredibly contentious it all is, particularly between the "rocks and clocks," see Foley et al., 2016. Among the better "clocks" studies, a recent Stanford paper argues for closer to 120 million years and finds species-specific gene expression in mature placentas, which speaks to the many ways placentas have evolved to meet their hosts' specific developmental plans (Knox and Baker, 2008).

93 firing unfertilized eggs down the fallopian tubes: Miller et al., 2022.

93 violent bit of in-family cannibalism: Chapman et al., 2013.

93 swim between the mother's two uteri: Tomita et al., 2018.

93 *Juramaia sinensis:* Luo et al., 2011.

94 *Protungulatum donnae:* O'Leary et al., 2013.

97 Roughly 1 in every 350 human girls: Grimbizis et al., 2001; Saravelos et al., 2008.

97 1 in 200 women are born: Saravelos et al., 2008.

97 Roughly 1 in 45 girls are born: Ibid.

98 1 out of every 10 girls are born: Ibid. Note that these are simply rounded estimates based mostly on findings in clinical settings, which are naturally vulnerable to sampling bias: more women who have problems with their reproductive organs have them examined, and women who do not are

unlikely to, which could skew both how commonly women with funky sex organs have difficulty making babies and/or enjoying sex *and* how common it really is in the general population (Chan et al., 2011).

98 1 in 4,500 girls born every year: Fontana et al., 2017.

99 as many as 20 percent of humans are homosexual: Coffman et al., 2017. Kinsey's original estimate from the mid-twentieth century was closer to 10 percent, but accurate self-reporting in the face of bias is incredibly tricky to obtain (ibid.).

100 homosexuality is a trait present from birth: I admit this is an anecdotal claim based on my knowledge of other scientists and how such things are talked about in the community, and I was unable to find an appropriate survey of the scientific community's beliefs on the matter. Still, the truly massive array of research into the biological underpinnings of homosexuality, bisexuality, and general queerity is, I believe, sufficient evidence here. For one recent review paper, see Bogaert and Skorska, 2020. The authors unfortunately note that studies on women subjects in these areas are woefully thin on the ground, as are studies of subjects who aren't cisgendered.

100 Homosexuality has been observed: Savolainen and Hodgson, 2016.

102 in 2021, a lab managed to keep mouse embryos alive: Aguilera-Castrejon et al., 2021. Note that the goal here was not to find a technological way out of our mammalian uterine hell but to develop powerful new ways of studying embryonic development without all the muss and fuss of actually having to impregnate living bodies, mouse or otherwise.

103 only a handful of species: Emera et al., 2012.

104 no particular training in anatomy: I wish this weren't common for American schools, but I'm afraid it's all too often true. Things have improved (a bit, in some places) since the 1990s in this regard, but I'm not going to sugarcoat this for you. In the United States, requirements for sexual education—whether to have it at all, how much professional training a person has to have to teach it, what the curriculum should be, how it's funded—are almost entirely left up to state governments rather than any federal law. As of 2022, only seventeen of the fifty states require that sexual education in public schools be medically accurate (AGI, 2022).

104 I'm having too many periods: Strassmann, 1997; Eaton et al., 1994.

104 greater risk for certain cancers: Eaton et al., 1994.

104 delaying pregnancy into my thirties could make my babies deformed: This is largely tied to increased risk of chromosomal abnormalities, which is far more significant over age forty than age thirty-five (Frederiksen et al., 2018), but don't forget the trials of pregnancy itself, which is simply harder for older bodies to bear, with most obstetric outcomes becoming an issue from age forty on (but most OBs will tell you anyone aged thirty-five and up should receive additional monitoring and care).

105 European women who fall pregnant: Or at least they get less of a happi-

ness boost, though that might be because of many miscarriages prior to first children in women's mid-thirties and later (Myrskylä and Margolis, 2014).

105 menstruation was a kind of anti-pathogen: Profet, 1993.

105 vagina doesn't seem particularly *less* loaded: Strassmann, 1996.

105 women's periods evolved as a social signal: Knight, 1991.

105 enhanced sexual drive while menstruating: Though such women may be "outliers," they certainly exist, and the mechanisms underlying human sexuality have repeatedly proven to be complex in the lab. The only reliable peak in sexual desire/motivation for normally fertile women lies around ovulation, with a decline toward menstruation, but while progesterone closely tracked the decline in desire toward the start of women's periods in one study, researchers couldn't locate a predictable measure for the *rise* in desire toward ovulation (Roney and Simmons, 2013). We have some idea what sex hormones are involved in turning women off, in other words, but less for what turns us on.

105 One ambitious fellow: Knight, 1991.

107 this trait is exceedingly rare: O'Leary, 2013.

108 an American woman was at the salon: Goldman, 2014. Please note that the cardiac surgeon Dr. Omar Lattouf performed a heroic surgery that saved this woman's life and deserves all praise for it. Please also note that in interviews about this case he repeatedly says he was motivated to succeed because he wanted to make sure that newborn baby had a mom. By all indications, he surely wanted his *patient* to live because she deserved to have life all on her own merits. But I'm concerned that whether or not a woman has children continues to be a motivating factor for saving her life. Call it a trolley problem, call it sexism, call it whatever you like, but it's true that pregnant women regularly suffer when clinicians hesitate to treat the woman over the fetus, which is a strong factor in maternal mortality rates, so these moments do not exist in a vacuum (MBRRACE-UK, 2016).

110 all of which increase risk: Duckitt and Harrington, 2005.

110 the rise of in vitro fertilization: Le Ray et al., 2012. Notably, older mothers who use donated eggs have a higher rate of preeclampsia, which may be an immunological issue. Think of organ donation: the fetus created with the mother's own genetic material is a *semi*-allogenic graft, while fetuses made of donated eggs are a *total* allogenic graft. Normally, a placenta made from the mother's own egg has to "persuade" the mother's immune system to think it's not a foreign body only some of the time. With a donated egg, the placenta presumably has more work to do to distract and deceive that mother's immune system (ibid.).

110 one in three women pregnant: Bergman et al., 2020. Notably, women who have preeclampsia in a singleton pregnancy have an increased risk of cardiovascular disease later, while women who have twin pregnancies do

not. That implies women who have preeclampsia with just one baby may also have underlying cardiovascular problems, whereas in twin pregnancies the preeclampsia is more likely a direct result of the immuno-active placenta and the extra burden of an extra-large pregnancy (ibid.).

110 Researchers have managed to isolate: Mutter and Karumanchi, 2008.

111 In 2011, a group of researchers in Haifa: Kliman et al., 2012.

112 "Let's say we're planning to rob a bank": Kliman, in Rabin, 2011; Kliman et al., 2012.

112 as PP13 wages its war: PP13 is now being targeted as a potential therapy for preventing preeclampsia in women with higher risk, as well as a potential measure to *predict* that risk, because women with low circulating levels of PP13 in the first trimester are more likely to develop preeclampsia later (Huppertz et al., 2013).

113 a conflict that every eutherian pregnancy: Haig, 2015.

113 women who never give birth: For *female*-typical autoimmune diseases this is particularly true. Parity actually increases your risk by 11 percent, and the effect is particularly strong if you miscarry (Jørgensen et al., 2012). The theory holds that in miscarriage more immune-triggering material may enter the mother's bloodstream, though it's also hard to disentangle the miscarriage from any preexisting conditions that might have contributed to both the miscarriage and the eventual autoimmune disease (ibid.).

113 your risk for certain kinds of cancers is lower: The data are all over the place for this. The latest research indicates that barring the first five years after your pregnancy, breast cancer risks do seem to be lower if you've ever been pregnant and breast-fed an infant, though the effect is quite small and doesn't kick in for a good long while (Nichols et al., 2020). More important, the amount and duration of breast-feeding directly correlates to the reduction in ovarian cancer risk (Babic et al., 2020). That might be because breast-feeding reduces the total number of menstrual cycles your ovaries endure; however, having a longer reproductive life means your lifetime risk for thyroid cancer is higher (Schubart at al., 2021).

114 The *safest* thing for a woman's body: This is so incredibly obvious to anyone who studies mammalian biology that it's difficult to know which papers would be best to cite. Let's put it this way: in the United States, the risk of death for a pregnant person is *fourteen times* the risk of death associated with legal, safe abortions (Raymond and Grimes, 2012). Or at least that number was true in 2012; unfortunately, now that many women in the United States have to travel significant distances and wait longer to receive legal, safe abortions (if that option is even realistically available to that person), one presumes the numbers are going to change. There's a big difference between a legal, safe abortion obtained at eight weeks and one obtained (at great expense and hardship) much later. That's because the duration of pregnancy has a direct relationship to the degree of risk—not simply the relatively minor risk of medical complications from the

later-stage abortion, but the far greater risk of simply being pregnant for longer.

114 Nearly all women suffer from muscular tears: Frohlich and Kettle, 2015. What's more, depending on one's risk profile, the pattern of long-term injury persists for at least a year and potentially for one's lifetime (Miller et al., 2015).

114 of those who do, 50 percent will die: Schantz-Dunn and Nour, 2009.

114 0.65 out of every 100,000 legal abortions: Kortsmit et al., 2021.

114 26.4 American women still die: Kassebaum et al., 2016.

115 that stat was as true in 1930: Ibid.

115 one in four maternal deaths: Schantz-Dunn and Nour, 2009. Please note, however, that another 30–50 percent of maternal deaths in sub-Saharan Africa are due to *unsafe abortions* (AGI, 1999; Henshaw et al., 1999). Some of those dead women and girls were also infected with HIV, some malaria, and some neither. But their immediate cause of death was complications from unsafe abortions. Given that legal restrictions for abortions do not reduce the number of abortions performed in those communities, but simply drive women who need them to whatever's available, including back-alley quacks and self-harm (Henshaw et al., 1999), what actually killed those women and girls were antiabortion laws. More on why that's both predictable and incredibly bizarre in the "Tools" and "Love" chapters.

CHAPTER 3 PERCEPTION

117 Rūmī: Rūmī, 1270/1927. It's probably impossible to properly convey the prosody of Rūmī's work in English—the Persian poetic tradition he was working in was something *sung* as much as spoken or read, so the meter of the original is deeply embedded in both the aesthetic experience and meaning-making for its listener. Nevertheless, as someone who thought deeply about human perception, I thought he'd serve well here.

119 The first plants to return were the ferns: Berry, 2020.

120 fruiting trees and their canopies: Carvalho et al., 2021; Benton et al., 2022. It's not true, however, that large fruiting trees only came about after the asteroid; in fact, angiosperms were mixed in with conifers and other types of trees during the late Cretaceous (Jud et al., 2018). Rather, massive spreading canopies of fruiting trees—the thing one usually pictures when one thinks of a forest canopy, and particularly when thinking of the evolution of primates—come later. As for the rise of angiosperms in general (not just forests, but all fruiting plants), that comes in the mid-Cretaceous, and might well have been *stressful* for mammals who weren't well-suited to a changing ecology, thereby favoring small-bodied insectivores (Grossnickle and Polly, 2013).

120 *Purgatorius*, the world's earliest known primate: Chester et al., 2015.

120 her many sisters throughout the Fort Union Formation: Van Valen and Sloan, 1965; Clemens, 2004.

120 her teeth were specialized: Wilson Mantilla et al., 2021.

120 From looking at her ankles: Chester et al., 2015.

121 Some scientists even think primate hands: Or at least, emphasizing balance and stability functions in the hind paws, which may free the forepaws for other tasks as needed (Patel et al., 2015). We know a lot more about primates in this regard than other tree mammals, but the trait is clearly present in other arboreal mammals, too, with the more fruit-heavy eaters like kinkajous having particularly dexterous, grasping hands (McClearn, 1992).

121 each aiding the other's success: Sussman, 1991; Rasmussen, 1990; Sussman et al., 2013; Benton et al., 2022.

121 plesiadapiforms, ancient primates: Purgi herself was most likely a basal plesiadapiform, but there's considerable debate around this still (Clemens, 2004). For our purposes, this doesn't matter so much: like many of the Eves, Purgi should be considered an exemplar for the Eve of our primate sensory array.

121 most especially the terminal branches: Sussman et al., 2013.

122 the greatest diversity of land animals: Benton et al., 2022.

122 the damn thing hits 125 decibels: Podos and Cohn-Haft, 2019. Bizarrely, this was produced in close range of a female, who quickly moved from her spot on a branch to avoid the full force of the noise (ibid.).

122 Howler monkeys can reach: Dunn et al., 2015. Please also note that the louder one's calls, the smaller one's testes (ibid.).

123 our move into the forest canopy: Coleman, 2009. Note, however, that social features can put their own pressure on hearing systems, particularly when you've got an increasingly social set of species like later primates (see Ramsier et al., 2012). We don't actually know how social Purgi and other basal primate-like Eves were, so it's hard to say when, exactly, this would become a factor.

123 Primates did both: Mitani and Stuht, 1998. This model is controversial, however, and when you look at *all* mammals with suitably tested hearing ranges, primates fall under the general curve (Heffner, 2004). Generally speaking, the smaller primates hear higher pitches (which also happen to be better for up-close communication), and the larger primates (particularly those who spend quite a lot of time on the ground) lose some of that sensitivity to higher pitches (Coleman, 2009). Ancient australopithecines, however, show a distinct transition to more human-type pitch sensitivity (same low end as chimps but more ability in the higher end) by the time savanna living was common, in a presumed adaptation to both social communication and ecological shift (Quam et al., 2015).

126 correspond to the standard pitch: Not to be outdone by the audible portion of maternal experience, it's also been shown that the *ultrasonic* por-

tions of baby cry also, quite without the mother's awareness, change the amount of oxygenated blood in the listening mother's breast—something that doesn't happen when the ultrasonic portions are left out (Doi et al., 2019). However, the mother does need to hear the audible portions for this to work; simply jazzing her with inaudible ultrasonic waves doesn't have the same effect (ibid.).

126 women's ears are better at hanging on: Pearson et al., 1995.

126 baby cries alarm women: Messina et al., 2015. Please note, however, that responding to infant cries is something *all* mammals do, and even deer have been known to respond to the cries of other species' babies, including our own (Lingle et al., 2014). So even if males respond less obviously in the human species, it's simply not the case that they've somehow escaped the *entirety of mammalian hard wiring* for the care of our young.

127 one recent study had subjects: Parsons et al., 2012.

127 Men are also far more likely to suffer: Gordon-Salant, 2005.

128 most men over the age of twenty-five: Ibid.

128 Middle-aged and older men also have more trouble: Dubno et al., 1984.

129 women have fewer high-volume, ear-damaging jobs: Like many such assumptions, this is increasingly undermined or complicated by results in the lab. One of the more recent studies demonstrates that even when exposed to equivalently dangerous noise levels, male subjects had considerably more hearing loss (Wang et al., 2021). So if men jackhammer more than women, they're also significantly more at risk of losing their hearing by doing so—which, to my mind, is another solid pitch for female drummers in any given band.

130 *Fallout 4:* The soundtrack is fantastic. And the gaming company, Bethesda, is famously good. Let's just say Boston is a bit less interesting than Tamriel.

131 Women's OAEs tend to be: And this "weakening" of the cochlear amplifier seems to be present across Mammalia, presumably due in part to exposure to prenatal androgens (McFadden, 2009).

131 it's not just true for human beings: Ibid.

131 the length ratio of the pointer finger: Gillam et al., 2008.

131 something more complex: van Hemmen et al., 2017.

131 more likely to have "masculinized" OAEs: McFadden et al., 2011.

132 In male-female fraternal twins: McFadden et al., 1996.

132 If you castrate a sheep later in life: McFadden, 2009.

133 "hyper-masculinity" driving the system: Williams et al., 2000.

133 stereotype that gay men still endure: Sadly, even to the level of *scientific* understanding of sexual orientation, biasing findings that favor a feminized model of male queerity over the hypermasculine model (Gorman, 1994).

135 these genes constitute as much as 2 percent: Firestein, 2001.

135 men's noses aren't as good: Cain, 1982.

136 actually-banana-scented pee of a pregnant female: Rosen et al., 2022. It also makes the male feel less pain, oddly enough, but it's unclear if that's simply a normal stress response or if it's uniquely beneficial: pregnant rodent mothers will violently attack males nearby, presumably in part because the males are known to be infanticidal.

136 the scent of male urine: Roberts et al., 2012.

137 the further along primate evolution you go: Barton, 2004.

137 the olfactory system had massively degraded: Yoder, 2014; Gilad et al., 2004.

138 none of the obvious nerves: Trotier et al., 2000. However, this region may still play a role in the *prenatal* phase of development, in yet another case of evolution's rule of innovating in the later stages of development but leaving deep structural patterns in early body building well enough alone (Smith et al., 2014).

139 talking to men at speed-dating events: Saxton et al., 2008.

139 high activation in their hypothalamus: Savic and Berglund, 2010.

139 higher levels of cortisol in their saliva: Wyart et al., 2007. This was androstadienone—a similar musky compound, also present in sweat, and naturally produced in the testes of human and pig alike. Importantly, not every human nose is even *able to smell* this at all (Keller et al., 2007), which naturally narrows any olfactory study of the stuff among human subjects.

139 If you test gay men for AND reactions: Savic et al., 2005. However, in one study utilizing PET scans, Swedish lesbians seem to process AND in olfactory networks rather than the anterior hypothalamus (Berglund et al., 2006).

139 women preferred the stinky pits of gay men: Sargeant et al., 2007. Importantly, subjects came from cultures in the U.K. where regular bathing and deodorant are the norm, and given that the complexity of pit odor is produced by both the body and the microbiome of the armpit in question (Bawdon et al., 2015), one's ongoing hygiene habits may well influence the sorts of smells one produces over time, as would one's dietary choices and a host of other influences. It's not, in other words, clear that straight women in the U.K. are more attracted to gay men because of some kind of differing physiology, but rather that they may prefer the sorts of body odors a gay male lifestyle tends to produce in its associated armpits. Or at least the lifestyles of the nine gay men who participated in this study. Hedonics is a messy business.

140 male-to-female transsexuals showed: Berglund et al., 2008. Note that these women did not self-identify as homosexual and also had not undergone gender-affirming surgery before the study was conducted. Similar results were produced in minors (both younger children and adolescents) in the Netherlands in 2014, although importantly these were children diagnosed with gender dysphoria; not all people in the trans community

have gender dysphoria, nor should the trans experience automatically be equated with a medical disorder or mental illness.

140 men tipping strippers more if they're ovulating: Miller et al., 2008.

140 smelly T-shirts of ovulating women: Lobmaier et al., 2018.

140 don't like the pit smells of menstruating women: Ibid., 2018; Roberts et al., 2008.

140 dislike the smell of a woman's tears: Gelstein et al., 2011. It also made men notably less horny, both in self-reporting and in the extremely sexy environment of an fMRI machine.

140 deodorant and birth control pills: Roberts et al., 2008; Lobmaier et al., 2018. The most famous, which largely kicked off the trend, was a 1995 Swiss study looking at subjects' MHC and whether olfaction helped select for immunologically compatible mates (Wedekind et al., 1995). Women who weren't on birth control preferred smelly T-shirts worn by men who were ostensibly more compatible—for two days, with no deodorant or soap—over similar shirts worn by less compatible men, while women on birth control did not show those preferences.

141 Women are better at detecting faint scents: Cain, 1982; Sorokowski et al., 2019; Cherry and Baum, 2020; Oliveira-Pinto et al., 2014.

141 this female advantage is present: Kass et al., 2017; Doty and Cameron, 2009. Because other species aren't quite able to *report* what they're smelling, much of this is behavioral work, though some progress has been made in studying mechanisms. In mice, for instance, females seem to transmit more signal to the olfactory bulb (Kass et al., 2017). One possible driver may be sexual selection: male scent markings in many mammals seem to be more complex than female markings (Blaustein, 1981), which would naturally give females an advantage if they're better able to discern the complex features of those scents. That's above and beyond the basic survival benefit afforded by females who are good at avoiding toxins, which, as I've mentioned, is particularly critical for the female placental.

142 A typical teenage male will grow a nose: Holton et al., 2014. This trait only kicks in around puberty; children's noses are roughly the same across the sexes (ibid.). The larger male nose is also independent of the generally slightly larger facial features (and body size) of males in general, and is thought to be tied to the more expensive muscle mass of post-pubertal male bodies. In many ways, the nose should be thought of as both a part of our sensory array and an extension of our lungs; adult male lungs are also, on average, a titch larger. More on respiration in the "Voice" chapter.

142 a lovely little mouse study: Kass et al., 2017.

143 a woman's sense of smell heightens: Doty and Cameron, 2009.

143 Being pregnant might have changed her ability: Or, rather, enhanced her conscious *attention* to the fact that she'd noticed the smell—there isn't a

lot of consistent evidence that women's baseline olfactory capabilities rise during pregnancy (in fact, during the third trimester she'll smell less well, likely because of a stuffy nose), but there is a wealth of both anecdotal and scientific evidence for women having stronger feelings of *disgust* for certain smells, rating many smells as less pleasant, and generally having strong emotional reactions to scents across pregnancy, but particularly in the first trimester (Cameron, 2014).

144 women's olfactory bulbs: Oliveira-Pinto et al., 2014.

144 Evolutionary scientists tend to think: The actual scientific support for this concept, however, remains controversial (Cameron, 2014). It does appear that having anosmia (an inability to smell) is associated with reduced nausea in pregnant women (Heinrichs, 2002), but more research is needed in this area. More clearly, a heightened *emotional* response to olfactory cues seems to happen in the pregnant body (Cameron, 2014), and so the links between pregnancy, nausea, and olfaction may lie instead in the brain itself, rather than the nasal passages (ibid.).

148 line up fossilized skulls: Barton, 2004.

150 stereopsis gives you really good 3-D vision: Heesy, 2009.

150 the more stereoscopic the eye placement: Barton, 2004.

150 Binocular, stereoscopic vision is a convergent trait: Heesy, 2009.

151 circadian cycle is embedded: And among these systems can be found multiple sex differences (Yan and Silver, 2016).

151 The way we digest food: Segers and Depoortere, 2021; Hoyle et al., 2017; Santhi et al., 2016.

152 women who work night shifts: Fernandez et al., 2020.

152 it doesn't affect their fertility: Perhaps this is due in no small part to the fact that unlike ovaries testes don't show much circadian influence (Kennaway et al., 2012).

154 change to dichromatism: The notion of nocturnality driving mammals' dichromacy is mostly due to Walls (1942). These ideas have since been complicated. For instance, there is some controversy around whether it was a total switch to daylight vision or rather an emphasis on *dim* light, like twilight and dawn or, occasionally, a full moon (Melin et al., 2013). It's also unclear if dichromacy was a basal state for mammals or if a change to nocturnality is what drove the shift (Jacobs, 1993).

154 The genes responsible for our red-green color vision: Hunt et al., 1998.

156 a bit like smelling a melon: Hiramatsu et al., 2009.

156 The groups with a mixture of color vision: Osorio and Vorobyev, 1996; Caine et al., 2010. Dim-light advantages for the dichromats also carry to insect foraging (Melin et al., 2007), while there's greater parity for fruit in bright-light conditions (Vogel et al., 2007), which may imply that not only fruit eating but time shifts for foraging *in general* can provide differing advantages to differently sighted group members, depending on that

species' environment. It's also possible dichromats simply spend longer foraging for the same amount of food, as they seem to do in zoos (ibid.).

157 the cochlea tamps down its amplifier: In auditory research, this is often called the "dinner party problem." But it's not just a matter of boosting one noise over another; differing sensory systems *also* influence one another as attention shifts. For example, paying attention to visual information makes the cochlea become less responsive; mammals really do "tune out" as needed (Delano et al., 2007; Marcenaro et al., 2021).

157 your pupils will dilate: Zokaei et al., 2019.

158 There are known sex differences: Heisz et al., 2013; Sammaknejad et al., 2017.

160 women who have the genetic predisposition: Jordan et al., 2010.

160 As many as 12 percent of all human girls: Ibid.

CHAPTER 4 LEGS

163 "We should go forth": Thoreau, 1862.

163 "some opinions on what it means": Lemire, 2009. Zilpah White was a former slave who lived near Thoreau's bean field in the woods by the pond. Unlike many former slaves who stayed in the same house that had enslaved them, their daily lives largely unchanged, Zilpah struck out on her own when she became legally free. But the soil by the pond was sandy. It didn't grow things well. And someone burned down her little house in 1813, so she had to rebuild that, too. She lived into her eighties by that pond, making brooms—paid for it this time, but not much (ibid.). Much has already been made of the ways Thoreau's mother and others (reminiscent of Hrdy's *Mothers and Others*) enabled his particular philosophical tramping on a little pond in Massachusetts—some of it good (Solnit, 2013; Shultz, 2015), some of it less so. But should we care more about the invisible labor of women and former slaves here? The lying? The wild racism? (Read for yourself.) No, I think what's *most* interesting in Thoreau is the notion that the Wild belongs to the masculine leg, ever striding, which becomes America's e'er-westward stride, when the *real* story of humanity's great walk is probably *northward and outward* and significantly taken in stride by a feminine, endurance-enabled leg—as I'll address in this chapter.

164 Soldiers call it "the suck": The phrase "welcome to the suck" is probably best known from the marketing campaign for Sam Mendes's 2005 film, *Jarhead*. While various versions of and references to the suck are used throughout the military, most I've spoken to (and read) think it took root in the early years of the twenty-first century in the war in Iraq (though the author of the book *Jarhead* is based on was a marine, included the phrase, but served instead in Saudi Arabia and Kuwait during the Gulf War in the

1990s). The author of the screenplay for Kathryn Bigelow's *Hurt Locker*, Mark Boal—himself an embedded journalist in Iraq in 2004—likewise used "the suck" to describe the daily realities of combat, presumably having picked up the phrase from the soldiers around him. For a general insider view of the Army Ranger School and its general alignment with the suck, see Lock, 2004.

164 Sixty percent of the people who quit: Spencer, 2016.

165 Griest is a woman: A cisgender woman, to boot. My portrait of Griest is based on a number of different journalists' interviews and her own public commentary during this period and later, most of which are available online. In particular, I leaned heavily on reportage done by *The New York Times*, CBS News, *The Washington Post*, the *Army Times*, private interviews with members of the military, and military reports to Congress on the issue of gender integration in the armed forces (for example, Oppel and Cooper, 2015; Kamarck, 2016; CBS News, 2015; Tan, 2016). In more recent news, Griest wrote an op-ed in 2021 in which she rejected a proposal for a change in standards to include more women in the military, which subjected her to no end of online bullying, including accusations that she'd somehow "internalized sexism" (Lamothe, 2021). But, for Griest, such changes would not only diminish her own accomplishment but jeopardize the battle readiness of combat troops (Griest, 2021). As for the notion of internalized sexism, please see the "Love" chapter herein, though I don't actually think Griest is participating in sexism by asking for standards to remain as they lie.

165 Life was good up in the trees: Fleagle, 2003.

166 global weather had changed: Senut et al., 2009.

166 Tearing a continent apart: Sepulchre et al., 2006; Pik, 2011; Wichura et al., 2015.

167 nearly 99 percent of our DNA: The latest numbers show our genome sharing 98.7 percent with bonobos and 96 percent with chimpanzees, but, given issues with overlap, insertions, and deletions, most folk in the know—the Smithsonian, for instance—generally say we share 99 percent with them both (Prüfer et al., 2012; Waterson et al., 2005; Mao et al., 2021).

167 hanging was better supported: Hunt, 2015.

168 Ardi is the best evidence: Lovejoy et al., 2009. Many papers were released essentially simultaneously on Ardi's discovery, a handful of which can be found in the bibliography.

170 That's what bunions are: Latimer, 2005.

170 a bit stilted and waddling: For an excellent overview of the current literature on gait mechanics and ancient hominins—what we know and what we *could* know from fossils of the pelvis and lower limb and contemporary anatomical experiments on living humans, see Warrener, 2017.

170 more knee replacements on women: Maradit Kremers et al., 2015.

Women also have more hip replacements, and while aging is an independent factor for these joint surgeries, women's hip replacement is more closely tied to age than the knee (that is, more older people have hip replacements and more older people are women, but even at younger ages more women have to get their knees worked on or replaced than men).

171 Relaxin levels are highest: And when levels stray from the normal, relaxin may even have a role in preterm birth (Weiss and Goldsmith, 2005).

172 Relaxin is found in all placental mammals: This is an understatement; its analogues are also found in fish, and the relaxin family of peptides should be considered wildly ancient in the animal kingdom. Like many useful molecules, one presumes relaxin's role in *placental* reproduction was largely a repurposing of existing systems. For instance, while it's now known to have a role in early pregnancy in marmosets, a big part of that role may be the growth of new blood vessels (Goldsmith et al., 2004).

173 women's spines have evolved differently: Whitcome et al., 2007.

175 That tugging is intimately tied: This is, of course, a simplification, and there are a few different complex systems at play, but the central concept here is that load bearing as mediated through muscle directly affects bone growth throughout the life span and that it's deeply flawed to try to think about the skeleton fully independently from the musculature it interacts with (Tagliaferri, 2015). It's also true, however, that male and female bones build themselves slightly differently, with typical male bones building more of an internal layer and a thinner outer layer, while female bones build a thicker cortical layer and a thinner core, effectively making female-typical bones more slender. This makes them particularly vulnerable to fracture as that outer layer thins after menopause. For a good overview of what's known about aging musculoskeletal systems, see Novotny et al., 2015.

177 upper body skeletal muscle: Round et al., 1999. For one recent study specifically on muscle effects in trans men receiving hormonal gender-affirming therapy, see Van Caenegem et al., 2015.

177 Chimps' metabolisms and muscle tissues: O'Neill et al., 2017.

178 Once you get to ultra-endurance distances: Ronto, 2021.

179 more expressed in young women's muscle cells: Maher et al., 2009; Maher et al., 2010.

180 she lived in a riverine forest: Cerling et al., 2010; Cerling et al., 2011; Louchart et al., 2009; White et al., 2009; White et al., 2010; Wolde-Gabriel et al., 2009.

180 We freed up our arms to carry weapons: This is well trod enough in the field, but I think an old analogy by David Pilbeam provides the most articulate critique: much as our dexterous hands didn't evolve *in order to* play violin, the hunting capacities of our bipedalism are likely a later add-on, with bipedalism itself essentially "preadaptive" in this regard—evolving

first in ways that most likely involve plant-focused feeding behaviors and only later utilized for things like running and hunting (Pilbeam, 1978).

181 we evolved to outrun nimble ungulates: Lieberman is probably the most vocal advocate of this theory among academics, and his arguments for the evolution of *endurance* running are particularly persuasive (Bramble and Lieberman, 2004). In the popular press, Christopher McDougall gained traction with his *Born to Run* in 2009, which highlighted the evolutionary theories around endurance running and bipedalism alongside his more contemporary stories. As for whether sweat was particularly necessary, Lieberman also proposes that hominin locomotion and cooling systems evolved separately and only became tied (ability to run and ability to sweat to cool off while running) later (Lieberman, 2015).

182 What a guy: This is a commonly known behavior among extant apes, chimps particularly. Jablonski makes an interesting twist for the "display" theory of bipedalism, however: in competitive environments, having regular threat displays (standing up, puffing your chest, being threatening on two legs) and regular appeasements might do well for ancient hominins. In simple terms, that means the difference between blustering and actual war. One costs lives; one costs pride (Jablonski and Chaplin, 1993).

182 would be game for such a trade: White et al., 2015. I admit it's odd to have Dr. Lovejoy be one of the very few scientists I name outside the notes. Given the sheer mass of scientists' work underlying these chapters, I purposefully chose to let the ideas speak for themselves rather than build a hero's story out of one lab or another. (Science is, at its core, a collaborative process, and many science writers are guilty of erasing that process in the desire to name heroes.) But given how very important Dr. Lovejoy's work is to the field, and given that I'm *disagreeing* with him—or at least gently pointing out the unlikelihood of ancient primate monogamy being linked with bipedalism (which is hardly the focus of his major contributions in *Ardipithecus*, but nevertheless a behavioral interpretation he leans on)—it seemed more respectful to name him directly. For further evidence of the difficulty involved in creating a mostly monogamous primate society, much less a *patriarchal* version of such a thing, see the "Love" chapter.

183 We were probably still covered in fur: The best argument I've heard for when we must have been hairless is the divergence of our head and body lice, which dates to roughly 190,000 years ago—leaving our nonhuman Eves rather furry *and* lousy, I'd say (Reed, 2007; Toups et al., 2011).

183 many females will barter sex for meat: Gomes and Boesch, 2009.

184 That's what chimps do now: Carvalho et al., 2012.

185 They're about endurance: Resistance to fatigue is a big part of how you take a chimp-like body of our ancient Eves and make her walk around all the time. But not that *walking* isn't the only thing we were necessarily doing that required endurance. For example, digging up tubers all day

long also requires the sort of resistance to fatigue and general metabolism that can produce sufficient endurance to make such activities worthwhile, and a number of extant human food-sourcing behaviors likewise lean on our species' unique ability to endure (Kraft et al., 2021).

185 Sergeant Major Colin Boley: Lemmon, 2015.

187 had significant muscle atrophy: Perhonen et al., 2001.

187 astronauts also have *conversion:* Fitts et al., 2001.

189 In studies from 1999 and 2001: Semmler et al., 1999; Sayers and Clarkson, 2001.

190 We may need to stop sooner: Interestingly, even a report from the U.S. Marines recently found evidence of this phenomenon, quite without meaning to: In a gender-integrated unit, the female soldiers suffered higher rates of musculoskeletal injury (40.5 percent versus 18.8 percent). However, while they were therefore more likely to be unavailable for training due to the injury, they also were out for *fewer days* than males who'd suffered injuries. In other words: more injuries, but faster recovery times (USAMEDCOM, 2020). The general availability for training, however, came out rather similarly in the end: 98.4 percent available for the males versus 96.8 percent for the females, implying a rather small difference in global readiness. For both sexes, most of these injuries were tied to load-bearing activities (ibid.).

191 34 percent of Ranger candidates: Tan, 2015, quoting Colonel David Fivecoat, commander of the Airborne and Ranger Training Brigade.

193 She offered to carry it for him: DVIDS, 2015; Oppel and Cooper, 2015.

193 There he was, completely broken: DVIDS, 2015. Two stories, in fact, exist from Rangers who completed the course with Griest and other women, both similarly reporting that women would take heavy loads from their fellow male rangers when they'd become too tired, notably when other *male* recruits simply couldn't. You can hear them for yourself in the DVIDS 2015 video, starting at minute marker 13:06.

194 "It's a weapon for us": Lemmon, 2021. For a more general overview of the situation of women and war among the Kurds, also see Sankey, 2018. The author is a scholar housed in the U.S. Air Force Air War College. Neither of these books will tell you everything you need to know about the Kurds, not by a long shot, and Sankey's entry glosses rather quickly over the Marxist influence on the rise of the PKK. But for those who want to dip a toe, they do put Rehana's story in perspective; the conflict of how one should deal with *gender*, you see, is very much a part of the conflict between Turkey and the Kurdish people. For a deeper analysis of how gender is enmeshed in the rise of the PKK, see Açik, 2013.

194 posting photos on Twitter: Rakusen et al., 2014; Silverman, 2015. There is also a large Twitter history related to Rehana, some of which includes Silverman's tweet thread about the Rehana story and how Carl Drott, the Swedish journalist who met the woman in the photograph, is pretty sure

she didn't kill a hundred members of ISIS, nor was she even a sniper. I, too, was unable to track down this ghost of a woman, and I hope deeply that she is well and living comfortably somewhere and is not, in fact, murdered, unlike her father.

For the citation work, meanwhile, it just so happened that while this book was going into copyediting, Elon Musk had just finalized his purchase of Twitter, so I hesitated to cite Twitter sources directly, instead leaning on secondary sources where I could. I admit I don't know what will become of the public record this odd little social media company has produced, and I'm not the only one worried about that: even the U.S. Library of Congress is struggling to keep up (Stokol-Walker, 2022).

195 By the end of 2015: Pellerin, 2015.

196 mixed-sex military groups' feelings: And on especially challenging cognitive tasks, mixed-sex groups in the U.S. Marines apparently perform better (MCCDC, 2015).

197 The entire American military has a problem: Morral et al., 2015. For one good argument about why gender-integrated basic training might be particularly good for reducing these issues across the military, see Lucero, 2018.

CHAPTER 5 TOOLS

199 "I would rather stand three times": The translation varies. It should probably read "stand three times with a shield," as in David Kovacs's translation, meaning she'd rather go to war (ostensibly on the ancient Greek version of the front lines, in tight battle) than give birth even once. While it's tempting to think Euripides had some prescient insight into postpartum depression here, or a deep intuitive knowledge of our Eves' suffering, I think it's instead a reminder that his central character—a woman!—should be understood as equally heroic (or at least equally experienced in danger) as the typical heroes Greek audiences would watch in other plays on the same stage. She's also a woman oppressed and scorned in a foreign land. As Countee Cullen writes of her in 1935, the fact that she takes her dead sons away on the chariot is as much "saving" them as depriving Jason of joy. "I will not leave my children here to die by other hands than mine," she says. "I will bury them myself where no hostile hand can dig them up to defile their little bones" (Cullen, 1935, 54, 61). So off she goes in her chariot, given to her by the gods, to later cure Herakles of *his* god-given madness, which had led him to murder his own wife and sons. There's just a lot of son murdering going on in Greece. Just a lot.

201 octopus uses tools with its tentacles: Finn et al., 2009.

201 Crows are avid tool users: Rutz et al., 2018.

201 mostly ate grasses and bugs: Notably, the presumed Habilis diet was less

varied than the diet of *Homo erectus*, implying a much broader food strategy involving a range of harder things and softer things, which speaks to both their remarkable ability to migrate and their general opportunistic omnivory (Ungar and Sponheimer, 2011).

201 dig up some kind of ancient turnip: Harmand et al., 2015.

201 a chimp is hunting: Pruetz et al., 2015.

202 lived in the grassy highlands of Tanzania: Leakey et al., 1964. Please note that there's current debate as to whether Habilis even belongs in *Homo*, or if she's instead an australopithecine, or even her own genus (Wood, 2014). Using Habilis as an exemplar species isn't meant to erase that debate, but is more an homage to Leakey and the obvious choice, together with Erectus, for Eves that had all the right pieces in place for the origins of human gynecology.

203 she was often prey: Brochu et al., 2010; Arriaza et al., 2021.

205 we're obviously more collaborative: I don't think it's possible to overstate the influence of Sarah Hrdy's work on the scientific understanding of alloparenting and why it matters so much for human evolution. Her book *Mothers and Others* (2009) is probably the most famous, and terribly readable—do spend the time if you haven't already. In fact, one of the reasons I focus on the pregnancies and post-birth recoveries of our Eves isn't simply that it's a topic that's been neglected thus far and has such an obvious effect on our species' evolution; it's also that what happens *after* the postnatal recovery period is already discussed so beautifully by Hrdy that I'd simply spend all my time waving her flag.

205 *before* our famously needy babies are born: Trevathan makes a bid for the obstetric dilemma (and bipedalism in general) as the starting gun for our race toward midwifery (and, as such, toward assisted birth becoming not only a basal hominin behavior but an imperative for complex rules around the birth, thereby setting the stage for implicit power structures around female bodies in hominin societies). It's a fascinating argument, but it admittedly attends primarily to the act of midwifery during labor and delivery, leaving most of what comes much before or after by the wayside (Trevathan, 1996). As any good ob-gyn will tell you, what happens in that moment is profoundly shaped by the prenatal care you've received prior, and, while one needs to be ready for any of a host of ways the birth could go sideways, even perfectly normal births are still quite dangerous in the days *after* the birth has already occurred.

205 Most of the features that make our reproduction: The problem of the narrow birth canal was in place, by most estimates, by the time Lucy was walking about her furry day, although unlike humans her babies at least didn't have to rotate quite as much to get out (Rosenberg, 1992; DeSilva et al., 2017; Laudicina et al., 2019). Neanderthals, meanwhile, had roughly the same difficult births as we did, though their mechanisms were a bit more primitive than ours (Weaver and Hublin, 2009). The

problem of the invasive placenta is harder to date, but may be tied to encephalization (growing brains are hungry) or climate instability (it's good to have fat babies in a wasteland), both of which have a bursty, extended history that begins some time after Lucy and her fellow australopithecines and really hits its peak by the time we arrive at *Homo* (Potts, 2012). For more on climates and our giant heads, see the "Brain" chapter.

205 But if you can't have babies: This is a self-evident statement. However, I'd also like to point out that maternal mortality is a particular problem for *primates* of many stripes: not only does early maternal death make it more likely for a female's offspring to die before reproducing, but should those offspring survive to have children of their own, they're *also* less likely to be able to produce offspring (that is, dead mom's grandbabies) that survive to reproductive adulthood (Zipple et al., 2020). In other words, the primate reproduction model leans so heavily on the mother that having the mother crippled and/or dead has even *more* of an effect on evolutionary fitness than it would in other species. That makes the clearly bonkers human reproductive system that much more unlikely. Without behavioral innovations around these problems, we'd be kaput.

206 Northern hairy-nosed wombat: Not to rag on these gals, but they also prefer to dig their burrows in very particular, sandy soil near ancient dry creek beds in Australia, and, even then, particularly in the spreading root ball of one particular tree (Queensland Government, 2021). Ever met somebody who'll basically date only one very specific, mostly nonexistent type of person? A person who complains about how annoying dating is and maybe has an extensive collection of bubble-gum wrappers, carefully ironed and kept in plastic, from the 1980s? This person doesn't really want to date someone. This person is, in fact, a closeted northern hairy-nosed wombat.

206 ancient specula made of iron: Hargest, 2020; Milne, 1907.

208 Human women, meanwhile, hover: What would this number be in ancient humanity? We have some windows, but they're complicated. For example, certain premodern human populations lacking modern medical care have low rates of death during childbirth (less than 3 percent), but this is a particularly narrow window to test, because it doesn't count death during pregnancy nor during the extended postpartum period (Lahdenperä et al., 2011). But, more important, even "premodern" populations (whether the Finn and Canadian represented in this study, or even hunter-gatherer communities studied elsewhere in the world) have *gynecology* of the sort I mean here: they have midwives and shared knowledge of female reproduction. They have long-established medical practices and pharmacology associated with female fertility. And, as I'll discuss in the "Love" chapter, they also have other behavioral interventions around female fertility—established cultural rules around when women are to

become pregnant and give birth (a part of the evolutionary perks of sexism). So that 3 percent represents what you can get to once you have gynecology at all. What the death and complication rates would have been at the dawn of the hominin line is hard to say, but if gynecology has an on-ramp of a sort among the australopithecines, it's certainly going to be well in gear by the time we arrive at Habilis, and by the time you arrive (much later) at *Homo sapiens*, it should probably be considered part of our species' basal social-behavioral suite.

209 a lemon-size hole: Wittman and Wall, 2007. Some have recently argued the obstetric dilemma used to be more variable than it is now, which might be an agriculture thing (we make bigger babies now because we have the food to do so) (Wells et al., 2012), but, even with some variability in effect stemming from direct obstruction in the birth canal, that doesn't really eliminate that it's a part of our species' broader reproductive woes (Haeusler et al., 2021), including all the various ways we have heart attacks and/or strokes and/or hemorrhages and/or kidney damage and/or liver failure and/or ongoing metabolic problems because of our clearly terrible human pregnancies rooted in our deeply invasive placentas (Abrams and Rutherford, 2011).

209 Chimps labor about forty minutes: Elder et al., 1931; Hirata et al., 2011.

210 about thirty-seven days more: Dunsworth et al., 2012. Thurber even aligns human pregnancy and lactation with arctic trekking and other "extreme metabolic activities" in order to calculate the maximum human threshold for metabolic load (Thurber et al., 2019). It's a convincing paper. It also provides an interesting perspective on why human women do so well in ultramarathons.

211 a black-and-white snub-nosed monkey: Ding et al., 2013.

211 the first clear evidence of active birth assistance: Pan et al., 2014.

212 staying quiet and away from the troop: Nishida et al., 1990. Notably, it's not just the *females* who might murder the newborn; even local males might snatch away the baby and eat it if she gives birth where others can see her (Nishie and Nakamura, 2018).

213 Dominant female chimps are known: Goodall, 1986, 1977, 2010; Pusey et al., 2008.

213 it helps them maintain their social position: Pusey and Schroepfer-Walker, 2013.

213 witness a bonobo give birth: Douglas, 2014.

214 researchers gathered three more: Demuru et al., 2018.

215 Mallard ducks are constantly raping each other: Prum, 2017. This is true of a good many duck species, in fact, though the mallard is probably the most talked about, in part because the species is so common and visible in American suburbs and their associated lakes and ponds and little rivers. Learning that *mallards* are prolific rapists feels, to an American mind, a

bit like browsing the U.S. registry of sex offenders to learn how many live in your neighborhood. You're welcome to try it, but fair warning, it won't make you happy: www.nsopw.gov.

215 You can see this sort of coevolution: Hosken et al., 2019.

216 gangs of dolphin males: These behaviors are typically called herding, which calls to mind the Australian cattle dog, but I'm afraid the reality is decidedly more rapey (Smuts and Smuts, 1993; Connor et al., 1992; Connor et al., 2022). Due to the obvious problems with direct observation, most dolphin sexual coercion is inferred after the fact, such as through tooth-rake marks (Scott et al., 2005). Meanwhile, the female dolphin's vagina seems to have evolved rape countermeasures of a sort, commonly seen in other species with a significant history of rape (Orbach, 2017).

217 the Bruce effect in the 1950s: Bruce, 1959. For a nice framework to understand both infanticide and the Bruce effect, see Zipple et al., 2019, which puts both under an umbrella of "male-mediated prenatal loss."

217 Rodents do it: Mahady and Wolff, 2002; de Catanzaro et al., 2021; Yoles-Frenkel et al., 2022.

217 Horses do it: Bartos et al., 2011.

217 Lions seem to do it: Bertram, 1977.

217 Even *primates* do it: Roberts et al., 2012.

217 consider the gelada: Ibid.

218 let the mare have sex with a familiar male: Bartos et al., 2011.

219 a *third* of foals aren't sired: Bowling and Touchberry, 1990.

220 5 percent of American rapes: Holmes et al., 1996.

220 Rates in other human communities: One Ethiopian study did report a much higher rate (17 percent), but it's unclear if that's because it involved an anonymous, self-reported questionnaire of high school students, which might not have controlled for things like how *many* sexual events were involved close to the rape and subsequent pregnancy (Mulugeta et al., 1998).

220 dropping to near zero: In other words, being raped doesn't make you less likely to become pregnant than if you'd had the sex consensually, nor does it make you *more* likely to become pregnant (as discussed at greater length in Fessler, 2003).

220 a woman who's having regular sex: Kenny and Kell, 2018. She'll also have a lowered risk of preeclampsia if she's lived with him—and had sex on a semi-regular basis during that time—for at least twelve months (Di Mascio et al., 2020).

221 it's not as predictable as the Bruce effect: Qu et al., 2017.

221 the chance that her fetus was fathered: This is the natural conclusion, given that forty-eight women are apparently raped *every* hour in the DRC (Peterman et al., 2011) and women refugees are significantly more likely to be victims of sexual violence compared with the general population

(Hynes and Lopes Cardozo, 2000). This is also true of the men and children in refugee populations, but of the adult men, much less (ibid.).

222 And many animals—including primates: Huffman, 1997; Fruth et al., 2014.

222 Mahale chimps, sick: Huffman, 1997; Huffman et al., 1997.

222 Similar sorts of self-medicating behaviors: Huffman, 1997.

223 red colobus monkeys eat the leaves: Wasserman et al., 2012.

224 *Homo erectus*, was one of the most successful: Potts and Teague, 2010.

225 We've found charred remains: Berna et al., 2012.

225 She invented the Acheulean tools: de la Torre, 2016.

226 minimum viable population (MVP): Shaffer, 1981. By definition, any species that's managed to survive geographic expansion and the resulting founder effect did somehow manage to reach their MVP in at least some of their new territories, and the strength of that effect can be tied to both recency and speed by examining genetic loss of diversity in current populations (Peter and Slatkin, 2015). For humanity, analyses like these are generally used to support current models for ancient human migrations out of Africa (Ramachandran et al., 2005).

226 as few as 50 percent of human pregnancies: March of Dimes, 2017. This number remains mysterious in no small part because it's likely only a few human pregnancies are actually *known* to the person carrying them, much less that person's clinician. If you count pregnancies that are *known* to have implanted, the miscarriage rate seems to be around 30 percent (Hertz-Picciotto and Samuels, 1988). It's also true that "pregnancy" only begins after successful implantation in the uterus; a fertilized embryo is not yet a pregnancy, even though development has already begun. If you count all the fertilized human embryos that briefly exist in human pelvises, the number would be significantly higher.

226 still has only a 9 percent chance: Wilcox et al., 2001.

229 the MVP for a reproductive group: Smith, 2014.

230 This phenomenon is called a founder effect: Amos and Hoffman, 2010.

233 Zika in women might as well be a different disease: Lee et al., 2021.

234 Pregnant women are three times: Schantz-Dunn and Nour, 2009.

234 protozoans accumulate in placental tissue: Ibid.; Fried and Duffy, 2017.

CHAPTER 6 BRAIN

241 "The little girl was sliding back": Jackson, 1959/2006. Like much of Jackson's work, the book is solidly about the Feminine Condition in Jackson's time. You're welcome to disagree, but a hell of a lot of people are going to fight you on that one. The insanity is America's, but the House is Hers. For what it's worth, Jackson used the money from the book's film rights to buy curtains, a piano, and a washing machine (Franklin, 2016). For the author, too, domesticity continued.

243 Some for their own growing brains: My portrait of the brainy hominin Eve leans heavily on research in early hominins as a prey species and the brain-eating habits of local carnivores (Brain, 1981; Hart and Sussman, 2005; Arriatza et al., 2021), as well as conversations with various scientists in the field, including Rick Potts at the Smithsonian. Though Brain's work focuses particularly on skull punctures in australopithecines by felids, later work also shows hyenas doing the same (Arriatza et al., 2021). And, of course, ancient hominins *did this, too,* transporting the heads of already-dead animals over long distances to crack open and share the brains back home (Ferraro et al., 2013). It's unclear whether later hominins, like Erectus, would have prey as often as the australopithecines, given their capacity for running and hunting, but the notion that our Eves grew increasingly large brains simply because they became so *dominant* in their local food web never sat right with me. It also seemed useful, right at the start, to point out how very *juicy* hominin brains would be to their predators: sopped with sugar, run through with blood and fat, and increasingly expensive to build and maintain. Big-brained hominins' bodies (potentially with a lot of adipose tissue to support that lengthy brain growth and buffer against inconsistent food supply) would also have been *rewarding* to any predator that managed to get their teeth into one.

243 The prefrontal cortex, in particular: Ruff et al., 1997. And it seems particularly on the *left* side, which may be a generalized trend for the ape lineage, making the later hominin Eves an extreme example of Ape Brains at Large (Smaers et al., 2011; Smaers et al., 2017). The human-type globular shape, however, continued to evolve as late as 100,000 to 35,000 years ago, which may be mysteriously tied to modern human behavior (Neubauer et al., 2018), but some features of the braincase may evolve independently from internal organization of the brain itself, making endocasts harder to interpret (Alatorre Warren, 2019). What is abundantly clear, however, is that the degree of encephalization radically increased across the hominin lineage.

243 the hungriest part of your body: How expensive? The modern human body dedicates roughly 20 to 25 percent of its resting metabolic rate to brain activity (Leonard et al., 2003).

246 For instance, the parts of the brain: Premachandran et al., 2020.

246 risk-taking behavior and general aggression: Trivers, 1972; Byrnes et al., 1999; Apicella et al., 2017; Campbell et al., 2021.

246 male mammals also have more androgen receptors: Goldstein et al., 2001; Sato et al., 2004; Dart et al., 2013. However, see Eliot's recent three-decade meta-analysis of sex dimorphism studies, which notes that while androgen receptors do vary in density in ways that are sexually dimorphic, the most robust finding across the sexes in humans is largely tied to size (Eliot et al., 2021).

246 less good at learning from subtle negative stimuli: Yokota et al., 2017.

247 diagnosed with anxiety disorders: Kessler et al., 2012.

247 human brains keep coming up the same: Eliot et al., 2021. In fact, the general cognitive functionality between the human sexes is so bewilderingly similar that some even thought to name it the "gender similarities hypothesis," the implications of which I explore in this chapter (Hyde, 2005).

248 the average adult human's metabolism: Pontzer et al., 2016.

249 everyone else's pressing concern: At least adults and children alike seem to think intellectual brilliance is a male thing (Storage et al., 2020).

250 slightly higher mean IQ than girls: Lynn and Kanazawa, 2011; Ellis et al., 2013.

250 your possible income range: McCall, 1977; Deary et al., 2007; Strenze, 2007; Griffiths et al., 2007. Note, however, that your adult *wealth* is decoupled from your IQ, while your income is not—meaning plenty of smart folk fail to save much money as they go along, and of course your parents' wealth continues to be the strongest predictor of your wealth as an adult (Zagorsky, 2007). Please also note the difficulty in excluding known eugenicists from short lists like these in an endnote—for example, I've deliberately excluded Richard Lynn's research, despite how often he's cited regarding the fertility issue. I do so, in fact, not for the sake of politics (though I'd obviously *personally* prefer that such ideas be wrong), but because problems frequently lie in both those studies' interpretation of data and in their frameworks, as shown in subsequent research and analysis by others (for example, Rojahn and Naglieri, 2006; Savage-McGlynn, 2012). For a usefully scientific critique of Lynn's work on IQ and fertility, please see Nicholas Mackintosh's review (Mackintosh, 2007). I can say that while it may be true that there's a link between fertility and IQ in the twentieth century, it doesn't appear to be easily disentangled from general socioeconomic influences or any of the other messy things that influence a woman's likelihood to have children at one age or another.

250 identical twins separated by adoption: Segal, 2000; Deary et al., 2009; Lee et al., 2010.

250 latest research proposes that it's closer to 80: Deary et al., 2009; Lee et al., 2010; Panizzon et al., 2014; Lean et al., 2018.

250 white Americans tend to have higher IQ scores: Dickens and Flynn, 2006.

251 if you test a large enough group: This idea is well trod. I don't mean to wade into the IQ controversy much in this book, but suffice to say that Black Americans have significantly increased their in-group average IQ relative to white Americans in the last half century, which undermines the idea that genetic drivers are the sole cause of between-group differences (Dickens and Flynn, 2006). No one who works in the science of human intelligence assumes IQ is driven purely by nature *or* nurture, but some amalgam of both. For instance, IQ seems oddly more heritable in adulthood than early infancy, but then declines again in very old age (Lee et al.,

2010). Maybe that's because, until recently, most of us didn't survive into old age, or maybe it's because the complicated factors that drive human senescence aren't entirely genetic, and aging is associated with cognitive decline (ibid.).

251 the most difference is at the tails: Deary et al., 2003; Johnson et al., 2008.

252 male test takers have far more variability: Deary et al., 2003; Johnson et al., 2008.

253 tests that involve spatial reasoning: Halpern et al., 2007.

253 Men tend to memorize paths more abstractly: Maguire et al., 1999. Another way of saying this, as applied to math, is that male and female test takers deal with certain kinds of math questions using different *strategies*, and you're more likely to find differences in scores where questions reward only certain strategies (Spelke, 2005).

253 if you include a tiny picture of a person: Tarampi et al., 2016. Importantly, this inclusion is also framed as a social task, advertised as being something women are better at, which potentially included stereotype threat as a factor in subjects' performance (ibid.).

253 the design of certain IQ test questions: This is counterintuitive, in part, because IQ tests are largely *designed* to produce gender-equal results, selecting questions in development that are more likely to produce gender-neutral scores and scuttling ones that are strongly gender-biased (Halpern et al., 2005). So it's a bit unclear why we've ended up where we are in today's data sets. The answer might be simpler than you think: why not simply assume there *are* some minor sex differences in brains, which naturally produce sex-differentiated *strategies* for certain cognitive tasks, with certain strategies more rewarded in current IQ test questions? Thinking about the mental rotation tasks, for example, and what mechanisms might undergird these sex differences, many point to the parietal lobe, which seems to have a different proportion of white to gray matter between the sexes. In one study, such proportions were, even within-sex, strongly linked to performance on mental rotation tasks (Koscik et al., 2009). But, as the authors note, it's likely a matter of strategy and efficiency: perhaps the males rotate the whole object in an imagined space, while females (or, rather, people with female-typical parietal lobe organization, who tend to be otherwise female) instead rotate the object piecemeal, as suited to their particular parietal wiring. The latter strategy is less efficient, but not necessarily less accurate, given enough time (ibid.; time vs. strategy discussed in greater detail by Peters, 2005; Halpern et al., 2007; and Voyer, 2011). Do we assume, then, that's a difference in baseline general intelligence? Or simply an odd little glitch in an otherwise sex-same functionality? And how, exactly, should one design an IQ test to allow for such quirks between the sexes, without quickly being accused of writing tests for Kurt Vonnegut's Harrison Bergeron?

254 many tests that have to do with language: Hirnstein et al., 2023; Halpern and LaMay, 2000.

254 when the tests involve writing: Adams and Simmons, 2019; Pargulski and Reynolds, 2017. Please note, however, that the effect sizes we're talking about here are small, particularly in IQ tests, and only show up more significantly on tests that aren't already controlling for gender difference, like national writing assessments (Reilly et al., 2019).

254 Boys, as a rule, do a bit more poorly: Halpern and LaMay, 2000. It may depend on how, exactly, the question is asked in these short-answer-format math problems—styles that favor mental rotation to determine a solution may favor a male-typical brain, regardless of how much writing that brain is then required to do—but again, across a broad swath of such things that require articulate writing skills as part of the tester's performance, boys seem to do a bit more poorly, particularly if the tester is required to write at length.

255 the differences in tested ability: Spelke, 2005.

255 few scientific studies that measure: One study from 2007, using undergraduates at the University of Arizona, established that both men and women speak an average of sixteen thousand words a day. However, the males had the wider spread, from one guy who spoke only about five hundred words, to a guy who said about forty-seven thousand (much to the chagrin, one presumes, of his friends) (Mehl et al., 2007).

255 women tend to speak less: Kendall and Tannen, 1997. Importantly, they also speak less at scientific conferences *in general*, but that may be because women are less likely to request a long-form talk session than men; both are equally likely to be awarded the slot (Jones et al., 2014).

256 women and girls are called on less: This is a well-known and replicated finding over the last forty years of research in this area. But it may have as much to do with professors' class structure as male *interruption*—though male voices tend to occupy the sonic space in a classroom roughly 1.6x compared to female voices, it's also true that they speak without raising their hands more often (Lee and McCabe, 2021).

256 when we listen to a conversation: Cutler and Scott, 1990.

256 girls produce their first words: Eriksson et al., 2012. Fascinatingly, four-year-old human boys also seem to have lower levels of FOXP2 protein in their left hemisphere cortex than in same-age girls, but, as with all such studies, larger subject pools would give better insight (Bowers et al., 2013).

256 in a recent large-scale international assessment: Hirnstein et al., 2023.

256 males scored *higher* on that test: Spelke, 2005.

257 Large data sets from the U.S. Department of Education: Peterson, 2018.

257 Men make up only 20 percent: Weiner, 2007, citing Nielsen Bookscan surveys in the United States, Canada, and U.K.

257 boys aren't great at either: Scheiber et al., 2015. Because writing isn't tested as often as other types of verbal skills, knowledge of this gap isn't as widely supported as other verbal ability gaps, but it deserves more attention (Reilly et al., 2019).

258 most human beings weren't even remotely literate: Despite global gains in education in the last twenty years, even basic literacy—the mere ability to read text in one's native language—remains at a stubborn 86 percent worldwide (UNESCO, 2014). If that seems low, consider that forty years ago it was 68 percent, which again might seem reasonable, until you realize that in the early 1800s roughly 12 percent of the adult human world had any ability to read and write, and the entire nineteenth century improved that number by only about 9 percent (UNESCO, 1953; UNESCO, 1957). Even Ancient Rome, considered among the more literate of ancient societies, seems to have had a literacy rate that never rose above 10 percent, and most of that was clustered in the cities (Harris, 1991). Reading is simply not something the human species has done much of, anywhere, ever. I'm saying this in an endnote in an already-long book, which is an odd thing to do, but it's still deeply true.

258 boys are two to three times more likely to be dyslexic: Rutter et al., 2004; Quinn and Wagner, 2015.

258 as of 2013, less than 20 percent of students: Quinn and Wagner, 2015.

258 the gap in reading ability widens: Reilly et al., 2013.

260 diagnosed with major depressive disorder: Baxter et al., 2014. Please note, however, that in *global* data, the gap between male and female MDD diagnoses seems to narrow as societies become more gender-equal, which could reflect both a reduction in female gender role stressors and women's access to birth control, which itself might work to reduce depression by controlling hormone cycling and, of course, reducing the number of times a woman's brain is at risk of postpartum depression (Seedat et al., 2009).

260 in data from the United States: Eaton et al., 2012.

262 Women who suffer from depression: Soares and Zitek, 2008.

262 Bipolar women will also sometimes report: Rasgon et al., 2003.

262 Young girls *don't* seem to be more depressive: Cyranowski et al., 2000.

262 Though women are about 12 percent more likely: Terlizzi and Norris, 2021.

263 Men are slightly more likely to be diagnosed with schizophrenia: Aleman et al., 2003.

263 Men are also more likely to have debilitating drug and alcohol addictions: NIDA, 2020. Women are just as likely to develop substance abuse disorders, in fact, but men have much higher rates of use and dependence (ibid.).

263 But men and women are equally likely to be diagnosed with OCD:

Though females tend to be diagnosed later in life, both are roughly equal in diagnosis and treatment outcomes (Mathes et al., 2019).

263 Among bipolar people, women patients: Arnold, 2003.

264 Women commit suicide roughly three times less often: Krysinska et al., 2017.

265 women having a more robust social support network: This is notoriously tricky to measure, of course, because it's all self-reported; saying someone has a strong social support network is as much about that person's *feelings* of support as the actual people potentially involved, and both depressive and suicidal patients are rather known for being bad at positively perceiving the opinions others may have of them. Nonetheless, it's true that one either does or doesn't have a friend to call, or family members, or significant others. And women generally report both higher feelings of social support and larger numbers of people involved in those groups. When they grow older, this is also strongly tied to how long they're likely to live (Shye et al., 1995).

265 suicidal mothers are less likely to try: Dehara et al., 2021. Sadly, this is not the case for mothers whose children are taken into Child Protective Services (Wall-Wieler et al., 2018). And among pregnant and postpartum women, whether or not that woman has experienced abuse is a strong risk factor for suicidal ideation—much more, in fact, than issues of social support, though women without much social support were still more likely to attempt to kill themselves than other mothers (Reid et al., 2022).

265 the overall feeling of "closeness": Machin and Dunbar, 2013.

266 traumatic brain injuries: However, in the small amount of published data available for TBIs properly analyzed for sex differences, it oddly seems women fare a bit worse (Farace and Alves, 2000), despite clinical opinion siding with the idea that women patients fare better and the overwhelming evidence that female rodents in the lab likewise do better on most measures. This paradox may have a few different things driving it: First, the few published studies that properly present and analyze for sex differences here may be too poor a selection—one meta-analysis in 2000 counted 8 (ibid.). Second, modern human women experience more menstrual cycles than they would have historically, which includes monthly withdrawals from peaks in progesterone.

266 What takes that male patient: Please note that this is primarily for *severe* TBI. Mild to moderate TBI actually seems to have worse outcomes for women patients (Gupte et al., 2019). Thus, there's a bit of a gap between clinical expectation and patient outcomes, perhaps due in part to the differences in post-concussion syndrome in women, which presents differently (and seems to persist for longer) than it does in men (ibid.). One key to these differences might be that in mild (concussive) cases of TBI, what's more at stake is less the widespread edema and cell death that

one sees in severe TBI, and more how local axons respond in the brain around and in the injury site, which might be influenced by the pre-injury structure there, which is now known to have sex differences (Dollé et al., 2018).

266 Male brains seem to suffer more extensive inflammation: This is particularly true for severe TBIs, seen in both rodent and human, and the link between sex differences seen in animal models and human clinical reports clears up a bit in meta-analyses, with the greatest effect seen in severe TBI (Caplan et al., 2017).

266 If you do terrible things to a rat's brain: O'Connor et al., 2005.

266 clinical trials for human beings are under way: Two such trials completed in the last decade show mixed results (Skolnick et al., 2014; Wright et al., 2014). It may depend when and how the progesterone is administered: quite soon after injury, with no hormonal follow-up, doesn't seem to help, and slightly increased the chance of a stroke (Wright et al., 2014). But that might be, in part, because giving a quick dose of progesterone with no follow-up may risk *withdrawal* from PG afterward, which itself puts brains at risk—as seen in menstruating women—of a number of bad brain outcomes, including mood destabilization. In fact, when a woman suffers from a TBI during the luteal phase, when progesterone is naturally high and then drops off, she'll fare worse than women who are on birth control (who have consistently high PG) and women who are in a different phase of their cycle (Wunderle et al., 2014). Annoyingly, menstruating women and girls are also more *likely* to suffer such injuries during the luteal phase, potentially due to more lax joints, which might be part of what's driving women patients' poor outcomes in the data (Wunderle et al., 2014). In sum, progesterone shows promise for severe TBI, but it matters when you give it, and how much, and for how long, and it might simply help more for male brains than female brains, in the end.

267 Estrogens do seem to stabilize: Sohrabji, 2007. This may be particularly important for out-of-hospital (rural/battlefield) stabilization after injury, with females (as usual) surviving better than males (Mayer et al., 2021).

267 Progesterone, too, seems to play a role: Roof and Hall, 2000; Sayeed and Stein, 2009. For a solid review paper highlighting confounds and the overall mixed results of such studies, see Caplan et al., 2017. Microglia may play an important role, perhaps under the influence of progesterone particularly. Sex hormones can't be dismissed out of hand, given that postpubertal girls and women have better survival for severe TBI than males of any age and prepubescent girls (ibid.). However, the age at which injury occurs doesn't indicate *only* hormonal status. While authors like Caplan tend to look at pre- and postpubescent pediatric groups to look for hormone signal, it's also true that developing brains respond to injury a bit differently precisely because they're on a developmental pathway, which is an independent confound (Arambula, 2019).

267 higher self-awareness in terms of their limitations: Turkstra et al., 2020; Rigon et al., 2016. They're also known to be more proactive in seeking medical care and rehab (Chan et al., 2016).

268 Take Parkinson's disease: Gillies et al., 2014.

270 an Eve of the modern hippopotamus: There are a number of these in the fossil record, but the most obvious case would be *Hippopotamus gorgops*, a massive beast roughly twice the size of the modern hippo. Interestingly, this ancient hippo and ancient hominin Eves seem to have migrated at similar times (roughly 1.9 mya), when increasing humidity made for bigger lakes and rivers, which made the Sahara possible to cross (Zhang et al., 2014; van der Made et al., 2017). Given how many humans today's hippos slaughter, however, it's not likely the two would have been *friends* (van der Made et al., 2017). In fact, Erectus also had the habit of butchering them, when possible (Hill, 1983; Lepre et al., 2011)—a far cry from the adorable baby hippos we admire and protect in today's zoos.

270 *Theropithecus oswaldi:* Diet was a major factor in their extinction, though not merely from climate change: competition with local ungulates also probably mattered, as there's only so much grass to go around (Cerling et al., 2013). The last surviving members of their genus, the geladas we met in the "Tools" chapter aren't as large and don't have as many ancient proto-antelope to contend with in the Ethiopian highlands. Please also note that our Eve Erectus may have *also* contributed to their extinction, given the large number of butchered juveniles found at Olorgesailie in Kenya (Shipman et al., 1981). Given that today's chimps commonly hunt other primates when the opportunity presents itself, it's not that much of a stretch to think that Erectus may have hunted the more vulnerable members of a local troop of grass-eating baboons. It's not that bigger brains *couldn't* have made our Eves more dominant in the local food web; it's more that hunting, on its own, seems unlikely to be the central driver.

271 About six to seven million years ago: I'm leaning heavily on Potts's work on variability selection here. In particular, see his paper with Faith from 2015.

273 when human beings are born: DeSilva and Lesnik, 2006.

273 Chimps come out of the womb: Ibid.

274 Our brains aren't done internally organizing: Goddings et al., 2019.

275 Boy babies, meanwhile, tend to squirm: Fausto-Sterling et al., 2012. Importantly, we're really talking about early infancy here, predominantly in the first four months of life. These sex differences in motor skills then seem to drop off, only to pick up again after twelve months, which may have as much to do with the child interacting with gendered expectations for play as it does the body itself (ibid.).

275 boy babies are more likely to be born slightly prematurely: Blencowe et al., 2012. Unfortunately, preterm babies also don't do as well if they

happen to be male, despite tending to weigh more, which is otherwise thought of as an advantage for preemies (Peacock et al., 2012).

276 certain kinds of autistic brains over-prune or under-prune some regions: This is a broad field, and what we presently call the autistic spectrum will likely end up representing many different disorders with a variety of underlying mechanisms. But for a solid paper specifically tying autism to synaptic pruning, see Tang et al., 2014.

276 also had these chimpy childhood patterns: One tooth study does indicate they may have entered adulthood roughly three years earlier than *Homo sapiens* (Smith et al., 2010), while another suggests they were better able to process supplemental foods by age two (Mahoney et al., 2021). As ape developmental patterns go, that puts Neanderthal childhoods somewhere in between the average chimp and the average human.

277 The adolescent "bloom" isn't nearly as prolific: Goddings et al., 2019.

278 key pathways getting extra myelinated: Chavarria et al., 2014; Genc et al., 2023.

278 males prune later and *faster:* De Bellis et al., 2001; Neufang et al., 2008.

279 Being handicapped in this way: Piantadosi and Kidd, 2016.

281 A pregnant woman's brain: Hoekzema et al., 2017; Hoekzema et al., 2022.

281 Similar things seem to happen: Barba-Müller et al., 2019.

281 the volume loss is most notable: Hoekzema et al., 2017; Barba-Müller et al., 2019.

287 "her dresses showed all the imagination": Watson, 2001.

287 two networks the brain uses: Koss and Gunner, 2018; Wadsworth et al., 2019.

288 A little bit of stress is good: Miller et al., 2007; Koss and Gunnar, 2018; Wadsworth et al., 2019.

288 Psychological research is pretty clear: Mrazek et al., 2011; Berger and Sarnyai, 2015.

288 she's not going to do as well: Johns et al., 2005.

288 they'll be worse at a test asking them to discern: Eisenberger and Lieberman, 2004.

288 If you tell Black subjects: Sellers et al., 2003.

288 In people who encounter threat every day: Wadsworth et al., 2019.

289 similar patterns in people who suffer from chronic pain: Ibid.

289 emotional and perceptive *detachment:* Eisenberger and Lieberman, 2003.

289 a kind of "psychological disengagement": Ibid.

291 between the amygdala and the hippocampus: Fish et al., 2020.

291 some structural differences in the olfactory bulb: Oliveira-Pinto et al., 2014.

291 "ambiguous genitalia": Johannsen et al., 2006.

292 rough-and-tumble play: Dinkle and Snyder, 2020.

292 some XX girls do, too: Ibid.

CHAPTER 7 VOICE

297 "History here is an oral tradition": Kapuscinski, 2001. I love much of this man's writing. His thoughts on "Africans" maybe less.

297 "So caught up was I": Solnit, 2001.

297 Someone found him in a heap: My portrait of this accident is drawn from a report on telemedicine in Vermont and upstate New York in 2001 (Rogers et al., 2001).

300 We might have started talking: Everett, 2017. Even more controversially, one recent paper decides *none* of the most popular anatomical restrictions on speech capabilities are particularly unique in the hominin line, and thereby pushes the dawn of speech back to over 20 million years (Boë et al., 2019). Others stick to the guns, perhaps well fed on Chomsky, and maintain human specialness, comfortable only with 200,000 years or so—after the dawn of our species but before strongly symbolic behavior.

300 Some think it was only 50,000 years ago: Lieberman, 2007.

301 The soonest hominins seemed to have: Barney et al., 2012.

302 innovation spread at a pace: Aiello and Dunbar, 1993; Dunbar, 1993, 1996.

302 What is humanity's earliest art all about?: Well, unless you count handprints (Sharpe and Van Gelder, 2006; Bednarik, 2008; Zhang et al., 2021; Fernández-Navarro et al., 2022). Using a reed filled with red ocher dust, a number of cultures have made ghostly hand impressions on the walls of ancient caves. Other times tiny handprints have been found in soft mud. We only recently discovered, however, that many of these hands likely belonged to *children*—some as young as two or three years old—which rather puts a damper on the theory that ancient art is all about hunting or god (Langley and Litster, 2018). While it might be a kind of religious christening ceremony or a case of including children in normally adult activities (that is, tantrum avoidance under duress), it might also have been a bit like what mothers do with young children today: *give them some art project to make it from breakfast to freakin' nap time.*

303 Many scientists think even *Homo erectus:* And because many big carnivores that lived alongside them, like the felids I portray in the opening scene of the "Brain" chapter, just so happen to leave large amounts of meat on their prey, the number of such creatures living alongside our hominin Eves in any given ancient Eden might well have provided enough such nutrition for creatures like Erectus (Pobiner, 2015).

303 The warning calls are so flexible: Outtara et al., 2009.

303 human subjects like listening to male voices: And masculine (i.e., lower-pitched) voices in women, too, in certain leadership roles (Anderson and Klofstad, 2012) but not necessarily in the case of dating, which is more culturally and individually shaped.

304 Philadelphia, 2016: Video of this event is available free online through PBS NewsHour at youtu.be/pnXiy4D_I8g.

305 Men have 10–12 percent more absolute lung volume: Bellemare et al., 2003.

306 the force of air in the human respiratory tract: Nishimura, 2006; MacLarnon and Hewitt, 1999; Lieberman, 2007; Ghanzafar and Rendall, 2009.

307 women's brains send more frequent impulse signals: That finer control may be part of why women's diaphragms also seem to be more protected from fatigue (Geary et al., 2019) and both may be part of why women do so well in endurance sports.

307 Hillary actually took *five* breaths: You can watch her do this, if you like, because video of this moment is widely available on the internet.

308 If you're a woman who uses your voice professionally: Hunter et al., 2011.

309 hominins lost their throat sacs: de Boer, 2012.

309 we had throat sacs until very recently: Ibid.

310 speech benefited from their absence: Ibid.

311 Infections of their laryngeal pouches: Lowenstine and Osborn, 2012.

312 larynx dropped lower in our throats: Lieberman, 2007. Please note that, as with many things in academia, the debate around whether or not the human vocal apparatus is the essential feature for the evolution of human speech is contested. W. Tecumseh Fitch (surely the best-named scientist in the field) disagrees with Lieberman as to whether macaques could produce speech sounds based on their anatomy—Fitch says yes and only neurological evolution holds them back, while Lieberman thinks the less-electric wetware still matters (Fitch et al., 2017). To some degree this is probably a simple intergenerational kerfuffle, but it also represents deeper divides in how scientists approach these questions: what can we learn from fossils (Lieberman) vs. living mammal physiology (Fitch) to model how our Eves evolved to make something as complex as human speech?

312 human males are able to hit bass notes: Fitch and Reby, 2001.

313 It also pumps its penis up and down: Ibid.

313 Higher-pitched women's voices: Zuckerman and Driver, 1989.

313 women with naturally lower-pitched voices: Ibid.

313 Across the menstrual cycle: Ryan and Kenny, 2009.

314 Those with more bothersome symptoms of PMS: Banai, 2017; Ryan and Kenny, 2009.

314 changes in their voices at menopause: Schneider et al., 2004.

315 The uterus is the strongest muscle: This is both for constricting pressure in general and by potential force compared to the muscle's *weight:* the average adult uterus weighs about forty ounces but can exert a force of up to four hundred newtons with each contraction during labor. That's about eighty-eight *pounds* of downward pressure.

315 Australopithecines used to have their hyoid bones: Capasso et al., 2008; Steele et al., 2013.

316 we started walking upright: Steele et al., 2013.

317 Girls are less likely to develop lisps: Black et al., 2015.

317 Our unique human genes have preprogrammed: So much has been written on this subject that providing a single citation feels silly. But if you want to read a nearly perfect book, see Steven Pinker's *The Language Instinct*. I may differ from his position on motherese, but otherwise I'm an obvious student of his work. An acolyte, really.

319 French babies cry in a rising melody: Mampe et al., 2009.

322 For fluency in a second language: Hartshorne et al., 2018.

322 zebra finch parents: Gobes et al., 2019; Chen et al., 2016.

323 what seems to be the most critical part: Friedmann and Rusou, 2015.

324 prone to using motherese: Piazza et al., 2017. However, in same-sex couples, male heterosexual couples show no disadvantage (or even any particular difference) in their child-directed speech patterns, though there were some differences between primary and secondary caregivers within-couple (Grinberg et al., 2022). Given that male babies are likely exposed to these speech patterns in their own infancy, there's no reason to assume adult men wouldn't be perfectly adept at using it. Perhaps the dominance of female use of motherese, then, reflects a more cultural norm: women are usually the primary caretakers of human babies and may therefore have more occasion to *practice*.

325 especially effective at getting the infant's attention: Slonecker et al., 2018.

325 Squirrel monkeys also call to their babies: Biben et al., 1989.

325 Even dolphin mothers communicate: King et al., 2016.

326 ones who hear a motherese-style song: Chen et al., 2016.

326 a very common feature of motherese across languages: Han et al., 2018. These mothers also change the degree to which they modify their speech as the child gets older, as is often seen in other language cultures that use motherese (Liu et al., 2009).

326 Babies whose mothers put more emphasis on vowels: Fascinatingly, this works for *computers*, too: a computer model presented with samples of speech was better at identifying the vowels when the samples were motherese (de Boer and Kuhl, 2003).

326 might help us distinguish different words in a string: Thiessen et al., 2005.

327 The typical range of pitches in motherese: Among British folk, meanwhile, "fatherese" differs a bit from "motherese" in that fathers manipulate the prosody of their speech to infants more than the mothers do (Shute and Wheldall, 1999).

330 For example, just by changing the tense: And that understanding can be interestingly language- and culture-specific. Speakers of Mandarin and English seem to understand time slightly differently, which may be in part because their use of their respective languages orients them to slightly different constructs of temporality (Boroditsky, 2001).

332 human language came about in fits and starts: As a number of prominent scientists have gently pointed out, anatomically modern humans were living in the world for a good long time without any particular evidence of symbolic culture. The body was there. The brain was there. The mouth and throat and tongue and hypoglossal nerve canal, all of it. But if symbolic culture is rooted in the sort of narrative cognition we associate with linguistic, symbolic story making, those Eves didn't have much to speak of.

332 deception is ancient: King, 2019.

CHAPTER 8 MENOPAUSE

337 *"And yet, and yet"*: Borges, 1962/2007.

337 "Damn, I got out of hand!": Borges, 1978. He seemed fond of this anecdote after she died, though the precise birthday seemed to change depending on the interview—sometimes ninety-five, sometimes ninety-eight. Borges's mother was quite famous as his companion; his memory, unfortunately, was not (Alifano, 1984).

337 No time for an old body's complaints: It seems at least one genetic mutation associated with our departure from Africa greatly increased the risk for osteoarthritis. It shortened our bone growth, which was great for colder climates, but rather crappy for longer-term wear and tear (Capellini et al., 2017). The same mutation was found in the Neanderthals and Denisovans (ibid.). By the time our Jericho Eve came about, the likelihood that she, too, might have carried this ancient risk of aching joints was significant—yet another way our Eves' journeys carry on in our bodies long after their Edens fade away.

338 The girl's mother had died: Floods were a regular problem for ancient Jericho and might well have been the reason the walls were built in the first place (Bar-Yosef, 1986).

338 Everyone seemed to want a job: There's a strong link between grieving and task fulfillment, after all, and in times of crisis the feeling that one has agency can be incredibly motivating *and* soothing (Riches and Dawson, 2000).

338 woman turned the child in the womb: Because so many human fetuses aren't yet head-down before thirty-seven weeks, breech births are a particular risk for preterm babies (Bergenhenegouwen et al., 2014), but they remain a problem—as they presumably were throughout human history—for women today. In the 1600s in London, quite a business was made of advice for midwifery, and the issue of how to deal with breech births was nearly always included (Walsh, 2014). These days, C-sections are generally recommended when possible, but the likelihood of the mother surviving such a thing in ancient Jericho would have been slim.

340 more prone to osteoporosis: Karlamangla et al., 2018. However, the

three-year period around the final menstrual cycle is when the most rapid bone loss occurs, setting the stage for later osteoporosis (ibid.).

341 But for scientists who study evolution: Ellis et al., 2018.

342 the beginnings of the grandmother hypothesis: Hawkes, 2003.

344 every year of being on high-dose hormonal birth control: de Vries et al., 2001.

345 The data did suggest that I wouldn't: Longitudinal studies of egg donors are a bit thin in the literature, of course, given how recently egg donation (and IVF in general) became popular. But the general theory holds that what is recruited during an egg-harvesting cycle is not "extra" eggs, but rather eggs that would otherwise be normally lost to atresia.

345 mammalian eggs may have an expiration date: Double-strand breaks may even be tied to ovarian aging, it turns out (Oktay et al., 2015), but given that some other long-lived animals continue to give birth into their old age (elephants, for example), human menopause seems less likely to be tied to an innate property of mammalian eggs.

347 Even chimps can give birth in their sixties: Thompson et al., 2007.

349 chimpanzees stop ovulating around age fifty: Hawkes and Smith, 2010; Herndon et al., 2012.

349 chimp grannies are very sexy: Muller et al., 2006.

349 primate ovaries age at similar rates: Hawkes and Smith, 2010; Alberts et al., 2013. Note that Alberts rightly finds that across a *larger* swath of primates, the most important signal is less that "primate ovaries clock out at age 50" (many species simply don't live that long, but also experience some amount of reproductive senescence at differing points) and more that human bodies radically outlive ovarian production: so long as she has normally functional ovaries, the rest of a woman's body ages at a much slower rate across the lifespan than those ovaries will.

352 Creating and maintaining a regular class: While extant hunter-gatherer societies have women who live into their postmenopausal years, the numbers are smaller. Consideration of "menopause" as both a physical and a social phenomenon requires different measures. However, it should be noted that because many of these women's lives are more *vigorous* over the life span, some of the menopausal symptoms women in agricultural/urban lifestyles suffer may actually be less of a bother. For instance, hot flashes are more likely to be a problem for women in New York than for women in hunter-gatherer communities (Freeman and Sherif, 2007), which might be tied to lower amounts of adipose tissue and/or general cardiovascular health.

353 Transient orca pods: Marsh and Kasuya, 1986.

353 postmenopausal orcas don't spend more time: Brent et al., 2015.

353 What the grandmothers *are* responsible for: Ibid.

354 an Afghani woman named Abedo: Ehsan, 2011.

355 "Modern-day youngsters in the police": Ibid.

356 aging-agriculture feedback loop: Austad, 1994. One important pressure I haven't mentioned—which could be a good argument for pushing the usefulness of menopause earlier—is the Last Glacial Maximum (LGM) associated with the last great ice age on our planet. Half of Europe was covered in glaciers. Climate changed back in Africa, too, and the Middle East, and really anywhere one happened to be on the planet. Human populations retreated to refugia along the Mediterranean (Posth et al., 2023). Climate changed pretty quicky and proved deadly for many. Though humans were hunter-gatherers then, the LGM provided ample opportunity for the usefulness of intergenerational know-how in severely challenging environments. How many of those potentially older females even *survived* then, I couldn't say. It's known that our global population contracted considerably in the face of climate change (ibid.), so somewhere after the ice receded (14,000 years ago) seems, to me, a better bet for when *societies* of elderly women would have a chance to get going.

360 Centenarians used to be unicorns: International statistics in this paragraph are taken from the United Nations' World Population Prospects report in 2015.

360 More than 80 percent of today's centenarians: Meyer, 2012.

360 Of the three people alive today: They are Maria Branyas, Fusa Tatsumi, and Edie Ceccarelli. When I finished writing this chapter, there were four, but Lucile Randon (aged 118) unfortunately died in January 2023. The best place to look for who's still around is actually the Gerontology Research Group, an international nonprofit that usefully publishes their database of verified supercentenarians online at grg.org/WSRL/TableE.aspx. Though many census-keepers exist in government bodies and likewise track these things, public reports aren't as regularly updated.

361 And it's true among our ape cousins: Bronikowski et al., 2011.

362 most people in today's industrialized countries: You et al., 2015. Please note that this simply wasn't the case for the majority of human history (Volk and Atkinson, 2013).

362 on the lookout for what *male* bodies do: Shaw et al., 2008; Maas and Appelman, 2010.

363 The male cardiovascular system: Mozaffarian et al., 2016.

363 men and boys who fell ill with COVID-19: Takahashi et al., 2020.

363 better modeled as a cardiovascular disease: Reynolds et al., 2020.

363 Lung disease is another one of the Big Three: Gordon and Rosenthal, 1999.

363 sex matters for the immune system: For one recent well-executed review paper on this subject, see Klein and Flanagan, 2016.

364 pregnant women were more susceptible: Smith et al., 2023.

364 women who are diagnosed with lung disease: Martinez et al., 2012.

365 one alcoholic drink a day: Lowry et al., 2016.

365 But, in general, more men get cancer: Dunford et al., 2017.

366 This turns out to be true *except* for about fifty: Ibid.
366 Across twenty-one different sorts of cancers: Ibid.
367 The ancient people of Jericho: Nigro, 2017; Kenyon, 1957.

CHAPTER 9 LOVE

376 two physical traits usually tied to polygyny: Plavcan and van Schaik, 1992; Lindenfors et al., 2007; Plavcan, 2001, 2012b.
377 Male chimps and bonobos: Plavcan, 2012b.
377 fossils going back 300 million years: Benoit et al., 2016.
377 Let's start with weight class: Plavcan, 2001.
378 male and female *Australopithecus:* Reno et al., 2003; Reno et al., 2010.
378 male canines keep getting smaller and smaller: Suwa et al., 2009; Plavcan, 2012a. Though it's possible the canines were already quite small in males by Ardi's time (Suwa et al., 2021), which implies at least some of this pattern of reduction seen in later hominins may be a kind of mosaic evolution (Manthi et al., 2012) or a problem of statistical method in prior studies (Suwa et al., 2021).
378 Tooth size seems to be modulated: Alvesalo, 2013. Exposure to androgens in the womb may also matter (Ribeiro et al., 2013).
378 The trend, if anything, is convergence: Plavcan, 2012a; Reno et al., 2010; Lovejoy, 2009.
379 Promiscuous primates have gigantic balls: Shultz, 1938; Anderson et al., 2007; Kappeler, 1997.
379 When mandrills compete with one another: Setchell and Dixson, 2001.
380 the thicker this seminal plug: Dixson and Anderson, 2002.
380 And it's only thick and sticky at first: Zaneveld et al., 1974. The "spunk" of spunk, in other words, is quite temporary.
380 When in contact with a woman's fertile cervical fluid: Suarez and Pacey, 2006.
381 but rape is incredibly rare: de Waal, 2022.
381 when it comes to sex, chimp males: Ibid.
382 Primatologists called this "mate guarding": Muller et al., 2007.
382 They might even chase a too-aggressive male: Tokuyama and Furuichi, 2016.
382 bonobos don't do a lot of mate guarding: Smuts and Smuts, 1993; de Waal, 2022.
382 Genetically, we're equally related to both: Mao et al., 2021.
384 species that commonly use rape: Brennan et al., 2007; Orbach et al., 2017; Brennan and Orbach, 2020.
384 human vaginas are only a tiny bit foldy: The folds we do have are called rugae and seem largely to be about having enough tissue to properly expand when we're aroused and a large, intruding member might be about to make an internal visit (or a large-headed baby might be about to

make a violent exit), which is also one of the better arguments for why the human penis is wider and longer than other apes' (Bowman, 2008).

385 One lab even created: Gallup et al., 2003.

386 The acidic vaginal environment: Brannigan and Lipshultz, 2008.

386 It also has a useful structure: Ulcova-Gallova, 2010.

386 more than *four times* as long to ejaculate: McLean et al., 2011; Reno et al., 2013.

386 The bigger they are: McLean et al., 2011.

387 Thus far, it looks like about 510 deletions: Suntsova and Buzdin, 2020.

387 One—an androgen receptor: McLean et al., 2011; Reno et al., 2013. For a recent overview of penile spines in primates, see Dixon, 2018.

388 Among mallard ducks, a raping drake: Snow et al., 2019.

388 The most popular argument: Lovejoy, 2009; Dixon, 2009; Leigh and Shea, 1995.

389 this is called "paternal uncertainty": Hrdy, 1979. In fact, in one recent model, the more likely the infanticide in a given primate community, the longer the female's sexual receptive period becomes (until, one presumes, you arrive at something like the chimpanzee or the bonobo, where sexual receptivity happens essentially all the time) (Rooker and Gavrilets, 2020).

391 Daughters inherit their social rank: Melnick and Pearl, 1987.

392 all bonobo females look out for the babies: Furuichi, 2011; Tokuyama and Furuichi, 2016; de Waal, 2022.

393 babies are certainly under threat: Alberts, 2018. It seems to strongly depend on local culture and demographic pressures, with some groups having as little as 2.3 percent of infant deaths tied to infanticide, and others anywhere between 38 percent and 70 percent (ibid.; Zipple et al., 2017). Infanticide is more reliably true among the chacma baboon, where males killing babies seems to be a deep part of male reproductive strategy (Palombit et al., 2000). Among all baboons, it's more likely for *immigrant* males to commit infanticide than residents (Alberts, 2018), and those cases are more reliably deterred by groups of resident *male* coalitions, who are presumably motivated by paternal uncertainty (Noë and Sluijter, 1990).

393 Primatologists have seen this many times: The literature on agonistic buffering and male infant-carrying runs pretty consistently from the 1970s forward and is frequently observed in the field. It's also present in other social primates: Gelada males also carry babies around as a strategy to decrease conflict (Dunbar, 1984) and macaques seem to do this agonistic buffering, as well (Deag and Crook, 1971). For macaques, this might be tied to a "care-then-mate" strategy, wherein a male helps a lactating female with childcare and then gets a chance for himself once she ovulates again (Ménard et al., 2001).

395 the monkey world's gender-flipped version: Such couplings will typically be concealed through one or another sort of deception, in order to avoid

the rage of the dominant male (le Roux et al., 2013), though he may tolerate it to some degree in exchange for the parenting help and/or assistance in defense of his harem (Snyder-Mackler et al., 2012).

395 the friendlier you are with a female's offspring: Smuts, 1985. Please note that males and females that form bonds also *live longer*, potentially because of the stress reduction of receiving extended grooming (Campos et al., 2020). In fact, they're 28 percent less likely to die, no matter their age (ibid.). There are, in other words, many perks to cross-sex relationships among primates, and not all of them are directly reproductive.

395 men are more likely to be abusers: Leemis et al., 2022; ONS, 2020; Hester, 2013.

396 if their societies were both matriarchal *and* matrilocal: Here I should tell you that if our hominin Eves were matrilocal, that would have been a very, very long time ago. Recent research indicates females traveled far more widely than males for many thousands of years back, which strongly indicates a patrilocal history during humanity's global expansion (Dulias et al., 2022), though males *also* traveled quite a bit throughout Eurasia, so different forces may be at play depending on the culture at the time (Goldberg et al., 2017). But, as I've mapped throughout this chapter, the shift from prior mating strategies to one with less male competition has evidence throughout the hominin line, so the start of the devil's bargain should be a very long time ago, indeed.

398 their helpful, collaborative, Nice Guy genes: It's hard to know which "genes" these would be, exactly, that lead a male to be more solicitous and friendly toward females and their offspring. But there was one famous case of a troop of baboons radically shifting from competitive to friendly behavior: a bunch of aggressive, dominant males died suddenly from infection, and the males that remained were friendlier. It didn't take long for the entire group's general behavior to shift to friendly collaboration over competition, which remained true in subsequent generations (Sapolsky and Share, 2004).

398 less likely to lose their offspring to infanticide: Lowe et al., 2018. This is particularly true if those males have higher social rank, and particularly during times of social instability in the troop (ibid.).

398 some chimp infanticide is committed by other *females:* Townsend et al., 2007.

400 But research confirms that women use that word: Bartlett et al., 2014; Armstrong et al., 2014.

407 most women do not stick to sex with only one man: Should I cite common sense? Formally, let's look to Kinsey, 1948, and simply assume the little scrap of history that the prior one thousand years represents doesn't bear much weight on the three-hundred-thousand-year history of our species.

409 Consistent condom use is remarkably low: Felisbino-Mendes et al., 2021; Fernandez-Esque et al., 2004; Kim and Cho, 2012; Harrison et al., 2008.

409 Latino men in the Miami area: Sastre et al., 2015.
409 the less promiscuous people in question: Newman et al., 2015.
410 The Centers for Disease Control has been tracking: CDC, 2015.
410 Syphilis rates tripled between 2012 and 2014 in Louisiana: Bowen et al., 2015; CDC, 2015.
411 The majority of chlamydia infections: Wiesenfeld et al., 2012.
412 Every year, roughly sixty-two million people: Kirkaldy et al., 2019.
412 gonorrhea has been around since the time of the Old Testament: Because gonorrhea doesn't leave skeletal traces, it's a bit harder to find—one is left more with ancient texts than hard evidence. One scraping of dental plaque, at least, places it solidly in the twelfth century (Warinner et al., 2014). For an amusingly thorough account of the search for ancient gonorrhea, see Flemming, 2019.
412 a vaccine for chlamydia: Abraham et al., 2019.
412 That's still true of today's well-studied: Howell, 1979/2017.
413 Maternal age is the single most predictive factor: Raj and Boehmer, 2013.
413 Girls married before they enter puberty: Nour, 2006.
414 Ancient Greece aimed closer to sixteen: Baber, 1934; McClure, 2020.
414 in Rome the younger brides: Frier, 2015. It's likewise true that even in medieval Europe, wherein stories of royals marrying off at twelve were common, the average age of marriage for a woman outside of the nobility was twenty to twenty-five (Shapland et al., 2015).
415 pregnant women and new mothers are dying: Hoyert, 2022.
416 Where are most pregnant American women dying?: Martin and Montagne, 2017.
417 How much money a child's parents have: Currie and Goodman, 2020.
417 financially supporting females: The impacts are seen in both large-scale developmental programs (Woetzel et al., 2015) and in data from the world of microloans (Quigley and Patel, 2022; Mahjabeen, 2008), but it's also reshaping how some economists model historical trends in the United States and Europe (Diebolt and Perrin, 2013).
418 When women in India were given the opportunity: Chattopadhyay and Duflo, 2004.
418 Similar trends can be seen: Hessami and da Fonseca, 2020.
419 When women are empowered in local governance: De Araujo and Tejedo-Romero, 2016; Stanić, 2023. In the United States Congress, female lawmakers also send roughly 9 percent more discretionary funds back to their home districts than their male colleagues (Anzia and Berry, 2011).
419 Many well-regarded economists: IMF, 2018.
420 formal education reliably increases: Autor, 2014. Though in the United States, the cost of student *debt* rather complicates the picture (ibid.).
420 For every additional year you educate a girl: Wodon et al., 2018.

420 The World Bank estimates: Caldwell traced this trend in historical data back in 1980. To dig a little deeper in the WHO data, see Pradhan, 2015.
421 more of Kerala's women are educated: Nair, 2010; Nussbaum, 2003.
421 the greater the number of girls: Wodon, 2018.
421 even in agricultural communities: Ibid.
422 If you have an IQ even fifteen points higher: Whalley and Deary, 2001.
423 In many Indian states, it's normal for young women: Hathi et al., 2021.
423 the average woman in India weighs *less:* Coffey, 2015.
425 Poor nutrition in early childhood: Northstone et al., 2012.
425 Malnourished babies tend to become adolescents: Galler et al., 2012.
425 Malnourished mothers are far more likely: Li et al., 2016.
426 they're far less likely to continue: Wodon et al., 2018.
427 women should be considered equal to men in all respects: Belo, 2009. I would like to mention, however, that he *also* wrote that, despite women's equal nature and general potential to men, "since women in these cities are not prepared with respect to any of the human virtues, they frequently resemble plants. Their being a burden upon the men in these cities is one of the causes of poverty" (Averroes, 1974). One is tempted to mention that women's unpaid labor is, of course, a *great* contribution to the ability of others to engage in paid labor, however educated and prepared in virtue those women may be—witness the ever-climbing cost of child care in the United States!—but modern ideas about poverty and income (and even money generally) are rather different from anything that existed in Averroes's time. And, again, Islam did rather better for women than Christianity in this period, by many measures.
427 Public schooling didn't take hold: Consider, for example, the Elementary Education Act of 1891, wherein England *only then* made primary education free. It cost about 10 shillings a head (Boos, 2013). Two years later, the British government bumped it up to age eleven or so, and decided it'd be a good idea for deaf and blind children, too. Dickens would be proud.
428 as of 1989, many Arab nations had become: United Nations, 2002, 67.
429 Social primates are pretty good: Vigilant and Groeneveld, 2012. For brilliant insights into why this matters for alloparenting, see Hrdy, 2009. For a nuanced insight into altruism and primates' "motivations," see de Waal and Suchak, 2010.
433 the Black Death killed a *third* of the population: Courie, 1972.
433 the Spanish flu in 1918: Barry, 2005.
433 Europeans developed the premodern middle class: Courie, 1972.
434 Cultures tend to respond to stress: Morris et al., 2011.

BIBLIOGRAPHY

Abraham, S., Juel, H. B., Bang, P., Cheeseman, H. M., Dohn, R. B., Cole, T., et al. (2019). Safety and immunogenicity of the chlamydia vaccine candidate CTH522 adjuvanted with CAF01 liposomes or aluminium hydroxide: A first-in-human, randomised, double-blind, placebo-controlled, phase 1 trial. *The Lancet Infectious Diseases*, 19(10), 1091–1100. doi:10.1016/S1473-3099(19)30279-8.

Abrams, E. T., and Rutherford, J. N. (2011). Framing postpartum hemorrhage as a consequence of human placental biology: An evolutionary and comparative perspective. *American Anthropologist*, 113, 417–430. doi:10.1111/j.1548-1433.2011.01351.x.

Absalon, D., and Ślesak, B. (2010). The effects of changes in cadmium and lead air pollution on cancer incidence in children. *Science of the Total Environment*, 408(20), 4420–4428. doi:10.1016/j.scitotenv.2010.06.030.

Açik, N. (2013). Redefining the role of women within the Kurdish national movement in Turkey in the 1990s. *The Kurdish Question in Turkey: New Perspectives on Violence, Representation, and Reconciliation*, edited by Cengiz Gunes and Welat Zydanlioglu. London: Routledge.

Adams, A. M., and Simmons, F. R. (2019). Exploring individual and gender differences in early writing performance. *Reading and Writing*, 32(2), 235–263. doi:10.1007/s11145-018-9859-0.

AGI (Alan Guttmacher Institute) (1999). *Sharing Responsibility; Women, Society and Abortion Worldwide*. New York: AGI.

AGI (Alan Guttmacher Institute) (2022). Sex and HIV education. www.guttmacher.org.

AGI (Alan Guttmacher Institute) (2023). Parental involvement in minors' abortions. www.guttmacher.org.

Aguilera-Castrejon, A., Oldak, B., Shani, T., Ghanem, N., Itzkovich, C., Slomovich, S., et al. (2021). Ex utero mouse embryogenesis from pre-gastrulation to late organogenesis. *Nature*, 593(7857), 119–124.

Ahmed, L. (1986). Women and the advent of Islam. *Signs*, 11(4), 665–691.

Aiello, L., and Dunbar, R. (1993). Neocortex size, group size, and the evolution of language. *Current Anthropology*, 34, 184–193.

Aiello, L. C., and Wheeler, P. (1995). The expensive-tissue hypothesis: The brain and the digestive system in human and primate evolution. *Current Anthropology*, 36(2), 199–221.

Aimé, C., André, J. B., and Raymond, M. (2017). Grandmothering and cognitive resources are required for the emergence of menopause and extensive post-reproductive lifespan. *PLOS Computational Biology*, 13(7), e1005631. doi:10.1371/journal.pcbi.1005631.

Akerlof, G. A., Yellen, J. L., and Katz, M. L. (1996). An analysis of out-of-wedlock childbearing in the United States. *The Quarterly Journal of Economics*, 111(2), 277–317. doi:10.2307/2946680.

Al Rawi, S., Louvet-Vallée, S., Djeddi, A., Sachse, M., Culetto, E., Hajjar, C., et al. (2011). Postfertilization autophagy of sperm organelles prevents paternal mitochondrial DNA transmission. *Science*, 334(6059), 1144–1147.

Alatorre Warren, J. L., Ponce de León, M. S., Hopkins, W. D., and Zollikofer, C. P. E. (2019). Evidence for independent brain and neurocranial reorganization during hominin evolution. *Proceedings of the National Academy of Sciences*, 116(44), 22115–22121. doi:10.1073/pnas.1905071116.

Alberts, S. C., Altmann, J., Brockman, D. K., Cords, M., Fedigan, L. M., Pusey, A., et al. (2013). Reproductive aging patterns in primates reveal that humans are distinct. *Proceedings of the National Academy of Sciences*, 110(33), 13440–13445. doi:10.1073/pnas.1311857110.

Alberts, S. C., and Altmann, J. (1995). Balancing costs and opportunities: Dispersal in male baboons. *The American Naturalist*, 145(2), 279–306. doi:10.1086/285740.

Alberts, S. C., and Fitzpatrick, C. L. (2012). Paternal care and the evolution of exaggerated sexual swellings in primates. *Behavioral Ecology*, 23(4), 699–706. doi:10.1093/beheco/ars052.

Albrecht, S., Lane, J. A., Marino, K., Al Busadah, K. A., Carrington, S. D., Hickey, R. M., and Rudd, P. M. (2014). A comparative study of free oligosaccharides in the milk of domestic animals. *British Journal of Nutrition*, 111(7), 1313–1328.

Aleman, A., Kahn, R. S., and Selten, J. P. (2003). Sex differences in the risk of schizophrenia: Evidence from meta-analysis. *Archives of General Psychiatry*, 60(6), 565–571. doi:10.1001/archpsyc.60.6.565.

Altmann, J., Gesquiere, L., Galbany, J., Onyango, P. O., and Alberts, S. C. (2010). Life history context of reproductive aging in a wild primate model. *Annals of the New York Academy of Sciences*, 1204, 127–38. doi:10.1111/j.1749-6632.2010.05531.x.

Alvesalo, L. (2013). The expression of human sex chromosome genes in oral and craniofacial growth. Pp. 92–107 in *Anthropological Perspectives on Tooth Morphology*, edited by G. R. Scott and J. D. Irish. Cambridge, U.K.: Cambridge University Press.

Ambrose, S. H. (2001). Paleolithic technology and human evolution. *Science*, 291(5509), 1748–1753.

American Cancer Society (2020). *Breast Cancer Facts and Figures 2019–2020.* Atlanta: American Cancer Society.

American Society of Plastic Surgeons (ASPS) (2021). *National Plastic Surgery Statistics Report, 2020.* ASPS National Clearinghouse of Plastic Surgery Procedural Statistics. www.plasticsurgery.org.

Amos, W., and Hoffman, J. I. (2010). Evidence that two main bottleneck events shaped modern human genetic diversity. *Proceedings of the Royal Society B: Biological Sciences*, 277(1678), 131–137. doi:10.1098/rspb.2009.1473.

Anderson, D. R., and Pempek, T. A. (2005). Television and very young children. *American Behavioral Scientist*, 48(5), 505–522. doi:10.1177/0002764204271506.

Anderson, M. J., Chapman, S. J., Videan, E. N., Evans, E., Fritz, J., Stoinski, T. S., et al. (2007). Functional evidence for differences in sperm competition in humans and chimpanzees. *American Journal of Physical Anthropology*, 134, 274–280.

Anderson, R. C., and Klofstad, C. A. (2012). Preference for leaders with masculine voices holds in the case of feminine leadership roles. *PLOS ONE*, 7(12), e51216. doi:10.1371/journal.pone.0051216.

André, G. I., Firman, R. C., and Simmons, L. W. (2022). The effect of genital stimulation on competitive fertilization success in house mice. *Animal Behaviour*, 190, 93–101. doi:10.1016/j.anbehav.2022.05.015.

Antón, S. C. (2003). Natural history of *Homo erectus*. *American Journal of Physical Anthropology*, 122(S37), 126–170.

Antón, S. C., Potts, R., and Aiello, L. C. (2014). Evolution of early *Homo:* An integrated biological perspective. *Science*, 345(6192), 1236828.

Anzia, S. F., and Berry, C. R. (2011). The Jackie (and Jill) Robinson effect: Why do congresswomen outperform congressmen? *American Journal of Political Science*, 55(3), 478–493. doi:10.1111/j.1540-5907.2011.00512.x.

Apicella, C. L., Crittenden, A. N., and Tobolsky, V. A. (2017). Hunter-gatherer males are more risk-seeking than females, even in late childhood. *Evolution and Human Behavior*, 38(5), 592–603.

Arambula, S. E., Reinl, E. L., El Demerdash, N., McCarthy, M. M., and Robertson, C. L. (2019). Sex differences in pediatric traumatic brain injury. *Experimental Neurology*, 317, 168–179. doi:10.1016/j.expneurol.2019.02.016.

Archibald, J. D., Zhang, Y., Harper, T., and Cifelli, R. L. (2011). Protungulatum, confirmed Cretaceous occurrence of an otherwise Paleocene eutherian (placental?) mammal. *Journal of Mammalian Evolution*, 18, 153–161.

Armstrong, E. A., Hamilton, L. T., Armstrong, E. M., and Seeley, J. L. (2014). "Good girls": Gender, social class, and slut discourse on campus. *Social Psychology Quarterly*, 77(2), 100–122. doi:10.1177/0190272514521220.

Arnold, L. M. (2003). Gender differences in bipolar disorder. *The Psychiatric Clinics of North America*, 26(3), 595–620. doi:10.1016/s0193-953x(03)00036-4.

Arriaza, M. C., Aramendi, J., Maté-González, M. Á., Yravedra, J., and Stratford, D. (2021). The hunted or the scavenged? Australopith accumulation by brown hyenas at Sterkfontein (South Africa). *Quaternary Science Reviews*, 273, 107252. doi:10.1016/j.quascirev.2021.107252.

Atsalis, S., Margulis, S. W., Bellem, A., and Wielebnowski, N. (2004). Sexual behavior and hormonal estrus cycles in captive aged lowland gorillas (*Gorilla gorilla*). *American Journal of Primatology*, 62, 123–132.

Austad, S. N. (1994). Menopause: An evolutionary perspective. *Experimental Gerontology*, 29(3), 255-263. doi:10.1016/0531-5565(94)90005-1.

Autor, D. H. (2014). Skills, education, and the rise of earnings inequality among the "other 99 percent." *Science*, 344(6186), 843–851. doi:10.1126/science.1251868.

Averroes. (1974). *Averroes on Plato's Republic*. Translated by R. Lerner. Ithaca, N.Y.: Cornell University Press. (Originally written in the twelfth century; surviving manuscripts in Hebrew translated thereafter.)

Baber, R. E. (1934). Marriage in ancient China. *The Journal of Educational Sociology*, 8(3), 131–140. doi:10.2307/2961796.

Babic, A., Sasamoto, N., Rosner, B. A., Tworoger, S. S., Jordan, S. J., Risch, H. A., et al. (2020). Association between breastfeeding and ovarian cancer risk. *JAMA Oncology*, 6(6), e200421. doi:10.1001/jamaoncol.2020.0421.

Bakshani, C. R., Morales-Garcia, A. L., Althaus, M., Wilcox, M. D., Pearson, J. P., Bythell, J. C., and Burgess, J. G. (2018). Evolutionary conservation of the antimicrobial function of mucus: A first defence against infection. *NPJ Biofilms Microbiomes*, 4, 14. doi:10.1038/s41522-018-0057-2.

Banai, I. P. (2017). Voice in different phases of menstrual cycle among naturally cycling women and users of hormonal contraceptives. *PLOS ONE*, 12(8), e0183462. doi:10.1371/journal.pone.0183462.

Bar-Yosef, O. (1986). The walls of Jericho: An alternative interpretation. *Current Anthropology*, 27(2), 157–162. doi:10.1086/203413.

Bar-Yosef, O., and Belfer-Cohen, A. (2001). From Africa to Eurasia—early dispersals. *Quaternary International*, 75(1), 19–28.

Barba-Müller, E., Craddock, S., Carmona, S., and Hoekzema, E. (2019). Brain plasticity in pregnancy and the postpartum period: Links to maternal caregiving and mental health. *Archives of Women's Mental Health*, 22(2), 289–299. doi:10.1007/s00737-018-0889-z.

Bardeen, C. G., Garcia, R. R., Toon, O. B., and Conley, A. J. (2017). On transient climate change at the Cretaceous-Paleogene boundary due to atmospheric soot injections. *Proceedings of the National Academy of Sciences*, 114(36), E7415–E7424. doi:10.1073/pnas.1708980114.

Barney, A., Martelli, S., Serrurier, A., and Steele, J. (2012). Articulatory capacity of Neanderthals, a very recent and human-like fossil hominin. *Philosophical Transactions of the Royal Society of London. Series B, Biological Sciences*, 367(1585), 88–102. doi:10.1098/rstb.2011.0259.

Barros, B. A., Oliveira, L. R., Surur, C. R. C., Barros-Filho, A. A., Maciel-Guerra, A. T., and Guerra-Junior, G. (2021). Complete androgen insensitivity syndrome and risk of gonadal malignancy: Systematic review.

Annals of Pediatric Endocrinology & Metabolism, 26(1), 19–23. doi:10.6065/apem.2040170.085.

Barry, J. M. (2005). *The Great Influenza: The Story of the Deadliest Pandemic in History*. New York: Penguin Books.

Bartlett, J., Norrie, R., Patel, S., Rumpel, R., and Wibberley, S. (2014). Misogyny on Twitter. www.demos.co.uk/files/MISOGYNY_ON_TWITTER.pdf.

Barton, R. A. (2004). Binocularity and brain evolution in primates. *Proceedings of the National Academy of Sciences*, 101(27), 10113–10115. doi:10.1073/pnas.0401955101.

Bartos, L., Bartošová, J., Pluháček, J., and Šindelářová, J. (2011). Promiscuous behaviour disrupts pregnancy block in domestic horse mares. *Behavioral Ecology and Sociobiology*, 65, 1567–1572. doi:10.1007/s00265-011-1166-6.

Bawdon, D., Cox, D. S., Ashford, D., James, A. G., and Thomas, G. H. (2015). Identification of axillary *Staphylococcus sp.* involved in the production of the malodorous thioalcohol 3-methyl-3-sufanylhexan-1-ol. *FEMS Microbiology Letters*, 362(16). doi:10.1093/femsle/fnv111.

Baxter, A. J., Scott, K. M., Ferrari, A. J., Norman, R. E., Vos, T., and Whiteford, H. A. (2014). Challenging the myth of an "epidemic" of common mental disorders: Trends in the global prevalence of anxiety and depression between 1990 and 2010. *Depression and Anxiety*, 31(6), 506–516. doi:10.1002/da.22230.

Bayle, P., Macchiarelli, R., Trinkaus, E., Duarte, C., Mazurier, A., and Zilhão, J. (2010). Dental maturational sequence and dental tissue proportions in the early Upper Paleolithic child from Abrigo do Lagar Velho, Portugal. *Proceedings of the National Academy of Sciences*, 107(4), 1338–1342. doi:10.1073/pnas.0914202107.

Beck, K. L., Weber, D., Phinney, B. S., Smilowitz, J. T., Hinde, K., Lönnerdal, B., et al. (2015). Comparative proteomics of human and macaque milk reveals species-specific nutrition during postnatal development. *Journal of Proteome Research*, 14(5), 2143–2157. doi:10.1021/pr501243m.

Bednarik, R. G. (2008). Children as Pleistocene artists. *Rock Art Research: The Journal of the Australian Rock Art Research Association (AURA)*, 25(2), 173–182.

Beery, A. K., and Zucker, I. (2011). Sex bias in neuroscience and biomedical research. *Neuroscience and Biobehavioral Reviews* 35(3), 565–572. doi:10.1016/j.neubiorev.2010.07.002.

Bekkering, S., Quintin, J., Joosten, L. A. B., van der Meer, J. W. M., Netea, M. G., and Riksen, N. P. (2014). Oxidized low-density lipoprotein induces long-term proinflammatory cytokine production and foam cell formation via epigenetic reprogramming of monocytes. *Arteriosclerosis, Thrombosis, and Vascular Biology*, 34(8), 1731–1738. doi:10.1161/ATVBAHA.114.303887.

Bellemare, F., Jeanneret, A., and Couture, J. (2003). Sex differences in thoracic dimensions and configuration. *American Journal of Respiratory and Critical Care Medicine*, 168(3), 305–312. doi:10.1164/rccm.200208-876OC.

Bellis, M. A., Downing, J., and Ashton, J. R. (2006). Adults at 12? Trends in puberty and their public health consequences. *Journal of Epidemiology and Community Health*, 60(11), 910–911. doi:10.1136/jech.2006.049379.

Belmaker, M. (2010). Early Pleistocene faunal connections between Africa and Eurasia: An ecological perspective. Pp. 183–205 in *Out of Africa I: The First Hominin Colonization of Eurasia*, edited by J. G. Fleagle, J. J. Shea, F. E. Grine, A. L. Baden, and R. E. Leakey. Dordrecht: Springer.

Belo, C. (2008). Some considerations on Averroes' views regarding women and their role in society. *Journal of Islamic Studies*, 20(1), 1–20. doi:10.1093/jis/etn061.

Ben-Dor, M., Gopher, A., Hershkovitz, I., and Barkai, R. (2011). Man the fat hunter: The demise of *Homo erectus* and the emergence of a new hominin lineage in the Middle Pleistocene (ca. 400 kyr) Levant. *PLOS ONE*, 6(12), e28689.

Benoit, J., Manger, P. R., and Rubidge, B. S. (2016a). Palaeoneurological clues to the evolution of defining mammalian soft tissue traits. *Scientific Reports*, 6(1), 25604. doi:10.1038/srep25604.

Benoit, J., Manger, P. R., Fernandez, V., and Rubidge, B. S. (2016b). Cranial bosses of *Choerosaurus dejageri* (Therapsida, Therocephalia): Earliest evidence of cranial display structures in Eutheriodonts. *PLOS ONE*, 11(8), e0161457.

Benson, R. B. J., Butler, R. J., Carrano, M. T., and O'Connor, P. M. (2012). Air-filled postcranial bones in theropod dinosaurs: Physiological implications and the "reptile"-bird transition. *Biological Reviews*, 87(1), 168–193.

Bentley, G. R. (2001). The Evolution of the human breast. *American Journal of Physical Anthropology*, 114(S32), 38–38.

Benton, M. J., Wilf, P., and Sauquet, H. (2022). The angiosperm terrestrial revolution and the origins of modern biodiversity. *New Phytologist*, 233, 2017–2035. doi:10.1111/nph.17822.

Berge, C., and Goularas, D. (2010). A new reconstruction of Sts 14 pelvis (*Australopithecus africanus*) from computed tomography and three-dimensional modeling techniques. *Journal of Human Evolution*, 58(3), 262–272.

Bergenhenegouwen, L. A., Meertens, L. J. E., Schaaf, J., Nijhuis, J. G., Mol, B. W., Kok, M., and Scheepers, H. C. (2014). Vaginal delivery versus caesarean section in preterm breech delivery: A systematic review. *European Journal of Obstetrics & Gynecology and Reproductive Biology*, 172, 1–6. doi:10.1016/j.ejogrb.2013.10.017.

Berger, M., and Sarnyai, Z. (2015). "More than skin deep": Stress neurobiology and mental health consequences of racial discrimination. *Stress*, 18(1), 1–10. doi:10.3109/10253890.2014.989204.

Berglund, H., Lindström, P., and Savic, I. (2006). Brain response to putative pheromones in lesbian women. *Proceedings of the National Academy of Sciences*, 103(21), 8269–8274. doi:10.1073/pnas.0600331103.

Berglund, H., Lindström, P., Dhejne-Helmy, C., and Savic, I. (2008). Male-to-female transsexuals show sex-atypical hypothalamus activation when smelling odorous steroids. *Cerebral Cortex*, 18(8), 1900–1908.

Bergman, L., Nordlöf-Callbo, P., Wikström, A. K., Snowden, J. M., Hesselman, S., Bonamy, A. K. E., and Sandström, A. (2020). Multi-fetal pregnancy, preeclampsia, and long-term cardiovascular disease. *Hypertension*, 76(1), 167–175. doi:10.1161/HYPERTENSIONAHA.120.14860.

Berna, F., Goldberg, P., Horwitz, L. K., Brink, J., Holt, S., Bamford, M., and Chazan, M. (2012). Microstratigraphic evidence of in situ fire in the Acheulean strata of Wonderwerk Cave, Northern Cape province, South Africa. *Proceedings of the National Academy of Sciences*, 109(20), E1215–E1220.

Berry, K. (2020). The first plants to recolonize western North America following the Cretaceous/Paleogene mass extinction event. *International Journal of Plant Sciences*, 182. doi:10.1086/11847.

Bertram, B. C. R. (1975). Social factors influencing reproduction in wild lions. *Journal of Zoology*, 177, 463.

Biben, M., Symmes, D., and Bernhards, D. (1989). Contour variables in vocal communication between squirrel monkey mothers and infants. *Developmental Psychobiology*, 22, 617–631. doi:10.1002/dev.420220607.

Black, L. I., Vahratian, A., Hoffman, H. J. (2015). Communication disorders and use of intervention services among children aged 3–17 years: United States, 2012. NCHS data brief, no 205. Hyattsville, Md.: National Center for Health Statistics.

Blackburn, D. G. (2015). Evolution of vertebrate viviparity and specializations for fetal nutrition: A quantitative and qualitative analysis. *Journal of Morphology*, 276(8), 961–990.

Blaustein, A. R. (1981). Sexual selection and mammalian olfaction. *The American Naturalist*, 117(6), 1006–1010. doi:10.1086/283786.

Blaustein, J. D. (2012). Animals have a sex, and so should titles and methods sections of articles. *Endocrinology*, 153(6), 2539–2540. doi:10.1210/en.2012-1365.

Blencowe, H., Cousens, S., Oestergaard, M. Z., Chou, D., Moller, A. B., Narwal, R., et al. (2012). National, regional, and worldwide estimates of preterm birth rates in the year 2010 with time trends since 1990 for selected countries: A systematic analysis and implications. *Lancet*, 379(9832), 2162–2172. doi:10.1016/S0140-6736(12)60820-4.

Blumenschine, R. J., Bunn, H. T., Geist, V., Ikawa-Smith, F., Marean, C. W., Payne, A. G., et al. (1987). Characteristics of an early hominid scavenging niche [and comments and reply]. *Current Anthropology*, 28(4), 383–407.

Blurton Jones, N. G. (2016). *Demography and Evolutionary Ecology of Hadza Hunter-Gatherers*. Cambridge, U.K.: Cambridge University Press.

Bobe, R., and Behrensmeyer, A. K. (2004). The expansion of grassland ecosystems in Africa in relation to mammalian evolution and the origin of the genus *Homo*. *Palaeogeography, Palaeoclimatology, Palaeoecology*, 207(3–4), 399–420.

Boccolini, C. S., de Carvalho, M. L., de Oliveira, M. I. C., and Pérez-Escamilla, R. (2013). Breastfeeding during the first hour of life and neonatal mortality. *Jornal de Pediatria*, 89(2), 131–136. doi:10.1016/j.jped.2013.03.005.

Bockting, W., Benner, A., and Coleman, E. (2009). Gay and bisexual identity development among female-to-male transsexuals in North America: Emergence of a transgender sexuality. *Archives of Sexual Behavior*, 38(5), 688–701. doi:10.1007/s10508-009-9489-3.

Bocquet-Appel, J.-P. (2011). The agricultural demographic transition during and after the agriculture inventions. *Current Anthropology*, 52(S4), S497–S510. doi:10.1086/659243.

Boë, L.-J., Sawallis, T. R., Fagot, J., Badin, P., Barbier, G., Captier, G., et al. (2019). Which way to the dawn of speech? Reanalyzing half a century of debates and data in light of speech science. *Science Advances*, 5(12), eaaw3916. doi:10.1126/sciadv.aaw3916.

Boffoli, D., Scacco, S. C., Vergari, R., Persio, M. T., Solarino, G., Laforgia, R., and Papa, S. (1996). Ageing is associated in females with a decline in the content and activity of the b-c1 complex in skeletal muscle mitochondria. *Biochimica et Biophysica Acta (BBA)—Molecular Basis of Disease*, 1315(1), 66–72.

Bogaert, A. F. (2015). Asexuality: What it is and why it matters. *The Journal of Sex Research*, 52(4), 362–379. doi:10.1080/00224499.2015.1015713.

Bogaert, A. F., and Skorska, M. N. (2020). A short review of biological research on the development of sexual orientation. *Hormones and Behavior*, 119, 104659. doi:10.1016/j.yhbeh.2019.104659.

Bonomi, A. E., Trabert, B., Anderson, M. L., Kernic, M. A., and Holt, V. L. (2014). Intimate partner violence and neighborhood income: A longitudinal analysis. *Violence Against Women*, 20(1), 42–58. doi.org/10.1177/1077801213520580.

Boos, F. S. (2013). Education and work: Women and the education acts. Pp. 141–60, 224–28 in *Berg Cultural History of Women in the Age of Empire*, edited by T. Mangum. Oxford: Berg Publishers.

Boquien, C.-Y. (2018). Human milk: An ideal food for nutrition of preterm newborn. *Frontiers in Pediatrics*, 6. doi:10.3389/fped.2018.00295.

Borges, J. L. (1978). Interview with César Hildebrandt. *Caretas*, Dec. 19, 1978. borgestodoelanio.blogspot.com.

Borges, J. L. (2007). *Labyrinths: Selected Stories and Other Writings*. Edited by D. A. Yates and J. E. Irby. New York: New Directions.

Borges, R., Johnson, W. E., O'Brien, S. J., Gomes, C., Heesy, C. P., and Antunes, A. (2018). Adaptive genomic evolution of opsins reveals that early mammals flourished in nocturnal environments. *BMC Genomics*, 19(1), 1–12. doi:10.1186/s12864-017-4417-8.

Boroditsky, L. (2001). Does language shape thought? Mandarin and English speakers' conceptions of time. *Cognitive Psychology*, 43(1), 1–22. doi:10.1006/cogp.2001.0748.

Boudová, S., Cohee, L. M., Kalilani-Phiri, L., Thesing, P. C., Kamiza, S., Muehlenbachs, A., et al. (2014). Pregnant women are a reservoir of malaria transmission in Blantyre, Malawi. *Malaria Journal*, 13(1), 506. doi:10.1186/1475-2875-13-506.

Boulinguez-Ambroise, G., Pouydebat, E., Disarbois, É., and Meguerditchian, A. (2022). Maternal cradling bias in baboons: The first environmental factor affecting early infant handedness development? *Developmental Science*, 25(1), e13179. doi:10.1111/desc.13179.

Bouty, A., Ayers, K. L., Pask, A., Heloury, Y., and Sinclair, A. H. (2015). The genetic and environmental factors underlying hypospadias. *Sexual Development*, 9(5), 239–259. doi:10.1159/000441988.

Bowen, V., Su, J., Torrone, E., Kidd, S., and Weinstock, H. (2015). Increases in incidence of congenital syphilis—United States, 2012–2014. *MMWR Morbidity and Mortality Weekly Report*, 64(44), 1241–1245.

Bowers, J. M., Perez-Pouchoulen, M., Edwards, N. S., and McCarthy, M. M. (2013). Foxp2 mediates sex differences in ultrasonic vocalization by rat pups and directs order of maternal retrieval. *The Journal of Neuroscience*, 33(8), 3276–3283. doi:10.1523/JNEUROSCI.0425-12.2013.

Bowling, A. T., and Touchberry, R. W. (1990). Parentage of Great Basin feral horses. *The Journal of Wildlife Management*, 54(3), 424–429. doi:10.2307/3809652.

Bowman, E. A. (2008). Why the human penis is larger than in the great apes. *Archives of Sexual Behavior*, 37, 361. doi:10.1007/s10508-007-9297-6.

Bradshaw, C. D. (2021). Miocene climates. Pp. 486–496 in *Encyclopedia of Geology*, edited by D. Alderton and S. A. Elias. 2nd ed. Oxford: Academic Press.

Brain, C. K. (1981). *The Hunters or the Hunted? An Introduction to African Cave Taphonomy*. Chicago: University of Chicago Press.

Bramble, D. M., and Lieberman, D. E. (2004). Endurance running and the evolution of *Homo*. *Nature*, 432(7015), 345–352. doi:10.1038/nature03052.

Brannigan, R., and Lipshultz, L. (2008). *The Global Library of Women's Medicine*. doi:10.3843/GLOWM.10316.

Braun, D. R., Harris, J. W. K., Levin, N. E., McCoy, J. T., Herries, A. I. R., Bamford, M. K., et al. (2010). Early hominin diet included diverse terrestrial and aquatic animals 1.95 Ma in East Turkana, Kenya. *Proceedings of the National Academy of Sciences*, 107(22), 10002–10007. doi:10.1073/pnas.1002181107.

Brawand, D., Wahli, W., and Kaessmann, H. (2008). Loss of egg yolk genes in mammals and the origin of lactation and placentation. *PLOS Biology*, 6(3), e63. doi:10.1371/journal.pbio.0060063.

Brenna, J. T., Salem, N., Jr., Sinclair, A. J., and Cunnane, S. C. (2009). Alpha-Linolenic acid supplementation and conversion to n-3 long-chain polyunsaturated fatty acids in humans. *Prostaglandins, Leukotrienes, and Essential Fatty Acids*, 80(2–3), 85–91. doi:10.1016/j.plefa.2009.01.004.

Brennan, P. L. R., and Orbach, D. N. (2020). Copulatory behavior and its relationship to genital morphology. Pp. 65–122 in *Advances in the Study of Behavior*, Vol. 52, edited by M. Naguib, L. Barrett, S. D. Healy, J. Podos, L. W. Simmons, and M. Zuk. Cambridge, Mass.: Academic Press.

Brennan, P. L. R., Prum, R. O., McCracken, K. G., Sorenson, M. D., Wilson, R. E., and Birkhead, T. R. (2007). Coevolution of male and female genital morphology in waterfowl. *PLOS ONE*, 2(5), e418. doi:10.1371/journal.pone.0000418.

Brent, L. J. N., Franks, D. W., Foster, E. A., Balcomb, K. C., Cant, M. A., and Croft, D. P. (2015). Ecological knowledge, leadership, and the evolution of menopause in killer whales. *Current Biology*, 25(6), 746–750. doi:10.1016/j.cub.2015.01.037.

Broadfield, D. C., Holloway, R. L., Mowbray, K., Silvers, A., Yuan, M. S., and Márquez, S. (2001). Endocast of Sambungmacan 3 (Sm 3): A new *Homo erectus* from Indonesia. *The Anatomical Record*, 262(4), 369–379.

Brochu, C., Njau, J., Blumenschine, R., and Densmore, L. (2010). A new horned crocodile from the Plio-Pleistocene hominid sites at Olduvai Gorge, Tanzania. *PLOS ONE* 5(2), e9333. doi:10.1371/journal.pone.0009333.

Brock, T. D. (1988). *Robert Koch: A Life in Medicine and Bacteriology*. Heidelberg: Springer Berlin.

Brody, S., and Krüger, T. H. (2006). The post-orgasmic prolactin increase following intercourse is greater than following masturbation and suggests greater satiety. *Biological Psychology*, 71(3), 312–315. doi:10.1016/j.biopsycho.2005.06.008.

Bronikowski, A. M., Altmann, J., Brockman, D. K., Cords, M., Fedigan, L. M., Pusey, A., et al. (2011). Aging in the natural world: Comparative data reveal similar mortality patterns across primates. *Science*, 331(6022), 1325–1328. doi:10.1126/science.1201571.

Brooks, R., Singleton, J. L., and Meltzoff, A. N. (2020). Enhanced gaze-following behavior in Deaf infants of Deaf parents. *Developmental Science*, 23(2), e12900. doi:10.1111/desc.12900.

Bruce, H. M. (1959). Exteroceptive block to pregnancy in the mouse. *Nature*, 184(4680), 105.

Brusatte, S., and Luo, Z. X. (2016). Ascent of the mammals. *Scientific American*, 314(6), 28–35. doi:10.1038/scientificamerican0616-28.

Bryan, D. L., Hart, P. H., Forsyth, K. D., and Gibson, R. A. (2007). Immunomodulatory constituents of human milk change in response to infant bronchiolitis. *Pediatric Allergy and Immunology*, 18(6), 495–502.

Bryant, G. A., and Haselton, M. G. (2009). Vocal cues of ovulation in human females. *Biology Letters*, 5(1), 12–15. doi:10.1098/rsbl.2008.0507.

Buchanan, F. F., Myles, P. S., and Cicuttini, F. (2009). Patient sex and its influence on general anaesthesia. *Anaesthesia and Intensive Care*, 37(2), 207–218. doi:10.1177/0310057X0903700201.

Bunn, H. T. (1981). Archaeological evidence for meat-eating by Plio-Pleistocene hominids from Koobi Fora and Olduvai Gorge. *Nature*, 291(5816), 574–577.

Burke, S. M., Cohen-Kettenis, P. T., Veltman, D. J., Klink, D. T., and Bakker, J. (2014). Hypothalamic response to the chemo-signal androstadienone in gender dysphoric children and adolescents. *Frontiers in Endocrinology*, 5, 60. doi:10.3389/fendo.2014.00060.

Butler, P. M., and Sigogneau-Russell, D. (2016). Diversity of triconodonts in the Middle Jurassic of Great Britain. *Palaeontologia Polonica*, 67, 35–65.

Byrnes, J. P., Miller, D. C., and Schafer, W. D. (1999). Gender differences in risk taking: A meta-analysis. *Psychological Bulletin*, 125(3), 367.

Cafazzo, S., Natoli, E., and Valsecchi, P. (2012). Scent-marking behaviour in a pack of free-ranging domestic dogs. *Ethology*, 118, 955–966. doi:10.1111/j.1439-0310.2012.02088.x.

Cain, W. S. (1982). Odor identification by males and females: Predictions vs. performance. *Chemical Senses*, 7(2):129–142.

Caine, N. G., Osorio, D., and Mundy, N. I. (2010). A foraging advantage for dichromatic marmosets (*Callithrix geoffroyi*) at low light intensity. *Biology Letters*, 6(1), 36–38. doi:10.1098/rsbl.2009.0591.

Caldwell, J. C. (1980). Mass education as a determinant of the timing of fertility decline. *Population and Development Review*, 225–255.

Callaway, E. (2012). Fathers bequeath more mutations as they age. *Nature*, 488(7412), 439–439. doi:10.1038/488439a.

Cameron, E. L. (2014). Pregnancy and olfaction: A review. *Frontiers in Psychology*, 5. doi:10.3389/fpsyg.2014.00067.

Cammarota, G., Ianiro, G., and Gasbarrini, A. (2014). Fecal microbiota transplantation for the treatment of Clostridium difficile infection: A systematic review. *Journal of Clinical Gastroenterology*, 48(8), 693–702.

Campbell, A., Copping, L. T., and Cross, C. P. (2021). *Sex Differences in Fear Response: An Evolutionary Perspective*. Cham: Springer.

Campos, F. A., Altmann, J., Cords, M., Fedigan, L. M., Lawler, R., Lonsdorf, E. V., et al. (2022). Female reproductive aging in seven primate species: Patterns and consequences. *Proceedings of the National Academy of Sciences*, 119(20), e2117669119. doi:10.1073/pnas.2117669119.

Campos, F. A., Villavicencio, F., Archie, E. A., Colchero, F., and Alberts, S. C. (2020). Social bonds, social status and survival in wild baboons: A tale of two sexes. *Philosophical Transactions of the Royal Society B: Biological Sciences*, 375(1811), 20190621. doi:10.1098/rstb.2019.0621.

Cantwell, R., Clutton-Brock, T., Cooper, G., Dawson, A., Drife, J., Garrod, D., et al. (2011). Saving mothers' lives: Reviewing maternal deaths to make motherhood safer: 2006–2008. The Eighth Report of the Confidential Enquiries into Maternal Deaths in the United Kingdom. *British Journal of Obstetrics and Gynaecology*, 118 (S1), 1–203. doi:10.1111/j.1471-0528.2010.02847.x.

Capasso, L., Michetti, E., and D'Anastasio, R. (2008). A *Homo erectus* hyoid bone: Possible implications for the origin of the human capability for speech. *Collegium Antropologicum*, 32(4), 1007–1011.

Capellini, T. D., Chen, H., Cao, J., Doxey, A. C., Kiapour, A. M., Schoor, M., and Kingsley, D. M. (2017). Ancient selection for derived alleles at a GDF5 enhancer influencing human growth and osteoarthritis risk. *Nature Genetics*, 49(8), 1202–1210. doi:10.1038/ng.3911.

Caplan, H. W., Cox, C. S., and Bedi, S. S. (2017). Do microglia play a role in sex differences in TBI? *Journal of Neuroscience Research*, 95(1–2), 509–517. doi:10.1002/jnr.23854.

Cardinale, D. A., Larsen, F. J., Schiffer, T. A., Morales-Alamo, D., Ekblom, B., Calbet, J. A. L., et al. (2018). Superior intrinsic mitochondrial respiration in women than in men. *Frontiers in Physiology*, 9. doi:10.3389/fphys.2018.01133.

Carlson, B. A., and Kingston, J. D. (2007). Docosahexaenoic acid, the aquatic diet, and hominin encephalization: Difficulties in establishing evolutionary links. *American Journal of Human Biology*, 19(1), 132–141. doi:10.1002/ajhb.20579.

Caro, T. M., Sellen, D. W., Parish, A., Frank, R., Brown, D. M., Voland, E., and Mulder, M. B. (1995). Termination of reproduction in nonhuman and human female primates. *International Journal of Primatology*, 16(2), 205–220. doi:10.1007/BF02735478.

Carr, L. E., Virmani, M. D., Rosa, F., Munblit, D., Matazel, K. S., Elolimy, A. A., and Yeruva, L. (2021). Role of human milk bioactives on infants' gut and immune health. *Frontiers in Immunology*, 12, 604080. doi:10.3389/fimmu.2021.604080.

Carrano, M. T., and Sampson, S. D. (2004). A review of coelophysoids (Dinosauria: Theropoda) from the Early Jurassic of Europe, with comments on the late history of the Coelophysoidea. *Neues Jahrbuch für Geologie und Paläontologie-Monatshefte* (9), 537–558. doi:10.1127/njgpm/2004/2004/537.

Carvalho, M. R., Jaramillo, C., de la Parra, F., Caballero-Rodríguez, D., Herrera, F., Wing, S., et al. (2021). Extinction at the end-Cretaceous and the origin of modern Neotropical rainforests. *Science*, 372(6537), 63–68. doi:10.1126/science.abf1969.

Carvalho, S., Biro, D., Cunha, E., Hockings, K., McGrew, W. C., Richmond, B. G., and Matsuzawa, T. (2012). Chimpanzee carrying behaviour and the origins of human bipedality. *Current Biology*, 22(6), R180–R181. doi:10.1016/j.cub.2012.01.052.

Casolini, P., Cigliana, G., Alema, G. S., Ruggieri, V., Angelucci, L., and Catalani, A. (1997). Effect of increased maternal corticosterone during lactation on hippocampal corticosteroid receptors, stress response and learning in offspring in the early stages of life. *Neuroscience*, 79, 1005–1012.

CBS News. (2015). First women to pass Ranger School recount milestone. Aug. 20, 2015. www.cbsnews.com.

Centers for Disease Control and Prevention (CDC). (2015). *Sexually Transmitted Disease Surveillance 2014*. Atlanta: U.S. Department of Health and Human Services.

Centers for Disease Control and Prevention (CDC). (2022). *Sexually Transmitted Disease Surveillance 2020*. Atlanta: U.S. Department of Health and Human Services. www.cdc.gov.

Centerwall, B. (1995). Race, socioeconomic status, and domestic homicide. *Journal of the American Medical Association*, 273(22), 1755–1758.

Cera, N., Vargas-Cáceres, S., Oliveira, C., Monteiro, J., Branco, D., Pignatelli, D., and Rebelo, S. (2021). How relevant is the systemic oxytocin concentration for human sexual behavior? A systematic review. *Sexual Medicine*, 9(4), 100370. doi:10.1016/j.esxm.2021.100370.

Cerling, T. E., Chritz, K. L., Jablonski, N. G., Leakey, M. G., and Manthi, F. K. (2013). Diet of *Theropithecus* from 4 to 1 Ma in Kenya. *Proceedings of the National Academy of Sciences*, 110(26), 10507–10512. doi:10.1073/pnas.1222571110.

Cerling, T. E., Levin, N. E., Quade, J., Wynn, J. G., Fox, D. L., Kingston, J. D., et al. (2010). Comment on the Paleoenvironment of *Ardipithecus ramidus*. *Science*, 328(5982), 1105–1105.

Cerling, T. E., Wynn, J. G., Andanje, S. A., Bird, M. I., Korir, D. K., Levin, N. E., et al. (2011). Woody cover and hominin environments in the past 6 million years. *Nature*, 476(7358), 51–56.

Chakradhar, S. and Ross, C. (2019). The history of OxyContin, told through unsealed Purdue documents. *Stat*, Dec. 3, 2019. www.statnews.com.

Chan, E. K., Timmermann, A., Baldi, B. F., Moore, A. E., Lyons, R. J., Lee, S. S., et al. (2019). Human origins in a southern African palaeo-wetland and first migrations. *Nature*, 575(7781), 185–189.

Chan, Y. Y., Jayaprakasan, K., Zamora, J., Thornton, J. G., Raine-Fenning, N., and Coomarasamy, A. (2011). The prevalence of congenital uterine anomalies in unselected and high-risk populations: A systematic review. *Human Reproduction Update*, 17(6), 761–771. doi:10.1093/humupd/dmr028.

Chance, M. R. A. (1996). Reason for externalization of the testis of mammals. *Journal of Zoology*, 239(4), 691–695. doi:10.1111/j.1469-7998.1996.tb05471.x.

Chapman, D. D., Wintner, S. P., Abercrombie, D. L., Ashe, J., Bernard, A. M., Shivji, M. S., and Feldheim, K. A. (2013). The behavioural and genetic mating system of the sand tiger shark, *Carcharias taurus*, an intrauterine cannibal. *Biology Letters*, 9(3), 20130003. doi:10.1098/rsbl.2013.0003.

Chattopadhyay, R., and Duflo, E. (2004). Women as policy makers: Evidence from a randomized policy experiment in India. *Econometrica*, 72(5), 1409–1443.

Chavarria, M. C., Sánchez, F. J., Chou, Y. Y., Thompson, P. M., and Luders, E. (2014). Puberty in the corpus callosum. *Neuroscience*, 265, 1–8. doi:10.1016/j.neuroscience.2014.01.030.

Chen, Y., Matheson, L. E., and Sakata, J. T. (2016). Mechanisms underlying the social enhancement of vocal learning in songbirds. *Proceedings of the National Academy of Sciences*, 113(24), 6641–6646. doi:10.1073/pnas.1522306113.

Cherry, J. A., and Baum, M. J. (2020). Sex differences in main olfactory system pathways involved in psychosexual function. *Genes, Brain and Behavior*, 19, e12618. doi:10.1111/gbb.12618.

Chester, S. G. B., Bloch, J. I., Boyer, D. M., and Clemens, W. A. (2015). Oldest known euarchontan tarsals and affinities of Paleocene *Purgatorius* to primates. *Proceedings of the National Academy of Sciences*, 112(5), 1487–1492. doi:10.1073/pnas.1421707112.

Choi, H., Dey, A. K., Priyamvara, A., Aksentijevich, M., Bandyopadhyay, D., Dey, D., et al. (2021). Role of periodontal infection, inflammation and immunity in atherosclerosis. *Current Problems in Cardiology*, 46(3), 100638. doi:10.1016/j.cpcardiol.2020.100638.

Clark, J. D., de Heinzelin, J., Schick, K. D., Hart, W. K., White, T. D., WoldeGabriel, G., et al. (1994). African *Homo erectus:* Old radiometric ages and young Oldowan assemblages in the Middle Awash Valley, Ethiopia. *Science*, 264(5167), 1907–1910.

Clark, P., ed. (2013). *The Oxford Handbook of Cities in World History.* Oxford: Oxford University Press.

Clayton, J. A., and Collins, F. S. (2014). Policy: NIH to balance sex in cell and animal studies. *Nature*, 509(7500), 282–283. doi:10.1038/509282a.

Clemens, W. A. (2004). *Purgatorius* (Plesiadapiformes, Primates?, Mammalia), a Paleocene immigrant into northeastern Montana: Stratigraphic occurrences and incisor proportions. *Bulletin of Carnegie Museum of Natural History*, 2004(36), 3–13.

Coenen, P., Huysmans, M. A., Holtermann, A., Krause, N., van Mechelen, W., Straker, L. M., and van der Beek, A. J. (2018). Do highly physically active workers die early? A systematic review with meta-analysis of data from 193 696 participants. *British Journal of Sports Medicine*, 52(20), 1320–1326. doi:10.1136/bjsports-2017-098540.

Coffey, D. (2015). Prepregnancy body mass and weight gain during pregnancy in India and sub-Saharan Africa. *Proceedings of the National Academy of Sciences*, 112(11), 3302–3307. doi:10.1073/pnas.1416964112.

Coffey, D., and Hathi, P. (2016). Underweight and pregnant: Designing universal maternity entitlements to improve health. *Indian Journal of Human Development*, 10(2),176–190.

Coffman, K. B., Coffman, L. C., and Ericson, K. M. M. (2017). The size of the LGBT population and the magnitude of antigay sentiment are substantially underestimated. *Management Science*, 63(10), 3168–3186. doi:10.1287/mnsc.2016.2503.

Cohan, A. B., and Tannenbaum, I. J. (2001). Lesbian and bisexual women's judgments of the attractiveness of different body type. *The Journal of Sex Research*, 38(3), 226–232. doi:10.1080/00224490109552091.

Colchero, F., Aburto, J. M., Archie, E. A., Boesch, C., Breuer, T., Campos, F. A., et al. (2021). The long lives of primates and the "invariant rate of ageing" hypothesis. *Nature Communications*, 12(1), 3666. doi:10.1038/s41467-021-23894-3.

Colchero, F., Rau, R., Jones, O. R., Barthold, J. A., Conde, D. A., Lenart, A., et al. (2016). The emergence of longevous populations. *Proceedings of the National Academy of Sciences*, 113(48), E7681–E7690. doi:10.1073/pnas.1612191113.

Coleman, M. N. (2009). What do primates hear? A meta-analysis of all known nonhuman primate behavioral audiograms. *International Journal of Primatology*, 30(1), 55–91.

Colman, R. J., Kemnitz, J. W., Lane, M. A., Abbott, D. H., and Binkley, N. (1999). Skeletal effects of aging and menopausal status in female rhesus macaques. *The Journal of Clinical Endocrinology & Metabolism*, 84(11), 4144–4148. doi:10.1210/jcem.84.11.6151.

Conard, N. J. (2015). Cultural evolution during the Middle and Late Pleistocene in Africa and Eurasia. Pp. 2465–2508 in *Handbook of Paleoanthropology*, edited by W. Henke and I. Tattersall. Berlin: Springer.

Conith, A. J., Imburgia, M. J., Crosby, A. J., and Dumont, E. R. (2016). The functional significance of morphological changes in the dentitions of early mammals. *Journal of the Royal Society Interface*, 13(124), 20160713. doi:10.1098/rsif.2016.0713.

Connor, R. C., and Smolker, R. (1996). "Pop" goes the dolphin: A vocalization male bottlenose dolphins produce during consortships. *Behaviour*, 133, 643e662.

Connor, R. C., Krützen, M., Allen, S. J., Sherwin, W. B., and King, S. L. (2022). Strategic intergroup alliances increase access to a contested resource in male bottlenose dolphins. *Proceedings of the National Academy of Sciences*, 119(36), e2121723119. doi:10.1073/pnas.2121723119.

Connor, R. C., Smolker, R. A., and Richards, A. F. (1992). Two levels of alliance formation among male bottlenose dolphins (*Tursiops sp.*). *Proceedings of the National Academy of Sciences*, 89(3), 987–990.

Coppa, G. V., Pierani, P., Zampini, L., Carloni, I., Carlucci, A., and Gabrielli, O. (1999). Oligosaccharides in human milk during different phases of lactation. *Acta Paediatrica* suppl., 88(430), 89–94. doi:10.1111/j.1651-2227.1999.tb01307.x.

Coppa, G.V., Zampini, L., Galeazzi, T., Facinelli, B., Ferrante, L., Capretti, R., and Orazio, G. (2006). Human milk oligosaccharides inhibit the adhesion to Caco-2 cells of diarrheal pathogens: *Escherichia coli*, *Vibrio cholerae*, and *Salmonella fyris*. *Pediatric Research*, 59, 377–382.

Coquerelle, M., Prados-Frutos, J. C., Rojo, R., Mitteroecker, P., and Bastir, M. (2013). Short faces, big tongues: Developmental origin of the human chin. *PLOS ONE*, 8(11), e81287.

Corvinus, G. (2004). *Homo erectus* in East and Southeast Asia, and the questions of the age of the species and its association with stone artifacts, with special attention to handaxe-like tools. *Quaternary International*, 117(1), 141–151.

Courie, L. W. (1972). *The Black Death and Peasant's Revolt.* London: Wayland.

Cox, C., Bergmann, C., Fowler, E., Keren-Portnoy, T., Roepstorff, A., Bryant, G., and Fusaroli, R. (2022). A systematic review and Bayesian meta-analysis of the acoustic features of infant-directed speech. *Nature Human Behaviour,* Oct. 3, 2022. doi:10.1038/s41562-022-01452-1.

Cox-York, K., Wei, Y., Wang, D., Pagliassotti, M. J., and Foster, M. T. (2015). Lower body adipose tissue removal decreases glucose tolerance and insulin sensitivity in mice with exposure to high fat diet. *Adipocyte*, 4(1), 32–43. doi:10.4161/21623945.2014.957988.

Cristia, A., Dupoux, E., Gurven, M., and Stieglitz, J. (2019). Child-directed speech is infrequent in a forager-farmer population: A time allocation study. *Child Development*, 90(3), 759–773. doi.org/10.1111/cdev.12974.

Croft, D. P., Brent, L. J. N., Franks, D. W., and Cant, M. A. (2015). The evolution of prolonged life after reproduction. *Trends in Ecology & Evolution*, 30(7), 407–416. doi:10.1016/j.tree.2015.04.011.

Crofton, E. J., Zhang, Y., and Green, T. A. (2015). Inoculation stress hypothesis of environmental enrichment. *Neuroscience Biobehavioral Review*, 49, 19–31. doi:10.1016/j.neubiorev.2014.11.017.

Cuckle, H. S., Wald, N. J., and Thompson, S. G. (1987). Estimating a woman's risk of having a pregnancy associated with Down's syndrome using her age and serum alpha-fetoprotein level. *British Journal of Obstetrics and Gynaecology*, 94(5), 387–402. doi:10.1111/j.1471-0528.1987.tb03115.x.

Cullen, C. (1935). *The Medea, and Some Poems.* New York: Harper & Brothers.

Cunnane, S. C., and Crawford, M. A. (2003). Survival of the fattest: Fat babies were the key to evolution of the large human brain. *Comparative Biochemistry and Physiology. Part A, Molecular & Integrative Physiology*, 136(1), 17-26. doi:10.1016/s1095-6433(03)00048-5.

Cunnane, S. C., and Crawford, M. A. (2014). Energetic and nutritional constraints on infant brain development: Implications for brain expansion during human evolution. *Journal of Human Evolution*, 77, 88–98. doi:10.1016/j.jhevol.2014.05.001.

Currie, J., and Goodman, J. (2020). Parental socioeconomic status, child health, and human capital. Pp. 239–248 in *The Economics of Education*, edited by S. Bradley and C. Green. 2nd ed. Elsevier.

Cutler, A., and Scott, D. (1990). Speaker sex and perceived apportionment of talk. *Applied Psycholinguistics*, 11(3), 253–272. doi:10.1017/S0142716400008882.

Cyranowski, J. M., Frank, E., Young, E., and Shear, M. K. (2000). Adolescent onset of the gender difference in lifetime rates of major depression: A theoretical model. *Archives of General Psychiatry*, 57(1), 21–27. doi:10.1001/archpsyc.57.1.21.

Daddona, M. (2018). Got Milk? How the iconic campaign came to be, 25 years ago. *Fast Company.* www.fastcompany.com.

Dahlberg, E. L., Eberle, J. J., Sertich, J. J. W., and Miller, I. M. (2016). A new earliest Paleocene (Puercan) mammalian fauna from Colorado's Denver Basin, U.S.A. *Rocky Mountain Geology*, 51(1), 1–22. doi:10.2113/gsrocky.51.1.1.

Dalene, K. E., Tarp, J., Selmer, R. M., Ariansen, I. K. H., Nystad, W., Coenen, P., et al. (2021). Occupational physical activity and longevity in working men and women in Norway: a prospective cohort study. *The Lancet Public Health*, 6(6), e386–e395. doi:10.1016/S2468-2667(21)00032-3.

Dalley, S. (1991). *Myths from Mesopotamia*. Oxford: Oxford University Press.

Dallman, M. F., Pecoraro, N., Akana, S. F., La Fleur, S. E., Gomez, F., Houshyar, H., et al. (2003). Chronic stress and obesity: A new view of "comfort food." *Proceedings of the National Academy of Sciences*, 100(20), 11696–11701.

Daly, S. E., and Hartmann, P. E. (1995). Infant demand and milk supply. Part 2: The short-term control of milk synthesis in lactating women. *Journal of Human Lactation*, 11, 27–37.

Danesh, J., Collins, R., and Peto, R. (1997). Chronic infections and coronary heart disease: Is there a link? *The Lancet*, 350(9075), 430–436. doi:10.1016/S0140-6736(97)03079-1.

Darnall, B. D., Stacey, B. R., and Chou, R. (2012). Medical and psychological risks and consequences of long-term opioid therapy in women. *Pain Medicine*, 13(9), 1181–1211.

Dart, D. A., Waxman, J., Aboagye, E. O., and Bevan, C. L. (2013). Visualising androgen receptor activity in male and female mice. *PLOS ONE*, 8(8), e71694. doi:10.1371/journal.pone.0071694.

Dawkins, R. (1982, 1999). *The Extended Phenotype*. Rev. ed. Oxford: Oxford University Press.

De Araujo, J. F. F. E., and Tejedo-Romero, F. (2016). Women's political representation and transparency in local governance. *Local Government Studies*, 42(6), 885–906. doi:10.1080/03003930.2016.1194266.

De Bellis, M. D., Keshavan, M. S., Beers, S. R., Hall, J., Frustaci, K., Masalehdan, A., et al. (2001). Sex difference in brain maturation during childhood and adolescence. *Cerebral Cortex*, 11(6), 552–557. doi:10.1093/cercor/11.6.552.

de Blok, C. J. M., Klaver, M., Wiepjes, C. M., Nota, N. M., Heijboer, A. C., Fisher, A. D., et al. (2017). Breast development in transwomen after 1 year of cross-sex hormone therapy: Results of a prospective multicenter study. *The Journal of Clinical Endocrinology & Metabolism*, 103(2), 532–538. doi:10.1210/jc.2017-01927.

de Boer, B. (2012). Loss of air sacs improved hominin speech abilities. *Journal of Human Evolution*, 62(1), 1–6.

de Boer, B., and Kuhl, P. K. (2003). Investigating the role of infant-directed speech with a computer model. *Auditory Research Letters On-Line (ARLO)*, 4, 129–134.

de Catanzaro, D., MacNiven, E., and Ricciuti, F. (1991). Comparison of the adverse effects of adrenal and ovarian steroids on early pregnancy in mice. *Psychoneuroendocrinology*, 16(6), 525–536. doi:10.1016/0306-4530(91)90036-S.

de Dreu, C. K. W., Greer, L. L., Handgraaf, M. J. J., Shalvi, S., Van Kleef, G. A., Baas, M., et al. (2010). The neuropeptide oxytocin regulates parochial altruism in intergroup conflict among humans. *Science*, 328(5984), 1408–1411. doi:10.1126/science.1189047.

de Goffau, M. C., Lager, S., Sovio, U., Gaccioli, F., Cook, E., Peacock, S. J., et al. (2019). Human placenta has no microbiome but can contain potential pathogens. *Nature*, 572(7769), 329–334. doi:10.1038/s41586-019-1451-5.

de Heinzelin, J., Clark, J. D., White, T., Hart, W., Renne, P., WoldeGabriel, G., et al. (1999). Environment and behavior of 2.5-million-year-old Bouri hominids. *Science*, 284(5414), 625–629.

de la Torre, I. (2016). The origins of the Acheulean: Past and present perspectives on a major transition in human evolution. *Philosophical Transactions of the Royal Society*, 371, 20150245. doi:10.1098/rstb.2015.0245.

de Vries, E., den Tonkelaar, I., van Noord, P. A. H., van der Schouw, Y. T., te Velde, E. R., and Peeters, P. H. M. (2001). Oral contraceptive use in relation to age at menopause in the DOM cohort. *Human Reproduction*, 16(8), 1657–1662. doi:10.1093/humrep/16.8.1657.

de Waal, F. (2022). *Different: Gender Through the Eyes of a Primatologist*. New York: W. W. Norton.

de Waal, F. B., and Suchak, M. (2010). Prosocial primates: Selfish and unselfish motivations. *Philosophical Transactions of the Royal Society*, 365(1553), 2711–2722. doi:10.1098/rstb.2010.0119.

Deag, J. M., and Crook, J. H. (1971). Social behaviour and "agonistic buffering" in the wild Barbary macaque Macaca sylvana L. *Folia primatologica*, 15(3–4), 183–200.

Deary, I. J., Johnson, W., and Houlihan, L. M. (2009). Genetic foundations of human intelligence. *Human Genetics*, 126(1), 215–232. doi:10.1007/s00439-009-0655-4.

Deary, I. J., Strand, S., Smith, P., and Fernandes, C. (2007). Intelligence and educational achievement. *Intelligence*, 35(1), 13–21.

Deary, I. J., Thorpe, G., Wilson, V., Starr, J., and Whalley, L. (2003). Population sex differences in IQ at age 11: The Scottish Mental Survey 1932. *Intelligence*, 31, 533–542. doi:10.1016/S0160-2896(03)00053-9.

Defense Visual Information Distribution Service (DVIDS). (2015). Video of Ranger students after Griest and Haver finishes the course. www.dvidshub.net/video/420406/ranger-course-student-panel.

DeGiorgio, M., Jakobsson, M., and Rosenberg, N. A. (2009). Out of Africa: Modern human origins special feature: Explaining worldwide patterns of human genetic variation using a coalescent-based serial founder model of migration outward from Africa. *Proceedings of the National Academy of Sciences of the United States of America*, 106(38), 16057–16062. doi:10.1073/pnas.0903341106.

Dehara, M., Wells, M. B., Sjöqvist, H., Kosidou, K., Dalman, C., and Sörberg Wallin, A. (2021). Parenthood is associated with lower suicide risk: A register-based cohort study of 1.5 million Swedes. *Acta psychiatrica Scandinavica*, 143(3), 206–215. doi:10.1111/acps.13240.

Delano, P. H., Elgueda, D., Hamame, C. M., and Robles, L. (2007). Selective attention to visual stimuli reduces cochlear sensitivity in chinchillas. *The Journal of Neuroscience*, 27(15), 4146–4153. doi:10.1523/JNEUROSCI.3702-06.2007.

Demuru, E., Ferrari, P. F., and Palagi, E. (2018). Is birth attendance a uniquely human feature? New evidence suggests that Bonobo females protect and support the parturient. *Evolution and Human Behavior*, 39(5), 502–510. doi:10.1016/j.evolhumbehav.2018.05.003.

Dennison, T., and Ogilvie, S. (2014). Does the European marriage pattern explain economic growth? *The Journal of Economic History*, 74(3), 651–693. doi:10.1017/S0022050714000564.

DeSilva, J., and Lesnik, J. (2006). Chimpanzee neonatal brain size: Implications for brain growth in *Homo erectus*. *Journal of Human Evolution*, 51(2), 207–212.

DeSilva, J. M. (2011). A shift toward birthing relatively large infants early in human evolution. *Proceedings of the National Academy of Sciences*, 108(3), 1022–1027.

DeSilva, J. M., Laudicina, N. M., Rosenberg, K. R., and Trevathan, W. R. (2017). Neonatal shoulder width suggests a semirotational, oblique birth mechanism in *Australopithecus afarensis*. *The Anatomical Record*, 300(5), 890–899.

Di Mascio, D., Saccone, G., Bellussi, F., Vitagliano, A., and Berghella, V. (2020). Type of paternal sperm exposure before pregnancy and the risk of preeclampsia: A systematic review. *European Journal of Obstetrics & Gynecology and Reproductive Biology*, 251, 246–253. doi:10.1016/j.ejogrb.2020.05.065.

Di Stefano, N., Ghilardi, G., and Morini, S. (2017). Leonardo's mistake: Not evidence-based medicine? *The Lancet*, 390(10097), 845. doi:10.1016/S0140-6736(17)32140-2.

Diamanti-Kandarakis, E., Bourguignon, J. P., Giudice, L. C., Hauser, R., Prins, G. S., Soto, A. M., et al. (2009). Endocrine-disrupting chemicals: An Endocrine Society scientific statement. *Endocrine Reviews*, 30(4), 293–342.

Diamond, J. (1995). Father's milk. *Discover*, 16(2), 82–87.

Dickens, W. T., and Flynn, J. R. (2006). Black Americans reduce the racial IQ gap: Evidence from standardization samples. *Psychological Science*, 17(10), 913–920. doi:10.1111/j.1467-9280.2006.01802.x.

Diebolt, C., and Perrin, F. (2013). From stagnation to sustained growth: The role of female empowerment. *American Economic Review*, 103(3), 545–549. doi:10.1257/aer.103.3.545.

Diez-Martín, F., Sánchez, P., Domínguez-Rodrigo, M., Mabulla, A., and Barba, R. (2009). Were Olduvai Hominins making butchering tools or battering tools? Analysis of a recently excavated lithic assemblage from BK (Bed II, Olduvai Gorge, Tanzania). *Journal of Anthropological Archaeology*, 28(3), 274–289. doi:10.1016/j.jaa.2009.03.001.

Ding, W., Yang, L., and Xiao, W. (2013). Daytime birth and parturition assistant behavior in wild black-and-white snub-nosed monkeys (*Rhinopithecus bieti*) Yunnan, China. *Behavioural Processes*, 94, 5–8. doi:10.1016/j.beproc.2013.01.006.

Dinkel, D., and Snyder, K. (2020). Exploring gender differences in infant motor development related to parent's promotion of play. *Infant Behavior & Development*, 59, 101440. doi:10.1016/j.infbeh.2020.101440.

Diogo, R., Molnar, J. L., and Wood, B. (2017). Bonobo anatomy reveals stasis and mosaicism in chimpanzee evolution, and supports bonobos as the most appropriate extant model for the common ancestor of chimpanzees and humans. *Scientific Reports*, 7(1), 608.

Dixson, A. F. (2018). Copulatory and Postcopulatory Sexual Selection in Primates. *Folia Primatologica*, 89(3–4), 258–286. doi:10.1159/000488105.

Dixson, A. L., and Anderson, M. J. (2002). Sexual selection, seminal coagulation and copulatory plug formation in primates. *Folia Primatologica*, 73(2–3), 63–69. doi:10.1159/000064784.

Dobolyi, A., Oláh, S., Keller, D., Kumari, R., Fazekas, E. A., Csikós, V., et al. (2020). Secretion and function of pituitary prolactin in evolutionary perspective. *Frontiers in Neuroscience*, 14, 621. doi:10.3389/fnins.2020.00621.

Dodge, B., Reece, M., and Herbenick, D. (2009). School-based condom education and its relations with diagnoses of and testing for sexually transmitted infections among men in the United States. *American Journal of Public Health*, 99(12), 2180–2182. doi:10.2105/AJPH.2008.159038.

Dollé, J. P., Jaye, A., Anderson, S. A., Ahmadzadeh, H., Shenoy, V. B., and Smith, D. H. (2018). Newfound sex differences in axonal structure underlie differential outcomes from in vitro traumatic axonal injury. *Experimental Neurology*, 300, 121–134. doi:10.1016/j.expneurol.2017.11.001.

Dong, X., Milholland, B., and Vijg, J. (2016). Evidence for a limit to human lifespan. *Nature*, 538, 257–259. doi:10.1038/nature19793.

Donovan, M. P., Iglesias, A., Wilf, P., Labandeira, C. C., and Cúneo, N. R. (2016). Rapid recovery of Patagonian plant-insect associations after the end-Cretaceous extinction. *Nature Ecology & Evolution*, 1(1), 0012. doi:10.1038/s41559-016-0012.

Donovan, M. P., Iglesias, A., Wilf, P., Labandeira, C. C., and Cúneo, N. R. (2018). Diverse plant-insect associations from the Latest Cretaceous and Early Paleocene of Patagonia, Argentina. *Ameghiniana*, 55(3), 303–338, 336. doi:10.5710/AMGH.15.02.2018.3181.

Dorak, M. T., and Karpuzoglu, E. (2012). Gender differences in cancer susceptibility: An inadequately addressed issue. *Frontiers in Genetics*, 3, 268. doi:10.3389/fgene.2012.00268.

Doty, R. L., and Cameron, E. L. (2009). Sex differences and reproductive hormone influences on human odor perception. *Physiology & Behavior*, 97(2), 213–228.

Douglas, P. H. (2014). Female sociality during the daytime birth of a wild bonobo at Luikotale, Democratic Republic of the Congo. *Primates*, 55(4), 533–542. doi:10.1007/s10329-014-0436-0.

Dreger., A. D. (1998). *Hermaphrodites and the Medical Invention of Sex*. Cambridge, Mass.: Harvard University Press.

Drewett, R., Bowen-Jones, A., and Dogterom, J. (1982). Oxytocin levels during breast-feeding in established lactation. *Hormones and Behavior*, 16(2), 245–248.

Drewnowski, A., Krahn, D. D., Demitrack, M. A., Nairn, K., Gosnell, B. A. (1992). Taste responses and preferences for sweet high-fat foods: Evidence for opioid involvement. *Physiology and Behavior*, 51, 371–379.

Drews, B., Roellig, K., Menzies, B. R., Shaw, G., Buentjen, I., Herbert, C. A., et al. (2013). Ultrasonography of wallaby prenatal development shows that the climb to the pouch begins in utero. *Scientific Reports*, 3(1), 1458. doi:10.1038/srep01458.

Dubno, J. R., Dirks, D. D., and Morgan, D. E. (1984). Effects of age and mild hearing loss on speech recognition in noise. *The Journal of the Acoustical Society of America*, 76(1), 87–96. doi:10.1121/1.391011.

Duckitt, K., and Harrington, D. (2005). Risk factors for pre-eclampsia at antenatal booking: Systematic review of controlled studies. *British Medical Journal*, 330(7491), 565.

Dulias, K., Foody, M. G. B., Justeau, P., Silva, M., Martiniano, R., Oteo-García, G., et al. (2022). Ancient DNA at the edge of the world: Continental immigration and the persistence of Neolithic male lineages in Bronze Age Orkney. *Proceedings of the National Academy of Sciences*, 119(8), e2108001119. doi:10.1073/pnas.2108001119.

Dunbar, R. (1993). Coevolution of neocortical size, group size, and language in humans. *Behavioral and Brain Sciences*, 16(4), 681–735.

Dunbar, R. (1996). *Grooming, Gossip, and the Evolution of Language*. London: Faber & Faber.

Dunbar, R. I. M. (1984). Infant-use by male gelada in agonistic contexts: Agonistic buffering, progeny protection or soliciting support? *Primates*, 25, 28–35.

Dunford, A., Weinstock, D. M., Savova, V., Schumacher, S. E., Cleary, J. P., Yoda, A., et al. (2017). Tumor-suppressor genes that escape from X-inactivation contribute to cancer sex bias. *Nature Genetics*, 49(1), 10–16. doi:10.1038/ng.3726.

Dunn, J. C., Halenar, L. B., Davies, T. G., Cristobal-Azkarate, J., Reby, D., Sykes, D., et al. (2015). Evolutionary trade-off between vocal tract and testes dimensions in howler monkeys. *Current Biology*, 25(21), 2839–2844. doi:10.1016/j.cub.2015.09.029.

Dunsworth, H., and Eccleston, L. (2015). The evolution of difficult childbirth and helpless hominin infants. *Annual Review of Anthropology*, 44, 55–69.

Dunsworth, H. M., Warrener, A. G., Deacon, T., Ellison, P. T., and Pontzer, H. (2012). Metabolic hypothesis for human altriciality. *Proceedings of the National Academy of Sciences*, 109(38), 15212–15216. doi:10.1073/pnas.1205282109.

Durex. (2007). The face of global sex 2007. www.durexnetwork.org.

Easter, M., and Freedman, A. (2020). Here's why breast milk isn't a good workout supplement for bodybuilders. *Men's Health*, Aug. 11, 2020. www.menshealth.com.

Eaton, N. R., Keyes, K. M., Krueger, R. F., Balsis, S., Skodol, A. E., Markon, K. E., et al. (2012). An invariant dimensional liability model of gender differences in mental disorder prevalence: Evidence from a national sample. *Journal of Abnormal Psychology*, 121(1), 282–288. doi:10.1037/a0024780.

Eaton, S. B., Pike, M. C., Short, R. V., Lee, N. C., Trussell, J., Hatcher, R. A., et al. (1994). Women's reproductive cancers in evolutionary context. *The Quarterly Review of Biology*, 69(3), 353–367. doi:10.1086/418650.

Egeland, C. P., Domínguez-Rodrigo, M., and Barba, R. (2007). The "home base" debate. Pp. 1–10 in *Deconstructing Olduvai: A Taphonomic Study of the Bed I Sites*. Dordrecht: Springer.

Ehsan, G. A. (2011). Female militia chief keeps peace in Helmand District. *Institute for War & Peace Reporting*, Sept. 7, 2011. iwpr.net.

Eid, R. S., Gobinath, A. R., and Galea, L. A. M. (2019). Sex differences in depression: Insights from clinical and preclinical studies. *Progress in Neurobiology*, 176, 86–102. doi:10.1016/j.pneurobio.2019.01.006.

Eisenberger, N. I., and Lieberman, M. D. (2004). Why rejection hurts: A common neural alarm system for physical and social pain. *Trends in Cognitive Science*, 8, 294–300.

Eisenberger, N. I., Lieberman, M. D., and Williams, K. D. (2003). Does rejection hurt? An FMRI study of social exclusion. *Science*, 302, 290–292.

Elder, J. H., Yerkes, R. M., and Cushing, H. W. (1936). Chimpanzee births in captivity: a typical case history and report of sixteen births. *Proceedings of the Royal Society B: Biological Sciences*, 120(819), 409–421. doi:10.1098/rspb.1936.0043.

Eliot, L., Ahmed, A., Khan, H., and Patel, J. (2021). Dump the "dimorphism": Comprehensive synthesis of human brain studies reveals few male-female differences beyond size. *Neuroscience & Biobehavioral Reviews*, 125, 667–697. doi:10.1016/j.neubiorev.2021.02.026.

Ellis, L., Hershberger, S., Field, E., Wersinger, S., Pellis, S., Geary, D., et al. (2013). *Sex Differences: Summarizing More Than a Century of Scientific Research*. New York: Psychology Press.

Ellis, S., Franks, D. W., Nattrass, S., Cant, M. A., Bradley, D. L., Giles, D., et al. (2018). Postreproductive lifespans are rare in mammals. *Ecology and Evolution*, 8(5), 2482-2494. doi:10.1002/ece3.3856.

Emberling, G. (2003). Urban social transformations and the problem of the "First City": New research from Mesopotamia. Pp. 254–268 in *The Social*

Construction of Ancient Cities, edited by M. Smith. Washington, D.C.: Smithsonian Institution Press.

Emera, D., Romero, R., and Wagner, G. (2012). The evolution of menstruation: A new model for genetic assimilation: Explaining molecular origins of maternal responses to fetal invasiveness. *BioEssays*, 34(1), 26–35. doi:10.1002/bies.201100099.

English, R., Lebovitz, Y., and Giffin, R. (2010). *Transforming Clinical Research in the United States: Challenges and Opportunities: Workshop Summary*. Washington, D.C.: National Academies Press.

Eriksson, M., Marschik, P. B., Tulviste, T., Almgren, M., Pérez Pereira, M., Wehberg, S., et al. (2012). Differences between girls and boys in emerging language skills: Evidence from 10 language communities. *British Journal of Developmental Psychology*, 30(2), 326–343.

Espenshade, T. J., Guzman, J. C., and Westoff, C. F. (2003). The surprising global variation in replacement fertility. *Population Research and Policy Review*, 22(5), 575–583. doi:10.1023/B:POPU.0000020882.29684.8e.

Euripedes. (1994). *Medea*. In *Cyclops; Alcestis; Medea*, edited and translated by David Kovas. Cambridge, Mass.: Harvard University Press.

Everett, D. (2017). *How Language Began: The Story of Humanity's Greatest Invention*. London: Profile Books.

Exton, M. S., Bindert, A., Kruger, T., Scheller, F., Hartmann, U., and Schedlowski, M. (1999). Cardiovascular and endocrine alterations after masturbation-induced orgasm in women. *Psychosomatic Medicine*, 61(3), 280–289.

FDA (2022). Breast Implants: Reports of Squamous Cell Carcinoma and Various Lymphomas in Capsule Around Implants: FDA Safety Communication. Sept. 8, 2022. www.fda.gov.

Fagundes, N. J. R., Ray, N., Beaumont, M., Neuenschwander, S., Salzano, F. M., Bonatto, S. L., and Excoffier, L. (2007). Statistical evaluation of alternative models of human evolution. *Proceedings of the National Academy of Sciences*, 104(45), 17614–17619. doi:10.1073/pnas.0708280104.

Fairweather, D. (2015). Sex differences in inflammation during atherosclerosis. *Clinical Medicine Insights. Cardiology*, 8(suppl. 3), 49–59. doi:10.4137/CMC.S17068.

Falk, D. (1980). A reanalysis of the South African australopithecine natural endocasts. *American Journal of Physical Anthropology*, 53(4), 525–539.

Falk, D. (1983). Cerebral cortices of East African early hominids. *Science*, 221(4615), 1072–1074.

Falk, D. (2004). Prelinguistic evolution in early hominins: whence motherese? *The Behavioral and Brain Sciences*, 27(4), 491–583. doi:10.1017/s0140525x04000111.

Falk, D., Zollikofer, C. P. E., Morimoto, N., and Ponce de León, M. S. (2012). Metopic suture of Taung (*Australopithecus africanus*) and its implications for hominin brain evolution. *Proceedings of the National Academy of Sciences*, 109(22), 8467–8470.

Farace, E., and Alves, W. M. (2000). Do women fare worse: A metaanalysis of gender differences in traumatic brain injury outcome, *Journal of Neurosurgery*, 93(4), 539–545.

Faria, J. B., Santiago, M. B., Silva, C. B., Geraldo-Martins, V. R., and Nogueira, R. D. (2022). Development of *Streptococcus mutans* biofilm in the presence of human colostrum and 3′-sialyllactose. *The Journal of Maternal-Fetal & Neonatal Medicine*, 35(4), 630–635. doi:10.1080/14767058.2020.1730321.

Farmer, C. G. (2020). Parental care, destabilizing selection, and the evolution of tetrapod endothermy. *Physiology*, 35(3), 160–176. doi:10.1152/physiol.00058.2018.

Fausto-Sterling, A., Coll, C. G., and Lamarre, M. (2012). Sexing the baby: Part 1—What do we really know about sex differentiation in the first three years of life? *Social Science & Medicine*, 74(11), 1684–1692. doi:10.1016/j.socscimed.2011.05.051.

Fauve-Chamoux, A. (2001). Marriage, widowhood, and divorce. Pp. 221–256 in *The History of the European Family*, Vol. 1: *Family Life in Early Modern Times, 1500–1789*, edited by D. I. Kertzer and M. Barbagli. New Haven: Yale University Press.

Felisbino-Mendes, M. S., Araújo, F. G., Oliveira, L. V. A., Vasconcelos, N. M., Vieira, M. L. F. P., and Malta, D. C. (2021). Sexual behaviors and condom use in the Brazilian population: Analysis of the National Health Survey, 2019. *Revista brasileira de epidemiologia (Brazilian Journal of Epidemiology)*, 24(suppl. 2), e210018. doi:10.1590/1980-549720210018.supl.2.

Felshman, J., and Schaffer, N. (1998). Sex and the single rhinoceros. *Chicago Reader*, Feb. 20, 1998, 24–27.

Fernandez, R. C., Moore, V. M., Marino, J. L., Whitrow, M. J., and Davies, M. J. (2020). Night shift among women: Is it associated with difficulty conceiving a first birth? *Frontiers in Public Health*, 8, 595943. doi:10.3389/fpubh.2020.595943.

Fernandez-Esquer, M. E., Atkinson, J., Diamond, P., Useche, B., and Mendiola, R. (2004). Condom use self-efficacy among U.S.- and foreign-born Latinos in Texas. *The Journal of Sex Research*, 41(4), 390–399.

Fernández-Navarro, V., Camarós, E., and Garate, D. (2022). Visualizing childhood in Upper Palaeolithic societies: Experimental and archaeological approach to artists' age estimation through cave art hand stencils. *Journal of Archaeological Science*, 140, 105574. doi:10.1016/j.jas.2022.105574.

Ferraro, J. V., Plummer, T. W., Pobiner, B. L., Oliver, J. S., Bishop, L. C., Braun, D. R., et al. (2013). Earliest archaeological evidence of persistent Hominin carnivory. *PLOS ONE*, 8(4), e62174. doi:10.1371/journal.pone.0062174.

Ferreira L. F. (2018). Mitochondrial basis for sex-differences in metabolism and exercise performance. *American Journal of Physiology: Regulatory, Integrative and Comparative Physiology*, 314(6), R848–R849. doi:10.1152/ajpregu.00077.2018.

Ferretti, M. T., Iulita, M. F., Cavedo, E., Chiesa, P. A., Schumacher Dimech,

A., Santuccione Chadha, A., et al. (2018). Sex differences in Alzheimer disease—the gateway to precision medicine. *Nature Reviews Neurology*, 14(8), 457–469. doi:10.1038/s41582-018-0032-9.

Fessler, D. M. T. (2003). Rape is not less frequent during the ovulatory phase of the menstrual cycle. *Sexualities, Evolution & Gender*, 5(3), 127–147. doi:10.1080/14616660410001662361.

Field, T. M., Cohen, D., Garcia, R., and Greenberg, R. (1984). Mother-stranger face discrimination by the newborn. *Infant Behavior and Development*, 7(1), 19–25. doi:10.1016/S0163-6383(84)80019-3.

Fildes, V. (1986). *Breasts, Bottles and Babies: A History of Infant Feeding*. Edinburgh: Edinburgh University Press.

Fildes, V. (1988). *Wet Nursing: A History from Antiquity to the Present*. Oxford: Basil Blackwell.

Finn, J. K., Tregenza, T., and Norman, M. D. (2009). Defensive tool use in a coconut-carrying octopus. *Current Biology*, 19(23), R1069–R1070. doi:10.1016/j.cub.2009.10.052.

Firestein, S. (2001). How the olfactory system makes sense of scents. *Nature*, 413(6852), 211–218. doi:10.1038/35093026.

Fish, A. M., Nadig, A., Seidlitz, J., Reardon, P. K., Mankiw, C., McDermott, C. L., et al. (2020). Sex-biased trajectories of amygdalo-hippocampal morphology change over human development. *NeuroImage*, 204, 116122. doi:10.1016/j.neuroimage.2019.116122.

Fitch, W. T. (2000). The evolution of speech: A comparative review. *Trends in Cognitive Sciences*, 4(7), 258–267. doi:10.1016/S1364-6613(00)01494-7.

Fitch, W. T. (2010). *The Evolution of Language*. Cambridge, U.K.: Cambridge University Press.

Fitch, W. T., and Giedd, J. (1999). Morphology and development of the human vocal tract: A study using magnetic resonance imaging. *The Journal of the Acoustical Society of America*, 106(3 Pt. 1), 1511–1522. doi:10.1121/1.427148.

Fitch, W. T., and Reby, D. (2001). The descended larynx is not uniquely human. *Proceedings of the Royal Society B: Biological Sciences*, 268(1477), 1669–1675.

Fitch, W. T., de Boer, B., Mathur, N., and Ghazanfar, A. A. (2017). Response to Lieberman on "Monkey vocal tracts are speech-ready." *Science Advances*, 3(7), e1701859. doi:10.1126/sciadv.1701859.

Fitts, R. H., Riley, D. R., and Widrick, J. J. (2001). Functional and structural adaptations of skeletal muscle to microgravity. *The Journal of Experimental Biology*, 204(18), 3201–3208. doi:10.1242/jeb.204.18.3201.

Fleagle, G. J. (2013). *Primate Adaptation and Evolution*. San Diego: Academic Press.

Fleagle, J. G., Shea, J. J., Grine, F. E., Baden, A. L., and Leakey, R. E. (2010). *Out of Africa I: The First Hominin Colonization of Eurasia*. Dordrecht: Springer.

Flemming, R. (2019). (The wrong kind of) gonorrhea in antiquity. In *The Hidden Affliction: Sexually Transmitted Infections and Infertility in History*, edited by S. Szreter. Rochester: University of Rochester Press.

Flores, A. R., Meyer, I. H., Langton, L., and Herman, J. L. (2021). Gender identity disparities in criminal victimization: National Crime Victimization Survey, 2017–2018. *American Journal of Public Health*, 111(4), 726–729. doi:doi.org/10.2105/AJPH.2020.306099.

Flynn, A., and Graham, K. (2010). "Why did it happen?" A review and conceptual framework for research on perpetrators' and victims' explanations for intimate partner violence. *Aggression and Violent Behavior*, 15, 239–251.

Foley, N. M., Springer, M. S., and Teeling, E. C. (2016). Mammal madness: Is the mammal tree of life not yet resolved? *Philosophical Transactions of the Royal Society B*, 371, 20150140. doi:10.1098/rstb.2015.0140.

Fontana, L., Gentilin, B., Fedele, L., Gervasini, C., and Miozzo, M. (2017). Genetics of Mayer-Rokitansky-Küster-Hauser (MRKH) syndrome. *Clinical Genetics*, 91, 233–246. doi:10.1111/cge.12883.

Foose, T. J., and Wiese, R. J. (2006). Population management of rhinoceros in captivity. *International Zoo Yearbook*, 40, 174–196.

Forrester, G. S., Davis, R., Mareschal, D., Malatesta, G., and Todd, B. K. (2019). The left cradling bias: An evolutionary facilitator of social cognition? *Cortex*, 118, 116–121. doi:10.1016/j.cortex.2018.05.011.

Foster, E. A., Franks, D. W., Mazzi, S., Darden, S. K., Balcomb, K. C., Ford, J. K. B., and Croft, D. P. (2012). Adaptive prolonged postreproductive life span in killer whales. *Science*, 337(6100), 1313. doi:10.1126/science.1224198.

Foster, J. A., and McVey Neufeld, K.-A. (2013). Gut-brain axis: How the microbiome influences anxiety and depression. *Trends in Neurosciences*, 36(5), 305–312. doi:10.1016/j.tins.2013.01.005.

Fowler, A., and Hohmann, G. (2010). Cannibalism in wild bonobos (*Pan paniscus*) at Lui Kotale. *American Journal of Primatology*, 72(6), 509–514. doi:10.1002/ajp.20802.

Fowler, A., Koutsioni, Y., and Sommer, V. (2007). Leaf-swallowing in Nigerian chimpanzees: Evidence for assumed self-medication. *Primates*, 48, 73–76. doi:10.1007/s10329-006-0001-6.

Franklin, B. (1745/1961). Old mistress apologue, 25 June 1745. Pp. 27–31 in *The Papers of Benjamin Franklin*, Vol. 3, *January 1, 1745, Through June 30, 1750*, edited by L. W. Labaree. New Haven: Yale University Press, 27–31.

Franklin, R. (2016). *Shirley Jackson: A Rather Haunted Life*. New York: Liveright.

Frederiksen, L. E., Ernst, A., Brix, N., Braskhøj Lauridsen, L. L., Roos, L., Ramlau-Hansen, C. H., and Ekelund, C. K. (2018). Risk of adverse pregnancy outcomes at advanced maternal age. *Obstetrics & Gynecology*, 131(3), 457–463. doi:10.1097/AOG.0000000000002504.

Fredriks, A. M., van Buuren, S., Fekkes, M., Verloove-Vanhorick, S. P., and Wit, J. M. (2005). Are age references for waist circumference, hip circumference and waist-hip ratio in Dutch children useful in clinical practice? *European Journal of Pediatrics*, 164(4), 216–222. doi:10.1007/s00431-004-1586-7.

Freedman, D. S., Khan, L. K., Serdula, M. K., Dietz, W. H., Srinivasan, S. R., and Berenson, G. S. (2002). Relation of age at menarche to race, time period, and anthropometric dimensions: The Bogalusa Heart Study. *Pediatrics*, 110(4), e43.

Freeman, E. W., and Sherif, K. (2007). Prevalence of hot flushes and night sweats around the world: A systematic review. *Climacteric*, 10(3), 197–214.

Freire, G. M. G., Cavalcante, R. N., Motta-Leal-Filho, J. M., Messina, M., Galastri, F. L., Affonso, B. B., et al. (2017). Controlled-release oxycodone improves pain management after uterine artery embolisation for symptomatic fibroids. *Clinical Radiology*, 72(5), 428.e421–428.e425. doi:10.1016/j.crad.2016.12.010.

Fried, M., and Duffy, P. E. (2017). Malaria during pregnancy. *Cold Spring Harbor Perspectives in Medicine*, 7(6), a025551. doi:10.1101/cshperspect.a025551.

Friedman, S. H., Horwitz, S. M., and Resnick, P. J. (2005). Child murder by mothers: A critical analysis of the current state of knowledge and a research agenda. *American Journal of Psychiatry*, 162(9), 1578–1587.

Friedmann, N., and Rusou, D. (2015). Critical period for first language: The crucial role of language input during the first year of life. *Current Opinion in Neurobiology*, 35, 27–34. doi:10.1016/j.conb.2015.06.003.

Frier, B. W. (2015). Roman law and the marriage of underage girls. *Journal of Roman Archaeology*, 28, 652–664.

Frohlich, J., and Kettle, C. (2015). Perineal care. *BMJ Clinical Evidence*, 2015, 1401.

Fruth, B., Ikombe, N. B., Matshimba, G. K., Metzger, S., Muganza, D. M., Mundry, R., and Fowler, A. (2014). New evidence for self-medication in bonobos: *Manniophyton fulvum* leaf- and stemstrip-swallowing from Lui-Kotale, Salonga National Park, DR Congo. *American Journal of Primatology*, 76(2), 146–158.

Fuller, D. Q., and Stevens, C. J. (2019). Between domestication and civilization: The role of agriculture and arboriculture in the emergence of the first urban societies. *Vegetation History and Archaeobotany*, 28(3), 263–282. doi:10.1007/s00334-019-00727-4.

Furuichi, T. (2011). Female contributions to the peaceful nature of bonobo society. *Evolutionary Anthropology*, 20(4), 131–142. doi:10.1002/evan.20308.

Gaillard, J.-M., and Yoccoz, N. G. (2003). Temporal variation in survival of mammals: A case of environmental canalization? *Ecology*, 84, 3294–3306. doi:10.1890/02-0409.

Galen. (1968). *On the Usefulness of Parts of the Body*. Translated by M. T. May. Ithaca, N.Y.: Cornell University Press.

Galler, J. R., Bryce, C. P., Waber, D. P., Hock, R. S., Harrison, R., Eaglesfield, G. D., and Fitzmaurice, G. (2012). Infant malnutrition predicts conduct problems in adolescents. *Nutritional Neuroscience*, 15(4), 186–192. doi:10.1179/1476830512Y.0000000012.

Gallup, G. G., Jr., Burch, R. L., Zappieri, M. L., Parvez, R. A., Stockwell, M. L., and Davis, J. A. (2003). The human penis as a semen displacement device. *Evolution and Human Behavior*, 24(4), 277–289.

Galmiche, M., Déchelotte, P., Lambert, G., and Tavolacci, M. P. (2019). Prevalence of eating disorders over the 2000–2018 period: A systematic literature review. *The American Journal of Clinical Nutrition*, 109(5), 1402–1413. doi:10.1093/ajcn/nqy342.

Gan, T. J., Glass, P. S., Sigl, J., Sebel, P., Payne, F., Rosow, C., and Embree, P. (1999). Women emerge from general anesthesia with propofol/alfentanil /nitrous oxide faster than men. *Anesthesiology*, 90(5), 1283–1287. doi:10.1097 /00000542-199905000-00010.

García-López de Hierro, L., Moleón, M., Ryan, P. G. (2013). Is carrying feathers a sexually selected trait in house sparrows? *Ethology*, 119(3), 199. doi:10.1111/eth.12053.

Gardner, A. S., Rahman, I. A., Lai, C. T., Hepworth, A., Trengove, N., Hartmann, P. E., and Geddes, D. T. (2017). Changes in fatty acid composition of human milk in response to cold-like symptoms in the lactating mother and infant. *Nutrients*, 9(9). doi:10.3390/nu9091034.

Garrett, E. C., Dennis, J. C., Bhatnagar, K. P., Durham, E. L., Burrows, A. M., Bonar, C. J., et al. (2013). The vomeronasal complex of nocturnal strepsirhines and implications for the ancestral condition in primates. *The Anatomical Record*, 296(12), 1881–1894.

Garrison, J. L., Macosko, E. Z., Bernstein, S., Pokala, N., Albrecht, D. R., and Bargmann, C. I. (2012). Oxytocin/vasopressin-related peptides have an ancient role in reproductive behavior. *Science*, 338(6106), 540–543. doi:10.1126/science.1226201.

Gavrilov, L. A., and Gavrilova, N. S. (2002). Evolutionary theories of aging and longevity. *The Scientific World Journal*, 2, 339–356. doi:10.1100/tsw.2002.96.

Geary, C. M., Welch, J. F., McDonald, M. R., Peters, C. M., Leahy, M. G., Reinhard, P. A., and Sheel, A. W. (2019). Diaphragm fatigue and inspiratory muscle metaboreflex in men and women matched for absolute diaphragmatic work during pressure-threshold loading. *Journal of Physiology*, 597, 4797–4808. doi:10.1113/JP278380.

Geary, D. C. (2000). Evolution and proximate expression of human paternal investment. *Psychological Bulletin*, 126, 55–77. doi:10.1037/0033 -2909.126.1.55.

Geiser, S. (2015). The growing correlation between race and SAT scores: New findings from California. *CSHE Research and Occasional Paper Series*, 15(10). cshe.berkeley.edu.

Geller, S. E., Koch, A. R., Roesch, P., Filut, A., Hallgren, E., and Carnes, M. (2018). The more things change, the more they stay the same: A study to evaluate compliance with inclusion and assessment of women and minorities in randomized controlled trials. *Academic Medicine*, 93(4), 630–635. doi:10.1097/acm.0000000000002027.

Gelstein, S., Yeshurun, Y., Rozenkrantz, L., Shushan, S., Frumin, I., Roth, Y., and Sobel, N. (2011). Human tears contain a chemosignal. *Science*, 331(6014), 226–230. doi:10.1126/science.1198331.

Genc, S., Raven, E. P., Drakesmith, M., Blakemore, S.-J., and Jones, D. K. (2023). Novel insights into axon diameter and myelin content in late childhood and adolescence. *Cerebral Cortex*, 33(10), 6435–6448. doi:10.1093/cercor/bhac515.

Gerkema, M. P., Davies, W. I., Foster, R. G., Menaker, M., and Hut, R. A. (2013). The nocturnal bottleneck and the evolution of activity patterns in mammals. *Proceedings of the Royal Society B: Biological Sciences*, 280(1765), 20130508. doi:10.1098/rspb.2013.0508.

Gershon, R., Neitzel, R., Barrera, M., and Akram, M. (2006). Pilot survey of subway and bus stop noise levels. *Journal of Urban Health*, 83, 802–812. doi:10.1007/s11524-006-9080-3.

Ghassabian, A., Vandenberg, L., Kannan, K., and Trasande, L. (2022). Endocrine-disrupting chemicals and child health. *Annual Review of Pharmacology and Toxicology*, 62, 573–594. doi:10.1146/annurev-pharmtox-021921-093352.

Ghazanfar, A. A., and Rendall, D. (2008). Evolution of human vocal production. *Current Biology*, 18(11), R457–R460.

Gilad, Y., Wiebe, V., Przeworski, M., Lancet, D., and Pääbo, S. (2004). Loss of olfactory receptor genes coincides with the acquisition of full trichromatic vision in primates. *PLOS Biology*, 2(1), e5. doi:10.1371/journal.pbio.0020005.

Gilardi, K. V. K., Shideler, S. E., Valverde, C. R., Roberts, J. A., and Lasley, B. L. (1997). Characterization of the onset of menopause in the rhesus macaque1. *Biology of Reproduction*, 57(2), 335–340. doi:10.1095/biolreprod57.2.335.

Gill, P. G., Purnell, M. A., Crumpton, N., Brown, K. R., Gostling, N. J., Stampanoni, M., and Rayfield, E. J. (2014). Dietary specializations and diversity in feeding ecology of the earliest stem mammals. *Nature*, 512(7514), 303–305. doi:10.1038/nature13622.

Gillam, L., McDonald, R., Ebling, F. J., and Mayhew, T. M. (2008). Human 2D (index) and 4D (ring) finger lengths and ratios: Cross-sectional data on linear growth patterns, sexual dimorphism and lateral asymmetry from 4 to 60 years of age. *Journal of Anatomy*, 213(3), 325–335. doi:10.1111/j.1469-7580.2008.00940.x.

Gillies, G. E., Pienaar, I. S., Vohra, S., and Qamhawi, Z. (2014). Sex differences in Parkinson's disease. *Frontiers in Neuroendocrinology*, 35(3), 370–384. doi:10.1016/j.yfrne.2014.02.002.

Glintborg, D., T'Sjoen, G., Ravn, P., and Andersen, M. S. (2021). Management of endocrine disease: Optimal feminizing hormone treatment in transgender people, *European Journal of Endocrinology*, 185(2), R49–R63.

Gobes, S. M. H., Jennings, R. B., and Maeda, R. K. (2019). The sensitive period for auditory-vocal learning in the zebra finch: Consequences of limited-model availability and multiple-tutor paradigms on song imitation. *Behavioural Processes*, 163, 5–12. doi:10.1016/j.beproc.2017.07.007.

Goddings, A. L., Beltz, A., Peper, J. S., Crone, E. A., and Braams, B. R. (2019). Understanding the role of puberty in structural and functional development of the adolescent brain. *Journal of Research on Adolescence*, 29(1), 32–53.

Goldberg, A., Günther, T., Rosenberg, N. A., and Jakobsson, M. (2017). Ancient X chromosomes reveal contrasting sex bias in Neolithic and Bronze Age Eurasian migrations. *Proceedings of the National Academy of Sciences*, 114(10), 2657–2662. doi:10.1073/pnas.1616392114.

Goldman, M. (2014). Amazing deliveries. *Emory Medicine Magazine*. Fall. emorymedicinemagazine.emory.edu.

Goldsmith, L. T., Weiss, G., Palejwala, S., Plant, T. M., Wojtczuk, A., Lambert, W. C., et al. (2004). Relaxin regulation of endometrial structure and function in the rhesus monkey. *Proceedings of the National Academy of Sciences*, 101(13), 4685–4689. doi:10.1073/pnas.0400776101.

Goldstein, J. M., Seidman, L. J., Horton, N. J., Makris, N., Kennedy, D. N., Caviness, V. S., Jr., et al. (2001). Normal sexual dimorphism of the adult human brain assessed by in vivo magnetic resonance imaging. *Cerebral Cortex*, 11(6), 490–497.

Gomes, C. M., and Boesch, C. (2009). Wild chimpanzees exchange meat for sex on a long-term basis. *PLOS ONE*, 4(4), e5116. doi:10.1371/journal.pone.0005116.

Goodall, J. (1977). Infant killing and cannibalism in free-living chimpanzees. *Folia Primatologica*, 28(4), 259–282.

Goodall, J. (1986). *The Chimpanzees of Gombe: Patterns of Behavior.* Cambridge, Mass.: Harvard University Press.

Goodall, J. (2010). *Through a Window: My Thirty Years with the Chimpanzees of Gombe*. Boston: Houghton Mifflin Harcourt.

Goodson, J. L., Schrock, S. E., Klatt, J. D., Kabelik, D., and Kingsbury, M. A. (2009). Mesotocin and nonapeptide receptors promote estrildid flocking behavior. *Science*, 325(5942), 862–866. doi:10.1126/science.1174929.

Gordon, H. S., and Rosenthal, G. E. (1999). The relationship of gender and in-hospital death: Increased risk of death in men. *Medical Care*, 37(3), 318–324.

Gordon-Salant, S. (2005). Hearing loss and aging: New research findings and clinical implications. *Journal of Rehabilitation Research & Development*, 42.

Goren-Inbar, N., Feibel, C. S., Verosub, K. L., Melamed, Y., Kislev, M. E., Tchernov, E., and Saragusti, I. (2000). Pleistocene milestones on the Out-of-Africa Corridor at Gesher Benot Ya'aqov, Israel. *Science*, 289(5481), 944–947.

Gorman, M. R. (1994). Male homosexual desire: Neurological investigations and scientific bias. *Perspectives in Biology and Medicine*, 38(1), 61–81.

Gould, S. J. (1993). A special fondness for beetles. *Natural History*, 1(102), 4.

Gowlett, J. A. J. (2016). The discovery of fire by humans: A long and convoluted process. *Philosophical Transactions of the Royal Society B: Biological Sciences*, 371(1696), 20150164.

Grant, J. M., Mottet, L. A., Tanis, J., Harrison, J., Herman, J. L., and Keisling, M. (2011). *Injustice at Every Turn: A Report of the National Transgender Discrimination Survey*. Washington, D.C.: National Center for Transgender Equality and National Gay and Lesbian Task Force.

Gray, L., Miller, L. W., Philipp, B. L., and Blass, E. M. (2002). Breastfeeding is analgesic in healthy newborns. *Pediatrics*, 109(4), 590–593. doi:10.1542/peds.109.4.590.

Graybeal, A., Rosowski, J. J., Ketten, D. R., and Crompton, A. W. (1989). Inner-ear structure in Morganucodon, an early Jurassic mammal. *Zoological Journal of the Linnean Society*, 96(2), 107–117.

Gredler, M. L., Larkins, C. E., Leal, F., Lewis, A. K., Herrera, A. M., Perriton, C. L., et al. (2014). Evolution of External Genitalia: Insights from reptilian development. *Sexual Development*, 8(5), 311–326. doi:10.1159/000365771.

Green, H., McGinnity, A., Meltzer, H., Ford, T., and Goodman, R. (2005). *Mental Health of Children and Young People in Great Britain, 2004*. Basingstoke: Palgrave Macmillan.

Greenblatt, R. B. (1972). Inappropriate lactation in men and women. *Medical Aspects of Human Sexuality*, 6(6), 25–33.

Griest, K. (2021). With equal opportunity comes equal responsibility: Lowering fitness standards to accommodate women will hurt the Army—and women. Modern War Institute at West Point, Feb. 25, 2021. mwi.usma.edu.

Grieve, K. M., McLaughlin, M., Dunlop, C. E., Telfer, E. E., and Anderson, R. A. (2015). The controversial existence and functional potential of oogonial stem cells. *Maturitas*, 82(3), 278–281.

Griffiths, C., McGartland, A., and Miller, M. (2007). A comparison of the monetized impact of IQ decrements from mercury emissions. *Environmental Health Perspectives*, 115(6), 841–847.

Griffiths, M. (1978). *Biology of the Monotremes*. New York: Academic Press.

Grimbizis, G. F., Camus, M., Tarlatzis, B. C., Bontis, J. N., and Devroey, P. (2001). Clinical implications of uterine malformations and hysteroscopic treatment results. *Human Reproduction Update*, 7(2), 161–174.

Grinberg, D., Levin-Asher, B., and Segal, O. (2022). The myth of women's advantage in using child-directed speech: Evidence of women versus men in single-sex parent families. *Journal of Speech, Language, and Hearing Research*, 65(11), 4205–4227. doi:10.1044/2022_JSLHR-21-00558.

Grossnickle, D. M., and Polly, P. D. (2013). Mammal disparity decreases during the Cretaceous angiosperm radiation. *Proceedings of the Royal Society B: Biological Sciences*, 280(1771), 20132110. doi:10.1098/rspb.2013.2110.

Grothe, B., and Pecka, M. (2014). The natural history of sound localization in mammals—a story of neuronal inhibition. *Front Neural Circuits*, 8, 116. doi:10.3389/fncir.2014.00116.

Gruber, M. I. (1989). Breastfeeding practices in biblical Israel and in old Babylonian Mesopotamia. *Journal of the Ancient Near Eastern Society (JANES)*, 19.

Gruss, L. T., and Schmitt, D. (2015). The evolution of the human pelvis: Changing adaptations to bipedalism, obstetrics and thermoregulation. *Philosophical Transactions of the Royal Society B: Biological Sciences*, 370(1663), 20140063.

Gulick, S. P. S., Bralower, T. J., Ormö, J., Hall, B., Grice, K., Schaefer, B.,et al. (2019). The first day of the Cenozoic. *Proceedings of the National Academy of Sciences*, 116(39), 19342–19351. doi:10.1073/pnas.1909479116.

Guo, Z., Johnson, C. M., and Jensen, M. D. (1997). Regional lipolytic responses to isoproterenol in women. *American Journal of Physiology-Endocrinology and Metabolism* 273(1), E108–E112. doi:10.1152/ajpendo.1997.273.1.E108.

Gupte, R. P., Brooks, W. M., Vukas, R. R., Pierce, J. D., and Harris, J. L. (2019). Sex differences in traumatic brain injury: What we know and what we should know. *Journal of Neurotrauma*, 36(22), 3063–3091.

Gurven, M., and Kaplan, H. (2007). Longevity among hunter-gatherers: A cross-cultural examination. *Population and Development Review*, 33, 321–365. doi:10.1111/j.1728-4457.2007.00171.x.

Haas, R., Watson, J., Buonasera, T., Southon, J., Chen, J. C., Noe, S., et al. (2020). Female hunters of the early Americas. *Science Advances*, 6(45), eabd0310. doi:10.1126/sciadv.abd0310.

Haeusler, M., and McHenry, H. M. (2004). Body proportions of *Homo habilis* reviewed. *Journal of Human Evolution*, 46(4), 433–465.

Haeusler, M., Grunstra, N. D. S., Martin, R. D., Krenn, V. A., Fornai, C., and Webb, N. M. (2021). The obstetrical dilemma hypothesis: There's life in the old dog yet. *Biological Reviews*, 96(5), 2031–2057. doi:10.1111/brv.12744.

Haggarty, P. (2004). Effect of placental function on fatty acid requirements during pregnancy. *European Journal of Clinical Nutrition*, 58, 1559–1570. doi:10.1038/sj.ejcn.1602016.

Haig, D. (2015). Maternal-fetal conflict, genomic imprinting and mammalian vulnerabilities to cancer. *Philosophical Transactions of the Royal Society B, Biological Sciences*, 370(1673), 20140178. doi:10.1098/rstb.2014.0178.

Haizlip, K. M., Harrison, B. C., and Leinwand, L. A. (2015). Sex-based differences in skeletal muscle kinetics and fiber-type composition. *Physiology*, 30(1), 30–39. doi:10.1152/physiol.00024.2014.

Halpern, D., Wai, J., and Saw, A. (2005). A psychobiosocial model: Why females are sometimes greater than and sometimes less than males in math achievement. Pp. 48–72 in *Gender Differences in Mathematics*, edited by A. M. Gallagher and J. C. Kaufman. Cambridge, U.K.: Cambridge University Press.

Halpern, D. F., and LaMay, M. L. (2000). *Educational Psychology Review*, 12(2), 229–246. doi:10.1023/a:1009027516424.

Halpern, D. F., Benbow, C. P., Geary, D. C., Gur, R. C., Hyde, J. S., and Gernsbacher, M. A. (2007). The science of sex differences in science and mathematics. *Psychological Science in the Public Interest*, 8(1), 1–51. doi:10.1111/j.1529-1006.2007.00032.x.

Hammer, M. L. A., and Foley, R. A. (1996). Longevity and life history in hominid evolution. *Human Evolution*, 11(1), 61–66. doi:10.1007/BF02456989.

Hammoud, H. (2006). Illiteracy in the Arab world. Paper commissioned for the EFA Global Monitoring Report 2006, *Literacy for Life*, UNESCO.

Hammurabi. (2250 BCE). The code of Hammurabi, king of Babylon. Trans. In: Harper, R. F. (1973). Mesopotamian pediatrics. *Episteme*, 7, 283–288.

Handelsman, D. J., Hirschberg, A. L., and Bermon, S. (2018). Circulating testosterone as the hormonal basis of sex differences in athletic performance. *Endocrine Reviews*, 39(5), 803–829. doi:10.1210/er.2018-00020.

Hansen, T., Pracejus, L., and Gegenfurtner, K. R. (2009). Color perception in the intermediate periphery of the visual field. *Journal of Vision*, 9(4), 26. doi:10.1167/9.4.26.

Hargest, R. (2020). Five thousand years of minimal access surgery: 3000 BC to 1850: Early instruments for viewing body cavities. *Journal of the Royal Society of Medicine*, 113(12), 491–496.

Harmand, S., Lewis, J. E., Feibel, C. S., Lepre, C. J., Prat, S., Lenoble, A., et al. (2015). 3.3-million-year-old stone tools from Lomekwi 3, West Turkana, Kenya. *Nature*, 521(7552), 310–315. doi:10.1038/nature14464.

Harris, L. J. (2010). Side biases for holding and carrying infants: Reports from the past and possible lessons for today. *Laterality*, 15, 56–135. doi:10.1080/13576500802584371.

Harris, W. V. (1991). *Ancient Literacy*. Cambridge, Mass.: Harvard University Press.

Harrison, A., Cleland, J., and Frohlich, J. (2008). Young people's sexual partnerships in KwaZulu-Natal, South Africa: Patterns, contextual influences, and HIV risk. *Studies in Family Planning*, 39(4), 295–308.

Harrison, D., Reszel, J., Bueno, M., Sampson, M., Shah, V. S., Taddio, A., et al. (2016). Breastfeeding for procedural pain in infants beyond the neonatal period. *Cochrane Database of Systematic Reviews* (10). doi:10.1002/14651858.CD011248.pub2.

Harrison, R. (2004). Physiological roles of xanthine oxidoreductase. *Drug Metabolism Reviews*, 36(2), 363–375. doi:10.1081/DMR-120037569.

Harrison, T. (2010). Apes among the tangled branches of human origins. *Science*, 327(5965), 532–534.

Hart, D., and Sussman, R. W. (2005). *Man the Hunted: Primates, Predators, and Human Evolution*. Boulder: Westview Press.

Hartshorne, J. K., Tenenbaum, J. B., and Pinker, S. (2018). A critical period for second language acquisition: Evidence from 2/3 million English speakers. *Cognition*, 177, 263–277. doi:10.1016/j.cognition.2018.04.007.

Hathi, P., Coffey, D., Thorat, A., and Khalid, N. (2021). When women eat last: Discrimination at home and women's mental health. *PLOS ONE*, 16(3), e0247065. doi:10.1371/journal.pone.0247065.

Hausfater, G., and Hrdy, S. B. (2017). *Infanticide: Comparative and Evolutionary Perspectives*. London: Routledge.

Hawkes, K. (2003). Grandmothers and the evolution of human longevity. *American Journal of Human Biology*, 15(3), 380–400. doi:10.1002/ajhb.10156.

Hawkes, K., and Coxworth, J. E. (2013). Grandmothers and the evolution of human longevity: A review of findings and future directions. *Evolutionary Anthropology*, 22(6), 294–302. doi:10.1002/evan.21382.

Hawkes, K., and Smith, K. R. (2010). Do women stop early? Similarities in fertility decline in humans and chimpanzees. *Annals of the New York Academy of Sciences*, 1204, 43–53. doi:10.1111/j.1749-6632.2010.05527.x.

Hawkes, K., O'Connell, J. F., and Blurton Jones, N. G. (1997). Hadza women's time allocation, offspring provisioning, and the evolution of long postmenopausal life spans. *Current Anthropology*, 38(4), 551–577. doi:10.1086/204646.

Hayden, E. C. (2010). Sex bias blights drug studies. *Nature* 464(7287), 332–333. doi:10.1038/464332b.

Heesy, C. P. (2009). Seeing in stereo: The ecology and evolution of primate binocular vision and stereopsis. *Evolutionary Anthropology*, 18, 21–35. doi:10.1002/evan.20195.

Heffner, R. S. (2004). Primate hearing from a mammalian perspective. *The Anatomical Record*, 281A, 1111–1122. doi:10.1002/ar.a.20117.

Heinrich, J. (2000). Women's health: NIH has increased its efforts to include women in research. *Report to Congressional Requesters.* Washington, D.C.: U.S. General Accounting Office.

Heinrichs, L. (2002). Linking olfaction with nausea and vomiting of pregnancy, recurrent abortion, hyperemesis gravidarum, and migraine headache. *American Journal of Obstetrics and Gynecology*, 186, S215–S219. doi:10.1067/mob.2002.123053.

Heisz, J. J., Pottruff, M. M., and Shore, D. I. (2013). Females scan more than males: A potential mechanism for sex differences in recognition memory. *Psychological Science*, 24(7), 1157–1163. doi:10.1177/0956797612468281.

Henn, B. M., Cavalli-Sforza, L. L., and Feldman, M. W. (2012). The great human expansion. *Proceedings of the National Academy of Sciences*, 109(44), 17758–17764. doi:10.1073/pnas.1212380109.

Henshaw, S. K., Singh, S., and Haas, T. (1999). *International Family Planning Perspectives*, 25, suppl. (Jan.), S30–S38.

Hernandez, T. L., Kittelson, J. M., Law, C. K., Ketch, L. L., Stob, N. R., Lindstrom, R. C., et al. (2011). Fat redistribution following suction lipectomy: Defense of body fat and patterns of restoration. *Obesity*, 19(7), 1388–1395. doi:10.1038/oby.2011.64.

Herndon, J. G., Paredes, J., Wilson, M. E., Bloomsmith, M. A., Chennareddi, L., and Walker, M. L. (2012). Menopause occurs late in life in the captive chimpanzee (*Pan troglodytes*). *Age*, 34(5), 1145–1156. doi:10.1007/s11357-011-9351-0.

Herrera, A. M., Shuster, S. G., Perriton, C. L., and Cohn, M. J. (2013). Developmental basis of phallus reduction during bird evolution. *Current Biology*, 23(12), 1065–1074. doi:10.1016/j.cub.2013.04.062.

Hertz-Picciotto, I., and Samuels, S. J. (1988). Incidence of early loss of pregnancy. *The New England Journal of Medicine*, 319(22), 1483–1484. doi:10.1056/NEJM198812013192214.

Herzog, D., Wegener, G., Lieb, K., Müller, M., and Treccani, G. (2019). Decoding the mechanism of action of rapid-acting antidepressant treatment strategies: Does gender matter? *International Journal of Molecular Sciences*, 20(4), 949. MDPI AG. doi:10.3390/ijms20040949.

Hesketh, T., Lu, L., and Xing, Z. W. (2011). The consequences of son preference and sex-selective abortion in China and other Asian countries. *Canadian Medical Association Journal*, 183(12), 1374–1377. doi:10.1503/cmaj.101368.

Hessami, Z., and da Fonseca, M. L. (2020). Female political representation and substantive effects on policies: A literature review. *European Journal of Political Economy*, 63, 101896. doi:10.1016/j.ejpoleco.2020.101896.

Hesselmar, B., Sjöberg, F., Saalman, R., Aberg, N., Adlerberth, I., and Wold, A. E. (2013). Pacifier cleaning practices and risk of allergy development. *Pediatrics*, 131(6), e1829–1837. doi:10.1542/peds.2012-3345.

Hester, M. (2013). Who does what to whom? Gender and domestic violence perpetrators in English police records. *European Journal of Criminology*, 10, 623–637.

Hewlett, B. S. (1991). *Intimate Fathers: The Nature and Context of Aka Pygmy Paternal Infant Care*. Ann Arbor: University of Michigan Press.

Hill, A. (1983). Hippopotamus butchery by *Homo erectus* at Olduvai. *Journal of Archaeological Science*, 10(2), 135–137. doi:10.1016/0305-4403(83)90047-X.

Hillstrom, C. (2022). The hidden epidemic of brain injuries from domestic violence. *The New York Times*, March 1, 2022. www.nytimes.com.

Hilton, C. B., Moser, C. J., Bertolo, M., Lee-Rubin, H., Amir, D., Bainbridge, C. M., et al. (2022). Acoustic regularities in infant-directed speech and song across cultures. *Nature Human Behaviour*, 6(11), 1545–1556. doi:10.1038/s41562-022-01410-x.

Hinde, K., and Milligan, L. A. (2011). Primate milk: Proximate mechanisms and ultimate perspectives. *Evolutionary Anthropology: Issues, News, and Reviews*, 20(1), 9–23. doi:10.1002/evan.20289.

Hinde, K., Skibiel, A. L., Foster, A. B., Del Rosso, L., Mendoza, S. P., and Capitanio, J. P. (2014). Cortisol in mother's milk across lactation reflects maternal life history and predicts infant temperament. *Behavioral Ecology*, 26(1), 269–281.

Hirai, A. H., Ko, J. Y., Owens, P. L., Stocks, C., and Patrick, S. W. (2021). Neonatal abstinence syndrome and maternal opioid-related diagnoses in the US, 2010–2017. *JAMA*, 325(2), 146–155. doi:10.1001/jama.2020.24991.

Hiramatsu, C., Melin, A. D., Aureli, F., Schaffner, C. M., Vorobyev, M., and Kawamura, S. (2009). Interplay of olfaction and vision in fruit foraging of spider monkeys. *Animal Behaviour*, 77(6), 1421–1426. doi:10.1016/j.anbehav.2009.02.012.

Hirata, S., Fuwa, K., Sugama, K., Kusunoki, K., and Takeshita, H. (2011).

Mechanism of birth in chimpanzees: Humans are not unique among primates. *Biology Letters*, 7(5), 686–688. doi:10.1098/rsbl.2011.0214.

Hirnstein, M., and Hausmann, M. (2021). Sex/gender differences in the brain are not trivial—A commentary on Eliot et al. *Neuroscience & Biobehavioral Reviews*, 130, 408–409. doi:10.1016/j.neubiorev.2021.09.012.

Hirnstein, M., Stuebs, J., Moè, A., and Hausmann, M. (2023). Sex/gender differences in verbal fluency and verbal-episodic memory: A meta-analysis. *Perspectives on Psychological Science*, 18(1), 67–90. doi:10.1177/17456916221082116.

Hoekzema, E., Barba-Müller, E., Pozzobon, C., Picado, M., Lucco, F., García-García, D., et al. (2017). Pregnancy leads to long-lasting changes in human brain structure. *Nature Neuroscience*, 20(2), 287–296. doi:10.1038/nn.4458.

Hoekzema, E., van Steenbergen, H., Straathof, M., Beekmans, A., Freund, I. M., Pouwels, P. J. W., and Crone, E. A. (2022). Mapping the effects of pregnancy on resting state brain activity, white matter microstructure, neural metabolite concentrations and grey matter architecture. *Nature Communications*, 13(1), 6931. doi:10.1038/s41467-022-33884-8.

Hoffmann, D. E. and Tarzian, A. J. (2001). The girl who cried pain: A bias against women in the treatment of pain. *The Journal of Law, Medicine & Ethics* 29(1), 13–27. doi:10.1111/j.1748-720x.2001.tb00037.x.

Holmes, M. M., Resnick, H. S., Kilpatrick, D. G., and Best, C. L. (1996). Rape-related pregnancy: Estimates and descriptive characteristics from a national sample of women. *American Journal of Obstetrics and Gynecology*, 175(2), 320–325. doi:10.1016/s0002-9378(96)70141-2.

Holton, N. E., Yokley, T. R., Froehle, A. W., and Southard, T. E. (2014). Ontogenetic scaling of the human nose in a longitudinal sample: Implications for genus *Homo* facial evolution. *American Journal of Physical Anthropology*, 153, 52–60. doi:10.1002/ajpa.22402.

Hopson, J. A. (1973). Endothermy, small size, and the origin of mammalian reproduction. *The American Naturalist*, 107(955), 446–452.

Horsup, A. (2005). Recovery plan for the northern hairy-nosed wombat (*Lasiorhinus krefftii*) 2004–2008. Department of Climate Change, Energy, the Environment, and Water, Australian Government. www.dcceew.gov.au.

Hosken, D. J., Archer, C. R., House, C. M., and Wedell, N. (2019). Penis evolution across species: Divergence and diversity. *Nature Reviews Urology*, 16(2), 98–106. doi:10.1038/s41585-018-0112-z.

Howell, N. (2017). *Demography of the Dobe !Kung*. New York: Routledge. (Original work published in 1979.)

Howie, P. W., and McNeilly, A. S. (1982). Effect of breast-feeding patterns on human birth intervals. *Journal of Reproduction and Fertility*, 65(2), 545–557. doi:10.1530/jrf.0.0650545.

Hoyert, D. L. (2022). Maternal mortality rates in the United States, 2020. *NCHS Health E-Stats*. doi:10.15620/cdc:113967.

Hoyle, N. P., Seinkmane, E., Putker, M., Feeney, K. A., Krogager, T. P.,

Chesham, J. E., et al. (2017). Circadian actin dynamics drive rhythmic fibroblast mobilization during wound healing. *Science Translational Medicine*, 9(415), eaal2774. doi:10.1126/scitranslmed.aal2774.

Hrdy, S. (2009). *Mothers and Others: The Evolutionary Origins of Mutual Understanding*. Cambridge, Mass.: Harvard University Press.

Hrdy, S. B. (1979). Infanticide among animals: A review, classification, and examination of the implications for the reproductive strategies of females. *Ethology and Sociobiology*, 1(1), 13–40.

Huffman, M. A. (1997). Current evidence for self-medication in primates: A multidisciplinary perspective. *American Journal of Physical Anthropology*, 104(S25), 171–200.

Huffman, M. A., Gotoh, S., Turner, L. A., Hamai, M., and Yoshida, K. (1997). Seasonal trends in intestinal nematode infection and medicinal plant use among chimpanzees in the Mahale Mountains, Tanzania. *Primates*, 38(2), 111–125. doi:10.1007/BF02382002.

Humphrey, L. L., Fu, R., Buckley, D. I., Freeman, M., and Helfand, M. (2008). Periodontal disease and coronary heart disease incidence: A systematic review and meta-analysis. *Journal of General Internal Medicine*, 23(12), 2079–2086. doi:10.1007/s11606-008-0787-6.

Hunley, K. L., Cabana, G. S., and Long, J. C. (2016). The apportionment of human diversity revisited. *American Journal of Physical Anthropology*, 160(4), 561–569. doi:10.1002/ajpa.22899.

Hunt, D. M., Dulai, K. S., Cowing, J. A., Julliot, C., Mollon, J. D., Bowmaker, J. K., et al. (1998). Molecular evolution of trichromacy in primates. *Vision Research*, 38(21), 3299–3306. doi:10.1016/S0042-6989(97)00443-4.

Hunt, K. D. (2015). Bipedalism. Pp. 103–112 in *Basics in Human Evolution*, edited by M. P. Muehlenbein. Boston: Academic Press.

Hunter, E. J., Tanner, K., and Smith, M. E. (2011). Gender differences affecting vocal health of women in vocally demanding careers. *Logopedics, Phoniatrics, Vocology*, 36(3), 128–136. doi:10.3109/14015439.2011.587447.

Huppertz, B., Meiri, H., Gizurarson, S., Osol, G., and Sammar, M. (2013). Placental protein 13 (PP13): A new biological target shifting individualized risk assessment to personalized drug design combating pre-eclampsia. *Human Reproduction Update*, 19(4), 391–405. doi:10.1093/humupd/dmt003.

Hyde, J. S. (2005). The gender similarities hypothesis. *American Psychologist*, 60, 581–592.

Hynes, M., and Lopes Cardozo, B. (2000). Observations from the CDC: Sexual violence against refugee women. *Journal of Women's Health & Gender-Based Medicine*, 9(8), 819–823. doi:10.1089/152460900750020847.

Iantaffi, A., and Bockting, W. O. (2011). Views from both sides of the bridge? Gender, sexual legitimacy and transgender people's experiences of relationships. *Culture, Health & Sexuality*, 13, 355–370. doi:10.1080/13691058.2010.537770.

Ilany, A., Holekamp, K. E., and Akçay, E. (2021). Rank-dependent social inheritance determines social network structure in spotted hyenas. *Science*, 373(6552), 348. doi:10.1126/science.abc1966.

Ingoldsby, B. B. (2001). The Hutterite family in transition. *Journal of Comparative Family Studies*, 32(3), 377–392. doi:10.3138/jcfs.32.3.377.

Ingram, M. (1990). *Church Courts, Sex and Marriage in England, 1570–1640*. Cambridge, U.K.: Cambridge University Press.

Insel, T. R. (2010). The challenge of translation in social neuroscience: A review of oxytocin, vasopressin, and affiliative behavior. *Neuron*, 65(6), 768–779.

Institute of Medicine (U.S.) Committee on Women's Health Research (2010). *Women's Health Research: Progress, Pitfalls, and Promise*. Washington, D.C.: National Academies Press.

International Monetary Fund (IMF) (2018). Pursuing women's economic empowerment. Report prepared for the meeting of G7 ministers and central bank governors, June 1–2, 2018, Whistler, Canada. www.imf.org.

Isaacson, W. (2004). *Benjamin Franklin: An American life*. New York: Simon & Schuster.

Ivell, R., Agoulnik, A. I., and Anand-Ivell, R. (2016). Relaxin-like peptides in male reproduction—a human perspective. *British Journal of Pharmacology*, 174, 990–1001. doi:10.1111/bph.13689.

Jablonski, N. G., and Chaplin, G. (1993). Origin of habitual terrestrial bipedalism in the ancestor of the Hominidae. *Journal of Human Evolution*, 24(4), 259–280. doi:10.1006/jhev.1993.1021.

Jackson, S. (2006). *The Haunting of Hill House*. New York: Penguin. (Original work published in 1959.)

Jacobs, G. H. (1993). The distribution and nature of colour vision among the mammals. *Biological Reviews of the Cambridge Philosophical Society*, 68(3), 413–471. doi:10.1111/j.1469-185x.1993.tb00738.x.

Jacobson-Dickman, E., and Lee, M. M. (2009). The influence of endocrine disruptors on pubertal timing. *Current Opinion in Endocrinology, Diabetes and Obesity*, 16(1), 25–30.

James, D., and Drakich, J. (1993). Understanding gender differences in amount of talk: Critical review of research. In *Gender and Conversational Interaction*, edited by D. Tannen. Oxford: Oxford University Press.

James, F. R., Wootton, S., Jackson, A., Wiseman, M., Copson, E. R., and Cutress, R. I. (2015). Obesity in breast cancer—What is the risk factor? *European Journal of Cancer*, 51(6), 705–720. doi:10.1016/j.ejca.2015.01.057.

Jannini, E. A., Fisher, W. A., Bitzer, J., and McMahon, C. G. (2009). Controversies in sexual medicine: Is sex just fun? How sexual activity improves health. *The Journal of Sexual Medicine*, 6(10), 2640–2648.

Jansen, S., Baulain, U., Habig, C., Weigend, A., Halle, I., Scholz, A. M., et al. (2020). Relationship between bone stability and egg production in genetically divergent chicken layer lines. *Animals*, 10(5), 850. doi:10.3390/ani10050850.

Jasieńska, G., Ziomkiewicz, A., Ellison, P. T., Lipson, S. F., and Thune, I.

(2004). Large breasts and narrow waists indicate high reproductive potential in women. *Proceedings of the Royal Society B: Biological Sciences*, 271(1545), 1213–1217. doi:10.1098/rspb.2004.2712.

Jeffery, P., Jeffery, R., and Lyon, A. (1989). *Labour Pains and Labour Power: Women and Childbearing in India*. London: Zed Books.

Jeffrey, R. (2004). Legacies of matriliny: The place of women and the "Kerala Model." *Pacific Affairs*, 77(4), 647–664.

Jha, S., and Parker, V. (2016). Risk factors for recurrent obstetric anal sphincter injury (rOASI): A systematic review and meta-analysis. *International Urogynecology Journal*, 27(6), 849–857. doi:10.1007/s00192-015-2893-4.

Ji, Q., Luo, Z.-X., Yuan, C.-X., and Tabrum, A. R. (2006). A swimming mammaliaform from the Middle Jurassic and Ecomorphological diversification of early mammals. *Science*, 311(5764), 1123–1127. doi:10.1126/science.1123026.

Johannsen, T. H., Ripa, C. P. L., Mortensen, E. L., and Main, K. M. (2006). Quality of life in 70 women with disorders of sex development. *European Journal of Endocrinology*, 155(6), 877–885.

Johns, M., Schmader, T., and Martens, A. (2005). Knowing is half the battle: Teaching stereotype threat as a means of improving women's math performance. *Psychological Science*, 16(3), 175–179. doi:10.1111/j.0956-7976.2005.00799.x.

Johnson, W., Carothers, A., and Deary, I. J. (2008). Sex differences in variability in general intelligence: A new look at the old question. *Perspectives on Psychological Science*, 3(6), 518–531. doi:10.1111/j.1745-6924.2008.00096.x.

Joint United Nations Programme on HIV/AIDS (UNAIDS) (2004). *Report on the Global HIV/AIDS Epidemic: 4th Global Report*. Geneva: UNAIDS.

Jones, H. E., Kaltenbach, K., Heil, S. H., Stine, S. M., Coyle, M. G., Arria, A. M., et al. (2010). Neonatal abstinence syndrome after methadone or buprenorphine exposure. *The New England Journal of Medicine*, 363(24), 2320–2331.

Jones, J. H. (2011). Primates and the evolution of long, slow life histories. *Current Biology*, 21(18), R708–717. doi:10.1016/j.cub.2011.08.025.

Jones, K. P., Walker, L. C., Anderson, D., Lacreuse, A., Robson, S. L., and Hawkes, K. (2007). Depletion of ovarian follicles with age in chimpanzees: Similarities to humans. *Biology of Reproduction*, 77(2), 247–251. doi:10.1095/biolreprod.106.059634.

Jones, T. M., Fanson, K. V., Lanfear, R., Symonds, M. R., and Higgie, M. (2014). Gender differences in conference presentations: A consequence of self-selection? *PeerJ*, 2, e627. doi:10.7717/peerj.627.

Joordens, J. C., Kuipers, R. S., Wanink, J. H., and Muskiet, F. A. (2014). A fish is not a fish: Patterns in fatty acid composition of aquatic food may have had implications for hominin evolution. *Journal of Human Evolution*, 77, 107–116. doi:10.1016/j.jhevol.2014.04.004.

Jordan, G., Deeb, S. S., Bosten, J. M., and Mollon, J. D. (2010). The dimensionality of color vision in carriers of anomalous trichromacy. *Journal of Vision*, 10(8), 12. doi:10.1167/10.8.12.

Jørgensen, K. T., Pedersen, B. V., Nielsen, N. M., Jacobsen, S., and Frisch, M. (2012). Childbirths and risk of female predominant and other autoimmune diseases in a population-based Danish cohort. *Journal of Autoimmunity*, 38(2–3), J81–J87. doi:10.1016/j.jaut.2011.06.004.

Jud, N. A., D'Emic, M. D., Williams, S. A., Mathews, J. C., Tremaine, K. M., and Bhattacharya, J. (2018). A new fossil assemblage shows that large angiosperm trees grew in North America by the Turonian (Late Cretaceous). *Science Advances*, 4(9), eaar8568. doi:10.1126/sciadv.aar8568.

Kachel, A. F., Premo, L. S., and Hublin, J. J. (2011). Grandmothering and natural selection. *Proceedings of the Royal Society B: Biological Sciences*, 278(1704), 384–391. doi:10.1098/rspb.2010.1247.

Kamarck, K. N. (2016). Women in Combat: Issues for Congress. Congressional Research Service, Summary, Dec. 13, 2016. fas.org.

Kaplan, M. (2012). Primates were always tree-dwellers. *Nature*. doi:10.1038/nature.2012.11423.

Kappeler, P. M. (1997). Intrasexual selection and testis size in strepsirhine primates. *Behavioral Ecology*, 8(1), 10–19. doi:10.1093/beheco/8.1.10.

Karastergiou, K., Smith, S. R., Greenberg, A. S., and Fried, S. K. (2012). Sex differences in human adipose tissues—the biology of pear shape. *Biology of Sex Differences*, 3(1), 13. doi:10.1186/2042-6410-3-13.

Karlamangla, A. S., Burnett-Bowie, S. M., and Crandall, C. J. (2018). Bone health during the menopause transition and beyond. *Obstetrics and Gynecology Clinics of North America*, 45(4), 695–708. doi:10.1016/j.ogc.2018.07.012.

Karras, R. M. (2012). *Unmarriages: Women, Men, and Sexual Unions in the Middle Ages*. Philadelphia: University of Pennsylvania Press.

Kass, M. D., Czarnecki, L. A., Moberly, A. H., and McGann, J. P. (2017). Differences in peripheral sensory input to the olfactory bulb between male and female mice. *Scientific Reports*, 7(1), 45851. doi:10.1038/srep45851.

Kassebaum, N. J., Barber, R. M., Bhutta, Z. A., Dandona, L., Gething, P. W., Hay, S. I., et al. (2016). Global, regional, and national levels of maternal mortality, 1990–2015: A systematic analysis for the Global Burden of Disease Study 2015. *The Lancet*, 388(10053), 1775–1812. doi:10.1016/S0140-6736(16)31470-2.

Kasuya, T., and Marsh, H. (1984). Life history and reproductive biology of the short-finned pilot whale, *Globicephala macrorhynchus*, off the Pacific coast of Japan. *Reports International Whaling Commission*, 6, 259–310.

Katz-Wise, S. L., Reisner, S. L., Hughto, J. W., and Keo-Meier, C. L. (2016). Differences in sexual orientation diversity and sexual fluidity in attractions among gender minority adults in Massachusetts. *Journal of Sex Research*, 53(1), 74–84. doi:10.1080/00224499.2014.1003028.

Kawada, M., Nakatsukasa, M., Nishimura, T., Kaneko, A., and Morimoto, N.

(2020). Covariation of fetal skull and maternal pelvis during the perinatal period in rhesus macaques and evolution of childbirth in primates. *Proceedings of the National Academy of Sciences*, 117(35), 21251–21257. doi:10.1073/pnas.2002112117.

Keele, K. D., and Roberts, J. (1983). *Leonardo da Vinci: Anatomical Drawings from the Royal Library, Windsor Castle*. New York: Metropolitan Museum of Art.

Keller, A., Zhuang, H., Chi, Q., Vosshall, L. B., and Matsunami, H. (2007). Genetic variation in a human odorant receptor alters odour perception. *Nature*, 449, 468–472. doi:10.1038/nature06162.

Kendall, S., and Tannen, D. (1997). Gender and language in the workplace. In *Gender and Discourse*, edited by R. Wodak. London: SAGE Publications. doi:10.4135/9781446250204.

Kennaway, D. J., Boden, M. J., and Varcoe, T. J. (2012). Circadian rhythms and fertility. *Molecular and Cellular Endocrinology*, 349(1), 56–61.

Kenny, L. C., and Kell, D. B. (2018). Immunological tolerance, pregnancy, and preeclampsia: The roles of semen microbes and the father. *Frontiers in Medicine*, 4, 239. doi:10.3389/fmed.2017.00239.

Kent, J., Mitoulas, L., Cox, D., Owens, R., and Hartmann, P. (1999). Breast volume and milk production during extended lactation in women. *Experimental Physiology*, 84(2), 435–447. doi:10.1111/j.1469-445X.1999.01808.x.

Kenyon, K. M. (1957). *Digging Up Jericho*. London: Ernest Benn.

Kermack, D. M., and Kermack, K. A. (1984). The evolution of mammalian sight and hearing. Pp. 89–100 in *The Evolution of Mammalian Characters*. Boston: Springer. doi:10.1007/978-1-4684-7817-4_6.

Kermack, K. A., Mussett, F., and Rigney, H. W. (1973). The lower jaw of Morganucodon. *Zoological Journal of the Linnean Society*, 53(2), 87–175.

Kermack, K. A., Mussett, F., and Rigney, H. W. (1981). The skull of Morganucodon. *Zoological Journal of the Linnean Society*, 71(1), 1–158.

Kessler, R. C., Petukhova, M., Sampson, N. A., Zaslavsky, A. M., and Wittchen, H. U. (2012). Twelve-month and lifetime prevalence and lifetime morbid risk of anxiety and mood disorders in the United States. *International Journal of Methods in Psychiatric Research*, 21(3), 169–184.

Khesbak, H., Savchuk, O., Tsushima, S., and Fahmy, K. (2011). The role of water H-bond imbalances in B-DNA substate transitions and peptide recognition revealed by time-resolved FTIR spectroscopy. *Journal of the American Chemical Society*, 133(15), 5834–5842. doi:10.1021/ja108863v.

Kielan-Jaworowska, Z., Cifelli, R., and Luo, Z.-X. (2005a). Distribution: Mesozoic mammals in time and space. Pp. 19–108 in *Mammals from the Age of Dinosaurs: Origins, Evolution, and Structure*. New York: Columbia University Press.

Kielan-Jaworowska, Z., Cifelli, R., and Luo, Z.-X. (2005b). The earliest-known stem mammals. Pp. 161–186 in *Mammals from the Age of Dinosaurs: Origins, Evolution, and Structure*. New York: Columbia University Press.

Killgore, W. D., and Yurgelun-Todd, D. A. (2007). The right-hemisphere and valence hypotheses: Could they both be right (and sometimes left)? *Social Cognitive and Affective Neuroscience*, 2(3), 240–250. doi:10.1093/scan/nsm020.

Kim, M. Y., and Cho, S. H. (2012). Affecting factors of contraception use among Korean male adolescents: Focused on alcohol, illicit drug, internet use, and sex education. *The Korean Journal of Stress Research*, 20(4), 267–277.

King, B. J. (2019). Deception in the wild. *Scientific American*, 321(3), 50–54. doi:10.1038/scientificamerican0919-50.

King, S. L., Guarino, E., Keaton, L., Erb, L., and Jaakkola, K. (2016). Maternal signature whistle use aids mother-calf reunions in a bottlenose dolphin, *Tursiops truncatus*. *Behavioural Processes*, 126, 64–70. doi:10.1016/j.beproc.2016.03.005.

Kinsey, A. C., Pomeroy, W. R., and Martin, C. E. (1948). *Sexual Behavior in the Human Male*. Philadelphia: W. B. Saunders Co.

Kirkcaldy, R. D., Weston, E., Segurado, A. C., and Hughes, G. (2019). Epidemiology of gonorrhoea: A global perspective. *Sexual Health*, 16(5), 401–411. doi:10.1071/SH19061.

Klein, R. G. (2009). *The Human Career: Human Biological and Cultural Origins*. Chicago: University of Chicago Press.

Klein, S. L., and Flanagan, K. L. (2016). Sex differences in immune responses. *Nature Reviews Immunology*, 16(10), 626–638. doi:10.1038/nri.2016.90.

Kliman, H. J., Sammar, M., Grimpel, Y. I., Lynch, S. K., Milano, K. M., Pick, E., et al. (2012). Placental protein 13 and decidual zones of necrosis: An immunologic diversion that may be linked to preeclampsia. *Reproductive Sciences*, 19(1), 16–30. doi:10.1177/1933719111424445.

Knaplund, K. S. (2008). The evolution of women's rights in inheritance. *Hastings Women's Law Journal*, 19, 3.

Knight, C. (1995). *Blood Relations: Menstruation and the Origins of Culture*. New Haven: Yale University Press.

Knox, K., and Baker, J. C. (2008). Genomic evolution of the placenta using co-option and duplication and divergence. *Genome Research*, 18(5), 695–705. doi:10.1101/gr.071407.107.

Ko, J. Y., Patrick, S. W., Tong, V. T., Patel, R., Lind, J. N., and Barfield, W. D. (2016). Incidence of neonatal abstinence syndrome—28 states, 1999–2013. *Morbidity and Mortality Weekly Report (MMWR)*, 65, 799–802. doi:10.15585/mmwr.mm6531a2.

Kolata, G. (2011). With liposuction, the belly finds what the thighs lose. *The New York Times*, April 30, 2011. www.nytimes.com.

Konner, M., and Worthman, C. (1980). Nursing frequency, gonadal function, and birth spacing among !Kung hunter-gatherers. *Science*, 207(4432), 788–791. doi:10.1126/science.73522.

Kortsmit, K., Mandel, M. G., Reeves, J. A., Clark, E., Pagano, P., Nguyen, A., et al. (2021). Abortion surveillance—United States, 2019. *MMWR Surveillance Summaries* 70(9), 1–29. doi:10.15585/mmwr.ss7009a1.

Koscik, T., O'Leary, D., Moser, D. J., Andreasen, N. C., and Nopoulos, P. (2009). Sex differences in parietal lobe morphology: Relationship to mental rotation performance. *Brain and Cognition*, 69(3), 451–459. doi:10.1016/j.bandc.2008.09.004.

Koss, K. J., and Gunnar, M. R. (2018). Annual research review: Early adversity, the hypothalamic-pituitary-adrenocortical axis, and child psychopathology. *Journal of Child Psychology and Psychiatry*, 59(4), 327–346. doi:10.1111/jcpp.12784.

Kovacs, C. S. (2001). Calcium and bone metabolism in pregnancy and lactation. *The Journal of Clinical Endocrinology & Metabolism*, 86(6), 2344–2348. doi:10.1210/jcem.86.6.7575.

Kraft, T. S., Venkataraman, V. V., Wallace, I. J., Crittenden, A. N., Holowka, N. B., Stieglitz, J., et al. (2021). The energetics of uniquely human subsistence strategies. *Science*, 374(6575), eabf0130.

Kreuer, S., Biedler, A., Larsen, R., Altmann, S., and Wilhelm, W. (2003). Narcotrend monitoring allows faster emergence and a reduction of drug consumption in propofol-remifentanil anesthesia. *Anesthesiology*, 99(1), 34–41. doi:10.1097/00000542-200307000-00009.

Krijgsman, W., Hilgen, F. J., Raffi, I., Sierro, F. J., and Wilson, D. S. (1999). Chronology, causes and progression of the Messinian salinity crisis. *Nature*, 400(6745), 652–655.

Kring, D. A., and Durda, D. D. (2002). Trajectories and distribution of material ejected from the Chicxulub Impact Crater: Implications for postimpact wildfires. *Journal of Geophysical Research: Planets*, 107(E8), 6–22.

Kroodsma, D., Hamilton, D., Sánchez, J. E., Byers, B. E., Fandiño-Mariño, H., Stemple, D. W., et al. (2013). Behavioral evidence for song learning in the suboscine bellbirds (*Procnias* spp.; Cotingidae). *The Wilson Journal of Ornithology*, 125(1): 1–14. doi:10.1676/12-033.1.

Kruepunga, N., Hikspoors, J., Mekonen, H. K., Mommen, G., Meemon, K., Weerachatyanukul, W., et al. (2018). The development of the cloaca in the human embryo. *Journal of Anatomy*, 233(6), 724–739. doi:10.1111/joa.12882.

Kruger, T. H. C., Leeners, B., Naegeli, E., Schmidlin, S., Schedlowski, M., Hartmann, U., et al. (2012). Prolactin secretory rhythm in women: Immediate and long-term alterations after sexual contact. *Human Reproduction*, 27(4), 1139–1143. doi:10.1093/humrep/des003.

Krysinska, K., Batterham, P. J., and Christensen, H. (2017). Differences in the effectiveness of psychosocial interventions for suicidal ideation and behaviour in women and men: A systematic review of randomised controlled trials. *Archives of Suicide Research*, 21(1), 12–32. doi:10.1080/13811118.2016.1162246.

Kuhl, P. K., Tsao, F. M., and Liu, H. M. (2003). Foreign-language experience in infancy: Effects of short-term exposure and social interaction on phonetic learning. *Proceedings of the National Academy of Sciences*, 100(15), 9096–9101.

Kuhn, T. (1970). *The Structure of Scientific Revolutions.* 2nd ed. Chicago: University of Chicago Press.

Kulczycki, A., and Windle, S. (2011). Honor killings in the Middle East and North Africa: A systematic review of the literature. *Violence Against Women*, 17(11), 1442–1464. doi:10.1177/1077801211434127.

Kumar, S., Filipski, A., Swarna, V., Walker, A., and Hedges, S. B. (2005). Placing confidence limits on the molecular age of the human-chimpanzee divergence. *Proceedings of the National Academy of Sciences*, 102(52), 18842–18847.

Kunz, C., Rodriguez-Palmero, M., Koletzko, B., and Jensen, R. (1999). Nutritional and biochemical properties of human milk, Part I: General aspects, proteins, and carbohydrates. *Clinical Perinatology*, 26(2), 307–333.

Kunz, C., Rudloff, S., Baier, W., Klein, N., and Strobel, S. (2000). Oligosaccharides in human milk: Structural, functional, and metabolic aspects. *Annual Review of Nutrition*, 20, 699–722.

Lahdenperä, M., Lummaa, V., Helle, S., Tremblay, M., and Russell, A. F. (2004). Fitness benefits of prolonged post-reproductive lifespan in women. *Nature*, 428, 178–181. doi:10.1038/nature02367.

Lahdenperä, M., Mar, K. U., and Lummaa, V. (2014). Reproductive cessation and post-reproductive lifespan in Asian elephants and pre-industrial humans. *Frontiers of Zoology*, 11, 54. doi:10.1186/s12983-014-0054-0.

Lahdenperä, M., Russell, A. F., Tremblay, M., and Lummaa, V. (2011). Selection on menopause in two premodern human populations: No evidence for the Mother Hypothesis. *Evolution*, 65(2), 476–489. doi:10.1111/j.1558-5646.2010.01142.x.

Lamothe, D. (2021). An army trailblazer set her sights on a new target. The reaction highlights a deep rift. *The Washington Post*, May 8, 2021. www.washingtonpost.com.

Lamvu, G., Soliman, A. M., Manthena, S. R., Gordon, K., Knight, J., and Taylor, H. S. (2019). Patterns of prescription opioid use in women with endometriosis: Evaluating prolonged use, daily dose, and concomitant use with benzodiazepines. *Obstetrics and Gynecology*, 133(6), 1120–1130. doi:10.1097/AOG.0000000000003267.

Langhammer, A., Johnsen, R., Gulsvik, A., Holmen, T. L., and Bjermer, L. (2003). Sex differences in lung vulnerability to tobacco smoking. *European Respiratory Journal*, 21(6), 1017–1023. doi:10.1183/09031936.03.00053202.

Langley, M. C., and Litster, M. (2018). Is it ritual? Or is it children? Distinguishing consequences of play from ritual actions in the prehistoric archaeological record. *Current Anthropology*, 59(5), 616–643.

Lara-Villoslada, F., Olivares, M., Sierra, S., Miguel Rodríguez, J., Boza, J., and Xaus, J. (2007). Beneficial effects of probiotic bacteria isolated from breast milk. *British Journal of Nutrition*, 98(S1), S96–S100. doi:10.1017/S0007114507832910.

Larison, J. R., Crock, J. G., Snow, C. M., and Blem, C. (2001). Timing of mineral sequestration in leg bones of white-tailed ptarmigan. *The Auk*, 118(4), 1057–1062. doi:10.1093/auk/118.4.1057.

Larsen, C. S. (2003). Equality for the sexes in human evolution? Early hominid sexual dimorphism and implications for mating systems and social behavior. *Proceedings of the National Academy of Sciences*, 100(16), 9103–9104.

Lassek, W. D., and Gaulin, S. J. (2007). Menarche is related to fat distribution. *American Journal of Physical Anthropology*, 133, 1147–1151. doi:10.1002/ajpa.20644.

Lassek, W. D., and Gaulin, S. J. (2008). Waist-hip ratio and cognitive ability: Is gluteofemoral fat a privileged store of neurodevelopmental resources? *Evolution and Human Behavior*, 29, 26–34. doi:10.1016/j.evolhumbehav.2007.07.005.

Latimer, B. (2005). The perils of being bipedal. *Annals of Biomedical Engineering*, 33(1), 3–6.

Laudicina, N. M., Rodriguez, F., and DeSilva, J. M. (2019). Reconstructing birth in *Australopithecus sediba*. *PLOS ONE*, 14(9), e0221871. doi:10.1371/journal.pone.0221871.

Le Ray, C., Scherier, S., Anselem, O., Marszalek, A., Tsatsaris, V., Cabrol, D., and Goffinet, F. (2012). Association between oocyte donation and maternal and perinatal outcomes in women aged 43 years or older. *Human Reproduction*, 27(3), 896–901. doi:10.1093/humrep/der469.

le Roux, A., Snyder-Mackler, N., Roberts, E. K., Beehner, J. C., and Bergman, T. J. (2013). Evidence for tactical concealment in a wild primate. *Nature Communications*, 4(1), 1462. doi:10.1038/ncomms2468.

Leakey, L. S. B., Tobias, P. V., and Napier, J. R. (1964). A new species of the genus *Homo* from Olduvai Gorge. *Nature*, 202(4927), 7–9.

Leakey, M. G., Spoor, F., Dean, M. C., Feibel, C. S., Anton, S. C., Kiarie, C., and Leakey, L. N. (2012). New fossils from Koobi Fora in northern Kenya confirm taxonomic diversity in early *Homo*. *Nature*, 488(7410), 201–204.

Lean, R. E., Paul, R. A., Smyser, C. D., and Rogers, C. E. (2018). Maternal intelligence quotient (IQ) predicts IQ and language in very preterm children at age 5 years. *Journal of Child Psychology and Psychiatry and Allied Disciplines*, 59(2), 150–159. doi:10.1111/jcpp.12810.

LeBlanc, S., and Barnes, E. (1974). On the adaptive significance of the female breast. *The American Naturalist* 108(962), 577–578.

Lee, J. J., and McCabe, J. M. (2021). Who speaks and who listens: Revisiting the chilly climate in college classrooms. *Gender & Society*, 35(1), 32–60. doi:10.1177/0891243220977141.

Lee, L. J., Komarasamy, T. V., Adnan, N. A. A., James, W., and Balasubramaniam, V. R. M. T. (2021). Hide and seek: The interplay between Zika virus and the host immune response. *Frontiers in Immunology*, 12, 750365. doi:10.3389/fimmu.2021.750365.

Lee, T., Henry, J. D., Trollor, J. N., and Sachdev, P. S. (2010). Genetic influences on cognitive functions in the elderly: A selective review of twin studies. *Brain Research Reviews*, 64(1), 1–13.

Leemis, R. W., Friar, N., Khatiwada, S., Chen, M. S., Kresnow, M., Smith, S. G., et al. (2022). *The National Intimate Partner and Sexual Violence Survey: 2016/2017 Report on Intimate Partner Violence*. Atlanta: National Center for Injury Prevention and Control, Centers for Disease Control and Prevention.

Leeners, B., Kruger, T. H. C., Brody, S., Schmidlin, S., Naegeli, E., and Egli, M. (2013). The quality of sexual experience in women correlates with post-orgasmic prolactin surges: Results from an experimental prototype study. *The Journal of Sexual Medicine*, 10(5), 1313–1319. doi:10.1111/jsm.12097.

LeGates, T. A., Kvarta, M. D., and Thompson, S. M. (2019). Sex differences in antidepressant efficacy. *Neuropsychopharmacology*, 44(1), 140–154. doi:10.1038/s41386-018-0156-z.

Leigh, S. R., and Shea, B. T. (1995). Ontogeny and the evolution of adult body size dimorphism in apes. *American Journal of Primatology*, 36(1), 37–60.

Leland, A. (2022). Deafblind communities may be creating a new language of touch. *The New Yorker*, May 12, 2022. www.newyorker.com.

Lemay, D. G., Lynn, D. J., Martin, W. F., Neville, M. C., Casey, T. M., Rincon, G., et al. (2009). The bovine lactation genome: Insights into the evolution of mammalian milk. *Genome Biology*, 10(4), R43. doi:10.1186/gb-2009-10-4-r43.

Lemmon, G. T. (2015). Meet the first class of women to graduate from Army Ranger School. *Foreign Policy*, Aug. 17, 2015. foreignpolicy.com.

Lemmon, G. T. (2021). *The Daughters of Kobani: A Story of Rebellion, Courage, and Justice*. New York: Penguin Press.

Leonard, W. R., Robertson, M. L., Snodgrass, J. J., and Kuzawa, C. W. (2003). Metabolic correlates of hominid brain evolution. *Comparative Biochemistry and Physiology Part A: Molecular & Integrative Physiology*, 136(1), 5–15. doi:10.1016/S1095-6433(03)00132-6.

Lepre, C. J., Roche, H., Kent, D. V., Harmand, S., Quinn, R. L., Brugal, J.-P., et al. (2011). An earlier origin for the Acheulian. *Nature*, 477(7362), 82–85. doi:10.1038/nature10372.

Leslie, P. W., Campbell, K. L., and Little, M. A. (1993). Pregnancy loss in nomadic and settled women in Turkana, Kenya: A prospective study. *Human Biology*, 65(2), 237–254.

Leutenegger, W. (1987). Neonatal brain size and neurocranial dimensions in Pliocene hominids: Implications for obstetrics. *Journal of Human Evolution*, 16(3), 291–296.

Levenson, M. (2019). Yes, killer whales benefit from grandmotherly love too. *The New York Times*, Dec. 10, 2019. www.nytimes.com.

Levertov, D. (1996). *Sands of the Well*. New York: New Directions.

Levine, M. E., Lu, A. T., Chen, B. H., Hernandez, D. G., Singleton, A. B., Ferrucci, L., et al. (2016). Menopause accelerates biological aging. *Proceedings of the National Academy of Sciences*, 113(33), 9327–9332. doi:10.1073/pnas.1604558113.

Lewkowski, M. D., Ditto, B., Roussos, M., and Young, S. N. (2003). Sweet taste and blood pressure-related analgesia. *Pain*, 106, 181–186.

Li, C., Zhu, N., Zeng, L., Dang, S., Zhou, J., and Yan, H. (2016). Effect of prenatal and postnatal malnutrition on intellectual functioning in early school-aged children in rural western China. *Medicine*, 95(31), e4161. doi:10.1097/MD.0000000000004161.

Libby, P., Ridker, P. M., and Maseri, A. (2002). Inflammation and atherosclerosis. *Circulation*, 105(9), 1135–1143. doi:10.1161/hc0902.104353.

Lieberman, D. E. (2012). Human evolution: Those feet in ancient times. *Nature*, 483(7391), 550–551.

Lieberman, D. E. (2015). Human locomotion and heat loss: An evolutionary perspective. *Comprehensive Physiology*, 5, 99–117.

Lieberman, P. (1993). On the Kebara KMH 2 hyoid and Neanderthal speech. *Current Anthropology*, 34(2), 172–175.

Lieberman, P. (2007). The evolution of human speech: Its anatomical and neural bases. *Current Anthropology*, 48(1), 39–66.

Liebovitz, A., and Sontag, S. (2000). *Women*. New York: Random House.

Lindenfors, P., Gittleman, J. L., and Jones, K. E. (2007). Sexual size dimorphism in mammals. Pp. 16–26 in *Sex, Size and Gender Roles: Evolutionary Studies of Sexual Size Dimorphism*, edited by D. J. Fairbairn, W. U. Blanckenhorn, and T. Székely. Oxford: Oxford University Press.

Lindsay, J. O., Whelan, K., Stagg, A. J., Gobin, P., Al-Hassi, H. O., Rayment, N., et al. (2006). Clinical, microbiological, and immunological effects of fructo-oligosaccharide in patients with Crohn's disease. *Gut*, 55(3), 348–355. doi:10.1136/gut.2005.074971.

Lindsay, S., Ansell, J., Selman, C., Cox, V., Hamilton, K., and Walraven, G. (2000). Effect of pregnancy on exposure to malaria mosquitoes. *Lancet*, 355(9219), 1972. doi:10.1016/S0140-6736(00)02334-5.

Lingle, S., and Riede, T. (2014). Deer mothers are sensitive to infant distress vocalizations of diverse mammalian species. *The American Naturalist*, 184(4), 510–522. doi:10.1086/677677.

Lipkind, D., Marcus, G. F., Bemis, D. K., Sasahara, K., Jacoby, N., Takahasi, M., et al. (2013). Stepwise acquisition of vocal combinatorial capacity in songbirds and human infants. *Nature*, 498(7452), 104–108. doi:10.1038/nature12173.

Liu, C., Weaver, D. R., Strogatz, S. H., and Reppert, S. M. (1997). Cellular construction of a circadian clock: Period determination in the suprachiasmatic nuclei. *Cell*, 91(6), 855–860. doi:10.1016/s0092-8674(00)80473-0.

Liu, G., Zhang, C., Wang, Y., Dai, G., Liu, S. Q., Wang, W., et al. (2021). New exon and accelerated evolution of placental gene Nrk occurred in the ancestral lineage of placental mammals. *Placenta*, 114, 14–21. doi:10.1016/j.placenta.2021.08.048.

Liu, H.-M., Tsao, F.-M., and Kuhl, P. (2009). Age-related changes in acoustic modifications of Mandarin maternal speech to preverbal infants and five-year-old children: A longitudinal study. *Journal of Child Language*, 36, 909–922. doi:10.1017/S030500090800929X.

Liu, Z., Yang, Q., Cai, N., Jin, L., Zhang, T., and Chen, X. (2019). Enigmatic differences by sex in cancer incidence: Evidence from childhood cancers. *American Journal of Epidemiology*, 188(6), 1130–1135. doi:10.1093/aje/kwz058.

Lloyd, J., Crouch, N. S., Minto, C. L., Liao, L.-M., and Creighton, S. M. (2005). Female genital appearance: "Normality" unfolds. *BJOG: An International Journal of Obstetrics & Gynaecology*, 112(5), 643–646. doi:10.1111/j.1471-0528.2004.00517.x.

Lobmaier, J. S., Fischbacher, U., Wirthmüller, U., and Knoch, D. (2018). The scent of attractiveness: Levels of reproductive hormones explain individual differences in women's body odour. *Proceedings of the Royal Society B: Biological Sciences*, 285(1886), 20181520. doi:10.1098/rspb.2018.1520.

Loring-Meier, S., and Halpern, D. F. (1999). Sex differences in visuospatial working memory: Components of cognitive processing. *Psychonomic Bulletin & Review*, 6, 464–471. doi:10.3758/BF03210836.

Louchart, A., Wesselman, H., Blumenschine, R. J., Hlusko, L. J., Njau, J. K., Black, M. T., et al. (2009). Taphonomic, avian, and small-vertebrate indicators of *Ardipithecus ramidus* habitat. *Science*, 326(5949), 66–66e64.

Lovejoy, C. O. (2009). Reexamining human origins in light of *Ardipithecus ramidus*. *Science*, 326(5949), 74–74e78.

Lovejoy, C. O., Simpson, S. W., White, T. D., Asfaw, B., and Suwa, G. (2009). Careful climbing in the Miocene: The forelimbs of *Ardipithecus ramidus* and humans are primitive. *Science*, 326(5949), 70–70e78.

Lovejoy, C. O., Suwa, G., Spurlock, L., Asfaw, B., and White, T. D. (2009). The pelvis and femur of *Ardipithecus ramidus:* The emergence of upright walking. *Science*, 326(5949), 71–71e76.

Lowe, A. E., Hobaiter, C., and Newton-Fisher, N. E. (2019). Countering infanticide: Chimpanzee mothers are sensitive to the relative risks posed by males on differing rank trajectories. *American Journal of Physical Anthropology*, 168(1), 3–9. doi:10.1002/ajpa.23723.

Lowenstine, L. J., and Osborn, K. G. (2012). Respiratory system diseases of nonhuman primates. *Nonhuman Primates in Biomedical Research*, 413–481. doi:org/10.1016/B978-0-12-381366-4.00009-2.

Lowery, C. M., Bralower, T. J., Owens, J. D., Rodríguez-Tovar, F. J., Jones, H., Smit, J., et al. (2018). Rapid recovery of life at ground zero of the end-Cretaceous mass extinction. *Nature*, 558(7709), 288–291. doi:10.1038/s41586-018-0163-6.

Lowry, S. J., Kapphahn, K., Chlebowski, R., and Li, C. I. (2016). Alcohol use and breast cancer survival among participants in the Women's Health Initiative. *Cancer Epidemiology, Biomarkers & Prevention*, 25(8), 1268–1273. doi:10.1158/1055-9965.Epi-16-0151.

Lu, Y.-F., Jin, T., Xu, Y., Zhang, D., Wu, Q., Zhang, Y.-K. J., and Liu, J. (2013). Sex differences in the circadian variation of cytochrome p450 genes and corresponding nuclear receptors in mouse liver. *Chronobiology International*, 30(9), 1135–1143. doi:10.3109/07420528.2013.805762.

Lucero, G. (2018). From sex objects to sisters-in-arms: Reducing military sexual assault through integrated basic training. *Duke Journal of Gender Law & Policy*, 26,1. scholarship.law.duke.edu.

Luders, E., and Kurth, F. (2020). Structural differences between male and female brains. Pp. 3–11 in *Handbook of Clinical Neurology*, vol. 175, edited by R. Lanzenberger, G. S. Kranz, and I. Savic. Elsevier.

Luo, S.-M., Schatten, H., and Sun, Q.-Y. (2013). Sperm mitochondria in reproduction: Good or bad and where do they go? *Journal of Genetics and Genomics*, 40(11), 549–556. doi:10.1016/j.jgg.2013.08.004.

Luo, Z., Lucas, S., Li, J., and Zhen, S. (1995). A new specimen of *Morganucodon oehleri* (Mammalia, Triconodonta) from the Liassic Lower Lufeng Formation of Yunnan, China. *Neues Jahrbuch für Geologie und Paläontologie Monatshefte*, 11, 671-680. doi:10.1127/njgpm/1995/1995/671.

Luo, Z. X. (2007). Transformation and diversification in early mammal evolution. *Nature*, 450(7172), 1011–1019. doi:10.1038/nature06277.

Luo, Z.-X., Yuan, C.-X., Meng, Q.-J., and Ji, Q. (2011). A Jurassic eutherian mammal and divergence of marsupials and placentals. *Nature*, 476(7361), 442-445. doi:10.1038/nature10291.

Lytle, S. R., Garcia-Sierra, A., and Kuhl, P. K. (2018). Two are better than one: Infant language learning from video improves in the presence of peers. *Proceedings of the National Academy of Sciences*, 115(40), 9859–9866. doi:10.1073/pnas.1611621115.

Maas, A. H., and Appelman, Y. E. (2010). Gender differences in coronary heart disease. *Netherlands Heart Journal*, 18, 598–603.

MacFadden, A., Elias, L., and Saucier, D. (2003). Males and females scan maps similarly, but give directions differently. *Brain and Cognition*, 53(2), 297–300.

Machin, A., and Dunbar, R. (2013). Sex and gender as factors in romantic partnerships and best friendships. *Journal of Relationships Research*, 4, E8. doi:10.1017/jrr.2013.8.

Mackintosh, N. J. (2007). Race differences in intelligence: An evolutionary hypothesis. *Intelligence*, 35(1), 94–96. doi:10.1016/j.intell.2006.08.001.

MacLarnon, A. M., and Hewitt, G. P. (1999). The evolution of human speech: The role of enhanced breathing control. *American Journal of Physical Anthropology*, 109, 341–363.

Macy, I. G., Hunscher, H. A., Donelson, E., and Nims, B. (1930). Human milk flow. *American Journal of Diseases in Childhood*, 39, 1186–1204.

Maessen, S. E., Derraik, J. G., Binia, A., and Cutfield, W. S. (2020). Perspective: Human milk oligosaccharides: Fuel for childhood obesity prevention? *Advances in Nutrition*, 11(1), 35–40.

Maguire, E. A., Burgess, N., and O'Keefe, J. (1999). Human spatial navigation: Cognitive maps, sexual dimorphism, and neural substrates. *Current Opinion in Neurobiology*, 9(2), 171–177. doi:10.1016/S0959-4388(99)80023-3.

Mahady, S., and Wolff, J. O. (2002). A field test of the Bruce effect in the monogamous prairie vole (*Microtus ochrogaster*). *Behavioral Ecology and Sociobiology*, 52, 31–37. doi:10.1007/s00265-002-0484-0.

Maher, A. C., Akhtar, M., Vockley, J., and Tarnopolsky, M. A. (2010). Women have higher protein content of beta-oxidation enzymes in skeletal muscle than men. *PLOS ONE*, 5(8), e12025.

Maher, A. C., Fu, M. H., Isfort, R. J., Varbanov, A. R., Qu, X. A., and Tarnopolsky, M. A. (2009). Sex differences in global mRNA content of human skeletal muscle. *PLOS ONE*, 4(7), e6335.

Mahjabeen, R. (2008). Microfinancing in Bangladesh: Impact on households, consumption and welfare. *Journal of Policy Modeling*, 30(6), 1083–1092. doi:10.1016/j.jpolmod.2007.12.007.

Mahoney, P., McFarlane, G., Smith, B. H., Miszkiewicz, J. J., Cerrito, P., Liversidge, H., et al. (2021). Growth of Neanderthal infants from Krapina (120–130 ka), Croatia. *Proceedings of the Royal Society B: Biological Sciences*, 288(1963), 20212079. doi:10.1098/rspb.2021.2079.

Maines, R. P. (1999). The technology of orgasm: "Hysteria," the vibrator, and women's sexual satisfaction. Baltimore: Johns Hopkins University Press.

Mammi, C., Calanchini, M., Antelmi, A., Cinti, F., Rosano, G. M., Lenzi, A., et al. (2012). Androgens and adipose tissue in males: a complex and reciprocal interplay. *International Journal of Endocrinology*, 2012, 789653. doi:10.1155/2012/789653.

Mampe, B., Friederici, A. D., Christophe, A., and Wermke, K. (2009). Newborns' cry melody is shaped by their native language. *Current Biology*, 19(23), 1994–1997. doi:10.1016/j.cub.2009.09.064.

Mano, R., Benjaminov, O., Kedar, I., Bar, Y., Sela, S., Ozalvo, R., et al. (2017). PD07-10 malignancies in male *BRCA* mutation carriers: Results from a prospectively screened cohort of patients enrolled to a dedicated male *BRCA* clinic. *Journal of Urology*, 197(4S), e131–e132. doi:10.1016/j.juro.2017.02.385.

Manthi, F. K., Plavcan, J. M., and Ward, C. V. (2012). New hominin fossils from Kanapoi, Kenya, and the mosaic evolution of canine teeth in early hominins. *South African Journal of Science*, 108, 1–9.

Mao, Y., Catacchio, C. R., Hillier, L. W., Porubsky, D., Li, R., Sulovari, A., et al. (2021). A high-quality bonobo genome refines the analysis of hominid evolution. *Nature*, 594, 77–81. doi:10.1038/s41586-021-03519-x.

Maradit Kremers, H., Larson, D. R., Crowson, C. S., Kremers, W. K., Washington, R. E., Steiner, C. A., et al. (2015). Prevalence of total hip and knee replacement in the United States. *The Journal of Bone and Joint Surgery*, 97(17), 1386–1397. doi:10.2106/JBJS.N.01141.

Marcenaro, B., Leiva, A., Dragicevic, C., López, V., and Delano, P. H. (2021). The medial olivocochlear reflex strength is modulated during a visual working memory task. *Journal of Neurophysiology*, 125(6), 2309–2321. doi:10.1152/jn.00032.2020.

March of Dimes (2017). Miscarriage. www.marchofdimes.org.

Marcobal, A., Barboza, M., Froehlich, J. W., Block, D. E., German, J. B., Lebrilla, C. B., and Mills, D. A., 2010. Consumption of human milk oligosaccharides by gut-related microbes. *Journal of Agricultural and Food Chemistry*, 58, 5334–5340.

Margulis, S. W., Atsalis, S., Bellem, A., Wielebnowski, N. (2007). Assessment of reproductive behavior and hormonal cycles in geriatric western Lowland gorillas. *Zoo Biology*, 26, 117–139.

Marine Corps Combat Development Command (MCCDC) (2015). Analysis of the integration of female Marines into ground combat arms and units. Quantico, Va., Aug. 27, 2015.

Marlowe, F. W. (2000). The patriarch hypothesis—an alternative explanation of menopause. *Human Nature*, 11, 27–42.

Marlowe, F. W. (2010). *The Hadza: Hunter-Gatherers of Tanzania*. Berkeley: University of California Press.

Marsh, H., and Kasuya, T. (1986). Evidence for reproductive senescence in female cetaceans. *Reports of the International Whaling Commission*, 8, 57–74.

Martin, L. J., Carey, K. D., and Comuzzie, A. G. (2003). Variation in menstrual cycle length and cessation of menstruation in captive raised baboons. *Mechanisms of Ageing and Development*, 124(8–9), 865–871. doi:10.1016/s0047-6374(03)00134-9.

Martin, N., and Montagne, R. (2017). U.S. has the worst rate of maternal deaths in the developed world. NPR, May 12, 2017. www.npr.org.

Martinez, C. H., Raparla, S., Plauschinat, C. A., Giardino, N. D., Rogers, B., Beresford, J., et al. (2012). Gender differences in symptoms and care delivery for chronic obstructive pulmonary disease. *Journal of Women's Health*, 21(12), 1267–1274.

Martínez, I., Arsuaga, J. L., Quam, R., Carretero, J. M., Gracia, A., and Rodríguez, L. (2008). Human hyoid bones from the middle Pleistocene site of the Sima de los Huesos (Sierra de Atapuerca, Spain). *Journal of Human Evolution*, 54(1), 118–124.

Martínez, I., Rosa, M., Quam, R., Jarabo, P., Lorenzo, C., Bonmatí, A., et al. (2013). Communicative capacities in Middle Pleistocene humans from the Sierra de Atapuerca in Spain. *Quaternary International*, 295, 94–101.

Masataka, N. (1992). Motherese in a signed language. *Infant Behavior and Development*, 15(4), 453–460. doi:10.1016/0163-6383(92)80013-K.

Mathes, B. M., Morabito, D. M., and Schmidt, N. B. (2019). Epidemiological and clinical gender differences in OCD. *Current Psychiatry Reports*, 21(5), 36. doi:10.1007/s11920-019-1015-2.

Mautz, B. S., Wong, B. B. M., Peters, R. A., and Jennions, M. D. (2013). Penis size interacts with body shape and height to influence male attractiveness. *Proceedings of the National Academy of Sciences*, 110(17), 6925–6930. doi:10.1073/pnas.1219361110.

Mauvais-Jarvis, F., Bairey Merz, N., Barnes, P. J., Brinton, R. D., Carrero, J.-J., DeMeo, D. L., et al. (2020). Sex and gender: Modifiers of health, disease, and medicine. *The Lancet*, 396(10250), 565–582. doi:10.1016/S0140-6736(20)31561-0.

Mayer, A. R., Dodd, A. B., Rannou-Latella, J. G., Stephenson, D. D., Dodd, R. J., Ling, J. M., et al. (2021). 17α-Ethinyl estradiol-3-sulfate increases survival and hemodynamic functioning in a large animal model of combined traumatic brain injury and hemorrhagic shock: A randomized control trial. *Critical Care*, 25(1), 428. doi:10.1186/s13054-021-03844-7.

Mazure, C. M., and Jones, D. P. (2015). Twenty years and still counting: Including women as participants and studying sex and gender in biomedical research. *BMC Women's Health*, 15, 94. doi:10.1186/s12905-015-0251-9.

MBRRACE-UK (Mothers and Babies: Reducing Risk through Audits and Confidential Enquiries across the UK) (2016). *Saving Lives, Improving Mothers' Care: Surveillance of Maternal Deaths in the UK 2012–14 and Lessons Learned to Inform Maternity Care from the UK and Ireland Confidential Enquires into Maternal Deaths and Morbidity 2009–14*. Maternal, Newborn and Infant Clinical Outcome Review Programme. www.npeu.ox.ac.uk.

McAuliffe, K., and Whitehead, H. (2005). Eusociality, menopause and information in matrilineal whales. *Trends in Ecology & Evolution*, 20, 650. doi:10.1016/j.tree.2005.09.003.

McCall, R. B. (1977). Childhood IQ's as predictors of adult educational and occupational status. *Science*, 197(4302), 482–483. doi:10.1126/science.197.4302.482.

McClearn, D. (1992). Locomotion, posture, and feeding behavior of kinkajous, coatis, and raccoons. *Journal of Mammalogy*, 73(2), 245–261. doi:10.2307/1382055.

McClellan, H. L., Miller, S. J., and Hartmann, P. E. (2008). Evolution of lactation: Nutrition v. protection with special reference to five mammalian species. *Nutrition Research Reviews*, 21, 97–116. doi:10.1017/s0954422408100749.

McClure, L. (2020). *Women in Classical Antiquity: From Birth to Death*. Hoboken: John Wiley & Sons.

McComb, K., Moss, C., Durant, S. M., Baker, L., and Sayialel, S. (2001). Matriarchs as repositories of social knowledge in African elephants. *Science*, 292(5516), 491–494. doi:10.1126/science.1057895.

McComb, K., Shannon, G., Durant, S. M., Sayialel, K., Slotow, R., Poole, J., and Moss, C. (2011). Leadership in elephants: The adaptive value of age. *Proceedings of the Royal Society B: Biological Sciences*, 278(1722), 3270–3276. doi:10.1098/rspb.2011.0168.

McFadden, D. (2009). Masculinization of the mammalian cochlea. *Hearing Research*, 252(1), 37–48. doi:10.1016/j.heares.2009.01.002.

McFadden, D. (2011). Sexual orientation and the auditory system. *Frontiers in Neuroendocrinology*, 32, 201–213. doi:10.1016/j.yfrne.2011.02.001.

McFadden, D., and Pasanen, E. G. (1998). Comparison of the auditory systems of heterosexuals and homosexuals: Click-evoked otoacoustic emissions. *Proceedings of the National Academy of Sciences*, 95(5), 2709–2713. doi:10.1073/pnas.95.5.2709.

McLean, C. Y., Reno, P. L., Pollen, A. A., Bassan, A. I., Capellini, T. D., Guenther, C., et al. (2011). Human-specific loss of regulatory DNA and the evolution of human-specific traits. *Nature*, 471(7337), 216–219. doi:10.1038/nature09774.

McPherron, S. P., Alemseged, Z., Marean, C. W., Wynn, J. G., Reed, D., Geraads, D., et al. (2010). Evidence for stone-tool-assisted consumption of animal tissues before 3.39 million years ago at Dikika, Ethiopia. *Nature*, 466(7308), 857–860.

McSweeney, J. C., Rosenfeld, A. G., Abel, W. M., Braun, L. T., Burke, L. E., Daugherty, S. L., et al. (2016). Preventing and experiencing ischemic heart disease as a woman: State of the science. *Circulation*, 133(13), 1302–1331. doi:10.1161/CIR.0000000000000381.

Mehl, M. R., Vazire, S., Ramírez-Esparza, N., Slatcher, R. B., and Pennebaker, J. W. (2007). Are women really more talkative than men? *Science*, 317(5834), 82–82. doi:10.1126/science.1139940.

Melin, A. D., Fedigan, L. M., Hiramatsu, C., Sendall, C. L., and Kawamura, S. (2007). Effects of colour vision phenotype on insect capture by a free-ranging population of white-faced capuchins, Cebus capucinus. *Animal Behaviour*, 73(1), 205–214.

Melin, A. D., Matsushita, Y., Moritz, G. L., Dominy, N. J., and Kawamura, S. (2013). Inferred L/M cone opsin polymorphism of ancestral tarsiers sheds dim light on the origin of anthropoid primates. *Proceedings of the Royal Society B: Biological Sciences*, 280(1759), 20130189. doi:10.1098/rspb.2013.0189.

Melnick, D. A., and Pearl, M. C. (1987). Cercopithecines in multimale groups: Genetic diversity and population structure. Pp. 121–134 in *Primate Societies*, edited by B. B. Smuts, D. L. Cheney, R. M. Seyfarth, R. W. Wrangham, and T. T. Struhsaker. Chicago: University of Chicago Press.

Ménard, N., von Segesser, F., Scheffrahn, W., Pastorini, J., Vallet, D., Gaci, B., et al. (2001). Is male-infant caretaking related to paternity and/or mating activities in wild Barbary macaques (*Macaca sylvanus*)? *Comptes Rendus de l'Académie des Sciences—Series III—Sciences de la Vie*, 324(7), 601–610.

Mencke, T., Soltész, S., Grundmann, U., Bauer, M., Schlaich, N., Larsen, R., and Fuchs-Buder, T. (2000). Time course of neuromuscular blockade after rocuronium: A comparison between women and men. *Anaesthesist*, 49, 609–612. doi:10.1007/s001010070077.

Messina, I., Cattaneo, L., Venuti, P., de Pisapia, N., Serra, M., Esposito, G., et al. (2015). Sex-specific automatic responses to infant cries: TMS reveals greater excitability in females than males in motor evoked potentials. *Frontiers in Psychology*, 6, 1909. doi:10.3389/fpsyg.2015.01909.

Meyer, J. (2012). *Centenarians: 2010*. Washington, D.C.: U.S. Department of Commerce, Economics and Statistics Administration, U.S. Census Bureau.

Miaskowski, G. (1997). Women and pain. *Critical Care Nursing Clinics of North America*, 9(4), 453–458. doi:10.1016/S0899-5885(18)30238-7.

Mika, K., Whittington, C. M., McAllan, B. M., and Lynch, V. J. (2022). Gene expression phylogenies and ancestral transcriptome reconstruction resolves major transitions in the origins of pregnancy. *eLife*, 11, e74297. doi:10.7554/eLife.74297.

Miller, E., Wails, C. N., and Sulikowski, J. (2022). It's a shark-eat-shark world, but does that make for bigger pups? A comparison between oophagous and non-oophagous viviparous sharks. *Reviews in Fish Biology and Fisheries*, 32, 1019–1033. doi:10.1007/s11160-022-09707-w.

Miller, G. (2010). The prickly side of oxytocin. *Science*, 328(5984), 1343–1343. doi:10.1126/science.328.5984.1343-a.

Miller, G., Tybur, J. M., and Jordan, B. D. (2007). Ovulatory cycle effects on tip earnings by lap dancers: Economic evidence for human estrus? *Evolution and Human Behavior*, 28, 375–381. doi:10.1016/j.evolhumbehav.2007.06.002.

Miller, G. E., Chen, E., and Zhou, E. S. (2007). If it goes up, must it come down? Chronic stress and the hypothalamic-pituitary-adrenocortical axis in humans. *Psychological Bulletin*, 133(1), 25–45. doi:10.1037/0033-2909.133.1.25.

Miller, J. M., Low, L. K., Zielinski, R., Smith, A. R., DeLancey, J. O., and Brandon, C. (2015). Evaluating maternal recovery from labor and delivery: Bone and levator ani injuries. *American Journal of Obstetrics and Gynecology*, 213(2), 188-e1. doi:10.1016/j.ajog.2015.05.001.

Milligan, L. A., and Bazinet, R. P. (2008). Evolutionary modifications of human milk composition: Evidence from long-chain polyunsaturated fatty acid composition of anthropoid milks. *Journal of Human Evolution*, 55(6), 1086–1095. doi:10.1016/j.jhevol.2008.07.010.

Milne, J. S. (1907). *Surgical Instruments in Greek and Roman Times*. Oxford: The Clarendon Press.

Mischkowski, D., Crocker, J., and Way, B. M. (2016). From painkiller to empathy killer: Acetaminophen (paracetamol) reduces empathy for pain. *Social Cognitive and Affective Neuroscience*, 11(9), 1345–1353. doi:10.1093/scan/nsw057.

Mitani, J. C., and Stuht, J. (1998). The evolution of nonhuman primate loud calls: Acoustic adaptation for long-distance transmission. *Primates*, 39(2), 171–182.

Mitchell, H. H., Hamilton, T. S., Steggerda, F. R., and Bean, H. W. (1945). The chemical composition of the adult human body and its bearing on the biochemistry of growth. *Journal of Biological Chemistry*, 158(3), 625–637.

Mogil, J. S. (2020). Qualitative sex differences in pain processing: Emerging evidence of a biased literature. *Nature Reviews Neuroscience*, 21, 353–365. doi:10.1038/s41583-020-0310-6.

Mogil, J. S., and Chanda, M. L. (2005). The case for the inclusion of female subjects in basic science studies of pain. *Pain*, 117(1–2), 1–5.

Molitoris, J., Barclay, K., and Kolk, M. (2019). When and where birth spacing matters for child survival: An international comparison using the DHS. *Demography*, 56(4), 1349–1370. doi:10.1007/s13524-019-00798-y.

Morgan, T. J. H., Uomini, N. T., Rendell, L., Chouinard-Thuly, L., Street, S. E., Lewis, H. M., et al. (2014). Experimental evidence for the co-evolution of Hominin tool-making teaching and language. *Nature Communications*, 6, 6029.

Morin, L. P., and Allen, C. N. (2006). The circadian visual system, 2005. *Brain Research Reviews*, 51(1), 1–60. doi:10.1016/j.brainresrev.2005.08.003.

Morral, A. R., Gore, K. L., and Schell, T. L., eds. (2015). *Sexual Assault and Sexual Harassment in the U.S. Military*. Vol. 2, *Estimates for Department of Defense Service Members from the 2014 RAND Military Workplace Study*. Santa Monica, Calif.: RAND Corporation. www.rand.org.

Morris, M. W., Mok, A., and Mor, S. (2011). Cultural identity threat: The role of cultural identifications in moderating closure responses to foreign cultural inflow. *Journal of Social Issues*, 67(4), 760–773.

Morrow, A. L., Ruiz-Palacios, G. M., Altaye, M., Jiang, X., Lourdes Guerrero, M., Meinzen-Derr, J. K., Farkas, T., et al. (2004). Human milk oligosaccharides are associated with protection against diarrhea in breast-fed infants. *Journal of Pediatrics*, 145, 297–303. doi:10.1016/j.jpeds.2004.04.054.

Morton, R. A., Stone, J. R., and Singh, R. S. (2013). Mate choice and the origin of menopause. *PLOS Computational Biology*, 9(6), e1003092. doi:10.1371/journal.pcbi.1003092.

Moss, C. J. (2001). The demography of an African elephant (*Loxodonta africana*) population in Amboseli, Kenya. *Journal of Zoology*, 255, 145–156.

Mossabeb, R., and Sowti, K. (2021). Neonatal Abstinence Syndrome: A call for mother-infant dyad treatment approach. *American Family Physician*, 104(3), 222–223.

Motlagh Zadeh, L., Silbert, N. H., Sternasty, K., Swanepoel, D. W., Hunter, L. L., and Moore, D. R. (2019). Extended high-frequency hearing enhances speech perception in noise. *Proceedings of the National Academy of Sciences*, 116(47), 23753–23759. doi:10.1073/pnas.1903315116.

Mowitz, M. E., Dukhovny, D., and Zupancic, J. A. (2018). The cost of necrotizing enterocolitis in premature infants. *Seminar in Fetal Neonatal Medicine*, 23, 416–419. doi:10.1016/j.siny.2018.08.004.

Mozaffarian, D., Benjamin, E. J., Go, A. S., Arnett, D. K., Blaha, M. J., Cush-

man, M., et al. (2015). Heart disease and stroke statistics—2015 update. *Circulation*, 131(4), e29–e322. doi:10.1161/CIR.0000000000000152.
Mozaffarian, D., Benjamin, E. J., Go, A. S., Arnett, D. K., Blaha, M. J., Cushman, M., et al. (2016). Heart disease and stroke statistics—2016 update: A report from the American Heart Association. *Circulation*, 133(4), e38–e360.
Mrazek, M. D., Chin, J. M., Schmader, T., Hartson, K. A., Smallwood, J., and Schooler, J. W. (2011). Threatened to distraction: Mind-wandering as a consequence of stereotype threat. *Journal of Experimental Social Psychology*, 47(6), 1243–1248. doi:10.1016/j.jesp.2011.05.011.
Muller, M. N., Kahlenberg, S. M., Emery Thompson, M., and Wrangham, R. W. (2007). Male coercion and the costs of promiscuous mating for female chimpanzees. *Proceedings of the Royal Society B: Biological Sciences*, 274(1612), 1009–1014. doi:10.1098/rspb.2006.0206.
Muller, M. N., Thompson, M. E., and Wrangham, R. W. (2006). Male chimpanzees prefer mating with old females. *Current Biology*, 16(22), 2234–2238. doi:10.1016/j.cub.2006.09.042.
Mulugeta, E., Kassaye, M., and Berhane, Y. (1998). Prevalence and outcomes of sexual violence among high school students. *Ethiopian Medical Journal*, 36(3), 167–174.
Munson, L., and Moresco, A. (2007). Comparative pathology of mammary gland cancers in domestic and wild animals. *Breast Disease*, 28, 7–21. doi:10.3233/bd-2007-28102.
Murray, C. M., Stanton, M. A., Lonsdorf, E. V., Wroblewski, E. E., and Pusey, A. E. (2016). Chimpanzee fathers bias their behaviour towards their offspring. *Royal Society Open Science*, 3(11), 160441. doi:10.1098/rsos.160441.
Mutter, W. P., and Karumanchi, S. A. (2008). Molecular mechanisms of preeclampsia. *Microvascular Research*, 75(1), 1–8. doi:10.1016/j.mvr.2007.04.009.
Myrskylä, M., and Margolis, R. (2014). Happiness: Before and after the kids. *Demography*, 51(5), 1843–1866. doi:10.1007/s13524-014-0321-x.
Nair, P. S. (2010). Understanding below-replacement fertility in Kerala, India. *Journal of Health, Population, and Nutrition*, 28(4), 405–412. doi:10.3329/jhpn.v28i4.6048.
Nakano, K., Nemoto, H., Nomura, R., Inaba, H., Yoshioka, H., Taniguchi, K., et al. (2009). Detection of oral bacteria in cardiovascular specimens. *Oral Microbiology and Immunology*, 24(1), 64–68. doi:10.1111/j.1399-302X.2008.00479.x.
National Institute on Drug Abuse (NIDA) (2022). Sex and gender differences in substance use. May 4, 2022. nida.nih.gov.
Nattrass, S., Croft, D. P., Ellis, S., Cant, M. A., Weiss, M. N., Wright, B. M., et al. (2019). Postreproductive killer whale grandmothers improve the survival of their grandoffspring. *Proceedings of the National Academy of Sciences*, 116(52), 26669–26673. doi:10.1073/pnas.1903844116.
Neubauer, S., Hublin, J.-J., and Gunz, P. (2018). The evolution of modern

human brain shape. *Science Advances*, 4(1), eaao5961. doi:10.1126/sciadv.aao5961.

Neufang, S., Specht, K., Hausmann, M., Güntürkün, O., Herpertz-Dahlmann, B., Fink, G. R., and Konrad, K. (2008). Sex differences and the impact of steroid hormones on the developing human brain. *Cerebral Cortex*, 19(2), 464–473. doi:10.1093/cercor/bhn100.

Newburg, D., Warren, C., Chaturvedi, P., Newburg, A., Oftedal, O., Ye, S., and Tilden, C. (1999). Milk oligosaccharides across species. *Pediatric Research*, 45(5), 745–745.

Newman, L., Rowley, J., Vander Hoorn, S., Wijesooriya, N. S., Unemo, M., Low, N., et al. (2015). Global estimates of the prevalence and incidence of four curable sexually transmitted infections in 2012 based on systematic review and global reporting. *PLOS ONE*, 10(12), e0143304. doi:10.1371/journal.pone.0143304.

Nicholls, H. (2012). Sex and the single rhinoceros. *Nature*, 485(7400), 566–569.

Nichols, H. B., Shoemaker, M. J., Cai, J., Xu, J., Wright, L. B., Brook, M. N., et al. (2019). Breast cancer risk after recent childbirth. *Annals of Internal Medicine*, 170(1), 22–30. doi:10.7326/m18-1323.

Nigro, L. (2017). *Beheaded Ancestors: Of Skulls and Statues in Pre-Pottery Neolithic Jericho*, 3–30. Conference paper. Scienze dell'Antichità, March 23, 2017.

Nilsson, I. E. K., Åkervall, S., Molin, M., Milsom, I., and Gyhagen, M. (2022). Severity and impact of accidental bowel leakage two decades after no, one, or two sphincter injuries. *American Journal of Obstetrics & Gynecology*. doi:10.1016/j.ajog.2022.11.1312.

Nishida, T., Corp, N., Hamai, M., Hasegawa, T., Hiraiwa-Hasegawa, M., Hosaka, K., et al. (2003). Demography, female life history, and reproductive profiles among the chimpanzees of Mahale. *American Journal of Primatology*, 59(3), 99–121.

Nishida, T., Takasaki, H., and Takahata, Y. (1990). Demography and reproductive profiles. In *The Chimpanzees of the Mahale Mountains*, edited by T. Nishida. Tokyo: University of Tokyo Press.

Nishie, H., and Nakamura, M. (2018). A newborn infant chimpanzee snatched and cannibalized immediately after birth: Implications for "maternity leave" in wild chimpanzee. *American Journal of Physical Anthropology*, 165(1), 194–199. doi:10.1002/ajpa.23327.

Nishimura, T. (2006). Descent of the larynx in chimpanzees: Mosaic and multiple-step evolution of the foundations for human speech. Pp. 75–95 in *Cognitive Development in Chimpanzees*, edited by T. Matsuzawa, M. Tomonaga, and M. Tanaka. Tokyo: Springer.

Nishimura, T., Mikami, A., Suzuki, J., and Matsuzawa, T. (2003). Descent of the larynx in chimpanzee infants. *Proceedings of the National Academy of Sciences*, 100(12), 6930–6933.

Nishimura, T., Mikami, A., Suzuki, J., and Matsuzawa, T. (2006). Descent of the hyoid in chimpanzees: Evolution of face flattening and speech. *Journal of Human Evolution*, 51(3), 244–254.

Nishimura, T., Tokuda, I. T., Miyachi, S., Dunn, J. C., Herbst, C. T., Ishimura, K., et al. (2022). Evolutionary loss of complexity in human vocal anatomy as an adaptation for speech. *Science*, 377(6607), 760–763. doi:10.1126/science.abm1574.

Noë, R., and Sluijter, A. A. (1990). Reproductive tactics of male savanna baboons. *Behaviour*, 113(1/2), 117–170. doi:10.1163/156853990X00455.

Norell, M. A., Wiemann, J., Fabbri, M., Yu, C., Marsicano, C. A., Moore-Nall, A., et al. (2020). The first dinosaur egg was soft. *Nature*, 583(7816), 406–410. doi:10.1038/s41586-020-2412-8.

Northstone, K., Joinson, C., Emmett, P., Ness, A., and Paus, T. (2012). Are dietary patterns in childhood associated with IQ at 8 years of age? A population-based cohort study. *Journal of Epidemiology and Community Health*, 66(7), 624–628. doi:10.1136/jech.2010.111955.

Norton, P., and Brubaker, L. (2006). Urinary incontinence in women. *The Lancet*, 367(9504), 57–67. doi:10.1016/S0140-6736(06)67925-7.

Nour, N. M. (2006). Health consequences of child marriage in Africa. *Emerging Infectious Diseases*, 12(11), 1644–1649. doi:10.3201/eid1211.060510.

Novotny, S. A., Warren, G. L., and Hamrick, M. W. (2015). Aging and the muscle-bone relationship. *Physiology*, 30(1), 8–16. doi:10.1152/physiol.00033.2014.

Nozaki, M., Mitsunaga, F., and Shimizu, K. (1995). Reproductive senescence in female Japanese monkeys (*Macaca Fuscata*): Age- and season-related changes in hypothalamic-pituitary-ovarian functions and fecundity rates. *Biology of Reproduction*, 52, 1250–1257. doi:10.1095/biolreprod52.6.1250.

Nunn, C. L. (1999). The evolution of exaggerated sexual swellings in primates and the graded-signal hypothesis. *Animal Behaviour*, 58(2), 229–246. doi:10.1006/anbe.1999.1159.

Nussbaum, M. C. (2003). Women's education: A global challenge. *Signs*, 29(2), 325–355.

O'Connell-Rodwell, C. E. (2007). Keeping an "ear" to the ground: Seismic communication in elephants. *Physiology*, 22(4), 287–294. doi:10.1152/physiol.00008.2007.

O'Connor, C. A., Cernak, I., and Vink, R. (2005). Both estrogen and progesterone attenuate edema formation following diffuse traumatic brain injury in rats. *Brain Research*, 1062(1), 171–174. doi:10.1016/j.brainres.2005.09.011.

O'Leary, M. A., Bloch, J. I., Flynn, J. J., Gaudin, T. J., Giallombardo, A., Giannini, N. P., et al. (2013). The placental mammal ancestor and the post-K-Pg radiation of placentals. *Science*, 339(6120), 662–667. doi:10.1126/science.1229237.

O'Neill, M. C., Umberger, B. R., Holowka, N. B., Larson, S. G., and Reiser, P. J. (2017). Chimpanzee super strength and human skeletal muscle evolution. *Proceedings of the National Academy of Sciences*, 114(28), 7343–7348. doi:10.1073/pnas.1619071114.

Ochieng, S. (2020). Child marriage in the US: Loopholes in state marriage laws perpetuate child marriage. *Immigration and Human Rights Law Review*, 2(1), 3.

Office for National Statistics (ONS). (2020). *Domestic Abuse Victim Characteristics, England and Wales: Year Ending March 2020*. www.ons.gov.uk.

Oftedal, O. T. (2002). The mammary gland and its origin during synapsid evolution. *Journal of Mammary Gland Biology and Neoplasia*, 7(3), 225–252. doi:10.1023/a:1022896515287.

Oftedal, O. T. (2012). The evolution of milk secretion and its ancient origins. *Animal*, 6(3), 355–368. doi:10.1017/S1751731111001935.

Oktay, K., Turan, V., Titus, S., Stobezki, R., and Liu, L. (2015). BRCA mutations, DNA repair deficiency, and ovarian aging. *Biology of Reproduction*, 93(3). doi:10.1095/biolreprod.115.132290.

Olesiuk, P. F., Bigg, M. A., and Ellis, G. M. (1990). Life history and population dynamics of resident killer whales (*Ornicus orca*) in the coastal waters of British Columbia and Washington State. *Report of the International Whaling Commission*, 12, 209–243.

Oliva, M., Muñoz-Aguirre, M., Kim-Hellmuth, S., Wucher, V., Gewirtz, A. D. H., Cotter, D. J., et al. (2020). The impact of sex on gene expression across human tissues. *Science*, 369(6509), eaba3066. doi:10.1126/science.aba3066.

Oliveira-Pinto, A. V., Santos, R. M., Coutinho, R. A., Oliveira, L. M., Santos, G. B., Alho, A. T., et al. (2014). Sexual dimorphism in the human olfactory bulb: Females have more neurons and glial cells than males. *PLOS ONE*, 9(11), e111733. doi:10.1371/journal.pone.0111733.

Oliveros, E., Martin, M., Torres-Espinola, F. J., Segura-Moreno, T., Ramirez, M., Santos-Fandila, A., et al. (2021). Human milk levels of 2´-fucosyllactose and 6´-sialyllactose are positively associated with infant neurodevelopment and are not impacted by maternal BMI or diabetic status. *Nutrition & Food Science*, 4, 024.

Olshansky, S. J., Carnes, B. A., and Grahn, D. (1998). Confronting the boundaries of human longevity: Many people now live beyond their natural lifespans through the intervention of medical technology and improved lifestyles—a form of "manufactured time." *American Scientist*, 86(1), 52–61.

Oppel, R. A., Jr., and Cooper, H. (2015). 2 graduating rangers, aware of their burden. *The New York Times*, Aug. 20, 2015. www.nytimes.com.

Orbach, D. N., Kelly, D. A., Solano, M., and Brennan, P. L. R. (2017). Genital interactions during simulated copulation among marine mammals. *Proceedings of the Royal Society B: Biological Sciences*, 284, 20171265. doi:10.1098/rspb.2017.1265.

Orbach, D. N., Marshall, C. D., Mesnick, S. L., and Würsig, B. (2017). Patterns of cetacean vaginal folds yield insights into functionality. *PLOS ONE*, 12(3), e0175037. doi:10.1371/journal.pone.0175037.

Osorio, D., and Vorobyev, M. (1996). Colour vision as an adaptation to frugivory in primates. *Proceedings of the Royal Society B: Biological Sciences*, 263, 593–599.

Ouattara, K., Lemasson, A., and Zuberbühler, K. (2009). Campbell's monkeys use affixation to alter call meaning. *PLOS ONE*, 4(11), e7808.

Oxenham A. J. (2018). How we hear: The perception and neural coding of sound. *Annual Review of Psychology*, 69, 27–50. doi:10.1146/annurev-psych-122216-011635.

Pagels, E. (2013). *Revelations: Visions, Prophecy, and Politics in the Book of Revelation*. New York: Penguin Books.

Palombit, R., Cheney, D., Fischer, J., Johnson, S., Rendall, D., Seyfarth, R., and Silk, J. (2000). Male infanticide and defense of infants in chacma baboons. Pp. 123–152 in *Infanticide by Males and Its Implications*, edited by C. Van Schaik and C. Janson. Cambridge, U.K.: Cambridge University Press. doi:10.1017/CBO9780511542312.008.

Palsson, O. S., Peery, A., Seitzberg, D., Amundsen, I. D., McConnell, B., and Simrén, M. (2020). Human milk oligosaccharides support normal bowel function and improve symptoms of irritable bowel syndrome: A multicenter, open-label trial. *Clinical and Translational Gastroenterology*, 11(12). doi:10.14309/ctg.0000000000000276.

Pan, W., Gu, T., Pan, Y., Feng, C., Long, Y., Zhao, Y., et al. (2014). Birth intervention and non-maternal infant-handling during parturition in a nonhuman primate. *Primates*, 55(4), 483–488. doi:10.1007/s10329-014-0427-1.

Panizzon, M. S., Vuoksimaa, E., Spoon, K. M., Jacobson, K. C., Lyons, M. J., Franz, C. E., et al. (2014). Genetic and environmental influences on general cognitive ability: Is g a valid latent construct? *Intelligence*, 43, 65–76.

Parada, M., Abdul-Ahad, F., Censi, S., Sparks, L., and Pfaus, J. G. (2011). Context alters the ability of clitoral stimulation to induce a sexually-conditioned partner preference in the rat. *Hormones and Behavior*, 59(4), 520–527. doi:10.1016/j.yhbeh.2011.02.001.

Parada, M., Chamas, L., Censi, S., Coria-Avila, G., and Pfaus, J. G. (2010). Clitoral stimulation induces conditioned place preference and Fos activation in the rat. *Hormones and Behavior*, 57(2), 112–118. doi:10.1016/j.yhbeh.2009.05.008.

Pargulski, J. R., and Reynolds, M. R. (2017). Sex differences in achievement: Distributions matter. *Personality and Individual Differences*, 104, 272–278. doi:10.1016/j.paid.2016.08.016.

Parish, A. R. (1994). Sex and food control in the "uncommon chimpanzee": How bonobo females overcome a phylogenetic legacy of male dominance. *Ethology and Sociobiology*, 15(3), 157–179. doi:10.1016/0162-3095(94)90038-8.

Parker, D. (2018). Kuhnian revolutions in neuroscience: The role of tool development. *Biology & Philosophy*, 33(3), 17. doi:10.1007/s10539-018-9628-0.

Parra-Peralbo, E., Talamillo, A., and Barrio, R. (2021). Origin and development of the adipose tissue, a key organ in physiology and disease. *Frontiers in Cell and Developmental Biology*, 9. doi:10.3389/fcell.2021.786129.

Parsons, C. E., Young, K. S., Parsons, E., Stein, A., and Kringelbach, M. L. (2012). Listening to infant distress vocalizations enhances effortful motor performance. *Acta Paediatrica*, 101(4), e189. doi:10.1111/j.1651-2227.2011.02554.x.

Partridge, E. A., Davey, M. G., Hornick, M. A., McGovern, P. E., Mejaddam, A. Y., Vrecenak, J. D., et al. (2017). An extra-uterine system to physiologically support the extreme premature lamb. *Nature Communications*, 8(1), 15112. doi:10.1038/ncomms15112.

Patel, B. A., Wallace, I. J., Boyer, D. M., Granatosky, M. C., Larson, S. G., and Stern, J. T., Jr. (2015). Distinct functional roles of primate grasping hands and feet during arboreal quadrupedal locomotion. *Journal of Human Evolution*, 88, 79–84. doi:/10.1016/j.jhevol.2015.09.004.

Patrick, S. W., Barfield, W. D., Poindexter, B. B., and Committee on Fetus and Newborn, Committee on Substance Use and Prevention (2020). Neonatal opioid withdrawal syndrome. *Pediatrics*, 146(5), e2020029074. doi:10.1542/peds.2020-029074.

Paulozzi, L. J., Erickson, J. D., and Jackson, R. J. (1997). Hypospadias trends in two US surveillance systems. *Pediatrics*, 100(5), 831–834. doi:10.1542/peds.100.5.831.

Pavard, S., Metcalf, C. J., and Heyer, E. (2008). Senescence of reproduction may explain adaptive menopause in humans: A test of the "Mother" hypothesis. *American Journal of Physical Anthropology*, 136, 194–203. doi:10.1002/ajpa.20794.

Pavlicev, M., Herdina, A. N., and Wagner, G. (2022). Female genital variation far exceeds that of male genitalia: A review of comparative anatomy of clitoris and the female lower reproductive tract in theria. *Integrative and Comparative Biology*, 62(3), 581–601. doi:10.1093/icb/icac026.

Pawłowski, B., and Żelaźniewicz, A. (2021). The evolution of perennially enlarged breasts in women: A critical review and a novel hypothesis. *Biological Reviews*, 96, 2794–2809. doi:10.1111/brv.12778.

Peacock, J. L., Marston, L., Marlow, N., Calvert, S. A., and Greenough, A. (2012). Neonatal and infant outcome in boys and girls born very prematurely. *Pediatric Research*, 71(3), 305–310. doi:10.1038/pr.2011.50.

Pearson, J. D., Morrell, C. H., Gordon-Salant, S., Brant, L. J., Metter, E. J., Klein, L. L., and Fozard, J. L. (1995). Gender differences in a longitudinal study of age-associated hearing loss. *The Journal of the Acoustical Society of America*, 97(2), 1196–1205. doi:10.1121/1.412231.

Peigné, S., de Bonis, L., Likius, A., Mackaye, H. T., Vignaud, P., and Brunet, M. (2005). A new machairodontine (Carnivora, Felidae) from the Late Miocene hominid locality of TM 266, Toros-Menalla, Chad. *Comptes Rendus Palevol*, 4(3), 243–253. doi:10.1016/j.crpv.2004.10.002.

Pellerin, C. (2015). Carter opens all military occupations to women. *DOD News, Defense Media Activity*, Dec. 3, 2015.

Pennington, P. M., and Durrant, B. S. (2019). Assisted reproductive technologies in captive rhinoceroses. *Mammal Review*, 49(1), 1–15. doi:10.1111/mam.12138.

Perhonen, M. A., Franco, F., Lane, L. D., Buckey, J. C., Blomqvist, C. G., Zerwekh, J. E., et al. (2001). Cardiac atrophy after bed rest and spaceflight. *Journal of Applied Physiology*, 91(2), 645–653. doi:10.1152/jappl.2001.91.2.645.

Peter, B. M., and Slatkin, M. (2015). The effective founder effect in a spatially expanding population. *Evolution*, 69(3), 721–734. doi:10.1111/evo.12609.

Peterman, A., Palermo, T., and Bredenkamp, C. (2011). Estimates and determinants of sexual violence against women in the Democratic Republic of Congo. *American Journal of Public Health*, 101(6), 1060–1067. doi:10.2105/AJPH.2010.300070.

Peters, M. (2005). Sex differences and the factor of time in solving Vandenberg and Kuse mental rotation problems. *Brain and Cognition*, 57, 176–184. doi:10.1016/j.bandc.2004.08.052.

Petersen, J. (2018). Gender difference in verbal performance: A meta-analysis of United States state performance assessments. *Educational Psychology Review*, 30(4), 1269–1281. doi.org/10.1007/s10648-018-9450-x.

Pfefferle, D., West, P. M., Grinnell, J., Packer, C., and Fischer, J. (2007). Do acoustic features of lion, *Panthera leo*, roars reflect sex and male condition? *The Journal of the Acoustical Society of America*, 121(6), 3947–3953. doi:10.1121/1.2722507.

Pfenning, A. R., Hara, E., Whitney, O., Rivas, M. V., Wang, R., Roulhac, P. L., et al. (2014). Convergent transcriptional specializations in the brains of humans and song-learning birds. *Science*, 346(6215), 1256846. doi:10.1126/science.1256846.

Phillips, D. (2018). As economy roars, army falls thousands short of recruiting goals. *The New York Times*, Sept. 21, 2018. www.nytimes.com.

Phinney, S. D., Stern, J. S., Burke, K. E., Tang, A. B., Miller, G., and Holman, R. T. (1994). Human subcutaneous adipose tissue shows site-specific differences in fatty acid composition. *The American Journal of Clinical Nutrition*, 60(5), 725–729. doi:10.1093/ajcn/60.5.725.

Photopoulou, T., Ferreira, I. M., Best, P. B., Kasuya, T., and Marsh, H. (2017). Evidence for a postreproductive phase in female false killer whales *Pseudorca crassidens*. *Frontiers in Zoology*, 14(1), 30. doi:10.1186/s12983-017-0208-y.

Piantadosi, S. T., and Kidd, C. (2016). Extraordinary intelligence and the care of infants. *Proceedings of the National Academy of Sciences*, 113(25), 6874–6879. doi:10.1073/pnas.1506752113.

Piazza, E. A., Iordan, M. C., and Lew-Williams, C. (2017). Mothers consistently alter their unique vocal fingerprints when communicating with infants. *Current Biology*, 27(20), 3162–3167.e3163. doi:10.1016/j.cub.2017.08.074.

Pik, R. (2011). Geodynamics: East Africa on the rise. *Nature Geoscience*, 4(10), 660–661.

Pilbeam, D. (1978). Major trends in human evolution. In *Current Argument on Early Man: Report from a Nobel Symposium*, edited by Lars-König Königsson. Oxford: Published on behalf of the Royal Swedish Academy of Sciences by Pergamon Press.

Pinker, S. (1994). *The Language Instinct*. New York: William Morrow.

Platt, L., Grenfell, P., Meiksin, R., Elmes, J., Sherman, S. G., Sanders, T., et al. (2018). Associations between sex work laws and sex workers' health: A systematic review and meta-analysis of quantitative and qualitative studies. *PLOS Medicine*, 15(12), e1002680. doi:10.1371/journal.pmed.1002680.

Plavcan, J. M. (2001). Sexual dimorphism in primate evolution. *American Journal of Physical Anthropology*, 116(S33), 25–53.

Plavcan, J. M. (2012a). Body size, size variation, and sexual size dimorphism in early *Homo*. *Current Anthropology*, 53(S6), S409–S423. doi:10.1086/667605.

Plavcan, J. M. (2012b). Sexual size dimorphism, canine dimorphism, and male-male competition in primates. *Human Nature*, 23(1), 45–67. doi:10.1007/s12110-012-9130-3.

Plavcan, J. M., and van Schaik, C. P. (1992). Intrasexual competition and canine dimorphism in anthropoid primates. *American Journal of Physical Anthropology*, 87(4), 461–477. doi:10.1002/ajpa.1330870407.

Plomin, R., and Deary, I. J. (2015). Genetics and intelligence differences: Five special findings. *Molecular Psychiatry*, 20(1), 98–108. doi:10.1038/mp.2014.105.

Plummer, T. W., Ditchfield, P. W., Bishop, L. C., Kingston, J. D., Ferraro, J. V., Braun, D. R., et al. (2009). Oldest evidence of toolmaking Hominins in a grassland-dominated ecosystem. *PLOS ONE*, 4(9), e7199. doi:10.1371/journal.pone.0007199.

Pobiner, B. L. (2013). Evidence for meat-eating by early humans. *Nature Education Knowledge*, 4(6), 1.

Podos, J., and Cohn-Haft, M. (2019). Extremely loud mating songs at close range in white bellbirds. *Current Biology*, 29(20), R1068–R1069. doi:10.1016/j.cub.2019.09.028.

Pollack, A. (2015). Breast milk becomes a commodity, with mothers caught up in debate. *The New York Times*, March 20, 2015.

Pontzer, H., Brown, M. H., Raichlen, D. A., Dunsworth, H., Hare, B., Walker, K., et al. (2016). Metabolic acceleration and the evolution of human brain size and life history. *Nature*, 533(7603), 390–392. doi:10.1038/nature17654.

Posth, C., Yu, H., Ghalichi, A., Rougier, H., Crevecoeur, I., Huang, Y., et al. (2023). Palaeogenomics of Upper Palaeolithic to Neolithic European hunter-gatherers. *Nature*, 615(7950), 117–126. doi:10.1038/s41586-023-05726-0.

Potts, R. (1984). Home bases and early hominids: Reevaluation of the fossil record at Olduvai Gorge suggests that the concentrations of bones and stone tools do not represent fully formed campsites but an antecedent to them. *American Scientist*, 72(4), 338–347.

Potts, R. (1986). Temporal span of bone accumulations at Olduvai Gorge and implications for early hominid foraging behavior. *Paleobiology*, 12(1), 25–31.

Potts, R. (1994). Variables versus models of early Pleistocene hominid land use. *Journal of Human Evolution*, 27(1), 7–24.

Potts, R. (2012). Environmental and behavioral evidence pertaining to the evolution of early Homo. *Current Anthropology*, 53(S6), S299–S317. doi:10.1086/667704.

Potts, R., and Faith, J. T. (2015). Alternating high and low climate variability: The context of natural selection and speciation in Plio-Pleistocene hominin evolution. *Journal of Human Evolution*, 87, 5–20. doi:10.1016/j.jhevol.2015.06.014.

Potts, R., and Shipman, P. (1981). Cutmarks made by stone tools on bones from Olduvai Gorge, Tanzania. *Nature*, 291(5816), 577–580.

Potts, R., and Teague, R. (2010). Behavioral and environmental background to "Out-of-Africa I" and the arrival of *Homo erectus* in East Asia. Pp. 67–85 in *Out of Africa I: The First Hominin Colonization of Eurasia*, edited by J. G. Fleagle, J. J. Shea, F. E. Grine, A. L. Baden, and R. E. Leakey. Dordrecht: Springer.

Powell, A., Shennan, S., and Thomas, M. G. (2009). Late Pleistocene demography and the appearance of modern human behavior. *Science*, 324(5932), 1298–1301.

Pradhan, E. (2015). Female education and childbearing: A closer look at the data. *World Bank Blogs*, Nov. 24, 2015. blogs.worldbank.org.

Prat, S. (2018). First hominin settlements out of Africa. Tempo and dispersal mode: Review and perspectives. *Comptes Rendus Palevol*, 17(1), 6–16. doi:10.1016/j.crpv.2016.04.009.

Premachandran, H., Zhao, M., and Arruda-Carvalho, M. (2020). Sex differences in the development of the rodent corticolimbic system. *Frontiers in Neuroscience*, 14. doi:10.3389/fnins.2020.583477.

Prendergast, B. J., Onishi, K. G., and Zucker, I. (2014). Female mice liberated for inclusion in neuroscience and biomedical research. *Neuroscience and Biobehavioral Reviews*, 40, 1–5. doi:10.1016/j.neubiorev.2014.01.001.

Profet, M. (1993). Menstruation as a defense against pathogens transported by sperm. *The Quarterly Review of Biology*, 68(3), 335–386. doi:10.1086/418170.

Pruetz, J. D., Bertolani, P., Ontl, K. B., Lindshield, S., Shelley, M., and Wessling, E. G. (2015). New evidence on the tool-assisted hunting exhibited by chimpanzees (*Pan troglodytes verus*) in a savannah habitat at Fongoli, Senegal. *Royal Society Open Science*, 2(4), 140507. doi:10.1098/rsos.140507.

Prüfer, K., Munch, K., Hellmann, I., Akagi, K., Miller, J. R., Walenz, B., et al. (2012). The bonobo genome compared with the chimpanzee and human genomes. *Nature*, 486(7404), 527-531. doi:10.1038/nature11128.

Prühlen, S. (2007). What was the best for an infant from the Middle Ages to Early Modern times in Europe? The discussion concerning wet nurses. *Hygiea Internationalis*, 6. doi:10.3384/hygiea.1403-8668.

Prum, R. O. (2017). *The Evolution of Beauty: How Darwin's Forgotten Theory of Mate Choice Shapes the Animal World—and Us*. New York: Doubleday.

Pusey, A., Murray, C., Wallauer, W., Wilson, M., Wroblewski, E., and Goodall, J. (2008). Severe aggression among female *Pan troglodytes schweinfurthii* at Gombe National Park, Tanzania. *International Journal of Primatology*, 29, 949–973. doi:10.1007/s10764-008-9281-6.

Pusey, A. E., and Schroepfer-Walker, K. (2013). Female competition in chimpanzees. *Philosophical Transactions of the Royal Society B: Biological Sciences*, 368(1631), 20130077. doi:10.1098/rstb.2013.0077.

Qu, F., Wu, Y., Zhu, Y. H., Barry, J., Ding, T., Baio, G., et al. (2017). The association between psychological stress and miscarriage: A systematic review and meta-analysis. *Scientific Reports*, 7(1), 1731. doi:10.1038/s41598-017-01792-3.

Quam, R., Martínez, I., Rosa, M., Bonmatí, A., Lorenzo, C., de Ruiter, D. J., et al. (2015). Early hominin auditory capacities. *Science Advances*, 1(8), e1500355. doi:10.1126/sciadv.1500355.

Queensland Government (2021). About northern hairy-nosed wombats. Updated Oct. 7, 2021. www.qld.gov.au.

Quigley, N. R., and Patel, P. C. (2022). Reexamining the gender gap in microlending funding decisions: the role of borrower culture. *Small Business Economics*, 59(4), 1661–1685. doi:10.1007/s11187-021-00593-3.

Quinn, J. M., and Wagner, R. K. (2015). Gender differences in reading impairment and in the identification of impaired readers: Results from a large-scale study of at-risk readers. *Journal of Learning Disabilities*, 48(4), 433–445. doi:10.1177/0022219413508323.

Rabin, R. C. (2011). Turncoat of placenta is watched for trouble. *The New York Times*, Oct. 18, 2011.

Raj, A., and Boehmer, U. (2013). Girl child marriage and its association with national rates of HIV, maternal health, and infant mortality across 97 countries. *Violence Against Women*, 19(4), 536–551. doi:10.1177/1077801213487747.

Rakusen, I., Devichand, M., Yildiz, G., and Tomchak, A. (2014). #BBCtrending: Who is the "Angel of Kobane"? BBC News, Nov. 3, 2014. www.bbc.com.

Ramachandran, S., Deshpande, O., Roseman, C. C., Rosenberg, N. A., Feldman, M. W., and Cavalli-Sforza, L. L. (2005). Support from the relationship of genetic and geographic distance in human populations for a serial founder effect originating in Africa. *Proceedings of the National Academy of Sciences*, 102(44), 15942–15947.

Ramsier, M. A., Cunningham, A. J., Finneran, J. J., and Dominy, N. J. (2012). Social drive and the evolution of primate hearing. *Philosophical Transactions of the Royal Society B: Biological Sciences*, 367(1597), 1860–1868. doi:10.1098/rstb.2011.0219.

Randall, L. (2015). *Dark Matter and the Dinosaurs*. New York: Ecco.

Rasgon, N., Bauer, M., Glenn, T., Elman, S., and Whybrow, P. C. (2003). Menstrual cycle related mood changes in women with bipolar disorder. *Bipolar Disorders*, 5(1), 48–52.

Rasmussen, D. T. (1990). Primate origins: Lessons from a neotropical marsupial. *American Journal of Primatology*, 22(4), 263–277. doi:10.1002/ajp.1350220406.

Ray, P. R., Shiers, S., Caruso, J. P., Tavares-Ferreira, D., Sankaranarayanan, I., Uhelski, M. L., et al. (2022). RNA profiling of human dorsal root ganglia reveals sex-differences in mechanisms promoting neuropathic pain. *Brain*, awac266. doi:10.1093/brain/awac266.

Raymond, E. G., and Grimes, D. A. (2012). The comparative safety of legal induced abortion and childbirth in the United States. *Obstetrics and Gynecology*, 119(2, Pt. 1), 215–219. doi:0.1097/AOG.0b013e31823fe923.

Rebuffé-Scrive, M. (1987). Regional adipose tissue metabolism in women during and after reproductive life and in men. *Recent Advances in Obesity Research*, 5, 82–91.

Rebuffé-Scrive, M., Enk, L., Crona, N., Lönnroth, P., Abrahamsson, L., Smith, U., and Björntorp, P. (1985). Fat cell metabolism in different regions in women. Effect of menstrual cycle, pregnancy, and lactation. *The Journal of Clinical Investigation*, 75(6), 1973–1976. doi:10.1172/JCI111914.

Rechlin, R. K., Splinter, T. F. L., Hodges, T. E., Albert, A. Y., and Galea, L. A. M. (2021). Harnessing the power of sex differences: What a difference ten years did not make. *bioRxiv*. doi:10.1101/2021.06.30.450396.

Reed, D. L., Light, J. E., Allen, J. M., and Kirchman, J. J. (2007). Pair of lice lost or parasites regained: The evolutionary history of anthropoid primate lice. *BMC Biology*, 5, 7. doi:10.1186/1741-7007-5-7.

Rehrer, C. W., Karimpour-Fard, A., Hernandez, T. L., Law, C. K., Stob, N. R., Hunter, L. E., and Eckel, R. H. (2012). Regional differences in subcutaneous adipose tissue gene expression. *Obesity*, 20(11), 2168–2173. doi:10.1038/oby.2012.117.

Reid, H. E., Pratt, D., Edge, D., and Wittkowski, A. (2022). Maternal suicide ideation and behaviour during pregnancy and the first postpartum year: A systematic review of psychological and psychosocial risk factors. *Frontiers in Psychiatry*, 13. doi:10.3389/fpsyt.2022.765118.

Reilly, D., Neumann, D. L., and Andrews, G. (2019). Gender differences in reading and writing achievement: Evidence from the National Assessment of Educational Progress (NAEP). *The American Psychologist*, 74(4), 445–458. doi:10.1037/amp0000356.

Reinert, A. E., and Simon, J. A. (2017). "Did you climax or are you just laughing at me?" Rare phenomena associated with orgasm. *Sexual Medicine Reviews*, 5(3), 275–281. doi:10.1016/j.sxmr.2017.03.004.

Reis, E. (2009). *Bodies in Doubt: An American History of Intersex*. Baltimore: Johns Hopkins University Press.

Reisman, T., and Goldstein, Z. (2018). Case report: Induced lactation in a transgender woman. *Transgender Health*, 3(1), 24–26. doi:10.1089/trgh.2017.0044.

Renaud, H. J., Cui, J. Y., Khan, M., and Klaassen, C. D. (2011). Tissue distribution and gender-divergent expression of 78 Cytochrome P450 mRNAs in mice. *Toxicological Sciences*, 124(2), 261–277. doi:10.1093/toxsci/kfr240.

Reno, P. L., McCollum, M. A., Meindl, R. S., and Lovejoy, C. O. (2010). An enlarged postcranial sample confirms *Australopithecus afarensis* dimorphism was similar to modern humans. *Philosophical Transactions of the Royal Society B: Biological Sciences*, 365(1556), 3355–3363.

Reno, P. L., McLean, C. Y., Hines, J. E., Capellini, T. D., Bejerano, G., and Kingsley, D. M. (2013). A penile spine/vibrissa enhancer sequence is missing in modern and extinct humans but is retained in multiple primates with penile spines and sensory vibrissae. *PLOS ONE*, 8(12), e84258. doi:10.1371/journal.pone.0084258.

Reno, P. L., Meindl, R. S., McCollum, M. A., and Lovejoy, C. O. (2003). Sexual dimorphism in *Australopithecus afarensis* was similar to that of modern humans. *Proceedings of the National Academy of Sciences*, 100(16), 9404–9409.

Reynolds, A. S., Lee, A. G., Renz, J., DeSantis, K., Liang, J., Powell, C. A., et al. (2020). Pulmonary vascular dilatation detected by automated transcranial Doppler in COVID-19 pneumonia. *American Journal of Respiratory and Critical Care Medicine*, 202(7), 1037–1039. doi:10.1164/rccm.202006-2219LE.

Rhone, A. E., Rupp, K., Hect, J. L., Harford, E. E., Tranel, D., Howard, M. A., III, and Abel, T. J. (2022). Electrocorticography reveals the dynamics of famous voice responses in human fusiform gyrus. *Journal of Neurophysiology*. doi:10.1152/jn.00459.2022.

Ribeiro, D. C., Brook, A. H., Hughes, T. E., Sampson, W. J., and Townsend, G. C. (2013). Intrauterine hormone effects on tooth dimensions. *Journal of Dental Research*, 92(5), 425–431. doi:10.1177/0022034513484934.

Rich, A. (1978). *The Dream of a Common Language*. New York: W. W. Norton.

Riches, G., and Dawson, P. (2000). *An Intimate Loneliness: Supporting Bereaved Parents and Siblings*. Maidenhead, U.K.: Open University Press.

Rigon, A., Turkstra, L., Mutlu, B., and Duff, M. (2016). The female advantage: Sex as a possible protective factor against emotion recognition impairment following traumatic brain injury. *Cognitive, Affective, & Behavioral Neuroscience*, 16(5), 866–875.

Rimbaud, A. (2011). *Illuminations*. Translated by John Ashbery. New York: W. W. Norton.

Riskin, A., Almog, M., Peri, R., Halasz, K., Srugo, I., and Kessel, A. (2012). Changes in immunomodulatory constituents of human milk in response to active infection in the nursing infant. *Pediatric Research*, 71(2), 220–225. doi:10.1038/pr.2011.34.

Roach, N. T., Hatala, K. G., Ostrofsky, K. R., Villmoare, B., Reeves, J. S., Du,

A., et al. (2016). Pleistocene footprints show intensive use of lake margin habitats by *Homo erectus* groups. *Scientific Reports*, 6, 26374.

Robbins, A. M., Gray, M., Basabose, A., Uwingeli, P., Mburanumwe, I., Kagoda, E., and Robbins, M. M. (2013). Impact of male infanticide on the social structure of mountain gorillas. *PLOS ONE*, 8(11), e78256. doi .org/10.1371/journal.pone.0078256.

Robert, M., and Chevrier, E. (2003). Does men's advantage in mental rotation persist when real three-dimensional objects are either felt or seen? *Memory and Cognition*, 31, 1136–1145. doi:10.3758/BF03196134.

Roberts, E. K., Lu, A., Bergman, T. J., and Beehner, J. C. (2012). A Bruce effect in wild geladas. *Science*, 335(6073), 1222–1225. doi:10.1126/science.1213600.

Roberts, S. A., Davidson, A. J., McLean, L., Beynon, R. J., and Hurst, J. L. (2012). Pheromonal induction of spatial learning in mice. *Science*, 338(6113), 1462–1465. doi:10.1126/science.1225638.

Robertson, D. S., McKenna, M. C., Toon, O. B., Hope, S., and Lillegraven, J. A. (2004). Survival in the first hours of the Cenozoic. *GSA Bulletin*, 116(5-6), 760–768. doi:10.1130/b25402.1.

Rocca, C. H., and Harper, C. C. (2012). Do racial and ethnic differences in contraceptive attitudes and knowledge explain disparities in method use? *Perspectives in Sexual and Reproductive Health*, 44(3), 150–158.

Rodriguez-Hart, C., Chitale, R. A., Rigg, R., Goldstein, B. Y., Kerndt, P. R., and Tavrow, P. (2012). Sexually transmitted infection testing of adult film performers: Is disease being missed? *Sexually Transmitted Diseases*, 987–992.

Rogers, F. B., Ricci, M., Caputo, M., Shackford, S., Sartorelli, K., Callas, P., Dewell, J., and Daye, S. (2001). The use of telemedicine for real-time video consultation between trauma center and community hospital in a rural setting improves early trauma care: Preliminary results. *The Journal of Trauma*, 51(6), 1037–1041. doi.org/10.1097/00005373-200112000-00002.

Rojahn, J., and Naglieri, J. A. (2006). Developmental gender differences on the Naglieri Nonverbal Ability Test in a nationally normed sample of 5–17 year olds. *Intelligence*, 34(3), 253–260. doi:10.1016/j.intell.2005.09.004.

Roney, J. R., and Simmons, Z. L. (2013). Hormonal predictors of sexual motivation in natural menstrual cycles. *Hormones and Behavior*, 63(4), 636–645. doi:10.1016/j.yhbeh.2013.02.013.

Ronto, P. (2021). The state of ultra running 2020. RunRepeat, Sept. 21, 2021. runrepeat.com.

Roof, K. A., Hopkins, W. D., Izard, M. K., Hook, M., and Schapiro, S. J. (2005). Maternal age, parity, and reproductive outcome in captive chimpanzees (*Pan troglodytes*). *American Journal of Primatology*, 67(2), 199–207. doi:10.1002/ajp.20177.

Roof, R. L., and Hall, E. D. (2000). Gender differences in acute CNS trauma and stroke: Neuroprotective effects of estrogen and progesterone. *Journal of Neurotrauma*, 17(5), 367–388. doi:10.1089/neu.2000.17.367.

Rooker, K., and Gavrilets, S. (2020). On the evolution of sexual receptivity in female primates. *Scientific Reports*, 10(1), 11945. doi:10.1038/s41598-020-68338-y.

Rose, L., and Marshall, F. (1996). Meat eating, Hominid sociality, and home bases revisited. *Current Anthropology*, 37(2), 307–338.

Rosenberg, K. R. (1992). The evolution of modern human childbirth. *American Journal of Physical Anthropology*, 35(S15), 89–124. doi:10.1002/ajpa.1330350605.

Rosowski, J. J., and Graybeal, A. (1991). What did Morganucodon hear? *Zoological Journal of the Linnean Society*, 101(2), 131–168. doi:10.1111/j.1096-3642.1991.tb00890.x.

Rothman, E. F., Exner, D., and Baughman, A. L. (2011). The prevalence of sexual assault against people who identify as gay, lesbian, or bisexual in the United States: A systematic review. *Trauma, Violence & Abuse*, 12(2), 55–66. doi:10.1177/1524838010390707.

Round, J. M., Jones, D. A., Honour, J. W., and Nevill, A. M. (1999). Hormonal factors in the development of differences in strength between boys and girls during adolescence: A longitudinal study. *Annals of Human Biology*, 26(1), 49–62.

Rowe, T. B., Macrini, T. E., and Luo, Z.-X. (2011). Fossil evidence on origin of the mammalian brain. *Science*, 332(6032), 955–957. doi:10.1126/science.1203117.

Rudder, C. (2014). *Dataclysm: Who We Are (When We Think No One's Looking)*. New York: Crown.

Ruff, C. B., Trinkaus, E., and Holliday, T. W. (1997). Body mass and encephalization in Pleistocene Homo. *Nature*, 387(6629), 173–176.

Ruhräh, J. (1925). *Pediatrics of the Past*. New York: Paul B. Hoeber.

Rūmī, J. M. (1270/1927). Spiritual couplets. In *Mathnawi of Jalalu'ddin Rumi, Edited from the Oldest Manuscripts Available with Critical Notes, Translation, and Commentary*, Volume III: *Containing the Text of the Third and Fourth Books*, translated and edited by R. A. Nicholson. Printed by Messrs. E. J. Brill, Leiden, for the Trustees of the "E. J. W. Gibb memorial" and published by Messrs. Luzac & Co., London, 1925–40.

Rutter, M., Caspi, A., Fergusson, D., Horwood, L. J., Goodman, R., Maughan, B., et al. (2004). Sex differences in developmental reading disability: New findings from four epidemiological studies. *Journal of the American Medical Association*, 291, 2007–2012.

Rutz, C., Hunt, G. R., and St. Clair, J. J. H. (2018). Corvid technologies: How do New Caledonian crows get their tool designs? *Current Biology*, 28(18), R1109–R1111. doi:10.1016/j.cub.2018.08.031.

Ryan, M., and Kenny, D. T. (2009). Perceived effects of the menstrual cycle on young female singers in the Western classical tradition. *Journal of Voice*, 23(1), 99–108. doi:10.1016/j.jvoice.2007.05.004.

Sammaknejad, N., Pouretemad, H., Eslahchi, C., Salahirad, A., and Alinejad, A. (2017). Gender classification based on eye movements: A processing effect during passive face viewing. *Advances in Cognitive Psychology*, 13(3), 232–240. doi:10.5709/acp-0223-1.

Sanger, T. J., Gredler, M. L., and Cohn, M. J. (2015). Resurrecting embryos of the tuatara, *Sphenodon punctatus*, to resolve vertebrate phallus evolution. *Biology Letters*, 11(10), 20150694. doi:10.1098/rsbl.2015.0694.

Sankey, M. D. (2018). *Women and War in the 21st Century: A Country-by-Country Guide*. Santa Barbara, Calif.: ABC-CLIO.

Sapolsky, R. M., and Share, L. J. (2004). A pacific culture among wild baboons: Its emergence and transmission. *PLOS Biology*, 2(4), e106. doi:10.1371/journal.pbio.0020106.

Saravelos, S. H., Cocksedge, K. A., and Li, T. C. (2008). Prevalence and diagnosis of congenital uterine anomalies in women with reproductive failure: A critical appraisal. *Human Reproduction Update*, 14, 415–429.

Sardella, R., and Werdelin, L. (2007). *Amphimachairodus* (Felidae, Mammalia) from Sahabi (latest Miocene-earliest Pliocene, Libya), with a review of African Miocene Machairodontinae. *Rivista Italiana di Paleontologia e Stratigrafia (Research in Paleontology and Stratigraphy)*, 113(1), 67–77.

Sarton, E., Olofsen, E., Romberg, R., den Hartigh, J., Kest, B., Nieuwenhuijs, D., et al. (2000). Sex differences in morphine analgesia: An experimental study in healthy volunteers. *Anesthesiology*, 93(5), 1245–6A. doi:10.1097/00000542-200011000-00018.

Sastre, F., De La Rosa, M., Ibanez, G. E., Whitt, E., Martin, S. S., and O'Connell, D. J. (2015). Condom use preferences among Latinos in Miami-Dade: Emerging themes concerning men's and women's culturally-ascribed attitudes and behaviours. *Culture, Health & Sexuality*, 17(6), 667–681. doi:10.1080s/13691058.2014.989266.

Sato, T., Matsumoto, T., Kawano, H., Watanabe, T., Uematsu, Y., Sekine, K., et al. (2004). Brain masculinization requires androgen receptor function. *Proceedings of the National Academy of Sciences*, 101(6), 1673–1678.

Savage-McGlynn, E. (2012). Sex differences in intelligence in younger and older participants of the Raven's Standard Progressive Matrices Plus. *Personality and Individual Differences*, 53(2), 137–141. doi:10.1016/j.paid.2011.06.013.

Savic, I., and Berglund, H. (2010). Androstenol—a steroid derived odor activates the hypothalamus in women. *PLOS ONE*, 5(2), e8651.

Savic, I., Berglund, H., and Lindström, P. (2005). Brain response to putative pheromones in homosexual men. *Proceedings of the National Academy of Sciences*, 102(20), 7356–7361.

Savolainen, V., and Hodgson, J. A. (2016). Evolution of homosexuality. In *Encyclopedia of Evolutionary Psychological Science*. Weekes-Shackelford, V., and Shackelford, T. (eds). Cham: Springer. doi:10.1007/978-3-319-16999-6_3403-1.

Saxton, M. (2009). The inevitability of Child Directed Speech. Pp. 62–86 in

Language Acquisition,edited by S. Foster-Cohen. London: Palgrave Macmillan UK.

Saxton, T. K., Lyndon, A., Little, A. C., and Roberts, S. C. (2008). Evidence that androstadienone, a putative human chemosignal, modulates women's attributions of men's attractiveness. *Hormones and Behavior*, 54(5), 597–601. doi:10.1016/j.yhbeh.2008.06.001.

Sayeed, I., and Stein, D. G. (2009). Progesterone as a neuroprotective factor in traumatic and ischemic brain injury. Pp. 219–237 in *Progress in Brain Research*, 175, edited by J. Verhaagen, E. M., Hol, I. Huitenga, J. Wijnholds, A. B. Bergen, G. J. Boer, and D. F. Swaab. Cambridge, Mass.: Elsevier.

Sayers, S. P., and Clarkson, P. M. (2001). Force recovery after eccentric exercise in males and females. *European Journal of Applied Physiology, 84*(1), 122–126. doi:10.1007/s004210000346.

Schaal, B., Doucet, S., Sagot, P., Hertling, E., and Soussignan, R. (2006). Human breast areolae as scent organs: Morphological data and possible involvement in maternal-neonatal coadaptation. *Developmental Psychobiology*, 48(2), 100–110.

Schantz-Dunn, J., and Nour, N. M. (2009). Malaria and pregnancy: A global health perspective. *Reviews in Obstetrics & Gynecology*, 2(3), 186–192.

Scheiber, C., Reynolds, M. R., Hajovsky, D. B., and Kaufman, A. S. (2015). Gender differences in achievement in a large, nationally representative sample of children and adolescents. *Psychology in the Schools*, 52, 335–348. doi:10.1002/pits.21827.

Schneider, B., van Trotsenburg, M., Hanke, G., Bigenzahn, W., and Huber, J. (2004). Voice impairment and menopause. *Menopause*, 11(2), 151–158.

Schneiderman, I., Zagoory-Sharon, O., Leckman, J. F., and Feldman, R. (2012). Oxytocin during the initial stages of romantic attachment: Relations to couples' interactive reciprocity. *Psychoneuroendocrinology*, 37(8), 1277–1285. doi:10.1016/j.psyneuen.2011.12.021.

Schouten, L. (2016). First woman enters infantry as army moves women into combat roles. *Christian Science Monitor*, April 28, 2016. www.csmonitor.com.

Schreiweis, C., Bornschein, U., Burguière, E., Kerimoglu, C., Schreiter, S., Dannemann, M., et al. (2014). Humanized Foxp2 accelerates learning by enhancing transitions from declarative to procedural performance. *Proceedings of the National Academy of Sciences*, 111(39), 14253–14258. doi:10.1073/pnas.1414542111.

Schubart, J. R., Eliassen, A. H., Schilling, A., and Goldenberg, D. (2021). Reproductive factors and risk of thyroid cancer in women: An analysis in the Nurses' Health Study II. *Women's Health Issues*, 31(5), 494–502. doi:10.1016/j.whi.2021.03.008.

Schulte, P., Alegret, L., Arenillas, I., Arz, J. A., Barton, P. J., Bown, P. R., et al. (2010). The Chicxulub asteroid impact and mass extinction at the Cretaceous-Paleogene boundary. *Science*, 327(5970), 1214–1218.

Schultz, A. H. (1938). The relative weights of the testes in primates. *Anatomical Record*, 72, 387–394.

Schultz, K. (2015). The moral judgments of Henry David Thoreau. *The New Yorker*, Oct. 19, 2015. www.newyorker.com.

Scott, C. (2006). *Translating Rimbaud's "Illuminations."* Exeter: University of Exeter Press.

Scott, E. M., Mann, J., Watson-Capps, J. J., Sargeant, B. L., and Connor, R. C. (2005). Aggression in bottlenose dolphins: Evidence for sexual coercion, male-male competition, and female tolerance through analysis of tooth-rake marks and behaviour. *Behaviour*, 142(1), 21–44. doi:10.1163/1568539053627712.

Scott, G. R., and Gibert, L. (2009). The oldest hand-axes in Europe. *Nature*, 461(7260), 82–85.

Scott, I., Bentley, G. R., Tovee, M. J., Ahamed, F. U., Magid, K., and Sharmeen, T. (2007). An evolutionary perspective on male preferences for female body shape. In *Body Beautiful: Evolutionary and Socio-cultural Perspectives*, edited by Swami, V., and Furnham, A. New York: Palgrave Macmillan.

Scott, R., director (2012). *Prometheus*. 20th Century Fox Home Entertainment.

Scutt, D., Lancaster, G. A., and Manning, J. T. (2006). Breast asymmetry and predisposition to breast cancer. *Breast Cancer Research*, 8(2), R14. doi:10.1186/bcr1388.

Sear, R., Mace, R., and McGregor, I. A. (2000). Maternal grandmothers improve nutritional status and survival of children in rural Gambia. *Proceedings of the Royal Society B: Biological Sciences*, 267(1453), 1641–1647. doi:10.1098/rspb.2000.1190.

Sear, R., Steele, F., McGregor, I. A., and Mace, R. (2002). The effects of kin on child mortality in rural Gambia. *Demography*, 39(1), 43–63. doi:10.1353/dem.2002.0010.

Seedat, S., Scott, K. M., Angermeyer, M. C., Berglund, P., Bromet, E. J., Brugha, T. S., et al. (2009). Cross-national associations between gender and mental disorders in the World Health Organization World Mental Health Surveys. *Archives of General Psychiatry*, 66(7), 785–795. doi:10.1001/archgenpsychiatry.2009.36.

SEER (2021). SEER*Explorer. Surveillance, Epidemiology, and End Results Program, National Cancer Institute. Accessed Sept. 27, 2021. seer.cancer.gov.

Segal, N. L. (2000). Virtual twins: New findings on within-family environmental influences on intelligence. *Journal of Educational Psychology*, 92(3), 442–448. doi:10.1037/0022-0663.92.3.442.

Segers, A., and Depoortere, I. (2021). Circadian clocks in the digestive system. *Nature Reviews Gastroenterology & Hepatology*, 18(4), 239–251. doi:10.1038/s41575-020-00401-5.

Sellers, R. M., Caldwell, C. H., Schmeelk-Cone, K. H., and Zimmerman, M. A. (2003). Racial identity, racial discrimination, perceived stress, and psychological distress among African American young adults. *Journal of Health and Social Behavior*, 44(3), 302–317.

Selvaggio, M. M., and Wilder, J. (2001). Identifying the involvement of multiple carnivore taxa with archaeological bone assemblages. *Journal of Archaeological Science*, 28(5), 465–470.

Semaw, S., Renne, P., Harris, J. W. K., Feibel, C. S., Bernor, R. L., Fesseha, N., and Mowbray, K. (1997). 2.5-million-year-old stone tools from Gona, Ethiopia. *Nature*, 385(6614), 333–336.

Semmler, J. G., Kutzscher, D. V., and Enoka, R. M. (1999). Gender differences in the fatigability of human skeletal muscle. *Journal of Neurophysiology*, 82(6), 3590–3593.

Senut, B., Pickford, M., and Ségalen, L. (2009). Neogene desertification of Africa. *Comptes Rendus Geoscience*, 341(8), 591–602.

Sepulchre, P., Ramstein, G., Fluteau, F., Schuster, M., Tiercelin, J.-J., and Brunet, M. (2006). Tectonic uplift and eastern Africa aridification. *Science*, 313(5792), 1419–1423.

Serdarevic, M., Striley, C. W., and Cottler, L. B. (2017). Sex differences in prescription opioid use. *Current Opinion in Psychiatry*, 30(4), 238–246. doi:10.1097/YCO.0000000000000337.

Sereno, P. C., Martinez, R. N., Wilson, J. A., Varricchio, D. J., Alcober, O. A., and Larsson, H. C. E. (2008). Evidence for avian intrathoracic air sacs in a new predatory dinosaur from Argentina. *PLOS ONE*, 3(9), e3303.

Seretis, K., Goulis, D. G., Koliakos, G., and Demiri, E. (2015). Short- and long-term effects of abdominal lipectomy on weight and fat mass in females: A systematic review. *Obesity Surgery*, 25(10), 1950–1958. doi:10.1007/s11695-015-1797-1.

Sergeant, M. J., Dickins, T. E., Davies, M. N., and Griffiths, M. D. (2007). Women's hedonic ratings of body odor of heterosexual and homosexual men. *Archives of Sexual Behavior*, 36(3), 395–401. doi:10.1007/s10508-006-9126-3.

Setchell, J. M., and Dixson, A. F. (2001). Changes in the secondary sexual adornments of male mandrills (*Mandrillus sphinx*) are associated with gain and loss of alpha status. *Hormones and Behavior*, 39(3), 177–184. doi:10.1006/hbeh.2000.1628.

Sevelius, J. (2009). There's no pamphlet for the kind of sex I have: HIV-related risk factors and protective behaviors among transgender men who have sex with nontransgender men. *Journal of the Association of Nurses in AIDS Care*, 20, 398–410. doi:10.1016/j.jana.2009.06.001.

Shaffer, M. L. (1981). Minimum population sizes for species conservation. *BioScience*, 31(2), 131–134. doi:10.2307/1308256.

Shanley, D. P., Sear, R., Mace, R., and Kirkwood, T. B. (2007). Testing evolutionary theories of menopause. *Proceedings of the Royal Society B: Biological Sciences*, 274(1628), 2943–2949. doi:10.1098/rspb.2007.1028.

Shao, Y., Forster, S. C., Tsaliki, E., Vervier, K., Strang, A., Simpson, N., et al.

(2019). Stunted microbiota and opportunistic pathogen colonization in caesarean-section birth. *Nature*, 574(7776), 117–121. doi:10.1038/s41586-019-1560-1.

Shapland, F., Lewis, M., and Watts, R. (2015). The lives and deaths of young medieval women: The osteological evidence. *Medieval Archaeology*, 59(1), 272–289. doi:10.1080/00766097.2015.1119392.

Sharpe, K., and Van Gelder, L. (2006). Evidence for cave marking by Palaeolithic children. *Antiquity*, 80(310), 937–947.

Shaw, L. J., Shaw, R. E., Merz, C. N., Brindis, R. G., Klein, L. W., Nallamothu, B., et al. (2008). Impact of ethnicity and gender differences on angiographic coronary artery disease prevalence and in-hospital mortality in the American College of Cardiology–National Cardiovascular Data Registry. *Circulation*, 117(14), 1787–1801. doi:10.1161/CIRCULATIONAHA.107.726562.

Shea, J. J. (2010). Stone Age visiting cards revisited: A strategic perspective on the lithic technology of early hominin dispersal. Pp. 47–64 in *Out of Africa I: The First Hominin Colonization of Eurasia*, edited by J. G. Fleagle, J. J. Shea, F. E. Grine, A. L. Baden, and R. E. Leakey. Dordrecht: Springer.

Shehab, A., Al-Dabbagh, B., AlHabib, K. F., Alsheikh-Ali, A. A., Almahmeed, W., Sulaiman, K., et al. (2013). Gender disparities in the presentation, management and outcomes of Acute Coronary Syndrome patients: Data from the 2nd Gulf Registry of Acute Coronary Events (Gulf RACE-2). *PLOS ONE*, 8(2), e55508. doi:10.1371/journal.pone.0055508.

Shen, G., Gao, X., Gao, B., and Granger, D. E. (2009). Age of Zhoukoudian *Homo erectus* determined with 26Al/10Be burial dating. *Nature*, 458(7235), 198–200.

Shipman, P., Bosler, W., and Davis, K. L. (1981). Butchering of giant geladas at an Acheulian site. *Current Anthropology*, 22(3), 257–268.

Shirazi, T., Renfro, K. J., Lloyd, E., and Wallen, K. (2018). Women's experience of orgasm during intercourse: Question semantics affect women's reports and men's estimates of orgasm occurrence. *Archives of Sexual Behavior*, 47(3), 605–613. doi:10.1007/s10508-017-1102-6.

Shubin, N. (2013). *The Universe Within: The Deep History of the Human Body*. New York: Vintage Books.

Shultz, S., Nelson, E., and Dunbar, R. I. M. (2012). Hominin cognitive evolution: Identifying patterns and processes in the fossil and archaeological record. *Philosophical Transactions of the Royal Society B: Biological Sciences*, 367(1599), 2130–2140.

Shute, B., and Wheldall, K. (1999). Fundamental frequency and temporal modifications in the speech of British fathers to their children. *Educational Psychology*, 19(2), 221–233. doi:10.1080/0144341990190208.

Shye, D., Mullooly, J. P., Freeborn, D. K., and Pope, C. R. (1995). Gender differences in the relationship between social network support and mortality: A longitudinal study of an elderly cohort. *Social Science & Medicine*, 41(7), 935–947. doi:10.1016/0277-9536(94)00404-H.

Siegel, R. L., Miller, K. D., Fuchs, H. E., and Jemal, A. (2022). Cancer statistics, 2022. *CA: A Cancer Journal for Clinicians*, 72(1), 7–33. doi:10.3322/caac.21708.

Silk, J. B., Alberts, S. C., and Altmann, J. (2004). Patterns of coalition formation by adult female baboons in Amboseli, Kenya. *Animal Behaviour*, 67(3), 573–582. doi:10.1016/j.anbehav.2003.07.001.

Silverman, C. (2015). Lies, damn lies, and viral content. Tow/Knight Report. Tow Center for Digital Journalism, Columbia University. doi:10.7916/D8Q81RHH.

Simoni-Wastila, L. J. (2000). The use of abusable prescription drugs: The role of gender. *Women's Health and Gender-Based Medicine*, 9(3), 289–297.

Singh, D., Dixson, B. J., Jessop, T. S., Morgan, B., and Dixson, A. F. (2010). Cross-cultural consensus for waist-hip ratio and women's attractiveness. *Evolution and Human Behavior*, 31(3), 176–181. doi:10.1016/j.evolhumbehav.2009.09.001.

Siraj, A., and Loeb, A. (2021). Breakup of a long-period comet as the origin of the dinosaur extinction. *Scientific Reports*, 11(1), 3803. doi:10.1038/s41598-021-82320-2.

Skjærvø, G. R., and Røskaft, E. (2013). Menopause: No support for an evolutionary explanation among historical Norwegians. *Experimental Gerontology*, 48(4), 408–413. doi:10.1016/j.exger.2013.02.001.

Skolnick, B. E., Maas, A. I., Narayan, R. K., Van Der Hoop, R. G., MacAllister, T., Ward, J. D., et al. (2014). A clinical trial of progesterone for severe traumatic brain injury. *New England Journal of Medicine*, 371(26), 2467–2476.

Slater, G. J. (2013). Phylogenetic evidence for a shift in the mode of mammalian body size evolution at the Cretaceous-Palaeogene boundary. *Methods in Ecology and Evolution*, 4(8), 734–744. doi:10.1111/2041-210X.12084.

Slon, V., Mafessoni, F., Vernot, B., de Filippo, C., Grote, S., Viola, B., et al. (2018). The genome of the offspring of a Neanderthal mother and a Denisovan father. *Nature*, 561(7721), 113–116. doi:10.1038/s41586-018-0455-x.

Slonecker, E. M., Simpson, E. A., Suomi, S. J., and Paukner, A. (2018). Who's my little monkey? Effects of infant-directed speech on visual retention in infant rhesus macaques. *Developmental Science*, 21, e12519. doi:10.1111/desc.12519.

Smaers, J. B., Gómez-Robles, A., Parks, A. N., and Sherwood, C. C. (2017). Exceptional evolutionary expansion of prefrontal cortex in great apes and humans. *Current Biology*, 27(5), 714–720. doi:10.1016/j.cub.2017.01.020.

Smaers, J. B., Steele, J., Case, C. R., Cowper, A., Amunts, K., and Zilles, K. (2011). Primate prefrontal cortex evolution: Human brains are the extreme of a lateralized ape trend. *Brain, Behavior and Evolution*, 77(2), 67–78. doi:10.1159/000323671.

Smith, C. (2014). Estimation of a genetically viable population for multigenerational interstellar voyaging: Review and data for project Hyperion. *Acta Astronautica*, 97, 16–29. doi:10.1016/j.actaastro.2013.12.013.

Smith, E. R., Oakley, E., Grandner, G. W., Ferguson, K., Farooq, F., Afshar, Y., et al. (2023). Adverse maternal, fetal, and newborn outcomes among pregnant women with SARS-CoV-2 infection: An individual participant data meta-analysis. *BMJ Global Health*, 8(1), e009495. doi:10.1136/bmjgh-2022-009495.

Smith, T., Laitman, J., and Bhatnagar, K. (2014). The shrinking anthropoid nose, the human vomeronasal organ, and the language of anatomical reduction. *The Anatomical Record*, 297. doi:10.1002/ar.23035.

Smith, T. M., Tafforeau, P., Reid, D. J., Pouech, J., Lazzari, V., Zermeno, J. P., et al. (2010). Dental evidence for ontogenetic differences between modern humans and Neanderthals. *Proceedings of the National Academy of Sciences*, 107(49), 20923–20928. doi:10.1073/pnas.1010906107.

Smuts, B. B. (1985). *Sex and Friendship in Baboons*. Hawthorne, N.Y.: Aldine Publishing Co.

Smuts, B. B., and Smuts, R. W. (1993). Male aggression and sexual coercion of females in nonhuman primates and other mammals: Evidence and theoretical implications. *Advances in the Study of Behavior*, 22(22), 1–63.

Snow, S. S., Alonzo, S. H., Servedio, M. R., and Prum, R. O. (2019). Female resistance to sexual coercion can evolve to preserve the indirect benefits of mate choice. *Journal of Evolutionary Biology*, 32(6), 545–558. doi:10.1111/jeb.13436.

Snyder-Mackler, N., Alberts, S. C., and Bergman, T. J. (2012). Concessions of an alpha male? Cooperative defence and shared reproduction in multi-male primate groups. *Proceedings of the Royal Society B: Biological Sciences*, 279(1743), 3788–3795. doi:10.1098/rspb.2012.0842.

Soares, C. N., and Zitek, B. (2008). Reproductive hormone sensitivity and risk for depression across the female life cycle: A continuum of vulnerability? *Journal of Psychiatry & Neuroscience*, 33(4), 331–343.

Sohrabji, F. (2007). Guarding the blood-brain barrier: A role for estrogen in the etiology of neurodegenerative disease. *Gene Expression*, 13(6), 311–319. doi:10.3727/000000006781510723.

Solnit, R. (2013). Mysteries of Thoreau: Unsolved. *Orion*, May/June 2013.

Sorge, R. E., Mapplebeck, J., Rosen, S., Beggs, S., Taves, S., Alexander, J. K., et al. (2015). Different immune cells mediate mechanical pain hypersensitivity in male and female mice. *Nature Neuroscience*, 18(8), 1081–1083.

Sorokowski, P., Karwowski, M., Misiak, M., Marczak, M. K., Dziekan, M., Hummel, T., and Sorokowska, A. (2019). Sex differences in human olfaction: A meta-analysis. *Frontiers in Psychology*, 10, 242.

Spelke, E. S. (2005). Sex differences in intrinsic aptitude for mathematics and science? A critical review. *American Psychologist*, 60, 950–958. doi:10.1037/0003-066X.60.9.950.

Spencer, J. (2016). The challenges of Ranger School and how to overcome them. Modern War Institute at West Point, April 12, 2016. mwi.usma.edu.

Spoor, F., Leakey, M. G., Gathogo, P. N., Brown, F. H., Anton, S. C., McDougall, I., et al. (2007). Implications of new early *Homo* fossils from Ileret, east of Lake Turkana, Kenya. *Nature*, 448(7154), 688–691.

Sramek, J. J., Murphy, M. F., and Cutler, N. R. (2016). Sex differences in the psychopharmacological treatment of depression. *Dialogues in Clinical Neuroscience*, 18(4), 447–457. doi:10.31887/DCNS.2016.18.4/ncutler.

St. John, J., Sakkas, D., Dimitriadi, K., Barnes, A., Maclin, V., Ramey, J., et al. (2000). Failure of elimination of paternal mitochondrial DNA in abnormal embryos. *The Lancet*, 355(9199), 200. doi:10.1016/S0140-6736(99)03842-8.

St.-Onge, M. P. (2010). Are normal-weight Americans over-fat? *Obesity*, 18(11), 2067–2068. doi:10.1038/oby.2010.103.

Stanić, B. (2023). Gender (dis)balance in local government: How does it affect budget transparency? *Economic Research-Ekonomska Istraživanja*, 36(1), 997–1014. doi:10.1080/1331677X.2022.2081232.

Stansfield, E., Fischer, B., Grunstra, N., Pouca, M. V., and Mitteroecker, P. (2021). The evolution of pelvic canal shape and rotational birth in humans. *BMC Biology*, 19(1), 224. doi:10.1186/s12915-021-01150-w.

Steele, J. (1999). Palaeoanthropology: Stone legacy of skilled hands. *Nature*, 399(6731), 24–25.

Steele, J., Clegg, M., and Martelli, S. (2013). Comparative morphology of the hominin and African ape hyoid bone, a possible marker of the evolution of speech. *Human Biology*, 85(5), 639–672.

Steele, T. E. (2010). A unique hominin menu dated to 1.95 million years ago. *Proceedings of the National Academy of Sciences*, 107(24), 10771–10772. doi:10.1073/pnas.1005992107.

Steen, S. J., and Schwartz, P. (1995). *Communication, Gender, and Power: Homosexual Couples as a Case Study*. Pp. 310–343 in *Explaining Family Interactions*, edited by M. A. Fitzpatrick and A. L. Vangelisti. London: SAGE Publications. doi:10.4135/9781483326368.

Steiper, M. E., and Young, N. M. (2006). Primate molecular divergence dates. *Molecular Phylogenetics and Evolution*, 41(2), 384–394.

Stevens, E. E., Patrick, T. E., and Pickler, R. (2009). A history of infant feeding. *The Journal of Perinatal Education*, 18, 32–39. doi:10.1624/105812409x426314.

Stewart, J. R. (1997). Morphology and evolution of the egg of oviparous amniotes. Pp. 291–326 in *Amniote Origins*, edited by S. S. Sumida and K. L. M. Martin. San Diego: Academic Press. doi:10.1016/B978-012676460-4/50010-X.

Stinson, L. F., Boyce, M. C., Payne, M. S., and Keelan, J. A. (2019). The not-so-sterile womb: Evidence that the human fetus is exposed to bacteria prior to birth. *Frontiers in Microbiology*, 10, 1124. doi:10.3389/fmicb.2019.01124.

Stockman, J. K., Hayashi, H., and Campbell, J. C. (2015). Intimate partner violence and its health impact on ethnic minority women [corrected]. *Journal of Women's Health*, 24(1), 62–79. doi:10.1089/jwh.2014.4879.

Stokol-Walker, C. (2022). Twitter's potential collapse could wipe out vast records of recent human history. *MIT Technology Review*, Nov. 11, 2022. www.technologyreview.com.

Storlazzi, C. D., Gingerich, S. B., van Dongeren, A., Cheriton, O. M., Swarzenski, P. W., Quataert, E., et al. (2018). Most atolls will be uninhabitable by the mid-21st century because of sea-level rise exacerbating wave-driven flooding. *Science Advances*, 4(4), eaap9741. doi:10.1126/sciadv.aap9741.

Strassmann, B. I. (1996). The evolution of endometrial cycles and menstruation. *The Quarterly Review of Biology*, 71(2), 181–220. doi:10.1086/419369.

Strassmann, B. I. (1997). The biology of menstruation in *Homo Sapiens*: Total lifetime menses, fecundity, and nonsynchrony in a natural-fertility population. *Current Anthropology*, 38(1), 123–129. doi:10.1086/204592.

Strenze, T. (2007). Intelligence and socioeconomic success: A meta-analytic review of longitudinal research. *Intelligence*, 35(5), 401–426.

Suarez, S. S., and Pacey, A. A. (2006). Sperm transport in the female reproductive tract. *Human Reproduction Update*, 12(1), 23–37.

Subramanian, S. (2020). *A Dominant Character: The Radical Science and Restless Politics of J. B. S. Haldane*. New York: W. W. Norton.

Suntsova, M. V., and Buzdin, A. A. (2020). Differences between human and chimpanzee genomes and their implications in gene expression, protein functions and biochemical properties of the two species. *BMC Genomics*, 21(7), 535. doi:10.1186/s12864-020-06962-8.

Surovell, T., Waguespack, N., and Brantingham, P. J. (2005). Global archaeological evidence for proboscidean overkill. *Proceedings of the National Academy of Sciences*, 102(17), 6231–6236.

Susman, R. L. (1994). Fossil evidence for early hominid tool use. *Science*, 265(5178), 1570–1573.

Susman, R. L. (2008). Brief communication: Evidence bearing on the status of *Homo habilis* at Olduvai Gorge. *American Journal of Physical Anthropology*, 137(3), 356–361.

Sussman, R. W. (1991). Primate origins and the evolution of angiosperms. *American Journal of Primatology*, 23(4), 209–223. doi:10.1002/ajp.1350230402.

Sussman, R. W., Rasmussen, D. T., and Raven, P. H. (2013). Rethinking primate origins again. *American Journal of Primatology*, 75(2), 95–106. doi:10.1002/ajp.22096.

Suwa, G., Kono, R. T., Simpson, S. W., Asfaw, B., Lovejoy, C. O., and White, T. D. (2009). Paleobiological implications of the *Ardipithecus ramidus* dentition. *Science*, 326(5949), 94–99.

Suwa, G., Sasaki, T., Semaw, S., Rogers, M. J., Simpson, S. W., Kunimatsu, Y., et al. (2021). Canine sexual dimorphism in *Ardipithecus ramidus* was nearly human-like. *Proceedings of the National Academy of Sciences*, 118(49), e2116630118. doi:10.1073/pnas.2116630118.

Swanson, K. W. (2016). Rethinking body property. *Florida State University Law Review*, 44, 193–259.

Swers, M. L. (2005). Connecting descriptive and substantive representation: An analysis of sex differences in cosponsorship activity. *Legislative Studies Quarterly*, 30(3), 407–433.

Tagliaferri, C., Wittrant, Y., Davicco, M. J., Walrand, S., and Coxam, V. (2015). Muscle and bone, two interconnected tissues. *Ageing Research Reviews*, 21, 55–70.

Takahashi, M., Singh, R. S., and Stone, J. (2017). A theory for the origin of human menopause. *Journal of Frontiers in Genetics*, Jan. 6, 2017. doi:10.3389/fgene.2016.00222.

Takahashi, T., Ellingson, M. K., Wong, P., Israelow, B., Lucas, C., Klein, J., et al. (2020). Sex differences in immune responses that underlie COVID-19 disease outcomes. *Nature*, 588(7837), 315–320. doi:10.1038/s41586-020-2700-3.

Tall, A. R., and Yvan-Charvet, L. (2015). Cholesterol, inflammation and innate immunity. *Nature Reviews. Immunology*, 15(2), 104–116. doi:10.1038/nri3793.

Tan, M. (2015). Ranger School: Many do-overs rare, not unprecedented. *Army Times*, Sept. 18, 2015. www.armytimes.com.

Tan, M. (2016). Meet the army's first female infantry officer. *Army Times*, April 27, 2016. www.armytimes.com.

Tang, G., Gudsnuk, K., Kuo, S.-H., Cotrina, M. L., Rosoklija, G., Sosunov, A., et al. (2014). Loss of mTOR-dependent macroautophagy causes autistic-like synaptic pruning deficits. *Neuron*, 83(5), 1131–1143. doi:10.1016/.

Tannen, D. (1990). *You Just Don't Understand: Women and Men in Conversation.* New York: William Morrow.

Tao, N., Wu, S., Kim, J., An, H., Hinde, K., Power, M., et al. (2011). Evolutionary glycomics: Characterization of milk oligosaccharides in primates. *Journal of Proteome Research*, 10, 1548–1557. doi:10.1021/pr1009367.

Tarampi, M. R., Heydari, N., and Hegarty, M. (2016). A tale of two types of perspective taking: Sex differences in spatial ability. *Psychological Science*, 27(11), 1507–1516. doi:10.1177/0956797616667459.

Terlizzi, E. P., and Norris, T. (2021). Mental health treatment among adults: United States, 2020. NCHS Data Brief, no. 419. Hyattsville, Md.: National Center for Health Statistics. doi:10.15620/cdc:110593.

The Chimpanzee Sequencing and Analysis Consortium (2005). Initial sequence of the chimpanzee genome and comparison with the human genome. *Nature*, 437(7055), 69–87. doi:10.1038/nature04072.

Thiessen, E. D., Hill, E. A., and Saffran, J. R. (2005). Infant-directed speech facilitates word segmentation. *Infancy*, 1(1), 53–71.

Thompson, M. E., Jones, J. H., Pusey, A. E., Brewer-Marsden, S., Goodall, J., Marsden, D., et al. (2007). Aging and fertility patterns in wild chimpanzees provide insights into the evolution of menopause. *Current Biology*, 17(24), 2150–2156. doi:10.1016/j.cub.2007.11.033.

Thurber, C., Dugas, L. R., Ocobock, C., Carlson, B., Speakman, J. R., and

Pontzer, H. (2019). Extreme events reveal an alimentary limit on sustained maximal human energy expenditure. *Science Advances*, 5(6), eaaw0341. doi:10.1126/sciadv.aaw0341.

Tian, X., Iriarte-Díaz, J., Middleton, K., Galvao, R., Israeli, E., Roemer, A., et al. (2006). Direct measurements of the kinematics and dynamics of bat flight. *Bioinspiration and Biomimetics*, 1, 10–18.

Tobias, P. V. (1965). *Australopithecus*, *Homo habilis*, tool-using and tool-making. *The South African Archaeological Bulletin*, 20(80), 167–192.

Tokuyama, N., and Furuichi, T. (2016). Do friends help each other? Patterns of female coalition formation in wild bonobos at Wamba. *Animal Behaviour*, 119, 27–35. doi:10.1016/j.anbehav.2016.06.021.

Tomaszycki, M., Cline, C., Griffin, B., Maestripieri, D., and Hopkins, W. D. (1998). Maternal cradling and infant nipple preferences in rhesus monkeys (*Macaca mulatta*). *Developmental Psychobiology*, 32, 305–312.

Tomita, T., Murakumo, K., Ueda, K., Ashida, H., and Furuyama, R. (2019). Locomotion is not a privilege after birth: Ultrasound images of viviparous shark embryos swimming from one uterus to the other. *Ethology*, 125, 122–126. doi:10.1111/eth.12828.

Tomori, C., Palmquist, A. E. L., and Sally, D. (2016). Contested moral landscapes: Negotiating breastfeeding stigma in breastmilk sharing, nighttime breastfeeding, and long-term breastfeeding in the US and the UK. *Social Science & Medicine*, 168, 178–185. doi:10.1016/j.socscimed.2016.09.014.Co.

Toth, N. (1985). The Oldowan reassessed: A close look at early stone artifacts. *Journal of Archaeological Science*, 12(2), 101–120.

Toups, M. A., Kitchen, A., Light, J. E., and Reed, D. L. (2011). Origin of clothing lice indicates early clothing use by anatomically modern humans in Africa. *Molecular Biology and Evolution*, 28(1), 29–32. doi:10.1093/molbev/msq234.

Townsend, S. W., Slocombe, K. E., Emery Thompson, M., and Zuberbühler, K. (2007). Female-led infanticide in wild chimpanzees. *Current Biology*, 17(10), R355–R356. doi:10.1016/j.cub.2007.03.020.

Trevathan, W. (2015). Primate pelvic anatomy and implications for birth. *Philosophical Transactions of the Royal Society B: Biological Sciences*, 370(1663), 20140065. doi:10.1098/rstb.2014.0065.

Trevathan, W. R. (1996). The evolution of bipedalism and assisted birth. *Medical Anthropology Quarterly*, 10(2), 287–290. doi:10.1525/maq.1996.10.2.02a00100.

Trinkaus, E. (2011). Late Pleistocene adult mortality patterns and modern human establishment. *Proceedings of the National Academy of Sciences*, 108(4), 1267–1271.

Trivers, R. L. (1972). Parental investment and sexual selection. Pp. 136–179 in *Sexual Selection and the Descent of Man*, edited by B. Campbell. London: Routledge. doi:10.4324/9781315129266-7.

Trotier, D., Eloit, C., Wassef, M., Talmain, G., Bensimon, J. L., Døving, K. B., and Ferrand, J. (2000). The vomeronasal cavity in adult humans. *Chemical Senses*, 25(4), 369–380. doi:10.1093/chemse/25.4.369.

Tschopp, P., Sherratt, E., Sanger, T. J., Groner, A. C., Aspiras, A. C., Hu, J. K., et al. (2014). A relative shift in cloacal location repositions external genitalia in amniote evolution. *Nature*, 516(7531), 391–394. doi:10.1038/nature13819.

Turkstra, L. S., Mutlu, B., Ryan, C. W., Despins Stafslien, E. H., Richmond, E. K., Hosokawa, E., and Duff, M. C. (2020). Sex and gender differences in emotion recognition and theory of mind after TBI: A narrative review and directions for future research. *Frontiers in Neurology*, 11, 59.

Ulcova-Gallova, Z. (2010). Immunological and physicochemical properties of cervical ovulatory mucus. *Journal of Reproductive Immunology*, 86(2), 115–121.

Underwood, M. A. (2013). Human milk for the premature infant. *Pediatric Clinics of North America*, 60(1), 189–207. doi:10.1016/j.pcl.2012.09.008.

Unemori, E. N., Lewis, M., Constant, J., Arnold, G., Grove, B. H., Normand, J., et al. (2000). Relaxin induces vascular endothelial growth factor expression and angiogenesis selectively at wound sites. *Wound Repair and Regeneration*, 8(5), 361–370. doi:10.1111/j.1524-475x.2000.00361.x.

UNESCO (1953). *Progress in Literacy in Various Countries: A Preliminary Study of Available Census Data Since 1900*. Paris: Firmin-Didot.

UNESCO (1957). *World Illiteracy at Mid-Century: A Statistical Study*. Paris: Buchdruckerei Winterthur AG.

UNESCO (2014). Adult and Youth Literacy. National Regional and Global Trends 1985–2015. unesdoc.unesco.org.

Ungar, P. S. (2012). Dental evidence for the reconstruction of diet in African early *Homo*. *Current Anthropology*, 53(S6), S318–S329.

Ungar, P. S., and Sponheimer, M. (2011). The diets of early hominins. *Science*, 334(6053), 190–193. doi:10.1126/science.1207701.

Ungar, P. S., Grine, F. E., Teaford, M. F., and El Zaatari, S. (2006). Dental microwear and diets of African early *Homo*. *Journal of Human Evolution*, 50(1), 78–95.

Ungar, P. S., Krueger, K. L., Blumenschine, R. J., Njau, J., and Scott, R. S. (2012). Dental microwear texture analysis of hominins recovered by the Olduvai Landscape Paleoanthropology Project, 1995–2007. *Journal of Human Evolution*, 63(2), 429–437. doi:10.1016/j.jhevol.2011.04.006.

UNICEF (2022). Child marriage. data.unicef.org/topic/child-protection/child-marriage/.

United Nations (2002). *Arab Human Development Report 2002*. United Nations Development Programme, Arab Fund for Economic and Social Development. www.miftah.org.

United Nations (2015). *World Population Prospects: The 2015 Revision, Key Findings and Advance Tables*. Department of Economic and Social Affairs, Population Division.

United Nations (2015). *The World's Women: Trends and Statistics*. United Nations, Department of Economic and Social Affairs, Statistics Division.

United States Department of Justice, Office of Justice Programs, Bureau of Justice Statistics (BJS) (2017). National Crime Victimization Survey, 2010–2016.

United States Food and Drug Administration (U.S. FDA) (1992). Letter of Approval for Ambien (zolpidem tartrate tablets), NDA 19-908. Letter addressed to Lorex Pharmaceuticals, Attn: Keith Rotenberg, PhD, dated April 21, 1992. Letter includes notes from review and suggested labeling. Included in "Approval Letter(s) and Printed Labeling" document in U.S. FDA public files. www.accessdata.fda.gov.

United States Food and Drug Administration (U.S. FDA) (2013). Risk of next-morning impairment after use of insomnia drugs; FDA requires lower recommended doses for certain drugs containing zolpidem (Ambien, Ambien CR, Edluar, and Zolpimist). *Drug Safety Communications*, Jan. 10, 2013. www.fda.gov.

Urashima, T., Asakuma, S., Leo, F., Fukuda, K., Messer, M., and Oftedal, O. T. (2012). The predominance of type I oligosaccharides is a feature specific to human breast milk. *Advances in Nutrition*, 3(3), 473S–482S.

Urashima, T., Saito, T., Nakamura, T., and Messer, M. (2001). Oligosaccharides of milk and colostrum in non-human mammals. *Glycoconjugate Journal*, 18(5), 357–371.

USAMEDCOM (2020). Soldier 2020: Injury Rates/Attrition Rates Working Group; Medical Recommendations. Briefing by LTG Patricia Horoho, Surgeon General and Commanding General, USAMEDCOM, June 24, 2015.

Van Caenegem, E., Wierckx, K., Taes, Y., Schreiner, T., Vandewalle, S., Toye, K., et al. (2015). Body composition, bone turnover, and bone mass in trans men during testosterone treatment: 1-year follow-up data from a prospective case-controlled study (ENIGI), *European Journal of Endocrinology*, 172(2), 163–171.

van Dam, M. J. C. M., Zegers, B. S. H. J., and Schreuder, M. F. (2021). Case report: Uterine anomalies in girls with a congenital solitary functioning kidney. *Frontiers in Pediatrics*, 9. doi:10.3389/fped.2021.791499.

van der Made, J., Sahnouni, M., and Kamel, B. (2017). Hippopotamus gorgops from El Kherba (Algeria) and the context of its biogeography. In *Proceedings of the II Meeting of African Prehistory, Burgos 15-16 April, 2015*, 135–169.

van Hek, M., Buchmann, C., and Kraaykamp, G. (2019). Educational systems and gender differences in reading: A comparative multilevel analysis. *European Sociological Review*, 35(2), 169–186. doi:10.1093/esr/jcy054.

van Hemmen, J., Cohen-Kettenis, P. T., Steensma, T. D., Veltman, D. J., and Bakker, J. (2017). Do sex differences in CEOAEs and 2D:4D ratios reflect androgen exposure? A study in women with complete androgen insensitivity syndrome. *Biology of Sex Differences*, 8(1), 11. doi:10.1186/s13293-017-0132-z.

van Schaik, C. P., Song, Z., Schuppli, C., Drobniak, S. M., Heldstab, S. A., and Griesser, M. (2023). Extended parental provisioning and variation in vertebrate brain sizes. *PLOS Biology*, 21(2), e3002016. doi:10.1371/journal.pbio.3002016.

van Valen, L., and Sloan, R. E. (1965). The earliest primates. *Science*, 150(3697), 743–745. doi:10.1126/science.150.3697.743.

Veale, D., Miles, S., Bramley, S., Muir, G., and Hodsoll, J. (2015). Am I normal? A systematic review and construction of nomograms for flaccid and erect penis length and circumference in up to 15,521 men. *BJU International*, 115(6), 978–986. doi:10.1111/bju.13010.

Velasco, E. R., Florido, A., Milad, M. R., and Andero, R. (2019). Sex differences in fear extinction. *Neuroscience & Biobehavioral Reviews*, 103, 81–108. doi:10.1016/j.neubiorev.2019.05.020.

Vellekoop, J., Sluijs, A., Smit, J., Schouten, S., Weijers, J. W. H., Sinninghe Damsté, J. S., and Brinkhuis, H. (2014). Rapid short-term cooling following the Chicxulub impact at the Cretaceous-Paleogene boundary. *Proceedings of the National Academy of Sciences*, 111(21), 7537–7541. doi:10.1073/pnas.1319253111.

Venn, O., Turner, I., Mathieson, I., de Groot, N., Bontrop, R., and McVean, G. (2014). Strong male bias drives germline mutation in chimpanzees. *Science*, 344(6189), 1272–1275.

Videan, E. N., Fritz, J., Heward, C. B., and Murphy, J. (2006). The effects of aging on hormone and reproductive cycles in female chimpanzees (*Pan troglodytes*). *Comparative Medicine*, 56(4), 291–299.

Vigilant, L., and Groeneveld, L. F. (2012). Using genetics to understand primate social systems. *Nature Education Knowledge*, 3(10), 87.

Vitetta, L., Chen, J., and Clarke, S. (2019). The vermiform appendix: An immunological organ sustaining a microbiome inoculum. *Clinical Science*, 133(1), 1–8. doi:10.1042/cs20180956.

Vogel, E. R., Neitz, M., and Dominy, N. J. (2006). Effect of color vision phenotype on the foraging of wild white-faced capuchins, *Cebus capucinus*. *Behavioral Ecology*, 18(2), 292–297. doi:10.1093/beheco/arl082.

Voland, E., Chasiotis, A., and Schiefenhovel, W. (2005). Grandmotherhood: A short overview of three fields of research of the evolutionary significance of the postgenerative female life. Pp. 1–17 in *The Evolutionary Significance of the Second Half of Female Life*, edited by E. Voland, A. Chasiotis, and W. Schiefenhovel. New Brunswick, N.J.: Rutgers University Press.

Volk, A. A., and Atkinson, J. A. (2013). Infant and child death in the human environment of evolutionary adaptation. *Evolution and Human Behavior*, 34(3), 182–192. doi:10.1016/j.evolhumbehav.2012.11.007.

von Stumm, S., and Plomin, R. (2015). Socioeconomic status and the growth of intelligence from infancy through adolescence. *Intelligence*, 48, 30–36. doi:10.1016/j.intell.2014.10.002.

Voyer, D. (2011). Time limits and gender differences on paper-and-pencil tests of mental rotation: A meta-analysis. *Psychonomic Bulletin & Review*, 18, 267–277. doi:10.3758/s13423-010-0042-0.

Voyer, D., and Voyer, S. D. (2014). Gender differences in scholastic achievement: A meta-analysis. *Psychological Bulletin*, 140(4), 1174–1204. doi:10.1037/a0036620.

Wade, L. (2020). An unequal blow. *Science*, 368(6492), 700. doi:10.1126/science.368.6492.700.

Wadsworth, M. E., Broderick, A. V., Loughlin-Presnal, J. E., Bendezu, J. J., Joos, C. M., Ahlkvist, J. A., et al. (2019). Co-activation of SAM and HPA responses to acute stress: A review of the literature and test of differential associations with preadolescents' internalizing and externalizing. *Developmental Psychobiology*, 61(7), 1079–1093. doi:10.1002/dev.21866.

Wald, C., and Wu, C. (2010). Biomedical research. Of mice and women: The bias in animal models. *Science*, 327(5973), 1571–1572. doi:10.1126/science.327.5973.1571.

Walker, A., and Leakey, R. E. (1993). *The Nariokotome* Homo erectus *Skeleton*. Cambridge, Mass.: Harvard University Press.

Walker, M. L., and Herndon, J. G. (2008). Menopause in nonhuman primates? *Biology of Reproduction*, 79(3), 398–406. doi:10.1095/biolreprod.108.068536.

Walker, R., Gurven, M., Hill, K., Migliano, A., Chagnon, N., De Souza, R., et al. (2006). Growth rates and life histories in twenty-two small-scale societies. *American Journal of Human Biology*, 18(3), 295–311. doi:10.1002/ajhb.20510.

Wall-Wieler, E., Roos, L. L., Brownell, M., Nickel, N., Chateau, D., and Singal, D. (2018). Suicide attempts and completions among mothers whose children were taken into care by child protection services: A cohort study using linkable administrative data. *Canadian Journal of Psychiatry*, 63(3), 170–177. doi:10.1177/0706743717741058.

Walls, G. L. (1942). *The Vertebrate Eye and Its Adaptive Radiation.* Bloomfield Hills, Mich.: Cranbrook Institute of Science. doi:10.5962/bhl.title.7369.

Walsh, K. P. (2014). Marketing midwives in seventeenth-century London: A re-examination of Jane Sharp's *The Midwives Book*. *Gender & History*, 26(2), 223–241.

Wamboldt, R., Shuster, S., and Sidhu, B. S. (2021). Lactation induction in a transgender woman wanting to breastfeed: Case report. *The Journal of Clinical Endocrinology and Metabolism*, 106(5), e2047–e2052. doi:10.1210/clinem/dgaa976.

Wang, Q., Wang, X., Yang, L., Han, K., Huang, Z., and Wu, H. (2021). Sex differences in noise-induced hearing loss: A cross-sectional study in China. *Biology of Sex Differences*, 12(1), 24. doi:10.1186/s13293-021-00369-0.

Warinner, C., Rodrigues, J. F., Vyas, R., Trachsel, C., Shved, N., Grossmann, J., et al. (2014). Pathogens and host immunity in the ancient human oral cavity. *Nature Genetics*, 46(4), 336–344. doi:10.1038/ng.2906.

Warren, M. (2018). Mum's a Neanderthal, Dad's a Denisovan: First discovery of an ancient-human hybrid. *Nature*, 560, 417–418. www.nature.com.

Warrener, A. G. (2017). Hominin hip biomechanics: Changing perspectives. *The Anatomical Record*, 300(5), 932–945. doi:10.1002/ar.23558.

Wasserman, M. D., Chapman, C. A., Milton, K., Gogarten, J. F., Wittwer, D. J., and Ziegler, T. E. (2012). Estrogenic plant consumption predicts red colobus monkey (*Procolobus rufomitratus*) hormonal state and behavior. *Hormones and Behavior*, 62(5), 553–562. doi:10.1016/j.yhbeh.2012.09.005.

Watson, J. D. (2001). *The Double Helix: A Personal Account of the Discovery of the Structure of DNA.* New York: Touchstone.

Weaver, T. D., and Hublin, J. J. (2009). Neandertal birth canal shape and the evolution of human childbirth. *Proceedings of the National Academy of Sciences*, 106(20), 8151–8156. doi:10.1073/pnas.0812554106.

Weber, G. W., Lukeneder, A., Harzhauser, M., Mitteroecker, P., Wurm, L., Hollaus, L.-M., et al. (2022). The microstructure and the origin of the Venus from Willendorf. *Scientific Reports*, 12(1), 2926. doi:10.1038/s41598-022-06799-z.

Wedekind, C., Seebeck, T., Bettens, F., and Paepke, A. J. (1995). MHC-dependent mate preferences in humans. *Proceedings of the Royal Society B: Biological Sciences*, 260(1359), 245–249. doi:10.1098/rspb.1995.0087.

Wedel, M. J. (2009). Evidence for bird-like air sacs in saurischian dinosaurs. *Journal of Experimental Zoology Part A: Ecological Genetics and Physiology*, 311A(8), 611–628.

Weiner, E. (2007). Why women read more than men. NPR, Sept. 5, 2007. www.npr.org.

Weiss, G., and Goldsmith, L. T. (2005). Mechanisms of relaxin-mediated premature birth. *Annals of the New York Academy of Sciences*, 1041(1), 345–350. doi:10.1196/annals.1282.055.

Wells, J. C., and Stock, J. T. (2007). The biology of the colonizing ape. *American Journal of Physical Anthropology*, 134 (S45), 191–222. doi:10.1002/ajpa.20735.

Wells, J. C., DeSilva, J. M., and Stock, J. T. (2012). The obstetric dilemma: An ancient game of Russian roulette, or a variable dilemma sensitive to ecology? *American Journal of Physical Anthropology*, 149 (S55), 40–71. doi:10.1002/ajpa.22160.

West, E., and Knight, R. J. (2017). Mothers' milk: Slavery, wetnursing, and Black and white women in the antebellum South. *Journal of Southern History*, 83(1): 37–68. doi:10.1353/soh.2017.0001.

Western, B., and Wildeman, C. (2009). The Black family and mass incarceration. *The Annals of the American Academy of Political and Social Science*, 621, 221–242. doi:10.1177/0002716208324850.

Whalley, L. J., and Deary, I. J. (2001). Longitudinal cohort study of childhood

IQ and survival up to age 76. *BMJ (Clinical Research Ed.)*, 322(7290), 819. doi:10.1136/bmj.322.7290.819.

Whitcome, K. K., Shapiro, L. J., and Lieberman, D. E. (2007). Fetal load and the evolution of lumbar lordosis in bipedal hominins. *Nature*, 450(7172), 1075–1078. doi:10.1038/nature06342.

White, K. J. C. (2002). Declining fertility among North American Hutterites: The use of birth control within a Dariusleut colony. *Social Biology*, 49, 1–2, 58–73.

White, T. D., Ambrose, S. H., Suwa, G., and WoldeGabriel, G. (2010). Response to Comment on the Paleoenvironment of *Ardipithecus ramidus*. *Science*, 328(5982), 1105.

White, T. D., Ambrose, S. H., Suwa, G., Su, D. F., DeGusta, D., Bernor, R. L., et al. (2009). Macrovertebrate paleontology and the Pliocene habitat of *Ardipithecus ramidus*. *Science*, 326(5949), 67–93.

White, T. D., Asfaw, B., Beyene, Y., Haile-Selassie, Y., Lovejoy, C. O., Suwa, G., and WoldeGabriel, G. (2009). *Ardipithecus ramidus* and the paleobiology of early Hominids. *Science*, 326(5949), 64–86.

White, T. D., Lovejoy, C. O., Asfaw, B., Carlson, J. P., and Suwa, G. (2015). Neither chimpanzee nor human, Ardipithecus reveals the surprising ancestry of both. *Proceedings of the National Academy of Sciences*, 112(16), 4877–4884.

White, U. A., and Tchoukalova, Y. D. (2014). Sex dimorphism and depot differences in adipose tissue function. *Biochimica et Biophysica Acta*, 1842(3), 377–392. doi:10.1016/j.bbadis.2013.05.006.

Wichura, H., Jacobs, L. L., Lin, A., Polcyn, M. J., Manthi, F. K., Winkler, D. A., et al. (2015). A 17-MY-old whale constrains onset of uplift and climate change in East Africa. *Proceedings of the National Academy of Sciences*, 112(13), 3910–3915.

Wiederman, M. W. (1997). The truth must be in here somewhere: Examining the gender discrepancy in self-reported lifetime number of sex partners. *The Journal of Sex Research*, 34(4), 375–386.

Wiesenfeld, H. C., Hillier, S. L., Meyn, L. A., Amortegui, A. J., and Sweet, R. L. (2012). Subclinical pelvic inflammatory disease and infertility. *Obstetrics and Gynecology*, 120(1), 37–43. doi:10.1097/AOG.0b013e31825a6bc9.

Wilcox, A. J., Dunson, D. B., Weinberg, C. R., Trussell, J., and Baird, D. D. (2001). Likelihood of conception with a single act of intercourse: Providing benchmark rates for assessment of post-coital contraceptives. *Contraception*, 63(4), 211–215. doi:10.1016/S0010-7824(01)00191-3.

Wilcox, A. J., Weinberg, C. R., O'Connor, J. F., Baird, D. D., Schlatterer, J. P., Canfield, R. E., et al. (1988). Incidence of early loss of pregnancy. *The New England Journal of Medicine*, 319(4), 189–194. doi:10.1056/NEJM 198807283190401.

Wilde, C. J., Prentice, A., and Peaker, M. (1995). Breast-feeding: Matching supply with demand in human lactation. *Proceedings of the Nutrition Society*, 54(2), 401–406. doi:10.1079/PNS19950009.

Wildt, D. E., Zhang, A., Zhang, H., Janssen, D. L., and Ellis, S. (2006). *Giant Pandas: Biology, Veterinary Medicine and Management.* Cambridge, U.K.: Cambridge University Press.

Williams, C. M., Peyre, H., Toro, R., and Ramus, F. (2021). Sex differences in the brain are not reduced to differences in body size. *Neuroscience & Biobehavioral Reviews*, 130, 509–511. doi:10.1016/j.neubiorev.2021.09.015.

Williams, T. J., Pepitone, M. E., Christensen, S. E., Cooke, B. M., Huberman, A. D., Breedlove, N. J., et al. (2000). Finger-length ratios and sexual orientation. *Nature*, 404(6777), 455–456. doi:10.1038/35006555.

Wilson Mantilla, G. P., Chester, S. G. B., Clemens, W. A., Moore, J. R., Sprain, C. J., Hovatter, B. T., et al. (2021). Earliest Palaeocene purgatoriids and the initial radiation of stem primates. *Royal Society Open Science*, 8(2), 210050. doi:10.1098/rsos.210050.

Winter, J. (2022). Why more and more girls are hitting puberty early. *The New Yorker*, Oct. 27, 2022. www.newyorker.com.

Winterbottom, M., Burke, T. A., and Birkhead, T. R. (2001). The phalloid organ, orgasm and sperm competition in a polygynandrous bird: The red-billed buffalo weaver (*Bubalornis niger*). *Behavioral Ecology and Sociobiology*, 50, 474–482.

Witt, C. (1995). Anti-essentialism in feminist theory. *Philosophical Topics*, 23(2), 321–344.

Wittman, A. B., and Wall, L. L. (2007). The evolutionary origins of obstructed labor: Bipedalism, encephalization, and the human obstetric dilemma. *Obstetrical & Gynecological Survey*, 62(11), 739–748. doi:10.1097/01.ogx.0000286584.04310.5c.

Wodon, Q., Montenegro, C., Nguyen, H., and Onagoruwa, A. (2018). *Missed Opportunities: The High Cost of Not Educating Girls.* Washington, D.C.: World Bank.

Woetzel, J., Madgavkar, A., Ellingrud, K., Labaye, E., Devillard, S., Kutcher, E., and Krishnan, M. (2015). The power of parity: How advancing women's equality can add $12 trillion to global growth. Shanghai: McKinsey Global Institute.

Woetzel, J., Madgavkar, A., Sneader, K., Tonby, O., Lin, D. Y., Lydon, J., and Gubieski, M. (2018). The power of parity: Advancing women's equality in Asia Pacific. Shanghai: McKinsey Global Institute.

Wolbers, K. A., and Holcomb, L. (2020). Why sign language is vital for all deaf babies, regardless of cochlear implant plans. The Conversation, Aug. 31, 2020. theconversation.com.

WoldeGabriel, G., Ambrose, S. H., Barboni, D., Bonnefille, R., Bremond, L., Currie, B., et al. (2009). The geological, isotopic, botanical, invertebrate, and lower vertebrate surroundings of *Ardipithecus ramidus*. *Science*, 326(5949), 65–65e65.

WoldeGabriel, G., Haile-Selassie, Y., Renne, P. R., Hart, W. K., Ambrose, S. H., Asfaw, B., et al. (2001). Geology and palaeontology of the Late Miocene Middle Awash valley, Afar rift, Ethiopia. *Nature*, 412(6843), 175–178. doi:10.1038/35084058.

Wood, B. (2012). Palaeoanthropology: Facing up to complexity. *Nature*, 488(7410), 162–163.

Wood, B. (2014). Human evolution: Fifty years after *Homo habilis*. *Nature*, 508, 31–33. doi:10.1038/508031a.

Wood, M. (2013). *Black Milk: Imagining Slavery in the Visual Cultures of Brazil and America*. Oxford: Oxford University Press.

World Health Organization (2009). SESSION 2, The physiological basis of breastfeeding. *Infant and Young Child Feeding: Model Chapter for Textbooks for Medical Students and Allied Health Professionals*. Geneva: World Health Organization. www.ncbi.nlm.nih.gov.

Wright, D. W., Yeatts, S. D., Silbergleit, R., Palesch, Y. Y., Hertzberg, V. S., Frankel, M., et al. (2014). Very early administration of progesterone for acute traumatic brain injury. *New England Journal of Medicine*, 371(26), 2457–2466. doi:10.1056/NEJMoa1404304.

Wu, Y., Wang, H., and Hadly, E. (2017). Invasion of ancestral mammals into dim-light environments inferred from adaptive evolution of the photo-transduction genes. *Scientific Reports*, 7, 46542.

Wunderle, M. K., Hoeger, K. M., Wasserman, M. E., and Bazarian, J. J. (2014). Menstrual phase as predictor of outcome after mild traumatic brain injury in women. *The Journal of Head Trauma Rehabilitation*, 29(5), E1.

Wyart, C., Webster, W. W., Chen, J. H., Wilson, S. R., McClary, A., Khan, R. M., and Sobel, N. (2007). Smelling a single component of male sweat alters levels of cortisol in women. *The Journal of Neuroscience*, 27(6), 1261–1265. doi:10.1523/jneurosci.4430-06.2007.

Wyatt, T. D. (2015). The search for human pheromones: the lost decades and the necessity of returning to first principles. *Proceedings of the Royal Society B: Biological Sciences*, 282(1804), 20142994. doi:10.1098/rspb.2014.2994.

Xiao, L., van't Land, B., Engen, P. A., Naqib, A., Green, S. J., Nato, A., et al. (2018). Human milk oligosaccharides protect against the development of autoimmune diabetes in NOD-mice. *Scientific Reports*, 8(1), 1–15.

Yalom, M. (1997). *History of the Breast*. New York: Knopf.

Yan, L., and Silver, R. (2016). Neuroendocrine underpinnings of sex differences in circadian timing systems. *The Journal of Steroid Biochemistry and Molecular Biology*, 160, 118–126. doi:10.1016/j.jsbmb.2015.10.007.

Yoder, A. D., and Larsen, P. A. (2014). The molecular evolutionary dynamics of the vomeronasal receptor (class 1) genes in primates: A gene family on the verge of a functional breakdown. *Frontiers in Neuroanatomy*, 8. doi:10.3389/fnana.2014.00153.

Yokota, S., Suzuki, Y., Hamami, K., Harada, A., and Komai, S. (2017). Sex differences in avoidance behavior after perceiving potential risk in mice. *Behavioral and Brain Functions*, 13(1), 9. doi:10.1186/s12993-017-0126-3.

Yoles-Frenkel, M., Shea, S. D., Davison, I. G., and Ben-Shaul, Y. (2022). The

Bruce effect: Representational stability and memory formation in the accessory olfactory bulb of the female mouse. *Cell Reports*, 40(8), 111262. doi:10.1016/j.celrep.2022.111262.

Yong, E. (2016). *I Contain Multitudes: The Microbes Within Us and a Grander View of Life*. New York: Ecco.

You, D., Hug, L., Ejdemyr, S., Idele, P., Hogan, D., Mathers, C., et al. (2015). United Nations Inter-agency Group for Child Mortality Estimation (UN IGME). Global, regional, and national levels and trends in under-5 mortality between 1990 and 2015, with scenario-based projections to 2030: A systematic analysis by the UN Inter-agency Group for Child Mortality Estimation. *Lancet*, 386(10010), 2275–2286.

Zagorsky, J. L. (2007). Do you have to be smart to be rich? The impact of IQ on wealth, income and financial distress. *Intelligence*, 35(5), 489–501. doi:10.1016/j.intell.2007.02.003.

Zaneveld, L. J. D., Tauber, P. F., Port, C., Propping, D., and Schumacher, G. F. B. (1974). Scanning electron microscopy of the human, guinea-pig and rhesus monkey seminal coagulum. *Reproduction*, 40(1), 223–225.

Zhang, D. D., Bennett, M. R., Cheng, H., Wang, L., Zhang, H., Reynolds, S. C., et al. (2021). Earliest parietal art: Hominin hand and foot traces from the middle Pleistocene of Tibet. *Science Bulletin*, 66(24), 2506–2515. doi:10.1016/j.scib.2021.09.001.

Zhang, X., and Firestein, S. (2007). Nose thyself: Individuality in the human olfactory genome. *Genome Biology*, 8(11), 230. doi:10.1186/gb-2007-8-11-230.

Zhang, X., Liu, Y., Liu, L., Li, J., Du, G., and Chen, J. (2019). Microbial production of sialic acid and sialylated human milk oligosaccharides: Advances and perspectives. *Biotechnology Advances*, 37(5), 787–800. doi:10.1016/j.biotechadv.2019.04.011.

Zhang, Z., Ramstein, G., Schuster, M., Li, C., Contoux, C., and Yan, Q. (2014). Aridification of the Sahara desert caused by Tethys Sea shrinkage during the Late Miocene. *Nature*, 513(7518), 401–404. doi:10.1038/nature13705.

Zhou, Z., and Zheng, S. (2003). The missing link in *Ginkgo* evolution. *Nature*, 423, 821–822. doi:10.1038/423821a.

Zipple, M. N., Altmann, J., Campos, F. A., Cords, M., Fedigan, L. M., Lawler, R. R., et al. (2021). Maternal death and offspring fitness in multiple wild primates. *Proceedings of the National Academy of Sciences*, 118(1), e2015317118. doi:10.1073/pnas.2015317118.

Zipple, M. N., Grady, J. H., Gordon, J. B., Chow, L. D., Archie, E. A., Altmann, J., and Alberts, S. C. (2017). Conditional fetal and infant killing by male baboons. *Proceedings of the Royal Society B: Biological Sciences*, 284, 20162561. doi:10.1098/rspb.2016.2561.

Zipple, M. N., Roberts, E. K., Alberts, S. C., Beehner, J. C. (2019). Male-mediated prenatal loss: Functions and mechanisms. *Evolutionary Anthropology*, 28(3):114–125. doi:10.1002/evan.21776.

Zittleman, K., and Sadker, D. (2002). Gender bias in teacher education texts: New (and old) lessons. *Journal of Teacher Education*, 53, 168–80.

Zivkovic, A. M., German, J. B., Lebrilla, C. B., and Mills, D. A. (2011). Human milk glycobiome and its impact on the infant gastrointestinal microbiota. *Proceedings of the National Academy of Sciences*, 108(suppl. 1), 4653–4658. doi:10.1073/pnas.1000083107.

Zokaei, N., Board, A. G., Manohar, S. G., and Nobre, A. C. (2019). Modulation of the pupillary response by the content of visual working memory. *Proceedings of the National Academy of Sciences*, 116(45), 22802–22810.

Zuckerman, M., and Driver, R. E. (1989). What sounds beautiful is good: The vocal attractiveness stereotype. *Journal of Nonverbal Behavior*, 13(2), 67–82. doi:10.1007/BF00990791.

INDEX

Page numbers in *italics* refer to illustrations.

A NOTE ABOUT THE AUTHOR

Cat Bohannon completed her PhD in 2022 at Columbia University, where she studied the evolution of narrative and cognition. Her writing has appeared in *Scientific American*, *Science*, *The Best American Nonrequired Reading*, and *The Georgia Review*, and on *The Story Collider*. This is her first book. She lives in the United States with her partner and their two offspring.

A NOTE ON THE TYPE

This book was set in Janson, a typeface long thought to have been made by the Dutchman Anton Janson, who was a practicing typefounder in Leipzig during the years 1668–1687. However, it has been conclusively demonstrated that these types are actually the work of Nicholas Kis (1650–1702), a Hungarian, who most probably learned his trade from the master Dutch typefounder Dirk Voskens. The type is an excellent example of the influential and sturdy Dutch types that prevailed in England up to the time William Caslon (1692–1766) developed his own incomparable designs from them.

Composed by North Market Street Graphics,
Lancaster, Pennsylvania

Printed and bound by Berryville Graphics,
Berryville, Virginia

Designed by Soonyoung Kwon